Hawai'i's Mauna Loa Observatory

Hawai'i's Mauna Loa Observatory

Fifty Years of Monitoring the Atmosphere

FORREST M. MIMS III

A Latitude 20 Book
University of Hawai'i Press
Honolulu

Publication of this book was made possible with funding from the
United States National Oceanic and Atmospheric Administration (NOAA)

Printed in the United States of America
17 16 15 14 13 12   6 5 4 3 2 1

Library of Congress Cataloging-in-Publication Data
Mims, Forrest M.
Hawai'i's Mauna Loa Observatory : fifty years of monitoring the atmosphere / Forrest M. Mims III.
p. cm.
"A latitude 20 book."
Includes bibliographical references and index.
ISBN 978-0-8248-3431-9 (hardcover : alk. paper)
1. Mauna Loa Observatory—History. 2. Meteorological stations—Hawaii—History. I. Title.
QC875.U62M386 2012
551.5072'3—dc23
2011036589

University of Hawai'i Press books are printed on acid-free
paper and meet the guidelines for permanence and durability
of the Council on Library Resources.

Designed by Mardee Melton
Printed by Sheridan Books, Inc.

Dedication

To Robert H. Simpson, the visionary who founded Hawai'i's Mauna Loa Observatory, Ralph Stair, who partnered with Simpson, and the scientists and staff who transformed a dream into one of the world's leading atmospheric research stations.

No pleasure is comparable to the standing upon the vantage ground of truth (a hill not to be commanded, and where the air is always clear and serene), and to see the errors and wanderings and mists and tempests in the vale below.

—The Roman poet and philosopher Lucretius as paraphrased in 1597 by Francis Bacon, the pioneer of the scientific method, in "On Truth" (*Essays by Francis Bacon,* New York: John B. Alden, 1885)

Contents

Foreword

By its title, this book is a historical account of half a century of observation, discovery, and research at Mauna Loa Observatory, perched on the upper slopes of the world's largest mountain mass. However, it encompasses much more—the accounts of earlier explorations, the hardships encountered, and hard-won accomplishments. In the telling, Forrest Mims' story dwells extensively on the lives and drives of explorers and the character of leadership, often influencing scientific progress and at times impacting national policies and governance issues, most notably the current controversies about global warming. In this comprehensive saga of Mauna Loa, the author's interests and concerns as an earth scientist and inventive technologist, complementing his literary talents as author of more than sixty books, are reflected in this historical account, crafted to read with the attractions and verve of a novel.

Mims traces the history of exploration and discovery on Mauna Loa from the late eighteenth century through the establishment of the present observatory and its half century of scientific achievements. As described, each exploration was different, both in motivation and the lessons learned along the way. And Mims' account of each exploration springs to life with numerous personal quotations from the participants themselves. All told, the volume is epochal in scope and character.

While each episode differs in original concept and outcome, the author's story presents surprising circumstances or incidents that tend to link or provide interesting contrasts between the succession of explorations. The narrative begins with a brief account of famed British explorer James Cook's expedition to Hawaii, where, at Kealakekua Bay in 1778, he lost his life during a skirmish with hostile natives before his exploration had made much progress. Cook's failed exploration, however, is interestingly linked to the highly successful Menzies' expedition through one of Cook's principals, George Vancouver.

Vancouver was also injured in Cook's fatal encounter, but he returned fifteen years later as leader of the succeeding expedition. Vancouver's expedition surgeon was a youthful but ambitious scientist named Menzies. The story describes how Archibald Menzies, a stubbornly determined youth with broad scientific interests, prevailed in his frequent disputes with the expedition leader and ultimately was allowed to organize his own assault team—a band of Hawaiians supplied by King Kamehameha after negotiations with Menzies. With these helpers, he climbed both Hualālai and Mauna Loa, making accurate measurements of summit elevations as well as scientific observations and meteorological records. In addition to Menzies unprecedented scientific achievements, Mims is careful to relate how this stubborn individual, often in quarrelsome disagreement with his peers, proved to be an attentively caring and compassionate leader in dealing with his team of Hawaiians during the rigors of the ascent and at the summit, a remarkable combination of grim determination and compassionate leadership in this successful exploration.

Following the author's brief discussion of two lesser explorations in the nineteenth century—first by James Macrae in 1825 and then David Douglas in 1834—the saga proceeds to the most ambitious of the early expeditions to Mauna Loa, the US Exploring Expedition, which receives full attention in this treatise. As a major expedition, it was conceived as and intended to be the successor to Thomas Jefferson's Lewis and Clark expedition of 1803. It was funded and charged by Congress to conduct a discovery mission to Antarctica, an exploration and mapping of various Pacific Ocean islands, and a full-fledged expedition to Hawai'i, including the scientific exploration and ascent of Mauna Loa. The expedition was well equipped, with a trained staff of more than three hundred. However, the responsibility for its success was placed in the hands of the US Navy's Captain Charles Wilkes, a competent scientist but regarded as a strict disciplinarian. Nevertheless, history records a long list of scientific achievements and unprecedented measurements under Wilkes' direction, including gravity at the summit altitude and variations of sound propagation with changes of altitude (or of air density).

The story of how the deserved recognition and acclaim that this expedition and its leader had earned were seriously blighted is one of the very sad episodes in exploration history. It was the unfortunate result of perceived shortcomings in Wilkes' management of the expedition and charges lodged against him by some of his disgruntled senior staff. As a result, Wilkes had to face courts-martial proceedings at the conclusion of the expedition. While most charges were dismissed, the damage had already been done. The blighted reputation of Wilkes continued to hang over the expedition and was not dissolved until nearly a century later. During the Pacific Science Congress of 1940, several distinguished speakers lauded the accomplishments and scientific successes of the Wilkes expedition and its leader, which succeeded in removing any stigma that remained. This unfortunate episode, however—in particular the means of its later resolution—forged a link, though subtle, to circumstances that contributed to the later establishment of both the Summit and Slope Observatories, as explained in the comment below and the personal note that follows.

In describing the exoneration of Wilkes and his expedition during the Pacific Science Congress of 1940, Mims quotes from one persuasive address by Francis W, Reichelderfer, the highly regarded naval officer and newly appointed chief of the US Weather Bureau (1938), who extolled the accomplishments of the Wilkes expedition and its leadership. It was an address that turned out to be an unanticipated connecting link to what was to follow eight years later, described in the personal note below.

At this point, Mims turns to the centerpiece of this historical volume: the development of today's Mauna Loa Observatory and its achievements. The connecting link mentioned above relates to Mims' account of this writer's role in establishing the Summit Observatory and the fortuitous circumstances that set the stage for establishing the permanent Slope Observatory. A few of the personal details I will recount here.

In January, 1948, less than a decade after his address of exoneration of the Wilkes Expedition, Francis W. Reichelderfer, my Washington boss and mentor, sent me to Honolulu on a four-year administrative mission ostensibly to coordinate the transfer of weather service offices in the Pacific, established by US forces during World War II, from the military to the US Weather Bureau. With some reluctance and misgivings, I accepted this reassignment, which seemingly would separate me from my primary interests in meteorology: the dynamical role of the tropics in breeding and ultimately the prediction of tropical cyclones. But Reichelderfer had a different outcome in mind for this assignment—set in motion by a prior negotiation of which I was unaware.

Soon after my arrival in Honolulu, Dr. Luna Leopold, distinguished geophysicist and water resources expert with the Pineapple Research Institute (PRI), introduced himself and offered (on behalf of PRI) to conduct me on an exploration of acquaintance with principal areas of Hawai'i of mutual concern. The exploration carried us to the more remote and less accessible regions of Kaua'i, O'ahu, Maui, and the Big Island of Hawai'i. It was indeed a trip of acquaintance and appreciation of the distinctive physical features, the wildlife, and the agriculture of these islands.

But despite the wonderment and individual uniqueness of each island, the eyes of this meteorologist were captivated by Mauna Loa—clearly the most promising venue for exploration and experimentation—its ever-so-gentle slopes extending upward almost to mid-troposphere, far above the prevailing "trade wind temperature inversion" and the day-to-day weather changes, including heavy rains trapped below; and above it a terrain promising minimal disruption to the flow of almost pristine air thousands of feet above the vegetation level and thousands of miles from the nearest sources of pollution.

Where else in the world could you find so idyllic a site for atmospheric observations and experimentation? Shortly afterward, my exuberant interview with a science reporter—and his account of it—stirred public and private interests in support of an observatory atop Mauna Loa and brought Tom Vance, director of Hawai'i Institutions, to my door a few days later with a proposal to accomplish it. Within weeks, a plan involving the Honolulu Weather Bureau, Tom Vance's Hawai'i Institutions, and the US Navy's Admiral William Radcliffe (CINCPAC) at Pearl Harbor was underway, in an effort that ended with the dedication of the Summit Observatory. The story of this venture is copiously detailed in chapter 4 of this volume.

Forrest Mims resumes his story of the Mauna Loa Observatory (MLO), a venture inspired and fueled as a cooperative effort involving federal and local agencies and other participants that led to the establishment of Mauna Loa's Summit Observatory, its joyous dedication, its two and a half years of operation, and the sad circumstances that forced its closure. Mims' account of how the MLO was reestablished several years later, however, reads like a fairy tale. As still another cooperative venture, the outcome hinged on the chance encounter of two scientists at a site thousands of miles removed from Mauna Loa—two people who quickly discovered a common scientific need. And in the course of a single exciting day of discussion and scheming, they agreed on cooperative plans of action to serve both needs. The plan succeeded: The Bureau of Standards and the Weather Bureau were persuaded to agree on a collaboration to reestablish the Mauna Loa Observatory, a larger structure, better equipped, and more habitable with a serviceable access road. Work on the project began soon afterward under the direction of Ralph Stair (Bureau of Standards), with the collaboration of Harry Wexler (Weather Bureau) and his staff. The new observatory, relocated on the slopes of Mauna Loa some two thousand feet below the summit, was dedicated and placed in operation little more than two years after the summit site was closed. It is yet another example of how the advancement of science came about almost solely through the pursuit of persuasive cooperation, and it is engagingly detailed by the author.

From the very day of its dedication, the Mauna Loa (Slope) Observatory was a creative factory of scientific achievements, including the initiation of its half century of records of daily weather variations and what became the world's longest record of atmospheric transmission, as well as the variable constituents of the atmosphere at the elevation of the observatory. The new observatory soon began accumulating a bevy of "firsts" among observatories, including the discovery of Asian dust and other pollutants transported across the Pacific to North America and the detection of volcanic plumes in the stratosphere emanating from El Chichón (1982) and Pinatubo (1991) using Lidar sensing. But by far the most globally significant discovery from Mauna Loa has been its long record of increasing carbon dioxide in the atmosphere that sparked scientific concerns and debates over its implications for global warming and international political controversies on what to do about it.

The final chapters detailing the activities and achievements of MLO decade by decade are a unique testimonial of what can and has been accomplished by exploiting the opportunities that MLO offers for scientific explorations and research, not only by a dedicated resident staff but in collaboration with talented scientists the world over attracted to MLO's unique facilities and venue. It is a facility whose achievements and cost-effectiveness are more than competitive with similar facilities in less suitable venues. In my view, it has only scratched the surface of opportunities that lie ahead, especially when means can be found to reestablish a summit facility as the prime site for atmospheric observation and measurements, supported by a hard-surface access road with year-around sustainability connecting it with the present Slope Observatory.

My earlier informal comment (quoted in chapter 15) notwithstanding, I personally remain as convinced now as I was in 1949 when planning the summit facility that the summit site, for many observations and measurements, is significantly superior to the one at a lower level. Air sampling at the summit should be far more representative of planetary flow and climate of the Hawai'i tropics than the slope site several thousand feet below, where the flow and climatic character is a product influenced—if not substantially controlled—by the mountain itself, bearing particulates and aerosols injected from the mountain. While the time for such an extension may not always seem propitious, it is never too early to undertake the planning.

Forrest M. Mims in this impressive treatise distinguishes himself as a historian well beyond that of his previous extensive publications. But by reputation and achievement, he remains *much more.* He has always been a concerned and talented participant in a wide range of earth sciences. This is still exhibited in his continuing collaboration at the Mauna Loa Observatory, his leadership in programs of science education, and his involvement as editor of the *Citizen Scientist.* And at home he continues to direct and expand activities at his own Geronimo Creek Observatory in south-central Texas.

ROBERT H. SIMPSON, PH.D.[1]
WASHINGTON, DC

1. Dr. Simpson, the founder of the Mauna Loa Observatory, received NOAA's Environmental Hero Award for Longtime Achievement on July 2, 2008. He was joined at the ceremony by his wife, Dr. Joanne Simpson (1923–2010), who was also a famed meteorologist, and his daughters Peggy Simpson and Lynn Gramzow. The award was presented by Dr. Richard Spinrad, assistant administrator for NOAA Research, who stated, "I don't need to overemphasize just how important the idea, the concept, and then the manifestation of that concept in terms of the development of the Mauna Loa Observatory truly are. We are extraordinarily dedicated to you, sir, for your foresight, your vision, and for making that happen."

Preface

Preparation for this book can be traced to 1989, when I began making instruments that measured the ozone layer, sunlight, haze, and water vapor for a column in *Scientific American* magazine. My dream was to someday visit Hawai'i's Mauna Loa Observatory (MLO) to calibrate my homemade instruments against their world-class ones. The opportunity came in August 1992, when I was invited to speak about my experience with *Scientific American* at a meeting of the American Scientific Affiliation at the University of the Nations in Kailua-Kona, Hawai'i.

The world standard instrument for measuring the ozone layer was at MLO that summer. The instrument was being operated by Gretchen "Grae" Roth of the University of Colorado. Judy Pereira, MLO's legendary secretary, put me in touch with Grae, and on three mornings she met me at my hotel at 3:45 a.m. for the 2.5-hour trip to the alpine station high on the slope of the world's biggest mountain.

During those first MLO visits, I confirmed my measurements in Texas of a calibration drift in ozone measurements by NASA's Nimbus-7 satellite. This led to my first refereed publication in *Nature*, and I was determined to make a return visit to MLO. Drs. Howard Malmstadt, John Kuhne, and Derek Chignell of the University of the Nations made this possible when they asked me to teach a short course about atmospheric science the following spring. This and other assignments have allowed me to calibrate at MLO every year since.

During the first several years, I drove to MLO every morning for a week or so. Those trips and the altitude change were very tiring, and, after beginning work for the GLOBE program, I received permission to stay at the observatory for ten days each spring. This solved the altitude acclimation problem and allowed me to better appreciate the remote, barren setting of the observatory and to experience nature at work high over the Pacific. Occasionally a thunderstorm arrived, and lightning bolts, accompanied by cracks of thunder, struck the mountain station. Some storms brought hail, which hammered the metal rooftops with a deafening roar and left snowlike drifts across the black lava. On spring days, dust and air pollution from Asia could sometimes be seen drifting by in layers above and below the observatory on its way to North America.

While staying at MLO, I learned much about its many scientific programs and its remarkable history. The first draft of this preface was typed into a laptop computer in the observatory's original building, where a noisy chorus of air sampling pumps was at work twenty-four hours a day. The nearby window provided a spectacular view of Mauna Kea, and I could almost sense the presence of the scientists who began working at MLO soon after the building was completed in June 1956. Their research transformed MLO into a world-class atmospheric monitoring station and established the tradition for collecting high-quality data that is very much evident today.

As the fiftieth anniversary of MLO approached, MLO station chief Dr. John Barnes and former director Dr. Russell Schnell asked if I was interested in writing a book about the observatory's history. In July 2005 the project was approved, NOAA issued a contract, and I began work. After spending months researching the history of science on Mauna Loa, I interviewed dozens of past and present MLO staff and many others associated with the observatory. NOAA provided complete access to the MLO files in Hilo, the mountain station, and Boulder, Colorado. Though the book touches on several controversial matters, including a look at both sides of the global warming debate, I was never told what to write or not to write.

The manuscript was reviewed at several stages by an ad hoc panel of some twenty-five past and present MLO staff and scientists associated with the observatory, all of whom were also interviewed. Their comments and suggestions vastly improved the book. The University of Hawai'i Press selected two experts to review the manuscript, the late Walter Steiger, Professor Emeritus of Physics and Astronomy at the University of Hawai'i at Manoa, and Volker Mohnen, Professor Emeritus of the Department of Earth and Atmospheric Sciences, State University of New York at Albany. Their suggestions greatly improved the manuscript. Dr. Steiger, who initiated professional astronomy in Hawai'i and who spoke at the dedication of MLO in 1956, also shared his personal adventures at MLO during interviews for the book.

All of those interviewed cheerfully tolerated my many inquiries. I am especially grateful to MLO founder Robert Simpson, who patiently guided me through the sequence of events that led to the development of the observatory. Many stories about their experiences on Mauna Loa were provided by each of the living former directors of MLO, including Howard Ellis, Lothar Ruenke, Rudolf Pueschel, Ronald Fegley, John Miller, Elmer Robinson, Russell Schnell, and current director John Barnes. Dr. Miller also provided some of the photographs in the book. Considerable assistance was provided by some of the early staff, especially Howard Ellis, Bernard Mendonca, Malcolm Keyes, John Chin, Barry Bodhaine, and longtime MLO secretary Judy Pereira. Major assistance was also provided by the recent and current staff, especially John Barnes, Darryl Kuniyuki, Steve Ryan, and Bob Uchida. Lena Ellis and Clara Mendonca, whose husbands Howard and Bernard staffed MLO during the budget crisis of 1964–1966, provided insight about the impact on family life of the long working hours at MLO during that time.

Essential assistance was provided by MLO's Preston Sato, who videotaped dozens of the interviews and made digital scans of many photos in the MLO archive. Frank Niepold of NOAA arranged for the transcripts of the video interviews that provided many of the quotations in the book. The late Leslie Pajo organized MLO's historical files before I began sorting through them.

In 1958, Charles Keeling began what became the most famous science program at MLO: the almost continuous measurement of the atmosphere's concentration of carbon dioxide. His sons Ralph Keeling, who has also become a leading atmospheric chemist, and historian Drew Keeling, together with the Scripps Oceanographic Institute Library, provided invaluable assistance in clarifying the elder Keeling's role at MLO.

Blake Vance, son of Thomas B. Vance, provided his father's memoir on the history of the Mauna Loa access road and related newspaper articles. MaryAnn Hogan of the National Institute of Standards and Technology provided several of the key papers by Ralph Stair published in the *Journal of Research of the National Bureau of Standards*. Gary Stair, son of Ralph Stair, provided several photographs made by his father. Leslie K. Overstreet and Erin Clements Rushing of the Smithsonian Institution Libraries provided a scan of a US Exploring Expedition (USExEx) map of Hilo that enabled me to locate the site of the USExEx's Observatory 1.

The book would not have been possible without the financial assistance provided by NOAA and the coordination provided by the late Dr. Dave Hofmann, former director of NOAA's Global Monitoring Division (GMD), his successor, Dr. James Butler, GMD deputy director and former MLO director Dr. Russell C. Schnell, and, especially, the past and current MLO staff.

Finally, acquisitions editor Keith Leber of the University of Hawai'i Press shepherded the publication process through its initial phases, editor Lee S. Motteler provided essential assistance with the spelling of Hawaiian names and the organization of the hundreds of references, and managing editor Ann Ludeman supervised the preparation of the book. In the end, I was responsible for writing the book and reviewing it for accuracy prior to publication. Therefore, any errors are solely my responsibility.

Forrest M. Mims III
Geronimo Creek Observatory, Texas

* * *

Unless otherwise credited, all photographs were taken by the author.

CHAPTER ONE

Earth's Biggest Mountain

An arc of some 132 islands and reefs sweeps 1,500 miles (2,400 km) across the Pacific Ocean (Plate 1). The chain is the Hawaiian Archipelago, and its major islands are among the most scenic and remote on Earth. As Mark Twain reminisced more than four decades after his 1866 visit, they are "The loveliest fleet of islands that lies anchored in any ocean" (Twain 1908).

The largest island in the fleet is Hawai'i, which is anchored twenty degrees north of the equator, more than 2,000 miles (3,200 km) from the nearest continental landmass. Known to its residents as the Big Island, Hawai'i has a provisional area of 4,028 square miles (10,433 square km). The area is provisional because the island is slowly growing as molten volcanic lava pours into the sea off the southeastern coast.

All the islands of Hawai'i are the summits of volcanic peaks that emerged from the sea floor over geologic time. The Big Island is formed from a cluster of five volcanoes, two of which exceed the elevation of Mount Everest when their height is measured from the ocean floor.

The highest of these volcanoes is Mauna Kea, which is usually translated from the Hawaiian as "White Mountain," but which some attribute to a Hawaiian deity. Mauna Kea, which is considered dormant, rises 13,796 feet (4,205 meters) over the sea. Its summit, which is home to some of the world's most important astronomical observatories, has become a popular sunset destination for tourists who rent four-wheel drive vehicles or who join professionally conducted tours.

The summit of Mauna Loa, the second highest mountain, is 13,679 feet (4,169 meters) above the Pacific. Perched high on its north slope is a cluster of buildings and domed structures known as the Mauna Loa Observatory. Very few tourists visit or even know about MLO. Maps and brochures label it a weather station, but MLO does much more than measure wind, temperature, and humidity. For more than half a century, scientists at MLO have carefully measured the respiration of Earth, the status of the ozone layer, the intensity of sunlight, and the clarity and composition of the free atmosphere over Hawai'i.

Today the Mauna Loa Observatory is widely regarded as one of the world's leading atmospheric monitoring observatories. A key factor behind its fame is its almost continuous record of measuring the increasing concentration of carbon dioxide in the atmosphere since 1958. This single measurement series has led to many international research programs about the role of carbon dioxide and other gases in warming the planet. The major reason for this success and that of many other measurement programs at MLO is the placement of the observatory on the upper slope of Mauna Loa in the middle of the Pacific, and an introduction to the great mountain, its vegetation, and its climate will explain why.

Mauna Loa

While Mauna Kea is slightly higher, Mauna Loa is far greater in volume and area. Indeed, Mauna Loa is believed to be the world's largest mountain. Its name means "Long Mountain," which well describes this huge pile of lava that forms the southern half of Hawai'i and extends 56 miles (90 km) from the base of its north slope to the southernmost point of the island.

On January 14, 2005, the snowcapped summits of Hawai'i's Mauna Kea and Mauna Loa (Plate 2) were captured by the Moderate Resolution Imaging Spectroradiometer (MODIS) on NASA's Aqua satellite (Earth Observatory 2005). The snow blocked access to the Mauna Loa Observatory on January 10–11. On January 12, physical scientist Steve Ryan and engineer Bob Uchida hiked through the snow to service the observatory (Barnes 2005).

The grand dimensions of Mauna Loa are belied by its gentle profile, which lacks a prominent peak and resembles a giant warrior's shield lying atop the island. The famous botanist and explorer David Douglas, who climbed Mauna Loa in 1834, perfectly described the mountain in his memoir:

> Mauna Loa appears, indeed, more like an elevated tableland than a mountain. It is a high, broad dome, formed by infinitude of layers of volcanic matter, thrown out from the many mouths of its craters. Vegetation does not exist higher than eleven thousand feet; there is no soil whatever, and no water. The lava is so porous that when the snow melts it disappears a few feet from the verge, the ground drinking it up like a sponge. . . .
>
> The summit of this extraordinary mountain is so flat, that from this point no part of the island can be seen, not even the high peaks of Mauna Kea, nor the distant horizon of the sea, though the sky was remarkably clear. It is a horizon of itself, and about seven miles in diameter. (Jarves 1843)

The north slope of Mauna Loa is usually in plain sight from the Saddle Road, which crosses the island between Hilo and Waimea. From Hilo the view of the mountain is often obscured by clouds, and the Hualālai volcano blocks the view from Kailua-Kona, the popular tourist destination on Hawai'i's west coast. When clouds do not block the view, Mauna Loa's shield can be seen in the distance over the coconut palms and canoes at Kawaihae Bay north of Kona.

Mauna Loa has erupted thirty-three times since 1843 (Lockwood and Lipman 1987), and it shows no sign of stopping anytime soon. During eruptions, the mountain can emit vast volumes of liquid rock and produce explosive eruptions that throw boulders of lava far out of its summit crater, the giant Moku'āweoweo. But these eruptions are gentle when compared to the explosive eruptions of volcanoes such as Mount Pinatubo in the Philippines, which in 1991 propelled vast columns of ash, steam, and gas high into the stratosphere. Instead, Mauna Loa's eruptions are characterized by flowing streams of molten lava, often punctuated by spectacular curtains of orange fire jetting up from one of the mountain's three rift zones. The last eruption occurred in 1984, when a river of lava advanced to within about 4 miles (6 km) of the outskirts of Hilo. On the way, it sliced through the power line that provides electricity to the Mauna Loa Observatory.

The Lava

Mauna Loa is principally composed of just two forms of lava that are chemically identical and differ only in their appearance and structure. One kind of lava is known as *pāhoehoe* (pronounced pah-hoy-hoy). It forms when a shallow flow of molten rock having little gaseous content stops moving downhill and cools in place. Thus, *pāhoehoe* lava can be likened to a homogenous sheet of black or dark gray rock poured across the landscape like honey on a pancake. Its relatively smooth surface preserves the surges, swirls, ripples, folds, eddies, and coils that once characterized the viscous lava as it oozed its way down the slope of the mountain. *Pāhoehoe* lava flows, which are easily walked upon,

Unless otherwise credited, all photographs taken by the author.

PLATE 1. Computer-generated image of the Hawaiian island chain by Nils Sparwasser, Thorsten Andresen, Stephan Reiniger, and Robert Meisner (2006). Hawai'i is in the foreground. Vertical dimensions are exaggerated 4.5 times. *(Copyright 2006 by the German Aerospace Center [DLR]. Used with permission.)*

PLATE 2. The snow-capped summits of Hawai'i's Mauna Kea (upper peak, north) and Mauna Loa (south) imaged by NASA's Aqua satellite on January 14, 2005 (Earth Observatory, 2005). The snow blocked access to the Mauna Loa Observatory on January 10–11. *(MODIS Rapid Response Team, Goddard Space Flight Center, NASA.)*

PLATE 3. The US Exploring Expedition landed at this unmarked beach on Hilo's Waiakea Peninsula in December 1841. This photograph was made from where the USExEx established its Observatory 1.

PLATE 4. Pu'u Huluhulu, a tree-covered cinder cone at the junction of Saddle Road and Mauna Loa Road.

PLATE 5. The Mauna Loa Observatory is the tiny white outline below the summit of Mauna Loa.

PLATE 6. The first several miles of the Mauna Loa Road were repaved in 2007.

PLATE 7. Portions of the Mauna Loa Road have been in disrepair for years.

PLATE 8. Arriving at Mauna Loa Observatory, 11,134 feet (3,394 meters) above sea level, at 6:22 a.m. on May 25, 2007.

PLATE 9. Bob Uchida of the Cloud Physics Observatory of the University of Hawai'i trims grass while Sharon Nagata, a Mauna Loa Observatory student employee, checks the rain gauge in this 1967 photograph. *(MLO archive.)*

PLATE 10. MLO meteorologist Bernard Mendonca at the 1969 eruption of Mauna Ulu. *(MLO archive.)*

PLATE 11. Bernard Mendonca and MLO interim secretary Janice Sias measure airborne particles during the 1969 eruption of Mauna Ulu. *(MLO archive.)*

PLATE 12. Janice Sias during a July 24, 1969, visit to the Mauna Loa Observatory. *(MLO archive.)*

PLATE 13. Judy Pereira of MLO, Lasaro R. Waipi of the weather bureau of Truk and the Caroline Islands, and a Volcanoes National Park ranger at the eruption of Mauna Ulu on December 30, 1969. *(Bernard Mendonca, MLO archive.)*

PLATE 14. Mauna Ulu's lava fountain reached a height of 1,000 feet (305 meters). *(Bernard Mendonca, MLO archive.)*

PLATE 15. An upslope view from about 1969 of the unpaved MLO road and power line east of Kūlani Mauka. *(MLO archive.)*

PLATE 16. John Chin checks the infrared gas analyzer in March 1972. *(MLO archive.)*

PLATE 17. Barry Bodhaine with a particle counter that identified sea-salt aerosols at MLO (shown here) and Cape Kumukahi in March 1972. *(Rudolf F. Pueschel, MLO archive.)*

PLATE 18. Bernard Mendonca checks chart recorders in the main building at MLO ca. 1972, including *(left, from top)* the microbarograph; the wind speed, gust, and direction recorder; and the operations recorder; *(right, from top)* the temperature and solar radiation recorders. *(MLO archive.)*

PLATE 19. The Eppley pyrheliometer *(left)* and Foskett near-infrared hygrometer were installed on sun-trackers mounted on the solar wall ca. 1972. *(MLO archive.)*

PLATE 20. Howard Ellis *(left)* and Rudolf Pueschel at the MLO office at the Cloud Physics Observatory ca. 1972. *(MLO archive.)*

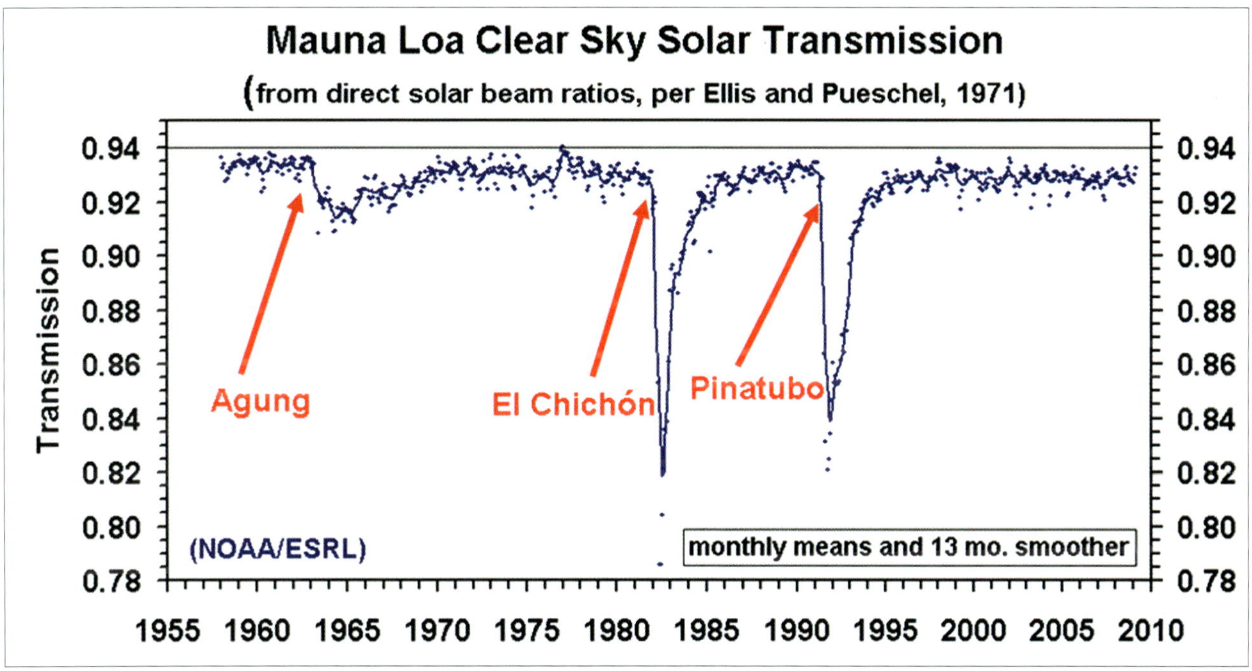

PLATE 21. The Ellis curve has tracked the direct sunlight at MLO since 1957. Aerosols from major volcanic eruptions cause the sharp dips in sunlight, and subtle reductions are caused by dust and air pollution from Asia. *(Ellsworth Dutton, NOAA/ESRL.)*

PLATE 22. MLO completed its move from the Cloud Physics Observatory to the Hilo Federal Building on July 1, 1976.

PLATE 23. Judy Pereira's office space in the Federal Building was a step up from the Cloud Physics Observatory, but parking was less convenient. *(John Miller, MLO archive.)*

PLATE 24. Prof. Kinsell L. Coulson at the keyboard of the computer that controlled his sky polarization system at MLO in 1977. *(John Miller, MLO archive.)*

are laced with cracks that provide shelter for ferns and mosses. The cracks reveal layers of color that indicate the presence of minerals that solidify at different temperatures.

The second kind of lava is called *'a'ā* (pronounced ah-ah), which arises when a gas-rich *pāhoehoe* flow slows to a crawl and becomes fragmented into clinkers. While *pāhoehoe* lava covers the landscape like the frosting on a cake, *'a'ā* resembles the cake. The 1984 *'a'ā* flows from Mauna Loa were from 6 to 23 feet (2 to 7 meters) thick (Volcano World 2005). According to the Hawaiian Volcano Observatory (HVO),

> Often what starts out as a pahoehoe flow may go through the transition to 'a'a when there is a change in conditions, such as a sudden steepening in slope, or just by the continuous loss of heat and gas as distance from the vent increases. (HVO 1999)

Both *pāhoehoe* and, especially, *'a'ā* lava are packed with vesicles formed when sulfur dioxide and other gases emerged from the lava, much like carbon dioxide forms a foam when it emerges from a carbonated beverage. *'A'ā* lava contains much more gas than the *pāhoehoe* form (Simkin 2000), and open bubbles in *'a'ā* form very sharp edges when they solidify at the surface of clinkers, where they quickly grind away the soles of shoes and boots and the tread of tires. *'A'ā* fields are very difficult to traverse and painfully unforgiving to those who fall onto the lava without the protection afforded by thick gloves and trousers. MLO is constructed over a field of *'a'ā* clinkers, which is why boardwalks are provided for the staff and visitors.

The Life Zones

The staff of the Mauna Loa Observatory travels through a sequence of climate zones as they drive between their office near sea level in Hilo to the mountain station 11,140 feet (3,400 meters) above the Pacific. The climate zones were succinctly described more than 140 years ago by Samuel Clemens, who toured Hawai'i in 1866 while serving as a correspondent for the *Sacramento Union*. Clemens had a fine view of Mauna Loa from the interisland schooner *Boomerang* as it approached Kailua-Kona on the western side of the Big Island. Writing under the name Mark Twain, he described several of the climate zones between the blue Pacific and the top of the great mountain:

> The rays of glittering snow and ice that clasped its summit like a claw looked refreshing when viewed from the blistering climate we were in. One could stand on that mountain (wrapped up in blankets and furs to keep warm), and while he nibbled a snowball or an icicle to quench his thirst he could look down the long sweep of its sides and see spots where plants are growing that grow only where the bitter cold of winter prevails; lower down he could see sections devoted to productions that thrive in the temperate zone alone; and at the bottom of the mountain he could see the home of the tufted coco palms and other species of vegetation that grow only in the sultry atmosphere of eternal summer. He could see all the climes of the world at a single glance of the eye, and that glance would only pass over a distance of four or five miles as the bird flies! (Twain 1866)

Twain was short on the distance, for the nearest beaches are 20 miles (32 km) as the bird flies from the summit of Mauna Loa. But he captured the essence of the dramatic changes between the stately coconut palms and the leafy banyans and bananas of Hilo to the vast wasteland of crunchy, boot-chewing lava at the Mauna Loa Observatory. Perhaps no place better illustrates the contrast between sea level and the desolation of the Mauna Loa Observatory than the small sandy beach decorated with tropical flowers and coconut palms on the west coast of Hilo's Waiākea Peninsula, where the US Exploring Expedition landed on the Big Island in December 1841 (Plate 3) prior to its three-week stay at the summit.

For more than half a century, the observatory staff has driven through the climate zones described by Twain. For the first dozen or so miles past Hilo, the way is lined with exotic trees and shrubs and luxurious display of tropical flowers. In the 1950s, the staff drove through the Kulani Prison to reach the rough road to MLO. Since 1963 they have left Hilo on Saddle Road, a curvy, scenic highway that links the east and west sides of the Big Island. The road follows the traditional Ala Mauna route, which in 1943 was developed into a cross-island gravel road by the Civilian Conservation Corps and the US Army Corps of Engineers in preparation for a possible Japanese invasion of the Big Island (Saddle Road 2005). Then as now, the road was known more for its many traffic accidents than the exotic scenery, a reputation that will likely change as major improvements are completed to the road.

Saddle Road climbs up from Hilo through what seems like a lush botanical garden. At an elevation of several thousand feet (around a kilometer), the large trees that shade the pavement give way to *'ōhi'a* trees that sometimes form a canopy over giant tree ferns as high as a house. The tall grass along both sides of the road is festooned with nonnative bamboo orchids bearing white blossoms tinged in lavender. Often this stretch of road is cloaked by orographic clouds that form as warm, moist trade winds rise into the cooler air of the saddle between Mauna Kea and Mauna Loa. This is rain forest country, where the annual rainfall is measured not in inches or centimeters but in feet. When the clouds are at the elevation of the road, visibility can fall to a few car lengths in the thick fog.

As the ferns are left behind, *'ōhi'a* trees continue to dominate the lavascape. They grow from near sea level and are the most common tree on the Big Island. Their flame-red lehua blossoms, each resembling a miniature lava fountain, announce their presence and more than make up for their dull green foliage.

Saddle Road is constructed over an old *pāhoehoe* lava flow from Mauna Loa. As the road climbs through about 4,800 feet (1,460 meters) and enters a broad expanse of bare lava, patches of shrubs and trees are scattered about. These are *kīpukas*—islands of vegetation growing from old lava flows that are surrounded by younger flows (Saddle Road 2005).

At an elevation of 6,560 feet (2,000 meters), the roads to the observatories on Mauna Loa and Mauna Kea branch, respectively, south and north from Saddle Road at Pu'u Huluhulu, an ancient cinder cone that long ago erupted from Mauna Loa's flank and was later surrounded by lava flows (Plate 4). This *pu'u* (lava hill) is an especially accessible and scenic example of a *kīpuka* that preserves a remnant of the dry koa forest that once covered much of the slopes of Mauna Kea at this elevation prior to its conversion to ranch land. Pu'u Huluhulu is surrounded by a fence to keep out feral sheep, goats, and pigs, and it features a network of trails leading to large koa trees at the summit. Also found on Pu'u Huluhulu are *māmane, 'iliahi* (sandalwood), and *'ōhi'a* trees and a wide variety of plants and shrubs, all of which provide a rare habitat for equally rare birds, including the endangered nēnē goose, Hawai'i's State Bird.

On a hot afternoon, the trees and plants growing from the bone-dry *'a'ā* lava of Pu'u Huluhulu seem to defy natural laws. How can any plant, much less large koa trees, survive in such a forboding place? The answer arrives with the trade wind orographic clouds that pour through the saddle and water the lava with cool showers or with mist and fog that paint plants and lava alike with a film of nourishing moisture.

On a clear day, when the clouds are below the elevation of the intersection of Saddle Road with the Mauna Loa road, the Mauna Loa Observatory is visible as a thin white rectangle high on the north slope of the big mountain (Plate 5). The observatory appears deceptively near, but it is another forty-five minutes or so away by car along a single-lane road (Plate 6). The first several miles of the road were repaved in 2007 and are a dream to drive compared to the previous condition of this section of the road.

From Pu'u Huluhulu to MLO, the vegetation becomes increasingly scarce. *'Ōhi'a* remains the dominant tree until the elevation reaches about 9,000 feet (2,740 meters), where they are reduced to miniature trees no larger than potted plants. Small ferns can be seen scattered about *pāhoehoe* lava flows, where they grow from cracks in the rock. In this region some portions of the road are in poor condition and must be driven with great care (Plate 7). Visible vegetation disappears long before the Mauna Loa Observatory is reached at 11,140 feet (3,400 meters) (Plate 8). Yet gray moss and a few small tufts of grass and fern were once found hiding in cracks in old *pāhoehoe* lava (Fosberg 1959).

The Climate

"The island of Hawaii, with a surface area of only 10,455 km^2, exhibits a spectacular range of climatic diversity comparable with that found on large continents." So began James Juvik, D. C. Singleton, and G. G. Clarke in their concise report on the climate of Hawai'i in *Mauna Loa Observatory: A 20th Anniversary Report* (1978). While the range of climate on the Big Island is indeed remarkable, the moderating influence of the Pacific Ocean and its tropical location combine to give much of Hawai'i a mild climate in which extremes of temperature, humidity, and wind are unusual. Drought, dust devils, thunderstorms, localized floods, and even hurricanes and tornadoes sometimes occur, but these weather extremes are unusual.

While the overall climate of Hawai'i is mild, the mountainous terrain of the major islands provides a wide variety of microclimates. The wet and dry extremes are especially evident along Saddle Road, for the windward, eastern side of Hawai'i is known for its abundant rainfall, while the leeward or *kona* side of the island is much drier and somewhat warmer. Saul Price, formerly Hawai'i's state climatologist and a US Weather Bureau scientist who worked closely with the original staff at the Mauna Loa Observatory, explained the difference in rainfall in an essay in the second edition of the *Atlas of Hawaii:*

> Over the open sea near Hawaii, rainfall averages between 25 to 30 inches a year. The islands themselves receive as much as 15 times this amount in some places and less than one-third of it in others. The cause of this remarkable variability, and of yearly totals that rival the greatest on earth, is principally the "orographic" (mountain-caused) rains which form within the moist trade-wind air as it moves in from the sea and overrides the steep and high terrain of the islands. . . . On the higher mountains the belt of maximum rainfall lies at only 2,000 to 3,000 feet, and amounts drop off rapidly with further elevation, so that the highest slopes are relatively dry. (Price 1983)

The transition between the rainy and dry microclimates of Hawai'i is a fact of life for the residents of the town of Waimea (or Kamuela) on the southeastern flank of the Kohala Mountains on the northwest portion of the Big Island. The east side of Waimea town is often sprinkled by cool showers that fall from a misty, cloud-filled sky. The west side is usually dry, warm, and sunny. The transition zone between the two weather patterns passes through the center of the town and can quickly flip between a cool, gentle shower and bright sunshine. The presence of both rain and sun often frames the town and nearby Mauna Kea within a brilliant rainbow.

A drive to the Mauna Loa Observatory may provide a similar weather experience, for the road often leads through heavy rain and thick fog before emerging into dry, clear air halfway or more up the mountain. The best description of the climate of Mauna Loa remains that provided by two meteorologists closely linked with the early history of the Mauna Loa Observatory, Saul Price and Jack Pales, the first director of MLO. They introduced a detailed account of the mountain's climate with a paragraph that perfectly summarizes its rather mellow nature:

> In keeping with its tropical maritime locale, the Observatory has a mild climate for the altitude. Severe weather is infrequent and the rigors of Alpine life virtually unknown. On the contrary, the brilliant skies, intense insolation, moderate temperature, and low humidity, encountered in so unusual and remote a setting, induce in most visitors a feeling of exhilaration and well being. (Price and Pales 1963)

While conditions on the mountain vary with the seasons, the changes are modest for an alpine site. For example, Price and Pales also reported that the mean maximum temperatures in February and June are, respectively, 46.6 and 58.8 degrees Fahrenheit. The mean minimums for the same months are, respectively, 28.9 and 37.1 F. During summer it is often possible to wear a short-sleeved shirt at the observatory, but this can be risky, for the sun's ultraviolet radiation is typically 20 percent more intense than at sea level.

Many of the scientific measurements made at MLO are much more influenced by the daily cycle of weather on the mountain than by annual variations in the climate. This is a result of a phenomenon called the *trade wind inversion layer* or simply the *trade inversion*. This is a layer of air typically 6,500 feet (2,000 meters) over the ocean that is warmer than the air below. Because the cool air below the inversion layer is denser than the warm air above, it is unable to rise through the inversion layer. In effect, the inversion layer serves as a cap that traps the cool air below, including its burden of moisture and any air pollution that might be present. The free atmosphere above the inversion layer is much drier and usually much cleaner than the air trapped below. Exceptions may occur where there is considerable airplane traffic, volcanic eruptions, or dust or smoke from distant sources.

Hilo is located on the northeast flank of Mauna Loa, where the giant volcano emerges from the Pacific Ocean. Trade winds from the east-northeast and the trade inversion prevail at Hilo during most of the year, and the inversion layer often plays the dominant role in Mauna Loa's daily weather pattern. As cool, moist Pacific air passes up and over the saddle that crosses the Big Island between Mauna Loa and Mauna Kea, orographic clouds are formed that, as noted earlier, often shroud the vegetation and the lower half of the road to the Mauna Loa Observatory in dense, wet fog. The eastern slopes of the lower half of the two mountains receive abundant rain from these clouds and are thick with vegetation. As noted by Price (1983), maximum rainfall occurs between about two to three thousand feet and decreases rapidly at higher elevations.

Conditions for monitoring the atmosphere at the Mauna Loa Observatory are best when the inversion layer is far down the slope of the mountain and the air at the observatory is extraordinarily dry and pristine. For weeks at a time the inversion layer will behave and remain tethered far below, in keeping with its altitude over the adjacent ocean. More commonly, between midmorning and early afternoon the inversion layer adjacent to the mountain will rise well above its usual height. No sensors are needed to notice the arrival of the layer, for one's face will feel the sudden chill that signifies the presence of cool, moist air from below even before the clouds arrive. Sometimes the cloud layer will cease rising just below the observatory and remain parked in place during the afternoon. When this occurs, the upper portion of Mauna Loa is invisible from below, while from an aircraft it resembles an island floating on a sea of clouds. More often, the rising clouds sweep by and then rise above the observatory, often preceded by cumulus clouds from the southwest that play overhead as their gossamer fragments and streamers engage in a complex ballet that is a delight to watch.

The observatory is immersed in cool, damp fog when the orographic layer arrives. Measurements of the ozone layer, solar ultraviolet, the atmosphere's optical depth (haze), and any other parameter involving sunlight are impossible when clouds block the sun. Instruments that sniff the air to measure its burden of particles, carbon dioxide, ambient (ground level) ozone, and a dozen other gases are unaffected by the clouds and moisture, and their pumps continue sucking in air to be measured.

Light rain often accompanies the clouds. Occasionally a thunderstorm arrives and throws lightning bolts at the power line poles and even the observatory's cluster of buildings and structures. Instruments and computers are sometimes damaged by lightning strikes. Small hail often accompanies these storms, and a strong storm can create quite a racket as the hail pounds on the metal roofs of the buildings. Some hailstorms can drop an inch or more of ice onto the observatory and transform its desolate environs into a chilly, wintry wonderland.

After a hailstorm, large expanses of black lava may remain cloaked in white ice until the next day. The hail around the observatory itself melts much faster, for the metal roofs and the asphalt of the parking area and road form a heat island that is warmer than the porous lava. This phenomenon helps explains the MLO garden, a random assortment of small plants scattered about the lava near the buildings and pavement. Some of the plants bear clusters of flowers that attract a few insects. Both the plants and the insects are ordinarily found far down the slope of the mountain.

The catchment systems on the roofs of several buildings at MLO capture water from the storms. When there is hail, the corrugated metal roof of the original building from 1956 serves as a mold for the sheet of ice that forms.

After the storm, the ice slides down toward the rain gutter and decorates it with an undulating, translucent border that eventually drips away.

By late afternoon, the cloudy flags of the inversion layer return to their normal elevation far down the slopes of the mountain. The sky quickly becomes as clear and dry as it was prior to the passage of the inversion layer.

The upper regions of Mauna Loa cool rapidly after sunset, and the freezing point can be reached even on spring and summer nights. A distinct change in wind direction is observed after sunset, as the breeze switches from the north or northeast to the south. Or at least that is its apparent direction, for this is a mountain, and one cannot be sure. Down in Hilo, the trade winds are still arriving from the Pacific. But Mauna Loa is so large that it creates its own weather. Dense, cool air from the free atmosphere above the island sinks over the mountain's summit and flows gently downward toward the observatory. So while the air appears to be arriving from the south at MLO, it is actually arriving from far above the mountain.

This is the downslope breeze favored by atmospheric chemists, who measure with precision almost unimaginably minuscule traces of gases in the air at MLO. They are especially interested in gases that are by-products of human activity, including carbon dioxide and methane, which have the ability to warm the earth, and the chlorofluorocarbons, which can cause both warming and destruction of the ozone layer that shields the earth from the sun's most dangerous ultraviolet rays. The clean air usually remains at MLO through midmorning, which is the best time to make measurements that require an ultraclear view of the sun. What happens thereafter is up to the inversion layer.

The Air Quality

On an especially clear day, a liter of air at the Mauna Loa Observatory may contain fewer than a dozen particles of matter larger than a micrometer across. Far below MLO, the situation is very different, for there both the chemistry of the sky and its burden of particles are dominated by an active volcanic vent on the slopes of Kīlauea that has been pouring molten rock into the Pacific off the southeast side of the island since 1982. When the lava plunges into the water, a huge plume of acidic steam erupts from the ocean and rises high in the sky to become a stream of cumulus clouds. A mist laden with bits of volcanic glass and hydrochloric acid falls from the steam cloud, and signs warn against staying in the area when the steam cloud is overhead. Those who stay might find that the mist falling from the cloud tastes a bit like lemonade.

The steam cloud causes far less air pollution than the lava itself, which emits vast quantities of sulfur dioxide gas. The gas reduces visibility only after it becomes tiny particles of sulfate upon which water vapor readily condenses. As the particles become coated with a film of liquid water, they gradually grow in size until they scatter light very efficiently. Eventually the sky is covered by a thick veil of gray. Residents of Hawai'i call this end product "vog" (from volcanic fog). On especially voggy days, the stuff looks much like smoke drifting across the lava and through the palm trees (Mims 2006). Vog increased significantly after March 12, 2008, when a new vent opened in the wall of Halema'uma'u, a large crater embedded in the summit crater of Kīlauea.

Why Place a World-Class Observatory on a Hawaiian Volcano?

The purpose of the Mauna Loa Observatory is to conduct long-term monitoring of sunlight and to probe, sample, count, and measure stuff in the air. The ozone layer high above the observatory is measured. Particles in the ambient air are collected and counted, and the concentration of a few dozen gases is measured with extraordinarily high accuracy. All this is done high in the free atmosphere over Hawai'i with minimal interference from vegetation and nearby human activity. Measurements made at MLO from as far back as 1956 provide a baseline against which subsequent measurements can be compared to look for hard evidence of changes in climate and the composition of the atmosphere. No better example of this is the observatory's measurements of the concentration of carbon dioxide, which began in March 1958 and have continued virtually uninterrupted for more than half a century.

Because carbon dioxide is an essential nutrient for plants, its concentration fluctuates around seven or eight parts per million (ppm) with the seasons. Thus, the measurements at MLO record the respiration of the earth. This oscillating seasonal signal is superimposed over the natural background level of carbon dioxide and its annual increase caused by the combustion of wood, coal, natural gas, and oil products. When the measurement of carbon dioxide was begun by Dr. Charles David Keeling in the spring of 1958, the concentration of carbon dioxide in the air at MLO was 316 ppm. By the summer of 2011, the concentration had reached 393.7 ppm (Scripps 2011), an increase of 24.6 percent.

Ordinarily an active volcano in Hawai'i would seem a completely unsuitable home for a baseline atmospheric and meteorological observatory. Indeed, as noted earlier, a lava flow from the 1984 eruption severed the power line to MLO for a few months. The observatory itself was undamaged and back in operation after a borrowed diesel generator was temporarily connected to the undamaged power line at the 8,300-foot (2,530-meter) level. Yet despite the presence of a lava diversion dike, the possibility remains that some or even all of the ongoing experiments and long-term observations conducted at the observatory might be damaged or destroyed by a future lava flow. This raises the obvious question about the wisdom of placing a world-class atmospheric and meteorological observatory on a mountain in Hawai'i when higher mountains with road access are available in the continental United States.

Consider that the summits of Colorado's Mount Evans and Pikes Peak are accessible by good roads, and neither mountain is an active volcano. Both mountains are higher than any of Hawai'i's peaks, with Mount Evans being 14,264 feet (4,348 meters) and Pikes Peak 14,110 feet (4,301 meters) (Forest Service 2006). Moreover, both of these mountains are convenient to the nerve centers for climate and atmospheric research in the United States, including MLO's parent organization, the Global Monitoring Division of the National Oceanic and Atmospheric Administration and the National Center for Atmospheric Research in Boulder.

A summer drive to the summits of Pikes Peak and Mount Evans will quickly demonstrate why these mountains are unsuited for the kind of baseline measurements made at the Mauna Loa Observatory. The most obvious reason is that the significant volume of automobile traffic on both mountains would contaminate delicate measurements of carbon dioxide, particulate matter, and sunlight. Pikes Peak is much worse in this regard, for, according to the City of Colorado Springs, in 2007 some 300,000 visitors made the drive to the summit along an unpaved, dusty road. The summit parking area can be filled by many scores of vehicles and hundreds of tourists enjoying the view. The road to Mount Evans is paved, and, according to the City of Denver, an average of 13,338 cars drive to Mount Evans each day it is open.

The proximity of vegetation is another issue. While the summits of both Pikes Peak and Mount Evans are above the timberline, upslope winds carry to the summits of both mountains the by-products of the forests far below. During July 1995, the summits of these mountains were evaluated by the author while making alpine measurements of solar ultraviolet and the clarity of the sky. The updraft at Pikes Peak was so strong that hang gliders rose above the summit as their pilots leapt from their takeoff points. The same updraft propelled across the summit a substantial assortment of dust particles, pine needles, detritus from the forest floor, and many insects. The breezy conditions also caused plumes of dust behind arriving and departing vehicles. The lesson is clearer than the air at the summit: Pikes Peak provides a superb view of the mountains and plains of Colorado, but it lacks the pristine conditions required for precise baseline measurements of the sun, sky, and atmosphere.

Regional air quality is an important consideration, even atop mountains like Pikes Peak and Mount Evans. Pollution from cities can reach the summits of both mountains, as can dust storms and smoke from forest fires.

Still another issue is security, for the delicate instruments of an atmospheric observatory are best placed in an isolated location secured against vandalism and theft. Even a well-meaning, curious tourist might inadvertently cause problems simply by leaving a fingerprint on the quartz dome of an ultraviolet monitor or by touching the moving shadowband of a sunlight radiometer.

Finally, winter snow accumulation makes access to the summits of mountains like Mount Evans and Pikes Peak unpredictable at best. The Pikes Peak road is open all year when snow has been cleared. The Mount Evans road is open from July 4 to Labor Day.

Mauna Loa bests Pikes Peak and Mount Evans in all these considerations. Snow only rarely blocks the road, and the giant mountain's vast fields of virtually lifeless lava isolate the observatory from the influences of dust, vegetation, and human activity. On a typical day, only several vehicles drive the observatory road. And thanks to the elevation of the observatory and the thousands of miles of open ocean that surround Hawai'i, the air over Mauna Loa is probably as clean as air can be in the Northern Hemisphere.

Yet even Mauna Loa is not the perfect site for an atmospheric observatory, for maintaining the road has long been a major problem, and there remains the threat of a future volcanic eruption. Moreover, the giant volcano leaks carbon dioxide and sulfur dioxide, both of which sometimes drift up to MLO and cause spikes and bumps over the normal background levels of these gases. Because the wind direction is carefully measured, eliminating interference caused by the volcano has long been a routine step in evaluating background measurements of these gases. As for the physical threat to the observatory posed by the volcano, according to the Hawaiian Volcano Observatory, 98 percent of the volcano's surface is covered with lava less than 10,000 years old (Lockwood 1995). Placing the observatory on an active volcano was a calculated risk made under the assumption that the site would provide many years of uninterrupted baseline measurements of the pristine sky over Hawai'i. It's a gamble that has paid off for more than half a century.

A Spring Day on the Mountain

First light on a typical spring day at the Mauna Loa Observatory begins when a thin orange glow emerges over the clouds overlying the Pacific east of Hilo while stars are still sprinkled across the black sky above. The air is cold, and sometimes there are traces of ice on the boardwalks between the buildings. The breeze from the south means clean air from the free troposphere is flowing down the slope of the big mountain.

The twilight glow gradually spreads along the horizon as it bulges upward into the darkness. Eventually an orange arch pushes high into the purplish sky, its base growing brighter by the minute as the eastern sky transitions from violet to blue. Thin, dark layers over the horizon are sometimes visible, especially during spring. They suggest the presence of Asian dust or air pollution, a phenomenon that was not reported during the early days at the observatory. While the twilight glow is rising in the eastern sky, the antitwilight glow is setting over the summit of Hualālai in the west. It forms a pink arc in the sky, and the crisp, clear air over Mauna Loa assigns it much more prominence than when it is observed from sea level. Looking back to the east, suddenly there is a brightening at the base of the arch in the sky, and within half a minute the upper edge of the solar disk bursts into view over the cloud deck cloaking the Pacific. It's sunrise at the Mauna Loa Observatory.

Arrival of the MLO Team

While the instruments that measure sunlight are awakening, many other instruments at the observatory have been working all night. These include an impressive array of meteorological instruments that continuously measure temperature, dew point, relative humidity, pressure, and wind speed and direction. More specialized are instruments that continuously count particles in the air and gas analyzers that measure carbon dioxide, ozone, sulfur dioxide, methane, and the CFCs and other halocarbons. All these instruments rely on noisy pumps that force air collected from the forty-meter tower or the Keeling Building mast through their sampling filters or chambers.

The window of time for the calibration of some of the instruments that stare directly at the sun begins around half an hour after sunrise. These are the instruments that track the sun and measure aerosols, the ozone layer, and water vapor in the atmosphere over MLO. Most are perched on a large deck over the NDACC (Network for the Detection of Atmospheric Composition Change) Building, the largest structure at MLO. Others are installed in telescope domes or on a concrete pad resting on the lava. They are best calibrated during the early morning during clean downslope conditions and when the position of the sun ranges from about 10 to 30 degrees over the horizon.

About an hour after sunrise, one of the MLO Solar Observatory staff arrives to begin a long day of work imaging the solar disk with specialized telescopes installed in a pair of domes south of the central buildings at MLO. Around 9:30, two or three staff members from the Mauna Loa Observatory arrive from Hilo. They bring supplies, replacement parts, empty air sampling flasks, and lunch. After unloading their vehicle and unlocking the original Mauna Loa building, now named the Keeling Building, the staff retrieve their clipboards and check sheets and quickly begins their rounds, checking dozens of instruments scattered around the site. Each staff member has specific responsibilities, and work on the mountain is generally limited to no more than three days a week. Therefore, while some instruments are checked every workday, others are checked two or three times a week. The schedule changes over time, so in the account that follows, all the tasks have been merged together for a hypothetical day.

More than a dozen solar instruments mounted on the solar deck over the NDACC Building must be checked and their optical surfaces cleaned. Special attention is given to the two Canadian Brewer spectrophotometers that measure the ozone layer and solar ultraviolet. While the data are sent directly to Canada's Atmospheric Environment Service via the Internet, a pair of computers in the NDACC Building below the deck permits the operation of the Brewers to be checked so that adjustments can be made when necessary.

Also in the NDACC Building are some of the ultrasensitive gas analyzers that sniff the atmosphere for carbon dioxide, methane, and a variety of other greenhouse gases that have the ability to trap infrared emitted by the earth and cause warming of the planet. Tall, heavy cylinders of calibration gas must be checked to be sure they contain an ample supply of gas. The freeze trap that removes water vapor from the air being sampled for carbon dioxide must be checked, for water vapor—by far the most potent of all the greenhouse gases—must be entirely removed to avoid artificially inflating the measurements.

In the Keeling Building are found instruments that capture or count particles in the air and measure the ambient ozone. There is also the Scripps Institution of Oceanography carbon dioxide analyzer. The original Model 70 Applied Physics Corporation Infrared Analyzer installed in the building in March of 1958 was retired a few years ago, much to the relief of a staff weary of the temperamental nature of the ancient—but accurate—vacuum tube instrument. Today Scripps is beginning its second half century of precisely measuring carbon dioxide with an analyzer as modern as the one used by NOAA in the NDACC Building. All these instruments are carefully inspected by the staff, and their operational status is checked off on the clipboard forms.

Varieties of instruments are mounted outside on the lava, and they too must be checked. This includes NASA's array of solar-powered, robotic sun photometers perched on a concrete pad west of the NDACC Building. Immediately south of the building is NOAA's GPS receiver system that measures the water vapor in a column through the atmosphere and automatically sends its data to Boulder via the Internet. While the GPS system needs no attention, the staff checks to make sure that current meteorological data are also being sent to Boulder, for when the temperature and wind data stop flowing, the GPS data are not plotted on the Web site.

In the midst of the checking and inspecting, one of the staff unlocks the tiny door of the Dobson dome and begins to make precise measurements of the ozone layer using an old Dobson spectrometer almost identical to the one that has been in the same location since 1957.

In the days before computers, the output of most instruments was automatically inscribed on rolls of chart paper. Today most data is sent directly to Boulder and various universities over the Internet. NASA's array of robotic sun photometers sends data directly to the Goddard Space Flight Center in Maryland via a satellite link. While the staff does not see much of the data unless they look for it on the Web, the recipients on the mainland are quick to inform them when there is a problem, which explains why the phone sometimes rings even before the staff arrives.

While the MLO mountain crew is at work, once a week a staff member in Hilo drives to the Coast Guard lighthouse at Cape Kumukahi to collect samples of air to be sent to the Global Monitoring Division in Boulder and the Scripps Institution of Oceanography in San Diego. Also once each week, an ozonesonde suspended from a weather balloon is launched from the National Weather Service office in Hilo to sample the ozone between sea level and the upper stratosphere.

Back on the mountain, as the morning progresses the cloud deck far below the observatory might remain locked in place, or it might begin rising upward. When the latter occurs, staff on the mountain will periodically glance down to the northeast to keep track of the rising clouds. Experienced MLO hands will also keep an eye on the sky to the southwest. The slope of the mountain blocks the view of any clouds in that direction until they suddenly spring into the sky and quickly swirl overhead. What happens next is unpredictable. The observatory may simply be covered by clouds. Or the cloud deck below may rise up and suddenly envelop the station in thick fog.

By noon, all the instruments have been inspected and the ozone layer has been measured two or more times. It's time for a lunch break and a chance to discuss any problems that might have been noted during the morning routine. Thereafter, the team works on troubled instruments, updates check sheets, replaces empty gas cylinders with full ones, and performs routine maintenance on everything from sophisticated analytical instruments and computer networks to vacuum pumps, air lines, and water pipes. When there is spare time, a timeworn shed or building might receive a fresh coat of paint.

Around 2:30, the MLO team packs shipping boxes with their cargo of fragile air-sampling flasks and a few bags of trash into their van and heads back to the Hilo office. When an occasional thunderstorm approaches the observatory, the staff pays special attention to the sound of thunder and may leave the station early to avoid the hazard posed by lightning (Plate 49).

The departure time is occasionally altered. Once every week or two the MLO mountain team stays past 4:00 to await the arrival of replacement gas cylinders and a large liquid nitrogen tank carried by a truck that also services the observatories atop Mauna Kea. The MLO staff assists the driver with unloading the heavy tanks and then helps him load the empty cylinders onto the truck.

Sunset

Even on days when the inversion layer rises near or above the observatory, the clouds have usually thinned significantly by sunset. As the sun descends toward the western horizon, the antitwilight glow begins to rise from the eastern horizon. Once a month, the simultaneous rising of the full moon and the deep pink antitwilight glow provides a spectacular sight in the eastern sky. During winter months, the shadow of Mauna Loa is visible over the north slope of the huge mountain, where it forms a purplish, triangular shadow in the antitwilight glow.

Turning back toward the west, occasionally the proverbial green flash can be seen just as the uppermost edge of the sun slips behind the cloud deck. The phenomenon is usually a brief green glow rather than the much rarer green flash.

If clouds are still present at or above the altitude of the observatory, they provide the backdrop for a strikingly beautiful display. Sometimes a series of sunsets occurs as the solar disk descends in sequence behind cloud layers stacked over the western horizon.

The colors in the twilight display provide clues about the clarity of the sky, for when the sun nears the horizon, its rays traverse a much longer path through the atmosphere than when it is high in the sky. The result is that the violet, blue, and green wavelengths are scattered away, leaving behind yellow, orange, and red (Minnaert 1940, 1993). Thus the sun reddens as it sets and illuminates clouds with increasingly orange and red wavelengths as it approaches the horizon. If there is dust in the sky, it too is reddened by the setting sun and may provide especially brilliant sunsets, unless there is so much of it that it absorbs more light than it scatters toward the observer. After the sun falls below the cloud deck, the twilight arch becomes increasingly red as it descends and forms a vivid crimson cap over the horizon.

If a lidar session has been scheduled to probe the stratosphere with a laser beam, after dark a thin green beam will be seen emerging from the roof of the NDACC Building. In his combined office and lab inside, MLO Station Chief Dr. John Barnes checks the power of the laser beam and slides a mouse across his desk to activate the program he developed to visualize the scattering of the beam by any aerosols that might be floating in the stratosphere high over MLO. Meanwhile, a scientist from the Jet Propulsion Laboratory (JPL) arrives to activate NASA's ultraviolet lidar, whose invisible beam measures the ozone layer while NOAA's lidar probes for aerosols and water vapor.

The lidar sessions end before midnight, the lights in the NDACC Building are switched off, and Dr. Barnes and the occasional visiting scientist retire to the two tiny bedrooms. By now sleep is very much on the mind of those who have been at the observatory since before sunrise. Sometimes the sleepy scientists are treated to a surprise light show while sliding into their sleeping bags, for, when the air is especially dry, streamers of crackling sparkles light up the interior of their bags.

Some visitors experience difficulty sleeping, especially during their first night or if they have a hint of a cold. The problem may be due as much to the dryness of the air as to the lack of oxygen, and a few spurts of saline solution in one's nostrils can make breathing much easier for the visitor in good health. Those with a cold or who experience altitude sickness should avoid staying overnight.

Having company while overnighting is helpful, for, after the lights are out, inexperienced visitors are sometimes startled when one of the resident Mauna Loa wolf spiders pays a visit or by the cacophony of sounds produced by the shutters, fan motors, air sampling pumps, and ventilators that drone away in the background or switch on and off at intervals. As Station Chief Barnes (2006) has observed, "Yes, at night if you are up there alone, it gets a little spooky."

Spooky indeed: Lonely scientists who realize they are the only human being spending the night on the world's largest mountain can turn their thoughts to other topics, like the huge crater looming a few thousand feet up the slope of the mountain. Will tonight be the night it erupts? Will there be any warning? Was former MLO director Russell Schnell serious when he said to grab the heavy Dobson spectrophotometer and run downhill should there be an eruption? Will molten lava melt the base of the forty-meter tower and vaporize the hapless scientist who flees to its top for safety? After the lonely scientist finally falls fitfully to sleep, the mountain might provide a reminder of who's really in charge in the form of a small earthquake that pushes the scientist's feet soundly against the wall to awaken him and initiate an even more animated cycle of spooky thoughts.

During its first decade, the observatory was manned around the clock, but today overnight stays are usually unnecessary, for, with the exception of the lidars, nearly all the instruments at the observatory are fully automated. Today staying overnight is beneficial for those who conduct occasional audits of the observatory's instruments and for those who must arise early to begin manual calibrations of sun photometers that are pointed toward the sun as it rises in the sky. The calibration session lasts only a few hours, but it's best performed when rested, not immediately after the tedious, bouncy drive from Hilo, often through thick fog. Staying overnight also provides the opportunity to experience the life of old-timers on the mountain back in the 1950s and 1960s, who observed and photographed the sunrises and sunsets, no two of which are alike. Sunsets watched through the crisp, cool air at MLO are visual treats that provide the perfect ending to a long day at the observatory. And there's no better way to begin the next day than walking carefully up the ice-coated steps of the solar radiation deck to watch the twilight glow slowly rising in the east while inhaling the pure downslope air flowing down from the summit.

The Old-Timers

The occasional scientist who stays at MLO for a week or ten days at a time will experience only some of what the old-timers did, for the routine on the mountain today is far less rigorous than when the observatory was manned twenty-four hours a day from 1957 through the mid-1960s. Back then it was essential to keep someone on the mountain to maintain the noisy diesel generators that provided electrical power and to fix the carbon dioxide analyzer when it ceased operating properly. Overnighting on the mountain became unnecessary only after the power line was installed in 1966. For decades the observatory has worked so well on its own that it is unmanned on weekends and holidays, and modern computer data acquisition systems long ago replaced nearly all the finicky pen-and-ink chart recorders.

The old-timers who used to stay on the mountain for four or more days at a time still recall the difficulties they faced. But they also remember how their well-being and work output improved after the first day on the mountain allowed them to become acclimated to the reduced oxygen at the observatory. The day crews that followed them don't have that advantage, and some report mild discomfort from the altitude change.

But even the old-timers at the observatory forty to fifty years ago faced few of the risks and challenges encountered by the ultimate old-timers who more than two centuries ago first ascended Mauna Loa for the purpose of making scientific observations and measurements. They had no roads, vehicles, or modern hiking boots. They had no electricity, and battery-powered flashlights had not been invented. Their only maps were the ones they drew, and GPS receivers were beyond imagination. Navigation was accomplished by dead reckoning, compass, and sextant. A barometer was considered portable when its delicate glass tube could be carried inside a padded leather case slung across the back along with a flask of mercury required for its operation. Back then there were no structures on the mountain, water holes were scarce, and food sources were nonexistent. Yet early scientific explorers managed to conduct pioneering research at the summit of the great mountain. They pioneered the science that has been performed every day at the Mauna Loa Observatory since June of 1956. Thus it is fitting that this history of today's Mauna Loa Observatory begins with the story of the scientific pioneers who led the way. We begin by traveling back to the year 1794, to join the first scientific expedition to the summit of the world's largest mountain.

References

Barnes, John. 2005. January 2005 activities report, Mauna Loa Observatory. Monthly Reports (CD-ROM), Hilo, Hawai'i.

———. 2006. Video interview, Hilo, Hawai'i (December).

Cheever, Rev. Henry T. 1856. *Life in the Sandwich Islands*. A. S. Barnes and Company, 115.

Earth Observatory. 2005. Snow on the summits of Hawaii. Earth Observatory, NASA. (Accessed at http://earthobservatory.nasa.gov/Newsroom/NewImages/images.php3?img_id=16796.)

Forest Service. 2006. Pike National Forest. US Department of Agriculture. (Accessed at http://www.fs.fed.us/r2/psicc/about/history/.)

Fosberg, F. R. 1959. Upper limits of vegetation on Mauna Loa, Hawaii. *Ecology* 40: 144–146.

Harpel, C., ed. 2006. Lava: Volcano world. Department of Geosciences, Oregon State University. (Accessed at http://volcano.und.edu/vwdocs/vwlessons/lava.html.)

Hawaiian Volcano Observatory. 1999. Pahoehoe and 'a'a lava flows. *Volcano Watch*, May 27. (Accessed at http://hvo.wr.usgs.gov/volcanowatch/1999/99_05_27.html.)

Jarves, James Jackson. 1843. *History of the Hawaiian or Sandwich Islands*. James Munroe and Co., 253. (Accessed at http://books.google.com.)

———. 2006. Mauna Loa, Earth's largest volcano. (Accessed at http://hvo.wr.usgs.gov/maunaloa/.)

Juvik, James O., D. C. Singleton, and G. G. Clarke. 1978. Climate and water balance on the island of Hawaii. In John Miller, ed., *Mauna Loa Observatory: A 20th Anniversary Report*, US Department of Commerce, 129–139. (Accessed at http://www.mlo.noaa.gov/pdf/

maunaloa20thanniversary/maunaloa20th101-158zip7.0later.pdf.)

Lockwood, J. P., 1995. Mauna Loa eruptive history—the preliminary radiocarbon record, Hawai'i. In J. M. Rhodes and J. P. Lockwood, eds., *Mauna Loa Revealed: Structure, Composition, History, and Hazards*. American Geophysical Union Monograph 92: 81–94. (Summary accessed at http://hvo.wr.usgs.gov/maunaloa/history/main.html.)

Lockwood, J. P., and P. W. Lipman. 1987. Holocene eruptive history of Mauna Loa Volcano. In R. W. Decker, T. L. Wright, and P. H. Stauffer, eds., *Volcanism in Hawaii*, US Geological Survey Professional Paper 1350, v. 1, 518. (Summary accessed at http://hvo.wr.usgs.gov/maunaloa/history/historytable.html.)

Mims, Forrest M. 2006. Hawaii's biggest air polluter: Forrest Mims' world of science. *The Citizen Scientist*. (Accessed at http://www.sas.org/tcs/weeklyIssues_2006/2006-10-20/mimsci/index.html.)

Minnaert, Marcel. 1940. *Light and Colour in the Open Air*. G. Bell and Sons. (Reprinted in 1954, Dover Publications; revised by Len Seymour and republished as *Light and Color in the Outdoors*, New York: Springer-Verlag, 1993.)

Price, Saul. 1983. Climate. In *Atlas of Hawaii*, 2nd ed. Honolulu: University of Hawai'i Press, 59–61.

Price, Saul, and Jack C. Pales. 1963. Mauna Loa Observatory: The first five years. *Monthly Weather Review* 91 (October): 665–680.

Saddle Road. 2005. Executive summary, S.5.3 biological resources. Central Federal Lands Highway Division, Federal Highway Administration. (Accessed at http://www.saddleroad.com/.)

Scripps. 2011. Data, in situ CO_2, Mauna Loa record. Scripps CO_2 Program, Scripps Institution of Oceanography. (Accessed at http://scrippsco2.ucsd.edu/.)

Simkin, T., and L. Siebert. 2002–2008. Global volcanism FAQs. Smithsonian Institution, Global Volcanism Program Digital Information Series, GVP-5. (Accessed at http://www.volcano.si.edu/faq/.)

Sparwasser, Nils, Thorsten Andresen, Stephan Reiniger, and Robert Meisner. 2006. Hawaii: The highest mountain on Earth. *Science* 313: 1730–1735. Visualization by the German Remote Sensing Data Center, German Aerospace Center. Landsat-7 imagery from the National Geophysical Data Center (NGDC), University of Maryland, and the Global Land Cover Facility (GLCF) of the US Geological Survey (USGS). Copyright 2006 by the German Aerospace Center (DLR). Used with permission.

Twain, Mark. 1866. At sea again. *Sacramento Daily Union*, August 18 (republished in *Roughing It*, 1871). (Accessed at http://www.fullbooks.com/Roughing-Itx82927.html.)

———. 1908. Letter to the Hawaiian Promotion Committee. In Albert Bigelow Paine, 1912, *Mark Twain: A Biography*, vol. 3, part 2, 1907–1910: 1474. (Accessed at http://www.gutenberg.org/files/2987/2987.txt and at http://books.google.com.)

Volcano World. (Accessed at http://volcanoworld.wordpress.com/.)

CHAPTER TWO

The First Scientific Expeditions (1794, 1825, and 1834)

The remoteness of the Hawaiian Islands preserved them from discovery by Europeans for a millennium or more after the arrival of Polynesians. There are no known written records of scientific observations on Mauna Loa prior to the arrival of Europeans. Yet the Hawaiians paved the way for expeditions to come by establishing the ancient ʻAinapō Trail to the eastern rim of the summit crater.

In 1778 the British naval officer Captain James Cook, the greatest explorer of his time, led the first confirmed European expedition that visited Hawaiʻi. The Hawaiian Archipelago was spotted during Cook's third major expedition when he was in command of the HMS *Resolution* and HMS *Discovery*. Cook named the chain the Sandwich Islands after his friend and sponsor, John Montagu, the Fourth Earl of Sandwich.

Cook and his men mapped the coast of Hawaiʻi and other islands in the archipelago. While a brief attempt by some of Cook's men to scale Mauna Loa was unsuccessful, Lieutenant James King used trigonometry to estimate that Mauna Loa's summit was 16,020 feet (4,883 meters) above the Pacific Ocean (Hawaiian Volcano Observatory 2001a). After completing their work at Hawaiʻi, Cook and four of his men perished at Kealakekua Bay on the island's west coast in 1779 during a conflict that arose after one of their boats was stolen.

When Cook's ships left Hawaiʻi without their commander, they had no idea that almost exactly fifteen years later another British expedition would depart from the same bay in a boat paddled by friendly Hawaiians. This expedition's goal was to ascend Hualālai and Mauna Loa and make the first scientific measurements from both of their summits.

Archibald Menzies (1794)

Archibald Menzies was a surgeon from Scotland who studied botany, chemistry, and medicine at the University of Edinburgh from 1771 to 1780. In 1782 Menzies became a surgeon in the Royal Navy, and in 1784 he was assigned to Halifax, Nova Scotia. In his spare time, Menzies collected botanical specimens and seeds for the famed naturalist Sir Joseph Banks, who was then president of the Royal Society and who had served as a botanist under Captain James Cook. In 1786 Menzies traveled to England, where he studied botany under Banks. The noted scientist recommended Menzies for service as surgeon aboard the *Prince of Wales*, which in October 1786 departed on a three-year voyage to the western coast of North America and China. During this voyage, Menzies spent a month collecting specimens at Nootka Sound on Vancouver Island (Stearn 2000).

Menzies' experience and Banks' recommendation perfectly qualified him for service as both surgeon and naturalist under George Vancouver, who was to command the British sloop HMS *Discovery* and the tender HMS *Chatham* on a circumnavigation of the world in the tradition of the grand scientific and exploration voyages of Captain Cook. Vancouver had been with Cook in 1779 when the famous explorer was killed at Kealakekua Bay, and he himself was injured during the skirmish. The planned voyage would include extended visits to the Pacific Coast of North America and the Sandwich Islands.

Discovery and *Chatham* left England on April 1, 1791, with Menzies performing a variety of scientific tasks in addition to his duties as the ship's surgeon. He kept a detailed record of the ship's coordinates, the temperature of the air and water, the barometric pressure, and the wind. A page in his journal reproduced by Walker (2005) and entitled "From Nootka to Monterrey" illustrates the meticulous nature of Menzies' meteorological observations in its record of changes in weather associated with barometric pressure. For example, on November 4, 1792, Menzies recorded, in the units of the time, that the pressure was 29.7 inches of mercury, the temperature of the air measured 65 degrees Fahrenheit, the wind was from the south, and there was a "moderate breeze with rain." The following day, the pressure rose to 30.3 inches, the temperature fell a degree, the wind changed from the west, and the sky was clear with a light breeze. Regular meteorological observations such as these prepared Menzies for what was to come.

Menzies' Portable Barometer

Menzies was especially intrigued by measurements of the pressure of the atmosphere provided by the barometer, an instrument invented in 1643 by the Italian scientist Evangelista Torricelli. The barometer functions as a scale that weighs the column of air over the instrument. Changes and the rate of changes in the pressure measured by a barometer provide valuable information about the movement of air masses and enable forecasts of the weather associated with the arrival and passage of low- and high-pressure masses of air. Besides being thoroughly acquainted with the relationship of atmospheric pressure and weather, Menzies knew that the barometer could also serve as an altimeter to register the elevation of mountain peaks, for the pressure of the atmosphere declines with altitude in a predictable fashion. This knowledge and his spirit of adventure led him to become the first person known to make scientific measurements from atop Hawai'i's Hualālai and Mauna Loa. His instruments consisted of a thermometer and a portable barometer.

The barometer was given to Menzies by Colonel Robert Gordon when *Discovery* stopped at the Cape of Good Hope. Gordon had used the instrument during his explorations of Africa "through many, many hundreds of leagues" (Menzies 1794a). Today both a thermometer and a barometer can be worn on the wrist inside a digital watch that provides the time with an accuracy only dreamed of in the eighteenth century. So the description of the "portable" barometer that Menzies penned in his journal is worth reading.

> The simplicity of its contrivance and the ease with which it is carried and managed may probably render it preferable on such occasions to a more expensive instrument, which in ascending pathless, rugged mountains is liable to be broke. I shall therefore subjoin a short account of it, leaving the reader to judge of the probable degree of accuracy of the observations made with it at different heights, both in this and subsequent journey[s] to the summit of Mauna Loa.
>
> It consists of a simple straight glass tube about three feet long, filled with mercury, which is secured by a small stopper and a piece of soft leather tied over the end of it. This tube, together with a brass scale of about the same length, divided and subdivided into inches and tenths, are carried in a small wooden case lined with cloth, where the scale is fastened, but the tube has a little play, and a few ounces of mercury are to be carried in a stone or wooden bottle, which completes the apparatus.
>
> For observing with this instrument, the mercury in the bottle is poured into a small open cup. After the stopper is taken out of the tube, the vacancy in it is to be filled up brimful with mercury, then with a finger

> placed over the mouth of the tube, that end of it is to be inverted into the mercury in the cup, taking care not to withdraw the finger until it is immersed. The mercury in the tube will then instantly fall down to its proper height according to the pressure or weight of the atmosphere on its external surface at the time and place of its observation. All that is further necessary to be done is to hold the tube upright and measure with the brass scale the exact length of the column of mercury in the tube above the surface of the mercury in the cup, which will give the true height of the mercury at each station.
>
> I had but only one tube which I was fortunate enough to preserve whole in my different journeys, but to guard against accidents several of these tubes may be closely packed in the same case by rolling each of them up in a piece of cloth, and it is almost needless to observe that in carrying them the stopper ends should be kept uppermost. This case may be slung on a man's back and carried with ease and safety over the most rugged mountains. (Menzies 1794a, 152–153)

The man's back referred to by Menzies was not always his, for he employed a servant to carry his portable barometer during at least part of his ascent of Mauna Loa.

The Ascent of Hualālai

Vancouver's expedition spent the winter of 1793–1794 in Hawai'i, where Menzies became acquainted with King Kamehameha I. On January 15, 1794, he met with the king to discuss his interest in ascending "a considerable mountain of conic form called Hualalai, situated a little to the southward of Kawaihae Bay" (Menzies 1794a).

Hualālai, one of the five principal volcanoes that form the island of Hawai'i, last erupted in 1801. Fortunately for Menzies, Hualālai was resting at the time of his stay in Hawai'i. King Kamehameha approved his proposal to climb the mountain and provided guides, security, and a double-hulled canoe to Menzies and his team from *Discovery,* which was anchored in Kealakekua Bay, where Captain Cook had been killed fifteen years earlier. In return, Menzies gave the king the key to his cabin on *Discovery,* which he suspected that the king wanted to use to entertain his lady friends (Menzies 1794a, 148). On January 16, Menzies and his team of shipmates and Native Hawaiians rowed north to Kailua Bay, where Kailua-Kona is located.

The expedition began its ascent on January 17 and reached the summit two days later, where Menzies was elated by the spectacular views of the other great mountains of Hawai'i and the Kona Coast below them (Menzies 1794a, 159–160). The summit of the giant dome of Mauna Loa loomed 2,900 feet (884 meters) over Hualālai. Within Menzies' field of view of the north slope of the great mountain was the future home of today's Mauna Loa Observatory, from which Hualālai is clearly visible when the inversion layer caps the clouds that otherwise rise over it.

The expedition enjoyed pleasant conditions at the summit for two days, during which some of the Hawaiians made offerings to the volcano and Menzies studied the various plants on the summit. He also found signs of recent eruptive behavior, and on both days he measured the atmospheric pressure and temperature.

> At both noons I observed the barometer on the top of the mountain. The first day it stood at 22 in. 40 pts., and the second day at 22 in. 44 points. So if we take the mean of these two observations, 22 in. 42 pts. [f]rom 30 in. 10 pts., what it stood at on the sea side, the remainder, 7 in. 62 pts. is the quantity it fell or sunk in our ascent of the mountain, which will give its height 8,062 feet above the level of the sea. (Menzies 1794a)

The elevation given here differs significantly from that in the same passage from the journal account that Menzies published thirty-five years later in the *Magazine of Natural History* (Menzies 1829), in which he gave an elevation of 8,457 feet (2,578 meters). Perhaps the latter number was corrected for the temperature, about which Menzies wrote in his journal account.

> But it is necessary to observe that the correction for temperature of the atmosphere has not been allowed for in

> this calculation, nor at any other station upon the mountain, which will make some difference in the result of the observations. (Menzies 1794a, 198)

The elevation of Hualālai given by the Hawaiian Volcano Observatory (2001b) is 8,271 feet (2,521 meters), which is 2.6 percent higher than what Menzies reported in his journal (8,062 feet) and 2.2 percent lower than what he reported in the same passage that he quoted in the *Magazine of Natural History* (8,457 feet). As demonstrated in the passage above, when Menzies made more than one measurement from the same location, he gave the mean as the result. Applying this approach to the uncorrected and corrected measurement for the elevation of Hualālai gives a mean of 8,259.5 feet (2,517.5 meters), which is within 0.1 percent of the actual elevation.

Menzies also measured the temperature at the summit and elsewhere on the mountain, both for its own merit and for correcting the barometer.

> The thermometer observed at the time was 65 degs. on the first day and 64 degs., on the second, and at seven on both mornings it was at 45 degs. But the first evening at sunset it was so low as 38 degs., and the second evening it stood at 41 degs., which showed in these instances that the evenings were colder than the morning, just the very reverse to what we experienced in the woody region of the mountain. (Menzies 1794a, 161)

After a difficult and failed attempt to reach Mauna Loa from Hualālai, Menzies and his party turned back around the south side of Hualālai and descended to the coast. They arrived in the vicinity of Kailua-Kona on January 25 and hired canoes to take them the final few miles back to Kealakekua Bay.

The Expedition to Mauna Loa

King Kamehameha was so pleased by the success of the Hualālai expedition that he agreed to a second attempt at climbing Mauna Loa during a meeting with Menzies on February 5 (Menzies 1794a, 175). The following day, Menzies and three crew members from *Discovery* and *Chatham* headed south from Kealakekua Bay with Chief Luhea and about twenty paddlers in a large double-hulled canoe. A second double-hulled canoe accompanied them.

The adventures experienced during the canoe trip to near the southernmost point of Hawai'i were recorded by Menzies in his journal (Menzies 1794a). After leaving the canoes at Pakini village on February 9, the expedition headed north along the southeastern slope of Mauna Loa, which forms most of the southern half of the island of Hawai'i. They passed through Kī'olokū and Punalu'u and reached a plantation called Kapāpala on February 12.

The next morning the expedition continued through the Kapāpala plantation until turning upward toward the summit of Mauna Loa, now being led by local guides who knew the way. From Menzies' narrative, the route was most likely the ancient 'Ainapō Trail. According to Bier's map of Hawai'i (1988), the trail proceeds northwest toward the summit crater from the present Hawai'i Belt Road (Highway 11) about 10 miles (16 km) southwest of the Kīlauea caldera.

An eruption of the Kīlauea volcano had made their time at Kapāpala unpleasant, and the conditions did not improve as the party headed up the mountain. Menzies wrote in his journal,

> We had the [Kīlauea] volcano to our right most part of this day and in the forenoon the smoke and ashes arising from it made the air very thick, which at times proved very tormenting to our eyes. (Menzies 1794a, 188–189)

Late on February 15, ten days after leaving Kealakekua Bay, Menzies and his party arrived within range of the summit. The party was reduced to about a dozen men, for most of the natives had begged to stay behind at the previous camp due to the cold and their lack of suitable clothing and shoes. Menzies recorded in his journal,

> As we had now taken up our abode at the lower edge of the snow, I observed the barometer at six in the evening, when it stood at 19 in. 80 pts., which in altitude is equal

> to 11,515 feet, and the thermometer at the same time was at 33 degrees.

They were about 325 feet (100 meters) higher and nearly due south of where the slope unit of the Mauna Loa Observatory would be constructed on the north slope of the mountain 162 years later. Menzies described the evening's sole nourishment in his journal:

> We managed to boil the chocolate in a tin pot over a small fire made of our walking sticks, and each had his share of it warm, with a small quantity of rum in it, before he went to bed. We had no other water than what we melted from the snow. (Menzies 1794a, 195)

More than a century and a half later, physicist Howard Ellis would spend many lonely nights sleeping inside two sleeping bags laid on the lava at the Mauna Loa Observatory at nearly the same altitude as the Menzies party, which had only slightly more primitive accommodations. As Menzies recorded in his journal,

> For our bed we made choice of a flat even rock on which we could all huddle close together . . . we raised a small parapet round it with the lava to break off the wind, which after sunset blew very keen and penetrating. . . . [A]fter spreading a mat on the bare lava rock . . . it was agreed we should all sleep together to keep ourselves warm, we joined together everything we had for our general covering, made pillows of the hard lava, and in this way passed the night tolerably comfortable.

Menzies wrote that the men slept very little during the still, chilly night, which was punctuated by the nearly full moon in a star-filled sky. He noted the absence of the sounds of insects to which they had become accustomed at lower elevations. The morning of February 16 arrived, and Menzies' journal preserves his account of how the summit was reached.

> Next morning, at sun-rise, the thermometer was at 26º and the air was exceedingly keen and piercing. We made a scanty meal on the remainder of our provision before we set out, but for want of fuel, had the greatest difficulty in getting the chocolate boiled, though we burned mats and everything we could think of. . . .
>
> About 11 in the forenoon we arrived at the mouth of an immense crater. . . . Mr. Howell's shoes being already cut and torn in pieces with the lava, and his strength being much exhausted with fatigue, he declined attempting the dreadful place. We therefore left him to the natives on the south side of it to wait our return, while Mr. Baker, Mr. McKenzie, myself and a servant who carried the barometer, crossed over this rugged hollow after a hard and persevering struggle, and by noon [February 16, 1794] got to the highest part of the mountain on the western brink of the great crater, where I observed the barometer and found the quicksilver stood at 18 in. 40 pts., and that on board the Discovery at Kealakekua Bay, observed at the same instant of time was found to be 30 in. 16 pts., so that the difference is 11 in. 76 pts., which will make the height of this mountain 13,634 feet above the level of the sea. But it is necessary to observe that the correction for temperature of the atmosphere has not been allowed for in this calculation, nor at any other station upon the mountain, which will make some difference in the result of the observations. The thermometer here was at 42 degrees. (Menzies 1794a, 195–198)

Today anyone with Web access can see the crater from very near where Menzies stood at the highest point on Mauna Loa, for there the Hawaiian Volcano Observatory has installed a sophisticated Web camera that provides a panoramic view of the giant Moku'āweoweo, the summit caldera. The Web site notes that "The images update every 10 minutes. At times, clouds may obscure visibility. The camera is subject to sporadic breakdown, and its remote location makes immediate repair impossible" (Hawaiian Volcano Observatory 2007).

Menzies summarized his ascent of Mauna Loa seven months later in one of the many letters he sent to Sir Joseph Banks during the voyage north from Hawai'i:

No. 7
North Bound September 8th, 1794

Sir Joseph,

My letter from San Diego California, in December last, by the way of New Spain, I trust, you have received, as it will acquaint you, with the progress of our transactions to that period. . . .

As no interruption took place during our stay at Awhyhee [Hawai‘i] of the most friendly intercourse with the Natives, I employed most part of my time in different excursions, to the interior parts of the Island; in one of these I reached the Summit of Mowna-roa [Mauna Loa] tho' not without hardships and great fatigue; I was how ever induced to persevere in this journey, which took upwards of a fortnight, from having a Portable Barometer, by which, I have been enabled to ascertain the height of that Mountain with sufficient exactness, as well as that of the lower line of snow—the lower line of Congelation—& the upper line of Vegetation, there being a series of corresponding observations, made at Karakakoaa [Kealakakua Bay] during my absence, & both Barometers carefully compared. By the lower line of congelation is here meant, where in our ascent, the Mercury in the Thermometer was first observed to sink, in the night time, below the freezing point, and which I found to happen, in this tropical region, so low, as 6490-feet which I dare say, will stagger the belief of some Theorists. Vegetation ceased at 10543-feet—the lower snow was met at 11595-feet and the height of the Mountain is 13634-feet above the level of the Sea but the correction for the Thermometer which is not been here allowed for, may make some trifling alteration. (Menzies 1794b)

The Elevation of Mauna Loa

Historical accounts of this first known ascent of Mauna Loa by a scientific expedition have concentrated on the remarkable accuracy of Menzies' measurement of the elevation of Mauna Loa. Scientists at the Hawaiian Volcano Observatory discussed various measurements of the height of the mountain above sea level and concluded that it is "Somewhere near 13,675 to 13,680 feet" (Hawaiian Volcano Observatory 2001a).

The uncorrected height given by Menzies in his journal and in his letter to Banks is "13,634 feet," only 43.5 feet less than the mean of the heights suggested by the Hawaiian Volcano Observatory (13,677.5 feet or 4,169 meters), a difference of only about –0.3 percent. Menzies' result might have been even more accurate than this, for the Hawaiian Volcano Observatory reports that the mountain subsided at least 61 centimeters (2 feet) after its 1984 eruption.

The significance of Menzies' measurement can be seen by comparing his result with the elevation measured by James King during Captain Cook's voyage to Hawai‘i in 1779. According to the Hawaiian Volcano Observatory, as noted above, King, "calculated the elevation as 16,020 feet (4,883 m), based on vertical angles to its summit from sea level." This measurement is 2,342.5 feet (714 meters) higher than the actual elevation, an error of +17 percent.

The high quality of Menzies' unprecedented pressure and temperature measurements made under exceedingly difficult conditions became a precedent for the precise scientific measurements that would be made by the next scientific expedition to the summit in 1840. The hardships, tensions, and uncertainties under which he performed some of his science were also prophetic of what was yet to take place on Hawai‘i's great mountain.

Assessing the Menzies Ascent of Mauna Loa

The Vancouver Expedition encircled the world and produced many more discoveries than some space shuttle missions. NASA even named one of its shuttles after the expedition's main ship, the *Discovery*. Much about the practical side of European science during the late eighteenth century can be gleaned from a careful review of the history of the expedition, which was highlighted by so many navigational achievements and scientific findings that Menzies' ascents of Hualālai and Mauna Loa were scarcely noticed in the half-million-word summary of the expedition prepared by Vancouver after his return. Yet the science conducted over the past half century at the Mauna Loa Observatory has as its earliest foundation the careful and precise observations that Menzies performed during his expeditions to the summits of Hualālai and Mauna Loa. Thus it is worthwhile to examine some of the parallels in the approach to science taken by Menzies and his modern-day counterparts on the mountain.

Like many of those who have worked at the Mauna Loa Observatory since the summit station was placed in operation in 1951, Menzies was a cross-disciplinarian. In addition to his formal training as a surgeon and a botanist, he possessed native skills in making measurements and observations and analyzing his findings. His care in doing so is apparent in the portions of his journal that describe his pressure and temperature measurements on the two mountains. (While it would be helpful to know how he corrected his pressure readings for temperature, no information on this could be found.)

Like Menzies before them, the scientists who have serviced Mauna Loa Observatory's facilities and its many instruments since the Weather Bureau began full-time observations in 1957 have also had a wide range of backgrounds, including chemistry, meteorology, physics, electrical engineering, and biology. Consider physical scientist Steve Ryan, the only scientist who has been assigned to all four of NOAA's climate monitoring observatories: Hawai'i, the South Pole, American Samoa, and Barrow, Alaska. Today Ryan manages a diverse range of programs at the Mauna Loa Observatory, including the monitoring of atmospheric sulfur dioxide, carbon dioxide, methane, and carbon monoxide. He also performs flask sampling, a process in which air at the observatory and from near sea level at Cape Kumukahi is collected in special containers for laboratory analysis. Ryan also launches sounding balloons that carry instruments known as ozonesondes into the stratosphere to measure the ozone layer. If Ryan is in the Hilo office of the Mauna Loa Observatory, he is probably working on a scientific paper or preparing for his next trip to the mountain. On the mountain he might be found connecting cylinders of calibration gas to an instrument, climbing to the top of the 120-foot (37-meter) meteorological tower to replace or install an air sampling hose, or using a hammer to repair a door frame. While Ryan is among the most versatile of Mauna Loa's scientists, virtually all those who have been assigned to MLO over the years have been assigned a wide variety of cross-disciplinary tasks.

Menzies' many years of hands-on field observations and measurements made him skeptical about the musings of theorists, as when he described in his letter to Sir Joseph Banks (quoted above) the frigid temperatures he measured during his ascent of Mauna Loa.

Vancouver, who risked lives and ships to collect data and conduct surveys, also disparaged theorists. An early account of the search for the Northwest Passage across North America that discussed Cook's discovery of the inlet near Anchorage, Alaska, that bears his name quoted Vancouver's assertion that a more thorough exploration of the inlet by Cook would

> have spared the theoretical navigators, who have followed him in their closets, the task of ingeniously ascribing to this arm of the ocean a channel, through which a north-west passage, existing according to their doctrines, might ultimately be discovered. (Anon. 1843, 125)

The scientists who followed Menzies to Mauna Loa in 1840 and those who staff the Mauna Loa Observatory today are experimentalists. They may employ the theoretician's statistical and mathematical tools to correct errors, process data, evaluate trends, and devise models that explain their findings. Yet in the end, as with Menzies before them, their research findings are based on precise physical measurements and observations. Archibald Menzies would no doubt be pleased by the monitoring philosophy espoused by the late Dr. David Hofmann (video interview with author, Boulder, Colorado, May 1, 2007), former director of NOAA's Global Monitoring Division (GMD), the organization responsible for MLO:

> I'm not a climate modeler; I'm not a climate scientist. I'm a person who makes atmospheric measurements, and the laboratory that I work for and the scientists in that laboratory pride themselves in making the best measurements in the world.

Most of the science conducted at today's Mauna Loa Observatory involves the collection of data over decades, with measurements of solar transmission going back to 1957 and carbon dioxide to 1958. Yet Menzies showed that even a brief time at the summit of Mauna Loa can yield a significant finding, and he demonstrated the value of simultaneous measurements at sea level and at the summit. Brief duration and simultaneous measurements at today's Mauna Loa Observatory have also yielded significant findings. For example, during the early 1960s, suites of identical sunlight measuring instruments were installed at MLO and at the Hilo office so that the effect of the atmosphere on the passage of sunlight through it could be better understood. A two-week comparison of many ozone measuring instruments at MLO in August 1995 provided important new information about the accuracy of measurements of the ozone layer (McPeters et al., 1999). Similarly, three days of observations of the ozone layer by the author using a new kind of instrument in 1992 confirmed a drift in ozone retrievals by NASA's ozone satellite (Mims 1993). And six days of observations by simple, self-contained instruments at MLO led to the discovery that scattering from cumulus clouds near the sun can increase solar ultraviolet at the surface by 15 percent or more (Mims and Frederick 1994).

Budget issues often dominate modern science, and the Vancouver Expedition is a reminder that financing has always been an issue, especially when the home office is half a world away. Scientists who work in remote places today might find of interest a letter to Lord Grenville of Britain's Home Office dated August 9, 1791, in which Captain Vancouver described his creative method for taking care of some necessary but unanticipated and unbudgeted expenses.

> Mr. Menzies having applied to me for some spirits for the preservation of the different natural curiosities he may fall in with, I have thought proper to add to the bill some breeding sheep, garden seeds, &c., &c., which I intend as presents to the different chiefs of the Sandwich Islands. . . . The amount of the things purchased here being about 334 rix dollars, I have taken the liberty of drawing on the Treasury; for which with a letter of advice have transmitted them attested vouchers, and your Lordship's giving directions for the same to be paid will oblige him who has the honour to be, &c. GEO VANCOUVER. (McNab 1908, 133)

Science usually involves human interaction at various levels, even when conducted from remote locations like MLO, and the Menzies' experience provides important lessons for scientists today. Just as the construction of the summit and slope Mauna Loa Observatories required formal approval by governors of the Territory of Hawai'i, Menzies' ascents of Hualālai and Mauna Loa were made with the permission and blessing of King Kamehameha I. During both expeditions, Menzies treated the Hawaiian bearers with respect, compensated them, and devised a clever way to provide bonuses to those who did the most work.

The journals of those who made long sea voyages include many stories about authoritarian sea captains, undisciplined men, and personality conflicts. By all accounts, Captain George Vancouver was certainly a disciplinarian, and this may account for the success of his expedition. During its four-and-a-half-year voyage, the longest in history, *Discovery* sailed 65,000 miles (104,600 km) with the loss of only six seamen, four of whom drowned, one being lost to disease, and one being poisoned.

One insubordinate crew member was sent home after being flogged, and his family connections led to problems for Vancouver. Another conflict arose between Menzies and Vancouver near the end of the voyage in July 1795. The incident was triggered when plants being cultured by Menzies died after a seaman failed to cover them during a storm. According to Walker (2005), "Menzies accused the man of negligence but Vancouver, an irascible man at the best of times, flew into a rage, pointing out that he had authorised the attendant's employment on other duties." The dispute caused Vancouver to demand that Menzies turn over all of his journals and drawings, which was within his rights. When Menzies refused, Vancouver had him arrested and in September received permission to court-martial him. As Walker relates, "At the beginning of October, however, when *Discovery* returned to the Thames, reason prevailed. Menzies apologized, and on October 24 Vancouver withdrew his demand for a court-martial."

In spite of their sharp differences, Menzies and Vancouver got along well enough for Menzies to complete the long voyage. When Menzies died in 1842, his obituary in the *Annual Register* noted the good relations between the two men:

> It was to this expedition that Mr. Menzies attached himself, and Captain Vancouver appears to have held him so much in estimation, that while in the preface to his "Voyage of Discovery," he omits to mention many more elderly and approved officers of the expedition, he especially devotes a space to commemorate the services of his able collaborateur, Mr. Menzies, and the friendship which he felt towards him. (Rivington and Rivington 1843, 253)

The significance of the altercation between Menzies and Vancouver goes beyond the dispute itself, for some of Vancouver's key documents from the voyage were later lost, and the same fate might have befallen Menzies' journals had he turned them over when Vancouver demanded them. As related by Walker (2005), many years later the Royal Meteorological Society discarded some of Menzies' weather records and many other old documents during a 1972 move. Fortunately, the documents were recovered by a bookseller, who offered to return them. The society declined the offer, and the bookseller then sold the records to the University of Texas. Nearly two decades later, officials of the Royal Meteorological Society learned about the lost records, and the University of Texas agreed to sell them back to the society. There are lessons here for properly safeguarding modern scientific records. Although some of the earliest data have gone missing, the Mauna Loa Observatory has an excellent record in this regard.

Finally, the most powerful lesson of the Menzies expedition to Mauna Loa is the sheer determination and grueling courage required to accomplish and survive the ascent. Those who followed exhibited similar determination and stamina, including two other Scottish explorers.

James Macrae (1825)

The Scottish botanist James Macrae accompanied Lord Byron (George Anson Byron), cousin of the poet, to Hawai'i aboard the HMS *Blonde* in 1825. There he collected a wide variety of plants, some of which bear his name. Macrae's specimens were probably delivered to Kew or the British Museum, but for nearly a century his personal journal about his trips to Kīlauea and the summit of Mauna Kea lay unpublished in the library of the Royal Horticultural Society, until William F. Wilson secured permission to publish the Hawai'i section of the journal. This was done in 1922, with helpful explanatory notes by Wilson (Macrae 1922), under the title *With Lord Byron at the Sandwich Islands in 1825: Being Extracts from the MS Diary of James Macrae, Scottish Botanist.* The book included informative background about the little town that was soon to host a major scientific expedition to the summit of Mauna Loa and which has long hosted today's Mauna Loa Observatory and served as home for its staff:

> Heddo [Hilo] Bay and its neighbourhood are very pleasant on account of its woody appearance and plentiful supply of water, and is far superior in scenery to any spot we had seen either at Mowee [Maui] or Woahoo [O'ahu], but like these is little cultivated, and only round about the native huts with patches of bananas, taro, tapa plants, etc. At the west end of the bay is a fair sized river [Wailuku River] with several waterfalls, convenient for watering ships, and in the centre of the bay is another subterranean one, which is dammed up a little distance from the sea and forms a large fresh water fish pond, tabooed for the use of the king and the chiefs. Toward the east end, not far from the shore and near a reef which runs partly across the bay, there is a small island thickly covered with low coco-nut trees, from which circumstance it is called Coco-nut Island.
>
> The whole of the E. side of Owhyee [Hawai'i], which is divided into two districts, belongs to Kaumanna and Pio. When at Heddo, their place of residence to receive the rents, is near the east side of the bay, and consists of

> no more than two huts, one of which is given to Lord Byron as a residence while here. (Macrae 1922, 46)

The hut assigned to Lord Byron was likely near or on the Waiākea Peninsula, the base of operations for history's greatest expedition to the summit of Mauna Loa in 1840–1841 and the site of most of today's tourist hotels in Hilo. Besides the background information about Hilo and its environs, Macrae's journal provides measurements of temperature on Mauna Kea and a concise description of the morning ascent and afternoon descent of the layer of clouds that denotes the inversion layer:

> The clouds generally rise on the mountains of Owhyee [Hawai'i] and the other islands in the morning and disperse towards evening. Rain often falls at night and also in the daytime some distance from the peaks, while on the coast the sun may be shining and there is no appearance of rain. (Macrae 1922, 57)

While Macrae did not climb Mauna Loa, he left behind a memorable observation of Hawai'i's two greatest mountains:

> May 4 [1825]. Saw the summit of Mouna Kaah [Mauna Kea] towering far above the clouds, and in places covered with snow. We also saw Mouna Roa [Mauna Loa] appearing not far off from the other, but without snow. The grand appearance of both these mountains as seen from on board the ship, upwards of 100 miles out at sea, would repay the traveler for coming so far on purpose to see them. (Macrae 1922, 7)

Macrae also left behind an intriguing observation about the climate of Hawai'i two centuries ago. Today both Mauna Kea and Mauna Loa are generally without snow during summer, but Macrae reported that this was not always the case.

> Mr. [John] Young this evening [June 13, 1825] gave me some account of Mr. Menzies' journey to Mouna Roah, next highest to Mouna Keah to which I am going. During the 26 years that Mr. Young has been on the island, he has never seen Mouna Kaah free from snow, but has not seen snow on Mouna Roah in summer, and on this he bases his theory of the greater height of Mouna Kaah. (Macrae 1922, 46)

David Douglas (1834)

While James Macrae's birthplace in Scotland is as yet unknown, Archibald Menzies was born in Perthshire. Five years after Menzies ascended Mauna Loa, David Douglas, another botanist destined for Hawai'i, was also born in Perthshire. While Douglas was more interested in Mauna Kea than in Mauna Loa, he managed to climb both mountains. This and other remarkable parallels in the lives of these two Scottish explorers and naturalists are well worth recalling as this chapter is ended.

Like Menzies before him, Douglas explored the Pacific Northwest, where in 1825 at Nootka Sound on Vancouver Island, British Columbia, he rediscovered a magnificent tree species first reported by Menzies in 1791 (Wier 1998). While it became known as the Douglas fir, Douglas recognized Menzies' priority of discovery by assigning the Latin name *Pseudetsiga menziessi*.

Douglas collected specimens from the Pacific Coast to Hudson Bay, often under primitive conditions and sometimes risking his life. He sent some seven thousand botanical species to organizations in Britain, about 13 percent of the species then known to the world (Szasz 2000), and he became one of the most famous scientists of his day. Like Menzies, Douglas did not always work well with others, a personality trait that is not unknown among other great scientists and explorers. But while Menzies' spat with Vancouver was temporary and eventually resolved, Douglas' temperament was more consistent. One historian has described him as "peevish and irritable" (Harvey 1947). Another wrote, "Douglas' years in the public limelight did not prove happy ones. Uncomfortable with all the attention, he became sour and difficult to work with. Even his friends tired of his extreme irritability" (Szasz 2000).

Perhaps Douglas' temperament explains why he was so successful in exploring lonely wildernesses largely on his own. He returned to remote areas again and again, even after falling down a ravine on one trip and on another nearly drowning and losing all his samples and supplies in a canoe accident. In 1834 he visited the island of Hawai'i and climbed both Mauna Loa and Mauna Kea. Like Menzies at Hualālai and Mauna Loa, Douglas made his ascent of Mauna Kea in winter. He reached the summit on January 19, forty years to the day after Menzies climbed Hualālai. Along the way he noted the vegetation and geology in great detail. He also measured the temperature with a mercury thermometer, and he measured the relative humidity. He recorded these observations along with descriptions of clear skies and clouds below.

In the account of his ascent of Mauna Loa, Douglas paid tribute to Archibald Menzies, who was both a surgeon and a botanist:

> The red-faced man, who cut off the limbs of men [Menzies had performed an amputation while in Hawai'i] and gathered grass, is still known here, and the people say that he climbed Mauna Loa. No one [except the American missionary Mr. Goodrich], however, has since done so, until I went up a short while ago. The journey took me seventeen days. On the summit of this extraordinary mountain is a volcano [the summit crater], nearly twenty-four miles in circumference, and at present in terrific activity. (Jarves 1843)

On a return visit to the flank of Mauna Kea, Douglas experienced what was either another exploration accident or, more likely, foul play (Hitchcock 1909) when he either fell or was thrown into a covered pit designed to capture wild cattle. Some accounts report that he was killed by a bull that had already fallen into the pit, while others say he was murdered (Fullard-Leo 2001/2002). The great naturalist had made several lifetimes of discoveries when he died at the age of thirty-four.

Six years after Douglas' untimely death, Charles Wilkes, the next great explorer of Mauna Loa, sailed into Hilo Bay and began organizing what remains history's greatest ascent of the world's most massive mountain.

References

Anon. 1843. Voyages round the world, from the death of Captain Cook to the present time. Edinburgh: Oliver & Boyd. (Accessed at http://books.google.com.)

Bier, James A. 1988. *Map of Hawai'i*, 4th ed. Honolulu: University of Hawai'i Press.

Fullard-Leo, Betty. 2001/2002. Who killed David Douglas? *Coffee Times*, fall–winter. (Accessed at http://www.coffeetimes.com/daviddouglas.htm.)

Harvey, Athelstan George. 1947. *Douglas of the Fir: A Biography of David Douglas, Botanist*. Cambridge, MA: Harvard University Press, 151.

Hawaiian Volcano Observatory. 2001a. How high is Mauna Loa above sea level? (Accessed at http://hvo.wr.usgs.gov/volcanowatch/2001/01_08_23.html.)

———. 2001b. Hualalai: Hawai'i's third active volcano. (Accessed at http://hvo.wr.usgs.gov/volcanoes/hualalai.)

———. 2007. Live panorama of Moku'āweoweo, the summit caldera of Mauna Loa volcano. (Accessed at http://hvo.wr.usgs.gov/cam2/.)

Hitchcock, Charles H. 1909. *Hawaii and Its Volcanoes*. Honolulu: Hawaiian Gazette Company.

Hofmann, David. 2007. Video interview with author, Boulder, Colorado, May 1, 2007.

Jarves, James Jackson. 1843. *History of the Hawaiian or Sandwich Islands*. James Munroe and Co., 252. (Accessed at http://books.google.com.)

Macrae, James. 1922. *With Lord Byron at the Sandwich Islands in 1825: Being Extracts from the MS Diary of James Macrae, Scottish Botanist*. Ed. Wm. F. Wilson, Honolulu. (Accessed at http://books.google.com.)

McNab, Robert, ed. 1908. *Historical Records of New Zealand*. Wellington. (Accessed at http://books.google.com.)

McPeters, R. D., D. J. Hofmann, M. Clark, L. Flynn, L. Froidevaux, M. Gross, B. Johnson, G. Koenig, X. Liu,

S. McDermid, T. McGee, F. Murcray, M. J. Newchurch, S. Oltmans, A. Parrish, R. Schnell, U. Singh, J. J. Tsou, T. Walsh, and J. M. Zawodny. 1999. Results from the 1995 stratospheric ozone profile intercomparison at Mauna Loa. *Journal of Geophysical Research* 104, D23: 30,505–30,514.

Menzies, Archibald. 1794a. *Journal of Archibald Menzies in Hawaii Nei—128 Years Ago*. W. F. Wilson, ed., 1920. Honolulu: The New Freedom. (Accessed at http://books.google.com.)

———. 1794b. Letter to Sir Joseph Banks (September 8, 1794). State Library of New South Wales, Sydney, Australia. (Accessed at http://www.sl.nsw.gov.au/banks/series_61/61_22.cfm.)

———. 1829. Some account of an ascent and barometrical measurement of Wha-ra-rai, a mountain in the island of Owhyhee. *Magazine of Natural History* (November): 435–442. (Accessed at http://books.google.com.)

Mims, Forrest M., III. 1993. Satellite monitoring error. *Nature* 361: 505.

Mims, Forrest M., III, and John E. Frederick. 1994. Cumulus clouds and UV-B. *Nature* 371: 291.

Rivington, J. G. F., and J. Rivington. 1843. *The Annual Register, or a View of the History and Politics of the Year 1842*. London: Thomas Curson Mansard. (Accessed at http://books.google.com.)

Stearn, William T. 2000. *Menzies, Archibald*. Dictionary of Canadian Biography Online, University of Toronto. (Accessed at http://www.biographi.ca/EN/ShowBio.asp?BioId=37679.)

Szasz, Ferenc Morton. 2000. Scots in the North American West, 1790–1917. Norman: University of Oklahoma Press: 45.

Walker, J. Malcolm. 2005. The weather observations of surgeon Menzies. Paper presented at the 85th annual meeting of the American Meteorological Society, San Diego, California. (Accessed at http://ams.confex.com/ams/pdfpapers/83619.pdf.)

Wier, Stuart. 1998. The Douglas fir of the southern Rocky Mountains. (Accessed at http://home.earthlink.net/~swier/DouglasFir.html.)

CHAPTER THREE

The United States Exploring Expedition (1840–1841)

Archibald Menzies and David Douglas demonstrated that scientists could scale Mauna Loa and bring back accurate measurements of temperature and pressure from the summit. They paved the way for the United States Exploring Expedition's establishment of the first observatory on Mauna Loa, a temporary camp and scientific facility at the summit of the great mountain that was manned from December 21, 1840, to January 13, 1841. That nineteenth-century observatory established the historical precedent for the summit facility of 1951–1954 that led to today's Mauna Loa Observatory.

Those who have visited or worked at MLO in modern times may have occasionally experienced some of the same problems and difficulties faced by their nineteenth-century predecessors, including funding shortfalls, broken instruments, severe weather, and altitude sickness. But they do not experience the astonishingly difficult conditions faced by an exploring party camped atop Mauna Loa for three weeks during the midst of winter, for the facilities and conditions at MLO today are luxurious by comparison. This is why the story of the US Exploring Expedition's stay atop Mauna Loa must be told in a book about the history of today's observatory.

While Archibald Menzies was writing about his ascent of Mauna Loa in 1794, the fledgling United States was distracted by domestic growing pains, serious disputes with Great Britain, and conflicts with the pirates of North Africa. After George Washington became president in 1789, a substantial fraction of the national budget went to pay tribute and ransoms to the Barbary pirates. In 1794, the US Congress and President Washington were planning the construction of six frigates to challenge pirates and to protect America from Great Britain. Voyages of exploration were not on the table, and the young republic's most significant claims to science were the private achievements of Thomas Jefferson and, especially, Benjamin Franklin.

The first major science expedition authorized by the United States followed the Louisiana Purchase of 1803, when President Jefferson sent Meriwether Lewis, William Clark, and their Corps of Discovery to the Pacific Ocean and back. Jefferson personally planned the expedition based on his considerable knowledge of geography and his hope that there might exist a northern passage between the Atlantic and Pacific Oceans. Jefferson had more in mind than surveying a route from Missouri to the Pacific Coast. From the outset, he assigned a high priority to the collection of scientific data about the new territory purchased from France, a territory so vast that it doubled the size of the United States.

On June 20, 1803, Jefferson wrote a 2,300-word letter of instructions to Lewis, who served as his personal secretary and who was to lead the expedition along with Clark. Jefferson's letter provided detailed instructions for the Corps of Discovery and directed that the expedition record its observations of plants, animals, minerals, and other natural resources. He also directed that the expedition keep a detailed record of

> climate as characterized by the thermometer, by the proportion of rainy, cloudy & clear days, by lightening [*sic*], hail, snow, ice, by the access & recess of frost, by the winds prevailing at different seasons, the dates at which particular plants put forth or lose their flowers, or leaf, times of appearance of particular birds, reptiles or insects. (Jefferson 1803)

The Lewis and Clark Expedition began its journey near the confluence of the Missouri and Mississippi Rivers in 1804 and returned in 1806 with numerous specimens and samples, as well as notebooks filled with data requested by Jefferson. Of special interest to modern scholars is the calibration of their thermometers and their measurements of temperature, which have been found to fall well within the modern range of temperature along the route of the expedition (Solomon and Daniel 2004). Their teacher was Jefferson himself, who had a lifelong habit of recording the daily temperature and whose careful measurements of the minimum and maximum temperature at Monticello from 1810 to 1816 were a close match to modern measurements (Mims 1993; Solomon et al., 2007). The success of the expedition made Lewis and Clark national heroes and established a precedent for more such explorations of unexplored regions of North America and beyond.

Several decades were to pass before the United States was to again sponsor a major scientific and exploration expedition. Meanwhile, there were too many competing interests with higher priorities. The progress of building a country was interrupted by war with Great Britain and persistent conflict with the Barbary pirates. The nation was busy building its capital in Washington and an infrastructure of roads and bridges.

Reynolds Challenges the Congress

In 1828, writer and explorer Jeremiah N. Reynolds enjoyed a friendly reception when he challenged the Congress "to explore the immense and unknown regions in the southern hemisphere" (Reynolds 1828). Reynolds had good reason to be heard, for he himself had voyaged south of the South Shetland Islands and had entered the icy realm of "the Antarctick." His enthusiasm for discovery and exploration had captured the notice of prominent citizens, including the poet, writer, and editor Edgar Allen Poe, who editorialized in support of Reynolds' ideas and published some of his work. Exactly two years after the defeat of Santa Ana by the Texian Army under Sam Houston at the Battle of San Jacinto, an account of Reynold's adventures was published in the *New York Mirror* (Reynolds 1836).

In April 1836, Reynolds took his audacious proposal directly to the Congress in an eloquent speech that proposed America mount an expedition on a scale that had never before been attempted. During his speech, Reynolds asked,

> Have we not shown, that this expedition is called for by national dignity and honor? Have we not shown, that our commanding position and rank among the commercial nations of the earth, makes it only equitable that we should take our share in exploring and surveying new islands, remote seas, and, as yet, unknown territory? Who so uninformed as to assert that all this has been done? Who so presumptuous as to set limits to knowledge, which, by a wise law of Providence, can never cease? As long as there is mind to act upon matter, the realms of science must be enlarged; and nature and her laws be better understood, and more understandingly applied to the great purpose of life. If the nation were oppressed with debt, it might, indeed, it would, still be our duty to do something, though the fact, perhaps, would operate as a reason for a delay of action. But have we any thing of this kind to allege, when the country is prosperous, without a parallel in the annals of nations? (Reynolds 1836)

The United States Exploring Expedition

The Congress answered Reynolds' challenge by appropriating $300,000 and authorizing the president to "send out a surveying and exploring expedition to the Pacific Ocean and the South Seas" (the *Globe* 1836). On May 14, 1836, President Andrew Jackson signed the bill into law. The venture would be known as the United States Exploring Expedition.

The appropriation for the four-year expedition was equivalent to $5,877,524 in 2010 dollars (Friedman 2011). Like the earlier Cook and Vancouver expeditions, the navigational, logistical, and scientific challenges faced by the US Exploring Expedition amounted to the equivalent of a manned space mission today. Thus, the modest size of the appropriation may be further placed in perspective by considering the cost of an unmanned satellite, which is typically more than $100 million. According to NASA, the average cost of a Space Shuttle mission was $450 million, an amount that does not include the cost of the shuttle itself.

The authorization for the expedition faced opposition in the House of Representatives, which eventually passed the bill by a vote of seventy-nine in favor and sixty-five against. The Senate handily approved the bill with a vote of twenty-six in favor and only three against (South-Pole 2007).

Organizing the Exploring Expedition, or the Ex. Ex. as it came to be known, proved contentious and controversial. Designated commanders resigned, most of the expedition's ships were changed, and there was a reduction in the scope of the expedition. Finally, after two years of preparation, on August 18, 1838, the three-masted flagship *Vincennes* led the Exploring Expedition's fleet of four ships and two tenders and its complement of officers, scientists, and sailors out of Hampton Roads, Virginia. In command was Lieutenant Charles Wilkes, a forty-year-old naval officer who had been sent to Europe in the summer of 1836 to purchase the navigational and scientific instruments for the expedition (Esch 1965). While Wilkes had considerable background as a nautical surveyor and possessed a strong scientific bent, he was a stern disciplinarian with no command experience.

The expedition's crew of 346 officers and men was supplemented with "a corps of scientific gentlemen," referred to by the crew as the "scientifics." The scientifics listed in the Ex. Ex. orders included the following men:

Horatio Hale, Philologist
Charles Pickering, Naturalist
Titian Ramsay Peale, Naturalist
Joseph P. Couthouy, Conchologist
James Dwight Dana, Mineralogist
William Rich, Botanist
Joseph Drayton, Draughtsman
Alfred T. Agate, Draughtsman
William Dunlop Brackenridge, Horticulturist (Wilkes 1845, vol. 1)

Couthouy had prior naval experience, and his displeasure over Wilkes' leadership style led to a falling out between the two men. Wilkes left him in Samoa before the expedition reached Hawai'i, but Couthouy was still to play a role in what was to come. Other scientifics included:

John G. Brown, Mathematical Instrument Maker
John W. W. Dyes, Assistant Taxidermist

Navy Secretary J. K. Paulding's detailed orders to Wilkes included a very specific science goal:

> Although the primary object of the Expedition is the promotion of the great interests of commerce and navigation, yet you will take all occasions, not incompatible with the great purposes of your undertaking, to extend the bounds of science, and promote the acquisition of knowledge. (Wilkes 1845, xxv–xxxii)

The fleet headed first for South America, where the scientifics began their observations and collections in earnest while Wilkes attended to surveying. So many items were collected during the voyage that eventually there was little room to store them. So the Ex. Ex. arranged to send things home on US ships they encountered. Over the nearly four-year duration of the voyage, some forty tons of materials were collected, much of which eventually became a major part of the initial collections for the Smithsonian Institution.

A principle goal of the Ex. Ex. was to sail to the South Pole, as proposed by Reynolds. After sailing 1,500 miles (2,400 km) along the coast of Antarctica now known as Wilkes Land, Wilkes confirmed that the land previously reported by Palmer was the size of a continent (Hobbs 1940). Later, he and the French explorer Dumont d'Urville both claimed credit for being the first to discover Antarctica.

Wilkes then turned toward New Zealand, where he had previously arranged to pick up the scientifics who had spent time touring portions of Australia and New Zealand during the Antarctic voyage. The expedition next visited the Fiji Islands, where Wilkes performed detailed surveys and became involved in various confrontations and even hostilities. A full description of these incidents is included in an account of the four-year voyage by Philbrick (2005). Hawai'i was the next destination.

The Ex. Ex. Arrives at Honolulu

When the Ex. Ex. arrived at Honolulu on September 30, 1840, Wilkes spent quality time with King Kamehameha III and Chiefess Kekāuluohi, the daughter of Kamehameha I who served as prime minister. The king wore a blue coat, white pantaloons, and a vest. Kekāuluohi wore a yellow silk dress and a tiara of yellow and red feathers that signified her high rank. Wilkes described her as "altogether one of the most remarkable-looking personages I have ever seen."

Wilkes spent the next few months learning about Hawaiian culture and politics, observing the sport of surfing, meeting again with the king and with many of the local missionaries, and preparing for an ascent of Mauna Loa on the island of Hawai'i. King Kamehameha III, like his predecessor Kamehameha I in 1794, authorized the expedition and the hiring of a few hundred Hawaiians to act as bearers. After ordering his fleet to survey various Pacific islands, Wilkes left Honolulu on December 3 in the *Vincennes* and sailed to Hilo with the goal of ascending the summit of Mauna Loa to make a variety of scientific measurements. He was accompanied by Dr. Gerrit C. Judd, a prominent missionary physician who was to later become prime minister of Hawai'i and who volunteered to assist during the Mauna Loa ascent as physician, interpreter, and manager of the Hawaiians hired as bearers.

Why Wilkes chose Mauna Loa instead of Mauna Kea was not uncovered during the preparation of this book, but a possible clue can be found in a report by the newspaper publisher and writer James Jackson Jarves (1843). Together with his wife Elizabeth, in 1840 Jarves founded Hawai'i's weekly *Polynesian* newspaper. While Elizabeth managed the newspaper, James traveled about the islands writing stories. In June 1840, some three months before the Ex. Ex. reached Honolulu, Jarves learned of a major volcanic eruption of Kīlauea on the Big Island.

> Having been long desirous of visiting Hawaii, as the largest and most interesting island, both from historical associations and its stupendous natural phenomena, I immediately determined to proceed to the scene of fire. Mr. J. P. Couthouy, a gentleman attached to the scientific corps of the United States Exploring Expedition, and another friend, joined me. A brig was just ready to sail for the leeward side of the island. Accordingly, on the afternoon of the 23d of June, we repaired on board. (Jarves 1843, 207)

Actually, Couthouy was no longer attached to the Ex. Ex., for he had been expelled by Wilkes in Samoa and had somehow managed to gain passage to Honolulu in advance of Wilkes and his expedition. Together with Jarves and his friend, the three planned an ascent of Mauna Loa after a visit to the erupting Kīlauea. Their mini-expedition was the first to make use of prison labor, a tradition that helped make possible today's Mauna Loa Observatory.

> The natives have no passion for high mountains, or cold weather. Good men were not to be procured at any rate; but to go we were determined, and, finally, by negotiating with the 'head man,' we procured twelve, and a guide, named Honoa, who professed to be acquainted with every path and route on the whole island. These

> fellows were all of the 'chain-gang' criminals, who had the choice of going with us, or continuing their labors on the public works. Their wages, I suppose, were to go towards the payment of fines. It was evident that to them it was but a choice of evils; for they manifested no enthusiasm in the cause. When they were all mustered, a more cut-throat looking set never made their appearance out of Newgate. The countenances of some were perfectly hideous, combining all of evil human features are capable of expressing. (Jarves 1843, 222)

Jarves' goal for the planned Mauna Loa ascent was to check and verify the contradictory measurements about the dimensions of the summit and its crater given by David Douglas in 1834, for "Considerable discrepancy seems to exist in regard to Mr. Douglas's measurements" (229). While Jarves and Couthouy had managed to climb Mauna Kea, their plans for Mauna Loa faced a serious challenge.

> July 6 [1840]. . . . The ascent of Mauna Loa was to us an object of great interest, in order to satisfy ourselves by actual observation of the accuracy of Douglas's account. . . . Although it was now midsummer, the weather continued cold and stormy. Upon assembling our men, we found them in a state of mutiny, refusing to go further, and threatening to leave us instanter. The day before, we had been too much occupied in exploring to have a care for them, and they had availed themselves of the opportunity to devour the remainder of the food. For five days they had averaged eleven pounds' weight of fish and poi to each man; a consumption which as many Esquimaux could not exceed. We had not anticipated vile ingratitude from kanakas on such a diet; but so it proved, and we were obliged to forego the ascent of Mauna Loa, and proceed at once to Hilo. The ascent of Mauna Loa was to us an object of great interest, in order to satisfy ourselves by actual observation of the accuracy of Douglas's account. (Jarves 1843, 252)

The Jarves-Couthouy ascent of Mauna Kea and their failed attempt to ascend Mauna Loa may have become known to Wilkes after his arrival in Honolulu several months later. If so, and in view of his rivalry with Couthouy, perhaps Wilkes' decision to climb Mauna Loa was influenced by his rival's failure.

From Hilo to Mauna Loa

On the evening of December 9, 1840, the *Vincennes* sailed into Hilo Bay on the Big Island of Hawai'i, the voyage from Honolulu having taken six days. They found the best anchorage to be in five fathoms of water over a soft mud bottom in the eastern side of the bay under the protection of the reef and due west of Coconut Island. From this anchorage, where the ship was to remain until March 5, 1841, Wilkes later wrote that

> Mauna Kea, distant about thirty-five miles, has the appearance of being by much the highest mountain; while Mauna Loa, distant sixty miles, and rounded at its summit to the shape of a regular dome, requires an effort of reason to satisfy the observer that it really has as great an elevation. . . . The appearance of this mountain is so deceptive, that one would not suppose it to have half its real altitude; and it might easily be passed unnoticed, so unpretending in its aspect. From Hilo, Mauna Loa looks as if one might walk over its smooth surface without difficulty; there is, indeed, so much optical deception in respect to this mountain, that it served to give us all great encouragement, and we set about making our preparations with a determination to succeed in the attempt to reach its highest summit. (Wilkes 1845, 122–123)

Buoyed by this optimistic prospect of the ascent that laid before them, Wilkes and his crew soon established a scientific observatory at the western base of Waiākea Point just north of where the Wailoa River empties into Hilo Bay. The site was slightly over a mile from the town of Hilo, which could be reached by walking along the sandy beach. Designated Observatory 1 on the expedition's chart of the harbor (Wilkes 1861), the facility was described by Lt. Colvocoresses in his journal:

> December 10th. This morning we sent the scientific instruments to the observatory. This [grass-roofed] building is situated on the south-eastern side of the bay; it is thirty feet long, by fifteen wide, and was, I am informed, erected expressly for our use by an order from the king. (Colvocoresses 1855)

Apparently there was more to Observatory 1, for a US Ex. Ex. chart (Wilkes 1861) shows a cluster of four structures facing Hilo Bay and backed by coconut palms. The chart indicates that Waiākea Point was lined by lava, with the exception of a small sand beach just north of Observatory 1 (Plate 3). Today there is a sand beach at the same location as that indicated on the old chart. This places Observatory 1 on a rise southeast of the beach and squarely within a small park designated as "Rakuen" on a metal plaque affixed to a large lava boulder. *Rakuen* is Japanese (楽園) for pleasure garden or paradise (Breen 2006). The plaque states that "The garden park is dedicated to all senior citizens of the Island of Hawaii." There is no mention of the Exploring Expedition's Observatory 1.

The park is less than 0.8 mile (1.25 km) from where the first Hilo office of today's Mauna Loa Observatory was established in 1958. It is directly behind the Suisan Fish Market at 85 Lihiwai Street and can be accessed by walking or driving west on Banyan Drive past the tourist hotels and crossing Lihiwai Street, where Banyan Drive terminates into a parking area lining the east side of the park. The sand beach is below the north end of the park (to the right when facing the bay) below a metal guard rail. Subsidence of the island and warming of the ocean have caused the sea level to rise an estimated 3.0 millimeters (0.12 inches) per year at Hilo since 1946, with negligible change since 1975 (Caccamise 2005, fig. 1). While the bay might have been somewhat shallower in 1840 than today, at low tide both the shoreline along either side of the mouth of the Wailoa River and the sand beach closely resemble what is depicted on the Wilkes chart. While the park is well above the bay, it is fully exposed to the deadly tsunamis that have ravaged the area from time to time, flooding the entire Waiākea Peninsula and much of Hilo.

Within a day of its arrival in Hilo Bay, the expedition had equipped Observatory 1 with chronometers and instruments designed to make meteorological and other observations simultaneous with those planned for the expedition to Mauna Loa, which today can be seen from the nearby beach. Many other instruments would make the trip up the mountain, including a pendulum apparatus capable of measuring gravity. Boats from the ship presumably landed at the sandy beach.

Organizing the expedition to the summit was a complicated affair that involved some two hundred Native Hawaiians as bearers. Eventually all was ready, and on December 14, 1840, the expedition assembled at Observatory 1 at Waiākea Point and, after several hours of mass confusion, departed for Mauna Loa. Wilkes was accompanied by his steward, cook, and servant and by American consul Peter Brinsmade. Wilkes and Brinsmade were each carried by four Hawaiians in umbrella-shaded sedan chairs arranged for by Judd (Philbrick 2003, 245).

About half a mile (0.8 km) southwest of Observatory 1, the party passed by the royal fishponds, from which, by order of the king, Wilkes' "table was constantly supplied with the fine fresh-water mullet that had been taken from the sea when small" (Wilkes 1845, 125). This suggests that the expedition had marched south from Observatory 1 along what is now Lihiwai Street and then becomes Manono Street. The expedition then veered southwest toward the fishponds along or near today's Park Road on its way to the old Volcano Road, which is today known as Kīlauea Avenue. The party must have passed very close to or even directly by the present office of the Mauna Loa Observatory at 1437 Kīlauea Avenue. Wilkes described the road as "tolerably good, although it scarcely admitted two persons to walk abreast" (126). He described his expedition as resembling a "vast caravan."

> [T]he party consisted of Lieutenant Budd, Passed Midshipman Eld, Midshipman Elliott, Mr. Brinsmade, Dr. Pickering, Mr. Brackenridge, Dr. Judd, myself, and ten men, including our servants from the ship . . . [and] two hundred bearers of burdens, forty hogs, a bullock and bullock hunter, fifty bearers of poi (native food), twenty-five with

> calabashes, of different sizes and shapes, from two feet to six inches in diameter. Some of the bearers had large and small panels of the portable house on their backs; others frying-pans or kettles; and others tents or knapsacks. Then there were the lame horses, which, instead of carrying their riders, were led by them; besides a very large number of hangers-on, in the shape of mothers, wives, and children, equaling in number all the bearers, all grumbling and complaining of their loads; so that wherever and whenever we stopped, confusion and noise ensued. I felt happy in not understanding the language, and of course was deaf to their complaints. (Wilkes 1845, 1:27)

By late afternoon, Wilkes had reached the outskirts of the Hilo District and reached a "hill [that] afforded a magnificent view of Hilo Bay, and of the surrounding country below." Lord Byron had passed this way in June 1825 on his way to the Kīlauea volcano (Macrae 1922, 66). By evening the expedition reached the Puna District on its way to Kīlauea's erupting crater. In three days, the caravan reached the Kīlauea Crater. Before he saw the crater itself, Wilkes caught sight of the ultimate goal:

> Mauna Loa burst upon us in all its grandeur. The day was extremely fine, the atmosphere pure and clear, except a few flying clouds, and this immense dome rose before us from a plain some twenty miles in breadth. I had not, until then, formed any adequate idea of its magnitude and height. The whole dome appeared of a bronze colour, and its uninterrupted smooth outline was relieved against the deep blue of a tropical sky. Masses of clouds were floating around it, throwing their shadows distinctly on its side. . . . I now, for the first time, felt the magnitude of the task I had undertaken. (Wilkes 1845, 130–131)

Back on the *Vincennes,* Lt. Colvocoresses described the expedition's plan:

> As soon as the instruments are up, Captain Wilkes and Messrs. Budd and Eld will proceed with the observations, and continue them until a sufficient number are obtained to form the data from which the proper results are to be obtained. Such are the objects of this enterprise, and we do most sincerely wish it all the success imaginable. No observations, we believe, have as yet been made by any one at so great an altitude with instruments like those with which the party are provided; and as all of them would be highly interesting and useful, it would give us particular pleasure to have the honor of making them secured to our own country.
>
> There are some, however, who are of the opinion that the whole affair will fall through from the fact of the natives not being able to stand cold, which is said, after an elevation of ten or twelve thousand feet; to be intense. (Colvocoresses 1855, 194–195)

The skeptics were nearly proven correct. After some preliminary explorations around Kīlauea, the party turned toward Mauna Loa the morning of December 18. The complaints heard from the first day of the journey were nothing compared to what was about to happen, for as they traveled west toward the great mountain that lay before them, the nights became cold and water was soon in such short supply that a gallon sold for $2, or $43.15 in 2010 dollars (Friedman 2011). Moreover, the lava over which they walked soon destroyed the leather sandals worn by most of the bearers. As noted by Wilkes, "there never was more difficult or unpleasant ground to travel over."

Wilkes, who assumed the title of "captain" for the duration of the Ex. Ex., had anticipated the difficulties to come, for Lt. Colvocoresses recorded in his journal that a letter from Wilkes had arrived the same day the expedition left Kīlauea for Mauna Loa:

> December 18th. Several letters have been received today from the Mauna Loa party, and among others, one from Captain Wilkes, addressed to the first Lieutenant, in which, he directs that fifty of the crew should be sent to him. One of the letters stated that about thirty of the natives had given out. . . .
>
> December 19th. At early dawn, Lieut. Alden and Mr. Sanford left here with the fifty men sent for by Captain Wilkes.

> It is justice due to those men to state, that not one of them waited to be ordered; they came forward and volunteered their services the moment they learned they were needed. (Colvocoresses 1855, 195–196)

The lack of water seriously threatened the expedition. The old 'Ainapō Trail apparently used by Menzies in 1794 passed by a number of water holes and was likely the intended route. However, the old trail began around 10 miles (16 km) southwest of Kīlauea, and the expedition's inexperienced guide missed the trail by taking a more direct route to the summit.

When it became obvious that their guide did not know his way, Wilkes himself "took the lead, compass in hand," and led the way toward the distant summit. For the benefit of stragglers who were carrying burdens, the branches of trees were broken off to serve as directional indicators, and fires were set at night. At an elevation of 6,071 feet (1,834 meters), two experienced guides arrived. "One of them," Wilkes reported, "was the celebrated bird catcher, called Keaweehu, who had been the guide of Lowenstern, and knew where water was to be obtained; but it was ten miles distant." Within a day or two, water was provided with Keawe'ehu's help. Keawe'ehu also confirmed that the expedition had indeed missed the trail.

Wilkes (1845) provides a full account of the ascent in volume 4 of his *Narrative of the U.S. Exploring Expedition*, which is relied on for much of what follows (page numbers in citations refer to this volume). To best appreciate the hardships endured by the expedition, we may rely on Wilkes himself as he describes a serious setback that occurred when he and the advance party reached an elevation of 13,190 feet (4,020 meters):

> The wind blew a strong gale from the southwest, and it was piercingly cold; the thermometer at 3 P.M. showed 25 [degrees]. For some time previous I had been obliged to keep the Kanakas [Native Hawaiians] before me, to prevent them from throwing their loads down and deserting; but I found them unable to go any further; being nearly naked, they were suffering much. Seeking a place of shelter under a high bank of clinkers, partly protected from the wind, I allowed them to deposit their loads, and gave them permission to return, upon which they seemed actually to vanish; I never saw such agility displayed by them before.
>
> As soon as the natives who were on the road saw those from the upper party coming down, they could no longer be induced to face the cold, and all deserted at once. The mountain became in consequence a scene of confusion; being strewn with instruments, boxes, pieces of the portable house, tents, calabashes, &c., which the natives had dropped.
>
> I now found myself with the guide and nine men, with nothing for a covering but the small tent used for the instruments, and the coming on of a snow-storm, made it very necessary to have something to protect us. The thermometer had gone down to 18 [degrees], and most of the men were much affected with the mountain sickness, with headache and fever, and were unable to do any thing. I felt quite unwell myself from the same cause, having a violent throbbing of the temples and a shortness of breath, that were both painful; and distressing. (149)

Three of the men were almost lost when they attempted to retrieve supplies. Later, an accumulation of snow on the roof of their improvised shelter caved in on Wilkes and the others during what was already a miserably cold night.

Pendulum Peak: The First Mauna Loa Observatory

After enduring additional hardships, on December 22 Wilkes and several others finally reached the eastern edge of the giant Moku'āweoweo Crater at the summit of Mauna Loa, just south of today's Mauna Loa summit cabin. Wilkes wrote, "The crater might be likened to an immense cauldron, boiling over the rim, and discharging the molten mass and scoriae which has floated on its top" (154).

That night they experienced near-hurricane-force winds that almost destroyed their tent. The next day they were joined by Drs. Judd and Pickering. While Pickering, who was one of the scientifics, descended into the crater, the others looked for a site to establish a more permanent camp for the planned stay on the summit. This survey revealed that the site selected the night before was the best location, and the men set to work constructing the observatory and camp, which they named Pendulum Peak and which history must regard as the first Mauna Loa observatory.

Some of the instruments arrived on Christmas Day, and Wilkes began expanding his measurements beyond the temperatures and pressures that had been recorded during the ascent. The men also began building a protective lava wall around a space large enough to hold the seven or so tents that were being carried up the mountain.

When Wilkes learned that some of his sailors were in difficulty, he and Dr. Judd hiked down to the 11,800-foot (3,597-meter) level and found their worst fears confirmed:

> There we found a large number of men in a temporary tent, lying on the panels of the portable houses: some of them were suffering from mountain-sickness, others vomiting; some had attacks of diarrhea, others had not got over their forced march, and showed me their bleeding feet and shoeless condition; all were looking half savage, with overgrown beards, dirty and ragged clothes—so totally different from their trim and neat appearance on board ship, that I was shocked at the change produced in so short a time. (157)

The sick were attended to by Judd, while Wilkes encouraged the men and boosted their morale. Over the next few days, more supplies arrived from below, and the men made leather sandals to replace their shoes that had been destroyed by the lava.

Meanwhile, Wilkes and Judd continued down to the rest stop known as the Recruiting Station at the 9,000-foot (2,743-meter) level, where they saw along the way much of the expedition's equipment scattered across the lava where it had been abandoned by the Hawaiians.

The next day, Wilkes and a sailor returned to the summit with a load of supplies and found that much progress had been made on the construction of the wall around the Pendulum Peak camp. Wilkes later wrote,

> We found the experiment of enclosing the camp in with a stone wall to succeed admirably, protecting us very much from the southwest wind. The temperature during the night fell to 17 [degrees]. On the 28th the day dawned with fine weather. At sunrise the effect of refraction was very similar to that before described. I was again struck with the apparent smallness of the diameter of the sun when seen in the horizon. The day continued beautifully clear, with a very strong wind from the westward. (159)

Wilkes spent the morning of December 29 setting up his delicate pendulum apparatus, with which he planned to measure the gravity at the summit. Finally, significant good news came with the arrival of the enthusiastic volunteers from the *Vincennes:*

> A short time after noon, Dr. Judd again joined us with the joyful news that the party from the ship had arrived, with sixty days' provisions for as many men. I now felt that through our own perseverance we should succeed in obtaining our wishes, for with this supply we could remain sufficiently long to effect my object in visiting the mountain. (160)

That day, the scientifics Pickering and Brackenridge quit the summit to make a difficult winter ascent of Mauna Kea. By December 30, the Pendulum Peak camp and its tents were completely encircled by a lava wall, and many scientific observations were begun. That afternoon Wilkes and Judd attempted to determine if they could find a spot from which the *Vincennes* might be visible. But, as previously reported by David Douglas, the massive dome of Mauna Loa formed its own horizon, and Hilo and its bay could not be seen. There, amidst the tranquility of the coconut palms at Observatory 1, Passed Midshipman William May, accompanied by his newly acquired Hawaiian girlfriend, made hourly weather measurements and observations that would later be compared with those made on the summit (Philbrick 2003). That evening, a special visitor arrived at Pendulum Peak to check on the expedition's progress.

> At night, on our return, we had a visit from the old guide, Keaweehu, the bird-catcher, who gave us the name of the terminal crater, as Moku-a-weo-weo, and that of the south of it as Pohakuohanalei. According to his statement, Moku-a-weo-weo emitted fire not long after Cook's visit, and again five years since, on the north side. . . . While he was in conversation with Dr. Judd, I thought it worth while to obtain a camera lucida sketch of him, as he sat wrapped in his tapa. (Wilkes 1845, 4:161)

On New Year's Day of 1841, Dr. Judd and two of the men went exploring in the crater, while Wilkes and the others worked on building a shielded, insulated tent for the pendulum. The day was clear, and that evening Wilkes observed the shadow of Mauna Loa against the eastern sky:

> This evening, at sunset, we had a beautiful appearance of the shadow of the mountain, dome-shaped, projected on the eastern sky: the colour, a light amethyst at the edges, increasing in intensity to a dark purple in the centre; it was as distinct as possible, and the vast dome seemed to rest on the distant horizon. The night was clear, with moonlight, the effect of which on the scene was beautiful: the clouds floating below us, with the horizon above them, reminded us of the icebergs and ice-fields of the Antarctic: the temperature lent its aid to the deception. (163)

Wilkes' sunset view of Mauna Loa's shadow can be observed from today's Mauna Loa Observatory during the winter months, when the spectacle is not blocked by the mountain itself. During summer months at the observatory, the shadow of the great mountain appears in the western sky at sunrise.

On the evening of January 2, Wilkes and several others experienced an attack of mountain sickness, for which the irreplaceable Dr. Judd provided them with remedies that Wilkes does not describe in his narrative. The next morning, all were again well, and they began a day of observations. The air was too dry for the dew point to be measured, but the sky below was painted with clouds that Wilkes describes in some detail. He also describes the static electricity that accompanies very dry days, which is familiar to those who have worked at or visited today's Mauna Loa Observatory on dry days. At such times, Wilkes, writes, "electricity was found to be easily excited: on moving any articles of dress from the person, it would envelope itself in sparks." Those who drive to MLO on a dry day experience this phenomenon while sliding across the seat to exit their vehicle, an exercise that generates a charge having sufficient electrical potential to produce a loud pop and an uncomfortable shock when a finger approaches a metal door handle. Those who spend the night in a sleeping bag at the observatory might notice popping and crackling sounds as they stir during the night. If they look down inside their bag, they will find that every movement is accompanied by a sparkling light show.

Despite the expedition's best efforts, the temperature of the carefully shielded tent that sheltered the pendulum's portable house varied unacceptably. On January 5, a large amount of dried grass ordered by Wilkes arrived from below. After grass was stuffed in the space between the tent and the pendulum house and spread across the lava floor, the temperature inside became stabilized at about 40 degrees. This allowed the pendulum to function properly, even though the temperature outside the lava wall varied from around 17 to 50 degrees each day. After the pendulum was cared for, Wilkes and Judd worked at improving the conditions in the tent they shared:

> Dr. Judd . . . invented a fire-place and chimney, which he built in one corner of the tent, and which occupied all the spare room we had. We thus were enabled to enjoy the comforts of what, on the top of Mauna Loa, we called a good fire. How good it was, may be understood when it is stated, that our allowance of fuel was three sticks of wood per day; and that the water froze within a couple of feet of the fire, when it was giving out the most heat. . . . The temperature of boiling water was again tried, and found 187 [degrees]; the barometer stood at 18.384 in. No dew point could be obtained. Electricity was developed in large sparks. (166)

FIGURE 3.1. This *kona* storm dumped snow on Mauna Loa on December 5, 2007.

After the summit party experienced several days of moderate circumstances, the mountain reminded them who was in control.

> On the 8th, we had a change to cold, raw, and disagreeable weather; snow began to fall, and a kona or southwest gale set in; the temperature fell soon to 20 [degrees].
>
> At 10 P.M., I was unable to proceed with the pendulum observations; for such was the fury of the storm, that the journeyman-clock, with a loud beat, although within three feet of my ear, could not be heard. I was indeed apprehensive that the whole tent, house, and apparatus would be blown over and destroyed. The barometer indicated but little change. This storm continued until sunrise of the 9th when it moderated. I have seldom experienced so strong a wind; it blew over and broke one of the barometers, although its legs had been guarded carefully by large stones; and the wind was so violent at times, that it was with difficulty we could keep our footing. We suffered the loss of three thermometers, by the frame being blown down on which they were fastened. (166–167)

A *kona* storm similar to the one experienced by Wilkes arrived at Mauna Loa the morning of December 5, 2007. It could be seen arriving at and enveloping the summit of the mountain from Coconut Island in Hilo Bay and the nearby sand beach at Observatory 1. The view of the storm from Coconut Island (Fig. 3.1) likely resembled that seen by the crew of the USS *Vincennes* in Hilo Bay when the storm described by Wilkes struck the US Exploring Expedition.

After the storm covered the mountain, a large thunderstorm rose over it, while Mauna Kea, framed by a rainbow, remained sunny. That morning, tourists visiting Volcano National Park found the view of Kīlauea Crater blocked by sheets of cold rain driven by powerful wind gusts. The storm remained over Mauna Loa the next day, when the Mauna Loa Observatory could occasionally be seen through the mist from Saddle Road. At one point, a small cloud plume emerged from directly over the observatory, a phenomenon presumably triggered by the warmth of the station's rooftops and pavement, which become much hotter than the surrounding lava. The clouds overhead appeared twisted and contorted by the interference to their passage caused by the big mountain. The storm left only after dumping snow on both Mauna Loa and Mauna Kea, breaking a power line at the observatory, and disabling the Webcam that provides images of the Moku'āweoweo Crater atop Mauna Loa.

Returning to the Wilkes Expedition, on January 9, 1841, several fearless members of the team made perilous measurements of the depths of the crater. The west side was found to have a depth of 784 feet (239 meters), which is less than that claimed by Douglas in 1834 but considerably deeper than its present depth of about 565 feet (172 meters).

The mountain again took charge the night of January 10, when, as Wilkes recorded,

> The temperature fell quickly to 16 [degrees], when for a few hours it blew a perfect hurricane. I thought the pendulum-house in great danger: it rocked to and fro, appearing at times to be lifted from the ground, and several of its staples were drawn out. Not a person in the camp could sleep until towards daylight, when the gale began to abate.
>
> These gales reminded me strongly of those we experienced along the Antarctic cruise. I regretted I had no anemometer to ascertain the direction, changes and forces of the wind. (168)

On January 11, Wilkes conducted a series of sound experiments by firing a small mortar brought to the summit. At sea level, Wilkes reported, the sound of the mortar was so loud "it was necessary to close the ears when standing within 20 feet. The sound could be heard six miles, and the report was equal to that of a large gun." But at the summit, Wilkes wrote, "we stood next to it without any precaution whatever, and the noise it there made was more like that of a squib." Yet down at the Recruiting Station some eight miles away, "the sound was loud and reverberatory."

Having completed his pendulum and magnetic measurements, on the twelfth of January Wilkes and Dr. Judd joined still another survey of the summit crater. Wilkes found the experience a good deal more dangerous than he anticipated.

> In traversing these fissures, we were in great danger, and experienced much difficulty in walking on the recent stream that seemed to have flowed from them, for the snow which covered the lava concealed the new and weak places. The idea of being precipitated down a chasm of one hundred and fifty or two hundred feet deep, was by no means agreeable. Our blood was occasionally stirred by breaking through with one leg or both; and I shall not soon forget my own descent into a vapour or steam bath, which on trial was found to be 169 [degrees], although only a few moments passed before I was out of danger. (169)

As they completed their work at midday, the team found that the leather sandals that protected their shoes were in such poor condition it was necessary to stop and repair them. Wilkes then notes, "The glare from the snow in the strong sunlight had now become exceedingly uncomfortable to the eyes, which was felt by several who were in company with us." He next describes a prospect that can often be seen from today's Mauna Loa Observatory:

> The weather was still and calm, and a deathlike stillness prevailed, which I dreaded to break, even by making a remark to my companions upon the splendor of the scene before us. The sight was surpassingly grand. In the distance, the island of Maui emerged from and broke the line of the deep-blue horizon, while its lower side was dimmed by a whitish haze that seemed to unite it to the island of Hawaii. The same haze enveloped the

> hills of Kohala on our right, and the western extremity of Hawaii, up whose sides a compact mass of white fleecy clouds was impelled by the sea breeze. To our right rose in bold relief Mauna Kea, covered with its snowy mantle; and at our feet was spread out, between the three great mountains, the black plain of lava, overhung by a dusky pall of clouds. All these features were so blended into each other by the mist, as to exhibit a tone of harmony that could hardly be conceived, considering the variety of forms, characters, and distances of the objects, and which seemed to blend earth, sea and sky into one. I can never hope again to witness so sublime a scene, to gaze on which excited such feelings that I felt relieved when I turned from it to engage in the duties that had called me to the spot. (170–171)

The haze mentioned by Wilkes was almost certainly caused by the eruption of Kīlauea, and a similar haze, or volcanic fog, has polluted the skies of Hawai'i since the latest eruption began in 1982. The locally familiar vog sometimes envelops Kohala and flows across the saddle between Mauna Kea and Mauna Loa, much as described by Wilkes.

The duty that called Wilkes to this point, the highest summit of Mauna Loa, was the measurement of the difference in height between Mauna Kea and Mauna Loa, for no one then knew with certainty which mountain was the highest. The measurement would be the first made from this lofty perch, but Wilkes was not the first person to employ scientific instruments at the site. Forty-seven years earlier, on February 17, 1794, Archibald Menzies stood at the same spot and measured the pressure using his portable barometer (see chapter 2). It is unknown if Wilkes reflected on this at the time, but we do know that he found himself overpowered more by the scene before him than by the challenge of the measurement he was about to make.

> The very idea of standing on the summit of one of the highest peaks in the midst of this vast ocean, in close proximity to a precipice of profound depth, overhanging an immense crater "outrageous as a sea," with molten rock, would have been exciting even to a strong man; but the sensation was overpowering to one already exhausted by breathing the rarified air and toiling over the lava which this huge cauldron must have vomited forth in quantities sufficient to form a dome sixty miles in diameter and nearly three miles in height. (171)

Almost exactly two years later, Mauna Loa would again erupt and send a massive flow of lava down to the saddle between the great mountain and Mauna Kea. One can only wonder how much more impressed Wilkes would have been had he fully understood the enormity of the mass of Mauna Loa formed by its countless eruptions. For we now know that the volume of the mountain is vastly greater than the portion that emerges from the Pacific. Indeed, Mauna Loa is the world's single largest mountain.

Wilkes found that Mauna Kea was indeed higher that Mauna Loa and that the highest cone of Mauna Kea was "one hundred and ninety-three feet above the place where I stood." They were then on the opposite side of the summit crater from Pendulum Peak camp, and their sandals had been worn out. It was with great difficulty that they managed to reach the camp, for walking across lava is exceedingly tedious, tiring and risky, even when wearing sturdy hiking boots. A seemingly gentle fall can bloody one's hands, and a misstep can sprain or break an ankle. In spite of the dangers of the nonexistent trail, they managed to arrive after sunset in a state of exhaustion.

A major problem faced Wilkes on his return to the Pendulum Peak camp. A large party of poorly clothed Hawaiians had arrived to help carry the gear back to the ship, for the expedition was scheduled to depart the following day. However, it was night, and there was no place for the men to stay. Wilkes decided to remove the clock and pendulum apparatus from the grass-insulated pendulum house, which the Hawaiians then gratefully occupied after being provided with ample food.

While the exhausted Wilkes provided for the comfort of the bearers, the effects of snow blindness began to take effect. The ultraviolet (UV) irradiance at today's Mauna Loa Observatory is the most intense of any site in the world where UV is regularly measured. It can be at least 20 percent more intense than at sea level and still greater at the summit. Snow reflects ultraviolet quite well, so the inflammation of the cornea known as snow blindness is a risk of hiking across snow-covered terrain on a sunny day without the protection of sunglasses or goggles, especially at a high-elevation site. Wilkes later wrote about his experience with snow blindness:

> During the time I was thus engaged, I began to feel as if cobwebs had passed over my face and eyes, and found the same feeling prevailed with two or three of the men who had accompanied me during the day. To this feeling succeeded excessive irritation and inflammation of the eyes and eyelids, brought on by exposure to the strong glare from the snow. Dr. Judd was kind enough to make various applications, but none of these produced any effect, and I felt forcibly the horror of probable blindness; indeed I was so for the Time, and notwithstanding all my fatigues, I passed a sleepless night in great pain. The night was stormy: the thermometer fell to 17. I, however, determined to leave the station in the morning, if I had to be led down the mountain, which I thought very probable. One consolation remained: my personal energies had not given way until every part of the objects of my ascent of Mauna Loa had been fully accomplished. (172–173)

Wilkes recovered enough during the night to carry on with the next day's plans. Several men were assigned to stay behind for a time to complete some additional surveying, and the camp was disassembled and prepared for transport down the mountain by the Hawaiians and sailors. Wilkes records the departure details in his narrative.

> At nine o'clock [on January 13, 1841], Dr. Judd, myself, and six of crew of the *Vincennes*, bade adieu to the walled village we had built. The men showed their delight at quitting this barren and desolate spot by three hearty cheers. It was no little gratification to me to be able to take my departure, after having successfully accomplished all the duties assigned to me here, without any serious mishap, except in the case of Longley [who suffered from exposure], although all the party had been more or less sufferers from the mountain-sickness. (177)

Prior to their departure, J. G. Clarke, a seaman from the *Vincennes*, chiseled into a lava block this inscription: "Pendulum Peak, 1841 U.S. Ex. Ex." The remains of the stone walls they left behind are still present today near the Mauna Loa cabin and may be the only remaining campsite of the U.S. Ex. Ex. within the Pacific (Hawai'i Volcanoes National Park 2004).

On January 16, Wilkes and his men had descended to about 4,000 feet (1,220 meters) above the ocean. There was no longer any altitude sickness induced by the rarified air of Mauna Loa, and they were much refreshed following the grueling expedition to the summit. They had spent twenty days on the summit making hundreds of meteorological and gravity measurements. Now, exactly one year after their first sighting of land at Antarctica, they were preparing to study the Kīlauea Crater.

After more adventures and close calls mapping and exploring Kīlauea, Wilkes and his men returned safely to the *Vincennes*, where they stayed another six weeks recuperating from the grueling Mauna Loa ascent and making a suite of meteorological observations from the ship, Observatory 1, and two additional observatories. Observatory 2 was situated on a hill just northwest of where the Wailuku River empties into Hilo Bay. Observatory 3 was located on a hill south of Hilo. Considerable time was invested in making gravity measurements using the pendulum apparatus that had been taken to the summit. On March 5, 1841, they set sail from Hilo for Honolulu, from where they departed for the Pacific Northwest Coast.

The Impact of the Exploring Expedition on the People of Hawai'i

The pursuit of science on the volcanoes of Hawai'i has long contributed to the economy of the Big Island. Charles Wilkes' narrative clearly establishes that the Ex. Ex. had an economic impact, for several hundred Hawaiians were employed as bearers for the expedition to Mauna Loa, and the expedition remained in Hilo for some three months, during which time food and other supplies were purchased from locals. The island was often visited by whaling ships, and their contributions were not always positive or desired, especially by the local missionaries and their flocks. One of the leading missionaries was the Reverend Titus Munson Coan, who is remembered both for his role in the dramatic conversion to Christianity of the high priest and priestess of the volcano and for his detailed reports of many volcanic eruptions (Coan 1882). Coan served in Hilo from 1835 to 1881 and was fully dedicated to his work with Native Hawaiians and new arrivals. He reflected on the visit of Charles Wilkes and the US Exploring Expedition in the autobiography that he wrote in his eightieth year:

> Here with an admirable corps of scientists [Wilkes] spent three months in explorations, measurements, observations, etc. Parties of officers and scientific gentlemen were detailed to visit different parts of the island, some to ascend the mountains, and some to survey the shore, making collections, drawings, and observations in all the branches of the natural history of the Islands. The commander called for 300 young and vigorous men to take him, with the materials of a wooden house and all the apparatus of a large observatory, with food, fuel, water, beds, etc., to the summit of Mauna Loa, where he and his attendants were to spend twenty or thirty days in taking observations.
>
> Other parties required large numbers of men to carry baggage, instruments, etc., and to act as guides and assistants in making surveys and collecting a large amount of specimens. (Coan 1882, 66)

Some of Coan's account differs from that of Wilkes, or at least what can be deduced from it.

> Parties of natives thus employed needed to be recruited often on account of fatigue and exhaustion, and for the lack of shoes and warm clothing to endure the hard travel and the rains, cold, and snows of the mountains. Some died of cold. It is supposed that about one thousand of our strongest men were brought into this service, and with small pay, during these three months. Some parties of men were required to travel and work on Sunday as on other days. All this had a demoralizing effect upon the poor natives. They had been accustomed to rest from all physical toil, and to worship on the Lord's day. Our congregations were much reduced in numbers. There was no little murmuring among the people at this new state of things, and for years the moral tone of the church and community could not be fully restored to its cheerful and normal state. (67)

Wilkes described the same hardships related by Coan. He also described the loss and apparent death of one Hawaiian, who became separated from his party during the descent down Mauna Loa. Coan was highly protective of his flock, and just three years before he had helped them recover from the 1837 tsunami that devastated Hilo with multiple walls of water. He was less concerned by the hardships endured by the Ex. Ex. on Mauna Loa than by those his people experienced, and he did not explain that the Hawaiians were hired and not forced to assist the expedition. Nor did he mention that virtually all of the bearers employed on the initial ascent deserted the Ex. Ex. and Wilkes had to send for help from the *Vincennes*. As for working on the Lord's day, Wilkes recorded his displeasure with the Rev. Coan when he learned that the missionary had delayed a Sunday delivery to Mauna Loa of a consignment of badly needed shoes and sandal leather for the sailors, who were reduced to going barefoot in freezing weather after their shoes were completely ruined by the sharp lava.

> In a note from Lieutenant Alden, he informed me that not more than half the men had shoes, and not more than that number were fit for duty, partly on that account and lameness, together with mountain-sickness. A supply of shoes, and sandals of raw hide, had been sent for, as the men had already worn out two pair. It was exceedingly provoking to learn that there was much delay in getting these articles and the provisions from the ships; which arose, as I was informed by letter, in consequence of the refusal of the Rev. Mr. Coan to allow the natives to set out early on Sunday morning; he required the officer to state that he believed our necessities were urgent before he would consent to the natives going. (Wilkes 1845, 4:160)

In the end, Wilkes wrote favorably about the missionaries of Hawai‘i, and there was no serious long-term impact on the community from the visit by the Ex. Ex. In spite of the misunderstandings and negative experiences he related, Coan was impressed by the scientifics and others associated with the expedition.

> This was a trial of faith, and a fan to winnow the church, but most of our Christians stood fast, and although it checked the progress of the revival, the loss to the church was less than might have been feared.
>
> The visit of the expedition to Hilo afforded us an opportunity to form an acquaintance with many worthy gentlemen, several of whom we met again in the United States in 1870–1871. Among these we met and received as a very welcome guest the then youthful James D. Dana, one of the scientific corps, now so distinguished in various departments of natural science, and honored as a Christian philosopher. The friendship then formed has been increased by years and can never wane. (Coan 1882, 67)

Several lessons can be drawn from the accounts by Wilkes and Coan. The visit of the Ex. Ex. under the permission of King Kamehameha III was an unprecedented event. Several prior voyagers had hired locals to assist them during ascents to the summit of Mauna Loa, but the Ex. Ex. ascent was a much bigger enterprise. No one realized how quickly the lava would destroy leather footwear designed for walking on the wood decks of ships. Nor did the explorers or their Hawaiian bearers fully anticipate the cold and the altitude sickness they would experience during the ascent. The Ex. Ex. succeeded admirably in spite of these unexpected difficulties and the total lack of rapid communications with those in Hilo.

The Scientific Legacy of Pendulum Peak

Charles Wilkes and the scientifics who accompanied him on the ascent of Mauna Loa conducted the first scientific measurements from the summit since the ascent by Archibald Menzies in 1794. Menzies climbed the mountain to measure its elevation and succeeded admirably. He also made temperature measurements and visual observations. Wilkes executed a far more ambitious plan that involved hourly measurements of temperature and regular observations of pressure, sunlight, atmospheric electricity, precipitation, and wind speed and direction. He and his team also measured the effect of the high altitude on sound propagation, surveyed and mapped the two summit craters, measured the gravity at the summit, and measured the difference in height between Mauna Loa and Mauna Kea. Wilkes also observed physiological reactions, including his own, to the high-altitude environment of the summit, including snow blindness, altitude sickness, and the increase in pulse rate that accompanied excitement at the summit (the increase ranged from thirty to forty beats per minute).

Wilkes also arranged that during the summit expedition hourly temperature measurements would be made at Observatory 1 at Waiākea Point near Hilo and from the *Vincennes*. Regular pressure measurements were also made. These data provided a valuable time series comparing the conditions at the summit and at sea level. The precision of the instruments and consistency of their use is described in the introduction to the Ex. Ex. volume on meteorology:

> The meteorological instruments used by the Expedition were procured in 1836 [by Charles Wilkes], from the best makers in England, France and Germany; others were prepared by Fisher, of Philadelphia, with great care and accuracy. They were all compared with the standard, with which they agreed. It was deemed essential at the beginning of the cruise, to insure a uniform mode of observation by all the officers; the method of procedure was pointed out to all who were to be engaged in the duty, and it is confidently believed, was adhered to throughout. Books of formulae were prepared for recording the observations. . . .
>
> The hours chosen for observation were those generally adopted, viz.: 3 a.m., 9 a.m., 3 p.m., and 9 p.m., according to civil time; when it was deemed desirable, more frequent observations were ordered. At the observatories, hourly observations were made, and throughout

the voyage, hourly observations of the temperature of the air and water were taken, both at sea and in port. A thermometer placed at the masthead was also observed at the hours above indicated; it was sheltered from accident by a small covering, which did not obstruct the free passage of air: this served to check the results on deck, proceeding either from reflection or the influence on the temperature arising from a crowded ship.

> The barometer for the use of the observatory was an ivory float gauge, made by Troughton & Simms; its diameter of tube was 0.35 in., and its error on the Royal Society's standard .007. This was considered the standard: it continued in use throughout the voyage, and was in perfect order at the return of the Expedition, having been used in the final experiments in Washington. It was read off to thousandth parts.
>
> We had several standard thermometers by Troughton & Simms, Dolland, and Jones, of Charing Cross, both for air and water, graduated on metal and ivory to half degrees, and easily read off to tenths; the length of scale 18 inches; also several self-registering spirit thermometers, among them, those of Six. The hygrometers in use were Pouillet (a capsule), Daniels, the silver cup, and the wetted bulb.
>
> The Photometer and Aetherioscope of Leslie were used.
>
> The rain gauge was that of Dewitt, a cone of nine inches in height by six at the base; it was measured by a rod graduated to hundredths, it was placed at the end of the spanker-boom, where it was free from the drip of the sails or rigging. The quantity of water that fell was measured immediately after rain.
>
> The force of the wind was noted in the usual way by the different observers. For the sake of space in the tables, the expressions have been changed into numbers. (Wilkes 1851, vi–vii)

Wilkes studied atmospheric electricity on the great mountain and reported that

> During the whole time that we were above the height of nine thousand feet, there were only one or two days in which the electrical excitement of the atmosphere was not apparent, and those were exceedingly damp; the electroscope, in fact, was in constant action during our stay. (Wilkes 1845, 4:178)

Every hour, the temperature of the air and the atmospheric pressure was recorded at the sea level observatory and, for a time, at Pendulum Peak. A black-bulb thermometer was employed to measure the temperature induced by its exposure to direct sunlight.

The care devoted to the instruments and the analysis of the data can be deduced from a comment by Wilkes in the introduction to the Ex. Ex. volume on meteorology:

> On the diagrams, the barometrical observations have been reduced to a uniform temperature (32 Fahr.), and to the level of the sea, and corrected for instrumental errors, as well as brought to the standard of the Royal Society, London, with which they were originally compared. (Wilkes 1851)

Some of the meteorological and geophysical measurements made by the Ex. Ex. on Mauna Loa and at Hilo can be compared with those made today; in particular, the elevations of Mauna Loa and Mauna Kea, the atmospheric pressure, the temperature measured by the shaded thermometers, and the temperature of the water in Hilo Bay.

For example, from January 31 to February 28, 1841, the mean temperature measured at Observatory 1 at Hilo was 70.93° F (21.6° C) (Wilkes 1851, 486). The mean February temperature from 1946 to 2006 about 0.8 mile (1.3 km) away at the Hilo National Weather Service station was 71.3° F (21.8° C) (WRCC 2006), a difference of +0.4° F (0.2° C), some of which may be attributed to the nearby aircraft runways at the Hilo station. Very close agreement is also found in the temperature of the water measured from the USS *Vincennes* in Hilo Bay near Coconut Island and at a modern site in the bay. The mean water temperature for February 1841 was 73.09° F (22.8° C), which is slightly warmer than the present February mean of 72° F (22.2° C), according to the National Oceanographic Data Center (2007). The difference of 1.1° F (0.6° C) should be considered provisional, for the depths at which the temperature was measured may have differed.

Of special interest would be a comparison of the temperature at the summit measured in December and January by Wilkes and that measured during the same time of the year at the Mauna Loa Summit Observatory from 1951 to 1954. The latter records, however, have not been located. Various meteorological data have been collected by astronomical observatories at the summit of Mauna Kea, which is only slightly higher than Mauna Loa, and Sara Couto da Silva (2006) analyzed some of the data in a dissertation. The mean temperature in January from 1994 to 2006 at the Canada-France-Hawai'i Telescope was 32.5° F (0.3° C), the mean at the UK Infrared Telescope was 32.2° F (0.1° C) and the combined mean was 32.25° F (0.14° C). From the summary by Guppy (1906), the mean measured by Wilkes at the summit of Mauna Loa from December 24, 1840, to January 12, 1841, was 38.5° F (3.6° C), but some of these data are the average of the daily minimum and maximum and may not be as reliable as the more numerous observations made from January 4 to 9, when the mean was 28.2° F (–2.1° C). Thus, the most that can be concluded from this very limited comparison is that the mean summit temperature atop Mauna Loa in January 1841 was somewhere between –2.1° C cooler to 3.6° C warmer than the January temperature atop Mauna Kea in recent years. Perhaps a more meaningful comparison can be produced when the Mauna Loa Summit Observatory data are located.

There is a significant difference in the measurement of the elevation of Mauna Loa by Menzies and by Wilkes. Menzies compared simultaneous pressure measurements from sea level and the summit to arrive at an elevation of 13,634 feet (4,156 meters), which is within 0.32 percent of the elevation given by the Hawaiian Volcano Observatory. Wilkes claimed an elevation of 13,440 feet (4,097 meters), which is within 1.74 percent of the correct elevation. The greater difference of the Wilkes measurement is unexplained and could be due to the absence of a temperature correction.

A comparison of many of the 1840–1841 measurements with modern data is limited or even prohibited by the absence of detailed information about the instruments and their use. Ramage (1978) examined this question for the *20th Anniversary Report of the Mauna Loa Observatory:*

> Lt. Wilkes had with him three mercurial barometers, thermometers, a Daniell silver cup hygrometer (Middleton 1969), and a Pouillet capsule hygrometer. (The expedition volume on physics, which was to have included details of the instruments, was never published. A search of the original papers failed to unearth the information [Reichelderfer 1940].) With them he measured "shade," temperature, and dewpoints but did not specify the method of exposure.

As Ramage notes, full details about the various instruments were to have been included in a scientific volume detailing the physical measurements during the expedition. Although Wilkes appealed for continued support from the Congress, funds were short, the Civil War had begun, and the physics volume was among several that were never approved for funding.

While Wilkes and his scientifics made many discoveries during their four-year voyage, they did not always offer explanations for them. Such is the case with Wilkes' description of the tropical temperature inversion. Ordinarily, temperature in the troposphere declines with height. However, the atmosphere over the tropical oceans exhibits what is known as the trade or trade wind temperature inversion that effectively divides the troposphere into a cool, damp region near the surface overlain by a very dry, warm layer of air (Riehl 1954). A trade inversion is often fixed in place over Hawai'i, and its presence is quite obvious to the scientists who work at today's Mauna Loa Observatory. The separation between the tops of the cumulus clouds and the warm, dry air above is referred to as the inversion layer, and on fair days the inversion layer remains well below MLO and its many instruments that require a clean, dry atmosphere for their precise measurements of sunlight and the clarity of the atmosphere. During such times, MLO is said to reside in the free troposphere. Often, daytime heating causes the inversion layer to rise up to MLO by midmorning, and when it arrives one's face can detect the sudden arrival of cool, moist air as well as any meteorological instrument. When the inversion layer rises higher that the observatory, the site is suddenly enveloped in dense cloud and fog. Later in the afternoon, the influence of sunlight wanes and the inversion layer usually retreats down below, leaving behind a clear blue sky. Ramage observed that Wilkes described the trade wind inversion long before its official discovery but failed to account for its origin:

> More than this, he described the diurnal variation of the inversion on the mountain. Arising as the afternoon upslope winds lifted moisture and clouds to Sunday Camp, sinking as the nocturnal downslope winds took over. At the lower camp, the temperature fell between 6:00 p.m. and 9:00 p.m. but at Sunday Camp, as the clouds once more retreated below, it rose. Wilkes was puzzled by his observations but unaccountably advanced no hypothesis to explain them. This brilliant man had come within a whisker of discovering the trade wind inversion more than 15 years before Piazzi-Smyth (1858). (In August 1856, wet/dry bulb thermometers in an instrument shelter strapped to the back of a mule were read every few hundred feet of an ascent and descent of Pico de Teide in the Canary Islands. Piazzi-Smyth identified a sharp temperature inversion at 2,000 ft which coincided with a strong lapse in humidity and the top of a dense layer of clouds (see also Riehl 1954). Ironically, Wilkes had already observed effects of the trade wind inversion not far from where Piazzi-Smyth made his discovery. "Whilst at Madeira [September 17–24, 1838] . . . we found the height of the vapour plain to be about 4,000 ft above the level of the sea . . . which corresponds to the highest point of cultivation. . . . There is little doubt that the vapour plain must have considerable influence upon the climate of Madeira."). . . .
>
> The party quit the summit on the 13th and reached Sunday Camp at 5:00 p.m., where they were "soon enveloped in mist." The great observer remained unaware of his near miss. "The vapour plain, or the height at which the clouds usually remained, was about 5,000 feet above the sea. . . . [They] were seldom seen above the height of 8,000 feet except during the stormy weather" (Ramage 1978).

While Wilkes may not have explained the inversion layer, he certainly experienced it and reported on it and its accompanying clouds in colorful detail as he encountered them again and again during his time on Mauna Loa. He also kept detailed notes about various atmospheric phenomena, including the aurora australis observed in the Southern Hemisphere and appearances of parhelia (sun dogs).

In the end, Wilkes' lapse in failing to explain the trade wind inversion was of little consequence in comparison to the results of his leadership failures and the concerns they raised among many members of the Ex. Ex. The concerns eventually became serious, and on the return of the Ex. Ex. to the United States, Wilkes was

> brought before a court martial to answer charges of illegal punishment [including whipping and solitary confinement] and scandalous conduct relating specifically to assuming the uniform and style of captain without a promotion and changing the Antarctic discovery logs to be able to supersede the French claim to discovery. The trial was more of an opportunity for revenge by the discontents of the voyage, by being able to tell tales of Wilkes' temper in a public forum. But Wilkes also had defenders who attested to his hard work and zeal for the success of the expedition. In the end he was convicted of two charges of improper punishment and sentenced to a public reprimand by the Secretary of the Navy. It was a light sentence, but it wounded Wilkes' pride. (Naval Historical Center 2004)

Philbrick (2004) and others have explored in detail the events that led to the charges against Wilkes, some of which were exaggerated or false and others that were witnessed by various officers, sailors, and scientifics. There can be no denying that Wilkes was both arrogant and abusive, and the details of his leadership style were related in detail in an unauthorized journal kept by a member of the Ex. Ex. (Reynolds et al. 2004). Neither can there be any denial of the immense accomplishments of the Ex. Ex. under Wilkes' leadership, such as it was. The unfortunate consequence of a leader so flawed and yet so successful was that Wilkes did not return as the hero he might have been, and the historic Exploring Expedition never received the full measure of fame that it surely deserved and might otherwise have earned.

Would a less authoritarian and determined leader have prevailed over the frigid challenges of the Antarctic coast and the summit of Mauna Loa? Mary Cooley, an instructor in geology at Mount Holyoke College, offered her opinion in a talk before the 1940 Centenary Celebration of the Wilkes Exploring Expedition, 1838–1842: "Probably no officer in the navy was better fitted to undertake the direction of such surveys than Lieutenant Charles Wilkes, the commander of the expedition."

Commander Francis W. Reichelderfer, chief of the US Weather Bureau from 1938 to 1963, also spoke at the 1940 Centenary Celebration and observed, "The most careful selection of illustrations can scarcely do justice to the vision of the men who, more than a century ago, planned this expedition in the interest of commerce and science, and no brief recital can pay due tribute to the patience, industry, courage and determination which carried the work to fruition years after the work of the Expedition."

The vision of those who proposed, led, and accompanied the United States Exploring Expedition was not lost on Reichelderfer, who eight years after the Centenary Celebration approved a proposal by Weather Bureau meteorologist Robert Simpson that the US Weather Bureau establish a small observatory on the summit of Mauna Loa. The planned observatory would be directly across the great Moku'āweoweo Crater from the Pendulum Peak camp of the US Exploring Expedition. For the first time in more than a century, there was hope that atmospheric conditions atop the world's largest mountain would once again be measured and studied.

References

Breen, Jim. 2006. EDICT2 Japanese-English electronic dictionary files. Electronic Dictionary Research & Development Group, Monash University. (Accessed at http://www.animelab.com/anime.manga/dictionary/ra/1120/40.)

Buerge, David M. 1987. The Wilkes Expedition in the Pacific Northwest. *Columbia Magazine* 1 (spring): 17.

Caccamise, Dana J., II, Mark A. Merrifield, Michael Bevis, James Foster, Yvonne L. Firing, Mark S. Schenewerk, Frederick W. Taylor, and Donald A. Thomas. 2005. Sea level rise at Honolulu and Hilo, Hawaii: GPS estimates of differential land motion. *Geophysical Research Letters* 32, L03607, doi:10.1029/2004GL021380.

Coan, Titus. 1882. Life in Hawaii: An autobiographic sketch of mission life and labors (1835–1881). Edward J. Coan, ed., 1997. (Accessed at http://www.soest.hawaii.edu/GG/HCV/COAN/V.html. See also http://www.books.google.com.)

Colvocoresses, Lieut. George Musalas. 1855. Four years in the government exploring expedition commanded by Captain Charles Wilkes, 5th ed. New York: J. M. Fairchild & Co. (Accessed at http://books.google.com/.)

Cooley, Mary E. 1940. The Exploring Expedition in the Pacific. *Proceedings of the American Philosophical Society* 82: 707. (Accessed at http://links.jstor.org/sici?sici=0003-049X(19400629)82%3A5%3C707%3ATEEITP%3E2.0.CO%3B2-X.)

Esch, Doris. 1965. Outfitting the United States Exploring Expedition: Lieutenant Charles Wilkes European assignment, August–November, 1836. *Proceedings of the American Philosophical Society* 109: 3, 159–172. (Accessed at http://links.jstor.org/sici?sici=0003-049X%2819650615%29109%3A3%3C159%3AOTUSEE%3E2.0.CO%3B2-L&size=LARGE.)

Friedman, S. Morgan. 2011. The inflation calculator. (Accessed at http://www.westegg.com/inflation/.)

The *Globe*. 1836. May 17: 2. Cited in Doris Esch, 1965, Outfitting the United States Exploring Expedition: Lieutenant Charles Wilkes European assignment, August–November, 1836, *Proceedings of the American Philosophical Society* 109, 3, 159–172. (Accessed at http://links.jstor.org/sici?sici=0003-049X%2819650615%29109%3A3%3C159%3AOTUSEE%3E2.0.CO%3B2-L&size=LARGE.)

Guppy, H. B. 1906. *A Naturalist in the Pacific between 1896 and 1899*. Vol. 2, Appendix, Note 61: Meteorological observations on the summit of Mauna Loa. London: Macmillan and Co.: 586–587.

Hawai'i Volcanoes National Park. 2004. Wildland fire management plan, chapter 3: Affected environment. (Accessed at http://www.nps.gov/archive/havo/manage/fire/eafire_affenv.htm.)

Jarves, James Jackson. 1843. Scenes and scenery of the Sandwich Islands and a trip through Central America: Being observations from my note-books during the years 1837–1842. Boston: James Munroe and Company. (Accessed at http://books.google.com/.)

Jefferson, Thomas. 1803. Instructions to Captain Lewis, June 20, 1803. In The letters of Thomas Jefferson: 1743–1826. (Accessed at http://grid.let.rug.nl/~usa/P/tj3/writings/brf/jefl155.htm.)

Macrae, James. 1922. With Lord Byron at the Sandwich Islands in 1825: Being extracts from the MS diary of James Macrae, Scottish Botanist. Wm. F. Wilson, ed., Honolulu. (Accessed at http://books.google.com.)

Middleton, W. E. K. 1969. *Invention of the Meteorological Instruments*. Baltimore: Johns Hopkins Press.

Mims, Forrest M., III. 1993. *Science Fair, Ticket to Your Future*. Fort Worth, Texas: Radio Shack: 4–5.

National Oceanographic Data Center. 2007. Water temperatures in degrees Fahrenheit, Hawaiian Island coast. (Accessed at http://www.nodc.noaa.gov/dsdt/cwtg/hawaii.html.)

Naval Historical Center. 2004. The Alfred Agate collection: The United States Exploring Expedition, 1838–1842. Department of the Navy. (Accessed at http://www.history.navy.mil/ac/exploration/wilkes/wilkes25.html.)

Philbrick, Nathaniel. 2003. *Sea of Glory: America's Voyage of Discovery—The U.S. Exploring Expedition, 1838–1842*. New York: Viking Penguin.

Ramage, C. S. 1978. The United States Exploring Expedition on Mauna Loa. In John Miller, ed., *Mauna Loa Observatory: A 20th Anniversary Report*, Washington, DC: National Oceanic and Atmospheric Administration, US Department of Commerce.

Reichelderfer, F. W. 1940. The contribution of Wilkes to terrestrial magnetism, gravity and meteorology. *Proceedings of the American Philosophical Society* 82: 583–600.

Reynolds, Jeremiah N. 1828. On the expediency of fitting out vessels of the navy for an exploration of the Pacific Ocean and South Seas. Second Session of the Nineteenth to the Second Session of the Twenty-first Congress, vol. 3, Naval Affairs, Washington, DC: Gale's & Seaton, 1860. (Accessed at http://www.du.edu/~ttyler/ploughboy/usexex363.htm.)

———. 1836. Address on the subject of a surveying and exploring expedition to the Pacific Ocean and South Seas. Delivered in the Hall of Representatives on the Evening of April 3, 1836. New York: Harper and Brothers. (Accessed at http://www.eapoe.org/works/CRITICSM/slm37r01.htm.)

Reynolds, William, Nathaniel Philbrick, and Thomas Philbrick. 2004. *The Private Journal of William Reynolds, United States Exploring Expedition, 1838–1842*. New York: Penguin Books.

Riehl, H. 1954. Tropical meteorology. New York: McGraw-Hill Book Co. (Accessed at the American Meteorological Society Glossary of Meteorology at com/glossary/browse?s=t&p=40.)

Silva, Sara Couto da. 2006. Climatological analysis of meteorological observations at the summit of Mauna Kea. University of Hawai'i and University of Lisbon. (Accessed at http://kiloaoloa.soest.hawaii.edu/guide/presentations/climate/MKclimatology.pdf.)

Solomon, Susan, and John S. Daniel, 2004. Lewis and Clark: Pioneering meteorological observers in the American West. *Bulletin of the American Meteorological Society* 85: 1273–1288.

Solomon, Susan, John S. Daniel, and Daniel L. Druckenbrod. 2007. Revolutionary minds. *American Scientist* 95. (Accessed at http://www.americanscientist.org/template/AssetDetail/assetid/55905.)

South-Pole. 2007. Charles Wilkes, 1798–1877. (Accessed at http://www.south-pole.com/p0000079.htm.)

Wilkes, Charles. 1845. United States Exploring Expedition during the years 1838, 1839, 1840, 1841, 1842. Vols. 1, 2, and 4. Philadelphia: Lea and Blanchard. (Accessed at http://www.sil.si.edu/digitalcollections/usexex/follow-02.htm.)

———. 1851. United States Exploring Expedition during the years 1838, 1839, 1840, 1841, 1842. Vol. 9, Meteorology. Philadelphia: C. Sherman. (Accessed at http://www.sil.si.edu/digitalcollections/usexex/follow-02.htm.)

———. 1861. United States Exploring Expedition during the years 1838, 1839, 1840, 1841, 1842. Vol. 23, Hydrography, Atlas, vol. 2. Philadelphia: C. Sherman. From a scan provided by Leslie K. Overstreet, curator of Natural-History Rare Books, and Erin Clements Rushing, New Media Office, Smithsonian Institution Libraries, Washington, DC.

WRCC. 2006. Hilo WSO AP 87. Hawaii: Period of record general climate summary—Temperature. Western Regional Climate Center. (Accessed at http://www.wrcc.dri.edu/cgi-bin/cliGCStT.pl?hihilo.)

CHAPTER FOUR

Robert Simpson's Dream

The Mauna Loa Summit Observatory (1951–1954)

The tough men of the US Exploring Expedition spent three grueling weeks at the Pendulum Peak camp on the summit of Mauna Loa during the winter of 1840–1841. Thereafter, Charles Wilkes' pioneering measurements of temperature, humidity, sunlight, and atmospheric electricity on the giant volcano were replaced by occasional eyewitness reports of the mountain's periodic eruptions. Several hearty men and at least one woman who ascended the flanks and even the summit of Mauna Loa provided the only data about the eruptive history of the mountain for more than half a century.

Among the earliest eruption observers was the Reverend Titus Coan, the highly regarded Hilo missionary who befriended James Dwight Dana when the US Exploring Expedition visited Hilo in preparation for its Mauna Loa ascent during 1840–1841. Coan and Dana did not always agree on the nature of Mauna Loa's eruptions, about which they engaged in friendly correspondence that might serve as a model for today's differences of opinion regarding various scientific matters. An example of this was given by Russell Apple (2000), who reported that Coan "hiked to and described almost every flank eruption of Mauna Loa between 1843 and 1880." Dana had become an editor of *American Journal of Science and Arts*, to which Coan submitted a detailed report about the 1843 eruption of Mauna Loa. Apple reports that Dana

> didn't always believe the testimony of the eyewitness. For instance, after Dana published Reverend Coan's graphic and detailed description of the source and flows of Mauna Loa's 1843 eruption, Dana followed with "Mr. Coan speaks of the lavas as flowing from an orifice in a broad stream down the mountain. It is probable that fissures opening to the fires below were continued at intervals along the course of the eruption, and that these afforded accession to the fiery flood. Any internal force sufficient to break through the sides of a mountain like Mauna Loa, must necessarily produce a linear fissure or a series of fissures, and not a single tunnel-like opening." (Apple 2000 quoting Dana 1852)

Apple explained how the dialogue between Coan and Dana inspired further exploration and study, the same motivators that drove Menzies and Wilkes to Mauna Loa and set the stage for the science that has been conducted at Mauna Loa down through the years:

> Arguments such as this could only be resolved by volcanologists doing the observing, but they served to publicize the need for further study. The Coan-Dana controversy helped raise questions that needed to be answered about volcanoes, thus helping pave the way for such institutions as the Hawaiian Volcano Observatory. (Apple 2000)

The establishment of the Hawaiian Volcano Observatory (HVO) in 1912 by Thomas A. Jaggar Jr. finally provided a foundation for the formal study of Kīlauea's volcanism. The US Weather Bureau approved the project and became a sponsor from 1919 to 1924 (Apple 2000). While the HVO's studies of Hawaiian volcanism attracted the interest of the scientific community, the press, and the general public, there was still little interest in the potential for meteorological studies atop the island's great volcanic mountains. Only a handful of hardy souls climbed Mauna Loa in the half century after the US Exploring Expedition, including the famous British traveler and author Isabella L. Bird (1875), who spent a frigid night at the snowy summit during the eruption of June 1874.

The most important nineteenth-century scientific visit to the summit of Mauna Loa after the US Ex. Ex. was by Henry Brougham Guppy, a British surgeon, botanist, and geologist. Guppy apparently had help in transporting his supplies and provisions to and from the summit, but he camped there alone from August 9 to 31, 1897. For twenty-three days, he made daily measurements of wind, precipitation, relative humidity, and temperature. In the appendix of volume 1 of *A Naturalist in the Pacific between 1896 and 1899*, Guppy lists his data, provides a detailed description of the wind and clouds, and gives details about his altitude sickness, static electricity, and other experiences on the summit (Guppy 1906, 588–592). He also describes his primary instruments:

> The instruments employed were a Sixe's maximum and minimum thermometer made by Negretti and Zambra, several unmounted thermometers, and a reference thermometer (with a Kew certificate) by the above named makers, which was used as a standard. The freezing point was also tested for all the instruments on the summit in melting powdered ice.

Guppy lacked instruments for measuring gravity, sound propagation, sunlight intensity, barometric pressure and other parameters pursued by the US Exploring Expedition. Nevertheless, he admirably demonstrated what can be achieved, measured, and published by a determined investigator working alone for three weeks on the unforgiving summit of Mauna Loa. It is likely that much could be learned by comparing his detailed measurements of temperature with those that were to come more than half a century later.

There was also scientific interest in Hualālai and Mauna Kea, both of which were climbed by Bird, Guppy, and others. In 1892, Prof. W. D. Alexander, astronomer E. D. Preston, and others ascended Mauna Kea for the purpose of carefully surveying its summit cones. This highly successful expedition was sponsored by a grant from the Bache fund, which numbered among its trustees the same James Dwight Dana who served as mineralogist on the US Exploring Expedition (Alexander 1892). Before the expedition, the young Dana had published *System of Mineralogy*, which he updated throughout his life and in which he classified minerals using a system still in use today (USGS 2007).

Like the Menzies and Wilkes expeditions, these and other nineteenth-century scientific studies on the summits of Mauna Loa and Mauna Kea were relatively brief. The permanent scientific observations begun in 1912 at the Hawaiian Volcano Observatory at Kīlauea were devoted to volcanism, and there was no organized effort to conduct meteorological observations on Mauna Loa or Mauna Kea.

Robert H. Simpson (1951)

After a century and a half of occasional, brief scientific visits to the summit of Mauna Loa begun in 1794 by Archibald Menzies, a small weather observatory at the summit was formally dedicated on December 12, 1951. The facility was made possible by an unlikely alliance of the Weather Bureau, the US Navy, the Park Service, prison laborers, and a man who wanted to develop a ski resort. This complex array of collaborations would never have occurred were it not for the dream of Robert H. Simpson, a determined Weather Bureau meteorologist who worked far more hours than required by his government position and who envisioned a weather station atop the highest point in the entire Pacific Basin.

During a 1989 interview with Edward Zipser for the American Meteorological Society and the University Corporation for Atmospheric Research (UCAR), Dr. Simpson discussed his life as a hurricane scientist and his work in Hawai'i. Simpson's background is well worth reviewing, for like Menzies, Wilkes, and others before him who made observations from atop Mauna Loa, a strong personality, flexibility, and a willingness to take risks made the difference between success and failure.

During the UCAR interview, Simpson traced his interest in weather to a boyhood experience that he never forgot. In 1919, when he was not quite seven years old, a major hurricane devastated his home town of Corpus Christi, Texas. The storm surge flooded parts of the city, including Simpson's house.

> The family had to swim, with me on my father's back, three blocks in near hurricane force winds to safe shelter in the courthouse, the only high building in the downtown area. A lot of what I saw frightened me but also supplied a fascination that left me with a lifelong interest in hurricanes. (Simpson 1989)

The hurricane devastated downtown Corpus Christi and killed from 250 to 500 people. Young Simpson and his family watched in horror as a man carrying a baby fell from a cistern floating down a flooded street below their vantage point in the courthouse. The man eventually surfaced but without the baby. He then quickly dove down into the water and was never seen again. When he was ninety-seven, Simpson vividly recalled the hurricane and described the experience as dreadful. It was a tragic experience that would eventually shape his career (personal communication with author, September 13, 2009).

Simpson's musical abilities earned him an invitation to attend Southwestern University in Georgetown, Texas, where he majored in physics and mathematics. He then attended Emory University, where in 1935 his studies of piezoelectricity helped earn him a master's degree in physics. Those were depression days, and jobs for physicists were scarce. So Simpson fell back on his musical abilities to secure a series of positions in high school music programs. But his career interests were in science, so in 1939 he took civil service examinations in an effort to qualify for positions with the Weather Bureau, Bureau of Standards, and Geological Survey. The Weather Bureau offered him a position at Brownsville, Texas, which he accepted even though the pay was only a third of what he earned as a high school music supervisor.

In 1941, Simpson accepted a six-month assignment as a meteorologist at Swan Island off Honduras in the west central Caribbean Sea, where he learned much about maintaining every aspect of a remote meteorological station. Each day the four observers recorded hourly weather observations, launched two radiosondes and four pilot balloons, and transmitted their reports by radio. The observers spent the rest of their time maintaining the facility's instruments, power generator, and water and sewage systems. Simpson recalls how the Swan Island assignment prepared him for hurricane research and general meteorology (personal communication with author, September 13, 2009). It also grounded him in the fundamentals required to establish a remote weather station, experience that would guide him a decade later as he pursued a dream of an even more remote weather station on the top of a mountain in the center of the Pacific Ocean.

During World War II, Simpson was assigned to a variety of Weather Bureau offices and became involved in serious hurricane research, including missions in which aircraft were flown into the giant storms. In 1947, he was transferred to the Weather Bureau headquarters in Washington, DC. The following year, Francis Reichelderfer, director of the Weather Bureau, asked Simpson to take over a project to consolidate weather operations in Hawai'i. Simpson at first rejected the assignment. His family wanted to stay in Washington, and he was concerned that still another move would jeopardize his marriage. But Reichelderfer persisted and, as Simpson recalled years later,

> After two weeks passed with no agreement reached, [Reichelderfer] suggested, "I think you should write yourself a letter for my signature stating the circumstances

> under which you will agree to take this assignment, which will probably be the biggest career opportunity you will have over the next ten years."

Simpson wrote a letter that specified the following:

> (1) while in Hawaii I would report directly to Reichelderfer for budget matters, personnel actions, research plans and activities, and for counsel and advice; (2) after four years, unless extended by mutual agreement, I would be returned to Washington into a job at or comparable to those of my peers who had been promoted while I was away. (Simpson 1989)

Reichelderfer approved and signed the letter, and Simpson left for Hawai'i in May 1948. There he occupied a position on the postwar Joint Meteorological Committee responsible for the transition of military weather operations to the Weather Bureau, a role that placed him on the staff of Admiral Arthur William Radford, commander in chief, Pacific. Years later, Simpson recalled that a benefit of his "membership on Admiral Radford's staff at Pearl Harbor resulted in the establishment of the Mauna Loa summit observatory" (1989). But this is getting ahead of the story.

Anticipating Simpson's Dream

During eighteen days of August 1920, almost exactly a year after the young Simpson and his family survived the Corpus Christi hurricane, a hundred scientists assembled in Honolulu for the first Pan-Pacific Scientific Conference. The conference addressed anthropology, biology, oceanography, climatology, and other scientific fields around and across the Pacific Ocean. The conference approved a series of resolutions on various matters, including three on meteorology that anticipated what was to come after Simpson arrived in Honolulu.

Resolution 6 was general in nature and explained the need for meteorological observations and encouraged their widespread introduction across the Pacific to better understand weather patterns over both land and sea:

> **Resolution 6. Meteorology**
> Investigations in meteorology, or the physics of the atmosphere designed to lead to an accurate scientific knowledge of atmospheric phenomena are of recognized importance. Very little is known of the behavior of the upper air over the land, and still less over the ocean. The fundamental aspects of these phenomena are exhibited in their simplest manner over the greatest of oceans, the Pacific. Hence it is necessary to make meteorological observation over the Pacific for the use in studying the more complex condition over the land. Moreover, the collection and prompt dissemination of marine meteorological data are of great benefit to humanity in carrying on its commerce and in weather forecasting which is now limited by a lack of synchronized uniform, meteorological data over great areas not within the customary track of vessels. Observation at the place of origin of typhoons, hurricanes, larger cyclonic and anticyclonic areas, as well as the development, dissipation, oscillation, and translation of the same, are essential to successful forecasting and the study of ocean meteorology. Moreover the meteorological survey of these ocean areas has practical value; therefore the governments bordering on the Pacific ocean are invited to carefully consider these matters with a view to increasing the number of meteorological vessel and land stations within the confines of this ocean and on its borders, especially the establishment of vessel reporting stations in somewhat fixed positions. In considering these matters, it is believed that special attention should be given to increasing the number of stations in the well known "centers of action." The Pan-Pacific Scientific Conference commends the ocean navigation companies and their masters of vessels for the valuable assistance they have rendered the meteorological services of the various stations, and urges them to further cooperate especially in the matter of transmitting their weather reports by radiograph as well as by mail. (Anon. 1920)

The next two resolutions were recommendations for specific actions. Resolution 7 called for the prompt resumption of observations at the meteorological station on

Macquarie Island, which the resolution erroneously stated had been interrupted by World War I. Macquarie Island is located halfway between Australia and Antarctica, and it briefly served as a rendezvous point for the ships of the US Exploring Expedition that managed to find it during the winter of 1839–1840. An Australian government report on the "History of Macquarie Island" recorded that

> In 1822 Captain Douglass, of the ship Mariner described it as "the most wretched place of involuntary and slavish exilium that can possibly be conceived; nothing could warrant any civilised creature living on such a spot. (Department of the Environment and Water Resources 2005)

How meteorology could be conducted at a site so wretched that no civilized creature could even live there has relevance to the history of the remote Mauna Loa Observatory. Accordingly, an inquiry requesting details was sent to the Antarctic Division of the Australian government. Dave Connell promptly replied that the Macquarie Island station provided data from 1911 to 1915. Unfortunately, the data from December 1913 to November 1914 were lost at sea when the relief ship carrying the data foundered (Van Rooy 1957). The ship was apparently the *Endeavour,* an Australian government research trawler that delivered supplies to the Macquarie Island meteorological station and then disappeared at sea on December 5, 1914 (Stone 2006). An extensive search by several ships failed to find any sign of the eighteen crew members.

This tragedy, which occurred in the cause of expanding scientific knowledge, sheds new light on the Pan-Pacific Scientific Conference resolution concerning Macquarie Island. The professional scientists at the meeting were conversing, dining, and relaxing in a pleasant and scenic setting in Honolulu to propose what should be done at the remote and dangerous island. None of the scientists would ever be stationed on the island, and the text of their resolution reveals that they had no idea why the vital observations at the island had come to an end. Moreover, while the resolution was well intentioned and scientifically justified, it offered no funds to carry out the task and no volunteers to brave "the most wretched place of involuntary and slavish exilium that can possibly be conceived."

The Macquarie Island weather station was not reopened until March 25, 1948, twenty-eight years after the passage of the resolution calling for its restoration. Thanks to improved technology and determined leadership, the station has provided regular observations ever since. Today anyone with access to the World Wide Web can visit the station and see the current weather conditions by means of the station's Webcam (www.aad.gov.au/asset/webcams/macca/default.asp).

This brings us to the second meteorologically related action recommendation of the Pan-Pacific Scientific Conference of 1920, Resolution 8, which called for the establishment of a "Meteorological Station on Mauna Loa":

> In view of the fact that Mauna Loa, Island of Hawaii, the highest accessible point in the central Pacific, offers exceptional opportunities for the exploration of the upper air, it is recommended that a station of the first order be established on its summit for continuous meteorological observations. (Anon. 1920)

Implementing this resolution posed challenges even greater than those in Resolution 6. Apparently none of the conferees in Honolulu volunteered to organize or man the station, and there was no provision to provide a budget for its founding and operation. The US Exploring Expedition had camped at the summit of Mauna Loa for twenty days during the winter of 1840–1841. But the ascent was exceedingly difficult, and all supplies, water, food, shelters, and instruments had to be carried by the expedition and its Hawaiian porters. Severe storms were experienced, and Lt. Wilkes and his men often suffered with altitude sickness and freezing conditions. Establishing a modern meteorological station at the summit would require a usable access road, instrumentation capable of surviving the elements, and a crew of hearty men capable of running the station. All this would require considerably more funds than a routine weather observatory and pose a nightmare of logistical problems, not the least of which was the likelihood that Mauna Loa might respond with one of its periodic eruptions.

Resolution 8 of the Pan-Pacific Scientific Conference of 1920 remained an unimplemented goal until Robert Simpson arrived in Honolulu. Simpson soon came to the same conclusion about Mauna Loa's potential for a weather station as had the conference that occurred twenty-eight years before his arrival, for, as he recalled during the UCAR interview,

> Shortly after arriving in Honolulu, I was interviewed by the science reporter of the *Honolulu Advertiser,* with the result that a front page article the next day proclaimed Mauna Loa to be an incomparable natural laboratory for study of atmospheric processes because of its smooth gentle slopes extending upward through the trade wind easterly to upper tropospheric westerlies where the atmosphere was free of pollutants, the nearest source being thousands of miles away. The article went on to suggest that an observatory near the summit would provide an unparalleled benchmark record for studying the structure of the atmosphere and of weather processes. (Simpson 1989)

The newspaper article and Simpson's membership on Admiral Radford's staff at Pearl Harbor triggered the implementation of Simpson's dream for Mauna Loa and the fulfillment of Resolution 8 of the first Pan-Pacific Scientific Conference. For Simpson was soon to lead the first effort to establish a permanent meteorological observatory at the summit of the mountain. Perhaps Simpson's Scottish heritage also played a role, for Archibald Menzies and David Douglas, two of Mauna Loa's earliest scientists, were from Scotland.

Fulfilling Simpson's Dream

Simpson's newspaper interview occurred at a fortuitous time, for on January 16, 1947, Hawai'i's territorial governor Ingram Stainback had announced plans for a "scenic highway" to the summit of Mauna Loa and that the road had reached "a densely forested area" at an elevation of 5,000 feet (1,524 meters) (Associated Press 1947). In his 1989 interview, Simpson discussed what happened after the 1948 newspaper interview was published:

> To my surprise, only two days later, I received a call from a man by the name of Tom Vance, Director of Institutions in Hawaii, who said, "I think we might be able to do business together in achieving our common interests in Mauna Loa. I have a prison camp at Kulani deep in the rain forest at the 5,000 feet level on Mauna Loa. The inmates there need a challenge to help equip them to be responsible citizens when they leave prison. They can help both in road building and in construction of the observatory building." Not until later did he confide that what he really wanted to do was establish a ski lodge at the summit to be operated by the inmates. This at first seemed a questionable way at best to go about the establishment of the observatory, but after further thought and consultation with Dr. Reichelderfer I decided to pursue the matter further.

Reichelderfer's endorsement came during a trip Simpson made to Washington. While there, Simpson also met with Reichelderfer's director of research, Harry Wexler (Simpson 1978). Thereafter, Simpson was on his own as he was granted permission for the project from the Department of the Interior, secured the loan for the necessary heavy equipment, and worked with Tom Vance to plan a road to the summit. Simpson's dream was underway.

Tom Vance and the Lions Club Pave the Way

Thomas B. Vance, who headed the Department of Public Institutions of the Territory of Hawai'i, had unknowingly prepared the way for the Mauna Loa Observatory by establishing the Kūlani Prison Camp, which sprang from the need to move the Waiākea Prison Camp from where prisoners were constructing the Hilo airport during World War II to a more remote location. Vance had to overcome considerable opposition in moving the prison to the site he selected, especially from various officials and the Hilo Women's Club (Vance 1969). Finally, the Kūlani Prison Camp was established on the eastern slope of Mauna Loa. The prison was connected to Highway 11 south of Hilo by a new road built under Vance's supervision. Governor Stainback, who visited the facility more than once, enthusiastically supported Vance's work, and the prison road was named the Stainback Highway following a 1950 resolution to that effect by the Hawai'i Lions Clubs.

After the Kūlani Prison Camp was well established, Vance began plans for extending the Stainback Highway past Kūlani toward, he hoped, the summit of Mauna Loa. Simpson's proposed summit observatory was but one of his goals, for Vance hoped to use prison labor for the road project as part of his rehabilitation strategy. Prisoners had served as bearers for James Jarves and J. P. Couthouy when they climbed Mauna Kea in 1840, and there was ample precedent for using prison labor to build a scientific facility in Hawai'i. Thomas A. Jaggar's first volcano observation site on the rim of the Kīlauea caldera, which led to the founding of the Hawaiian Volcano Observatory, was hand-dug by prison laborers in 1912.

> The diggers were prisoners of the Territory of Hawaii, sentenced to a term of hard labor (HVO record book for 1912 . . .). Jaggar was not above using free labor, prison or otherwise, to help stretch his limited funds. Their prison camp was nearby at what is now Kilauea Military Camp in Hawaii Volcanoes National Park. The prisoners dug through almost six feet of volcanic ash and pumice to a layer of thick pahoehoe lava—a firm base for the concrete piers on which seismometers would be anchored. (Apple 2000)

Vance lacked money to build the road, and territorial law blocked his plans to use prison labor. But his membership in the Honolulu Lions Club provided a potential solution to both issues. The law allowed prison labor to be used for nonprofit community service organizations, and the Lions supported a variety of charities and needs. Howard Ellis, who knew Tom Vance and who served on the staff of the slope unit of the Mauna Loa Observatory for more than two decades, summed up what happened next in his monograph, *The Observatory: On Their Own* (Ellis 1988):

> He went to the Lions Clubs of Hawaii and the Hilo Lions [Club] was specifically chartered to build a road to the summit of Mauna Loa. The Lions held their convention in Honolulu in the spring and came up with $300 for the project.

The $300 donation, which amounts to $2,714 in 2010 dollars (Friedman 2011), followed a presentation by Vance at the annual Hawai'i Lions Club convention in Honolulu in the spring of 1949 (Vance 1969). As for direct involvement by the Lions Club, the 1949 convention designated the Hilo Lions Club as the official sponsor of the road, and this met the requirement of territorial law. Thus, the Lions Club would supervise the road project, their grant would provide money for fuel and supplies, and the Kūlani Prison Camp would provide the necessary labor. The Lions were enthusiastic about their involvement and even shared Vance's goals for a recreational facility on Mauna Loa. As the press reported, "The Lions Club of Hilo is planning to erect a snowline lodge . . . for skiers, skaters and other lovers of winter sports and recreation" (Shiramizu 1951).

Vance supplemented the Lions Club grant by selling lava rocks from Mauna Loa for a dollar each, or $9.05 in 2010 dollars (Friedman 2011). The Hilo Women's Club, which had opposed the establishment of the Kūlani Prison, was won over after Vance gave the members a tour of the facility and then escorted them to jeeps and trucks driven by prisoners for a bouncy ride along the newly extended Stainback Highway above the prison. The women, who didn't realize that Vance and a detachment of guards were following the prisoner-led convoy, enjoyed the adventure of the experience, but they suggested that the road should be renamed the "Strainback Highway" (Vance 1969). They joined with Vance in proposing "Gardens of the World," which reporter Hugh Lytle (1951) described as including "flora of all climates from that of the tropics to the arctic, growing along the highway as it soars from sea level to the snow."

The Lions Club, Vance, and the prison workers eventually constructed 9 miles (14.5 km) of road beyond Kūlani across the north slope of Mauna Loa before the $300 ran out (Vance 1969). This placed the end of the road at what came to be known as Kūlani Mauka at an elevation of about 8,300 feet (2,530 meters).

Meanwhile, Vance began work on a major new cross-island road project, the Konahilo Road, which the prison road-building crew accessed by making a right turn at Kūlani Mauka and cutting a spur 4.25 miles (6.8 km) downslope toward Saddle Road. By January 1951 he had established Camp Jaggar, from where he had begun cutting a perfectly straight road toward Kealakekua on the Kona side of the island at a mile per month (Shiramizu 1951). At 7.75 miles (12.5 km), the road project came to a halt after Camp Jaggar burned. The abandoned Konahilo Road eventually became known as "The Road to Nowhere."

By February 20, 1951, the road to the summit had reached 9,100 feet (2,774 meters), which was about 3 miles (5 km) beyond the Kūlani Mauka junction. On that day, Charles C. Smith, superintendant of the Kūlani Prison, appeared before the Hawai'i County Board of Supervisors and its chairman, James Kealoha, to request the use of three county dump trucks to assist in sealing the rough Mauna Loa summit road with cinders. Smith told the officials that a group of scientists would arrive on March 10 "to test the feasibility of establishing a laboratory at the summit of the volcanic mountain for the study of geophysics." He then said, "The higher we can push the road by that time, the better will be our chances for the laboratory" (*Honolulu Advertiser* 1951).

Smith's request was approved, and on March 9, 1951, Tom Vance and James Kealoha met with the visiting scientists in Hilo (Fig. 4.1). While the details of the visit have yet to be found, it is likely that the visiting scientists were taken on a tour of the newly extended road.

Meanwhile, Robert Simpson was looking for the road-building equipment necessary to continue the road to the summit.

> The big question was how to acquire the road building equipment to construct the access road. To explore the options, I sought out my colleagues on the JMC/PAC at Pearl Harbor. I explained what we wanted to do and told them, "I can get the help of the Geological Survey at the Volcano Observatory to help stake out a roadway likely to be safe from lava flows, and can provide from my regional funds the cash to purchase building materials and equipment for the observatory. The territory (of Hawaii) can supply the labor to construct the building and to build the road. However, we don't have the equipment to build the access road. Can you help us?" (Simpson 1989)

Simpson recalls that his military colleagues on the committee were intrigued by his plan:

> The Navy member of the Committee suggested the Navy might be able to help, and offered to introduce me to Admiral Radford to see if he would approve it. When I explained the plan to Radford, he was quite interested. (Personal communication with author, January 24, 2008)

> He agreed to loan the project two big road building CATS and ship them to Hilo, providing others would pick up the cost of fuel and spare parts. I quickly agreed the Weather Bureau would do so. Over my signature as member of the JMC/PAC, the Navy supplied the two road building CATS, and therewith a project was set in motion supported officially only by a handshake or two, to establish an observatory at the summit of Mauna Loa just below the caldera of Makeoweoweo [*sic*] Crater at an altitude of 13,453 feet. It was a small but auspicious beginning. (Simpson 1989)

The handshake agreements also included "the unfunded blessings of the Weather Bureau and . . . a preliminary endorsement of the Department of the Interior to enter National Park grounds for this purpose" (Miller 1978). There was good reason for Simpson to insert "unfunded" and "preliminary" into his recollection of the early status of the project.

Simpson's dream was well on the way to fruition when progress was threatened by a roadblock even greater than the lava boulders and fissures that blocked the way to the

FIGURE 4.1. The Mauna Loa site survey party on March 9, 1951, included *(left to right)* Charles M. Woffinden, US Weather Bureau; W. P. Mordy, head, Meteorology Div., Pineapple Research Co.; Tom Vance, County of Hawai'i; Leon Sherman, UCLA; James Kealoha, chairman, County of Hawai'i; Mr. Youman, US Weather Bureau; James Steiner, US Weather Bureau, Hilo office. *(MLO archive.)*

summit. The proposed observatory and access road would occupy land under the jurisdiction of the National Park Service, whose staff vigorously opposed the plan. Tom Vance provides details in his unpublished memoir:

> The Volcano National Park officials were expressing adamant opposition to the Mauna Loa summit road as an entrance to the summit extension of the Park via the back door. The Park Service said that public access to this portion would require supervisory personnel from the Park, personnel for which there were then no funds. The *Honolulu Star Bulletin* supported the position of the Park Service but the *Honolulu Advertiser* supported the plans of the Weather Bureau and the Kulani Prison Camp, proof that there is not collusion in the news media to consolidate their attack on an enterprise. Each

> of the two leading Honolulu daily papers took an independent position and went its own way. In one of the meetings at Park headquarters, the Assistant Superintendent urgently requested authority to go to Washington so as to try to put a stop to the whole affair. Such authority was never granted. (Vance 1969)

Robert Simpson had much to do with the reason the assistant superintendant was not granted authority to visit Washington. The Menzies expedition of 1794 and the US Exploring Expedition found a much friendlier reception from the Hawaiian leadership of their day than Simpson did, and this would not be the last attempt to block scientific studies from the summits of Hawai'i's major volcanoes. But Simpson persevered. As he recalled during his 1989 interview,

> The real fight came, however, after getting final approval from Dr. Reichelderfer, in persuading the Secretary of the Interior, Oscar Chapman, to allow us to use National Park land to build the road and erect the observatory building. This was a tough job. But with Reichelderfer's help, we finally got approval, after strong objections by Chapman's staff.

The Road to the Summit

Simpson recently provided new details about the construction of what came to be known as the Mauna Loa access road:

> Tom Vance and his staff supervised the construction. I visited the scene of construction and reviewed its progress perhaps half a dozen times between travels mandated by responsibilities elsewhere in the Pacific. I am reasonably sure that the road construction to the summit began near the 8,300 ft level [Kūlani Mauka]. The climbs to the summit on foot to stake out the routing of the road, and to select the observatory site began near that level after driving a 4-wheel drive vehicle from the prison to that level. (Simpson 2009)

Simpson pursued the project with the same intensity he had pursued his meteorology career, but hiking to the summit of Mauna Loa was a good deal more difficult than his usual duties. During one of his hikes to the summit, the mountain reminded him to slow down.

> During the climb I did have quite an elevated pulse rate, which concerned me, and I stopped frequently to let it settle down from 150 to 100 before proceeding. The only pain I experienced was after I got back to sea level when I had a corker of a headache. Two days later my physician examined me and said "Forget it; you're OK." (Personal communication with author, January 24, 2008)

Simpson had assistance in planning the summit access road:

> Gordon MacDonald, Director of the Volcano Observatory, made a number of trips to the summit with me and on one occasion with Charles Woffinden [see Fig. 4.1], Head of the Weather Forecast Office at Honolulu, to stake out the access road. (Simpson 1989)

Howard Ellis provided details about the construction of the summit access road in his monograph (Ellis 1988):

> Mr. Vance enjoyed describing his technique for building the road to the summit. He headed out with a small party aboard the two bulldozers [Fig. 4.2], loaded with bedding, food and other provisions. On the lava fields he sent a man out uphill with a hand-held level. The man would keep walking as long as he could sight back to the crew on the bulldozers.
>
> The machine operators kept the blades up. Their tracks left a fairly good road. They planned to back-blade on their way down and to work over some of the particularly bad spots. For meals and night time rest, they simply stopped wherever they were.
>
> There was a network of lava tubes along the route that the road took. Such a tube results from the flow of lava that courses down a channel between two previous flows and solidifies on top and continues to flow under-

FIGURE 4.2. In 1949 the US Navy loaned these two earth-moving machines for the construction of a rough road from Kūlani Mauka to the summit of Mauna Loa. *(MLO archive.)*

> neath. When the source stops, the last of the lava flows out the bottom leaving the empty tube.
>
> The lighter of the two bulldozers went first and the operator developed a special talent for feeling by the "seat of his pants." Whenever he felt the ground shake excessively and in a certain way, he jumped from his machine until checks were made.
>
> On at least one occasion the bulldozer fell into a tube after the operator had jumped. This delayed progress only until the crew could get extra equipment and men from the base camp, get the machine out, and fill up the caved-in portion of the road.
>
> Mr. Henry Auwae, an instructor at Kulani, stayed with all the road building on Mauna Loa from the beginning and did most of the work and operational decision making. He said that they were out on the summit road project almost six months. (Ellis 1988)

The road crew followed Vance's alignment method as they continued the road in an almost straight line across the north slope of Mauna Loa. A change in plans occurred about 1.2 miles (1.9 km) beyond the site that would later be selected for today's Mauna Loa Observatory. As Ellis notes in his history of the Mauna Loa Observatory, "it became obvious that they would have to go all the way around the top of the mountain before they could reach it if they continued with this method." Ellis records what happened next:

> Mr. Auwae backed up a bit and made about a 100-degree turn to the left (east) and went back across the north face of the mountain for about five miles until the alignment with Hilo was reached and then switched back and forth, picking his own way to the summit.
>
> There were such cracks and awful features that he sensibly gave up on some starts. He tried to skirt close to cinder cones because of the good road surfacing material found there. It is a wonder that they made it without a serious accident.

Aerial imagery generally affirms this account. At 1.2 miles (1.9 km) beyond the present site of the Mauna Loa

Observatory, the road turns from southwest to due south and directly up the slope of the mountain for 270 feet (82 meters). The road then switches back to the southeast. After 2 miles (3.2 km), the road switches back to the southwest.

Eventually the road reached 12,200 feet (3,719 meters), which was close enough to the summit to finalize Simpson's selection of a site for the new observatory. Representatives from the Weather Bureau, Vance's Territorial Department of Institutions, and the National Park Service gathered at the Kūlani Prison for their third attempt to reach the summit. Ellis' monograph gives the date as September 6, 1951. The *Honolulu Advertiser* (September 11, 1951) reported that the party included James W. Steiner of the Hilo office of the Weather Bureau; Charles M. Woffinden and Arthur W. Youmans of the Honolulu office of the Weather Bureau; Harry Brisbane, Ishmael Manus, Donald J. Murray, and A. M. Huronymous of the Department of Institutions; Thomas E. Carpenter, landscape architect with the San Francisco office of the National Park Service; and Nash Castro and Ray Bolen of the Hawaiʻi National Park.

The *Honolulu Advertiser* reported that the party gathered for a predawn breakfast at the prison before heading up the newly constructed road by truck. At the end of the road, the party hiked some 2.5 miles (4 km) to a point about 1,000 feet (305 meters) northwest of the summit crater. Unlike the fresh lava flows they traversed at many points while driving the new road, here the lava was ancient and much less likely to be overrun in a future eruption. The newspaper reported that the elevation was about "13,450 feet," which is quite close to the elevation of 13,453 feet (4,100 meters) given online by Google Earth™ (2009). The paper reported that "A large white flag was laid out on the ground as a marker for aerial photographs." The party also laid out a route for the remainder of the road to the summit.

The bulldozer crews soon resumed work on the road, and two switchbacks later the road neared the north pit of the summit crater. The men then continued another 1.5 miles (2.4 km) to the site selected for the Mauna Loa Summit Observatory. While bulldozers can cut through lava with surprising efficiency, a drive along today's Mauna Loa road illustrates the many challenges faced by the road crew as they drove their two bulldozers across the rough lava.

There's a story behind an upslope turnoff about 2.5 miles (4 km) from today's observatory at an elevation of 10,232 feet (3,119 meters). The origin of this road, which is clearly visible in aerial imagery, was explained by Ellis (1988):

> When the [road construction crew] reached the top, a big snow storm came and covered the ground deeply for as far as they could see. They couldn't find their road and might have been trapped, but Mr. Auwae acted quickly and loaded his crew on the two machines and headed direct for Hilo through the snow. They intercepted their first summit road about three miles down from the 11,000 foot level. This road that they made while coming down in the snow turned out to not only be straighter and about one third the distance as the first, but an all-around better one with an easier slope.

The Mauna Loa Summit Observatory

On October 18, 1951, the *Honolulu Advertiser* reported Thomas Vance's announcement that the Mauna Loa road had finally reached the summit. The following day, an article in the *Honolulu Star-Bulletin*, noted that inmates from the Kūlani Prison had poured the concrete foundation for the new observatory's building, which they were constructing from plywood. The Weather Bureau provided $500 for the building and the instruments (Ellis 1988). Robert Simpson discussed the summit building in his 1989 interview:

> Plans were finally completed to put [up] a small observatory building, with heat and a cooking facility, a bunk for overnight visits, and oxygen to supplement the thin air, in addition to the autographic equipment and probe systems to record wind, temperature, pressure, and rainfall.

The completed building was hauled in sections to the summit in a large truck and installed on the prisoner-built concrete slab, which remains on the mountain to this day. Aerial imagery suggests that the slab measures 13 x 15 feet (4 x 4.6 meters). Photographs of the building suggest it measured 8 x 10 feet (2.4 x 3 meters), dimensions that are supported by press reports. A roof topped with corrugated metal extended about 3 feet (0.9 meters) out from the building's walls. Cables connected the east and west sides of the roof to three eyebolts installed in both sides of the concrete slab. The little building's base was secured to the slab by eighteen sturdy bolts, making it much more capable of withstanding hurricane-force winds than the tents of the US Exploring Expedition during its stay at the summit for nearly three weeks in the winter of 1840–1841.

Shortly after the building was completed, meteorologist Joachim Kuettner accompanied Robert Simpson to the summit of Mauna Loa to visit the new facility. On November 27, 1951, the *Honolulu Star-Bulletin* reported that

> Mauna Loa began to tick Friday [November 23]—not volcanically but statistically—when the recently installed weather observatory started recording temperature, wind and sunshine data. The measuring equipment was installed by James Steiner, head of the local U.S. weather bureau, who hiked to the mountain top with two other weather bureau men, James Felgren, liaison inspector for the Pacific area from Honolulu, and Donald Yager, general maintenance technician.

The new observatory had almost as many instruments as the US Exploring Expedition used at the opposite side of the summit 110 years before. The modern instruments were automated and required attention only every two weeks. The *Honolulu Star-Bulletin* article reported that the instruments included a hygrothermograph for recording temperature and humidity, a wind gauge for recording wind direction and speed, a sunshine switch for recording the duration of sunshine, and a rain recorder. The observatory was located very near where Henry Guppy had camped on the summit for three weeks in 1897, which suggests a future comparison of the temperature measured by Guppy and by the automated observatory fifty-four years later.

After more than three years of planning and hard work, Simpson's dream was finally in place at the summit of Mauna Loa, and meteorological data was being recorded just as proposed by a resolution of the 1920 Pan-Pacific Scientific Conference. Simpson had learned from the road-building project that promotion was important to the success of the mission. He therefore organized a dedication ceremony for the summit observatory to be held at the site in December 1951. He invited various dignitaries, including Secretary of Interior Oscar Chapman and Governor Oren E. Long, and he chartered a US Navy plane for the trip from Honolulu for $270.

At 7:30 a.m. on December 12, 1951, some twenty-five people departed from Volcano House on the rim of the Kīlauea Crater for the arduous trip to the summit of Mauna Loa. Secretary Chapman turned down the trip to the dedication. Instead, he wanted to visit Volcanoes National Park, and Governor Long asked Tom Vance to escort Chapman on a tour of the park. Back on the mountain, the dedication group was driven in sedans through Kūlani Prison and then ushered into heavy-duty military transports near Kūlani Mauka for the rough ascent to the summit. Robert Simpson led the dedication group, which, he later recalled, included

> Washington representatives from the Departments of Commerce, Interior and Navy, as well as a number of scientists from universities and government laboratories. All were transported to the summit in four-wheel drive vehicles. The dedication by Governor [Long] took place during a light snowfall and was followed by an open-pit barbecue around a blazing bonfire. (Simpson 1989)

Howard Ellis (1988) reported in his monograph that

> Dr. Simpson started the logbook and described the day as: cloudy, temperature 32 [degrees] F, wind SW 15 mph with occasional snow flurries. . . . Mr. Smith, who provided most of the labor and worked out the logistics details, said: "Hell, this place is so high and cold that I can boil water for coffee all day and stick my hand inside it and not get burned."

In a chapter for the *20th Anniversary Report of the Mauna Loa Observatory*, Simpson closed his report on the dedication by noting that

FIGURE 4.3. The Mauna Loa Observatory was dedicated under an overcast sky in frigid conditions on December 12, 1951. MLO founder Robert Simpson *(center, with safari hat)* and Saul Price *(foreground, back to camera)* were among those who spoke. *(MLO archive.)*

> The excursion in four-wheel-drive vehicles to the summit, punctuated by frequent stops for rests, for accommodation to the thin air, and for a few deep breaths from the "walk around" oxygen bottles borrowed from military supplies, was completed without incident. (Miller 1978)

The dedication ceremony was more akin to a frigid adventure than a usual building dedication. Howard Pierce, a reporter for the *Hilo Tribune-Herald,* described the view of the snow-capped Mauna Kea and the patches of snow at his feet, which some of the participants made into snowballs. Photographs of the event (Fig. 4.3) show Robert Simpson, wearing a safari hat, amidst a group of obviously cold men huddled together against the front side of the tiny observatory building. Newspaper reporter Hugh Lytle (1951) described "the group of scientists and interested individuals as they stood before a bleak instrument shelter in a temperature slightly below freezing." Lytle's report includes a detailed account of Simpson's address to the group:

> When something happens here, it can have an effect on the other side of the world two weeks hence. . . . This point is the highest accessible by car, above sea level,

FIGURE 4.4. The Mauna Loa Observatory *(right)* at the summit of the mountain included an outdoor instrument shelter *(far left)*, wind instrument mast *(center)*, and precipitation collector *(center)*. The observatory was abandoned in June 1954. *(MLO archive.)*

that can be reached anywhere except in the continental regions. From sea level to 13,000 feet is reached in less than 30 air line miles.

This is the only place in the world, where such a site is accessible, where the atmosphere is of a marine nature and free of continental influences. The climate all the way up the mountain side is determined by a marine, tropical air mass. Because of this we can make measurements which will be representatives of the free atmosphere in the tropics, and to make continuous recordings of these. Previously the only way we could do it was to send up sounding balloons.

Simpson described the instruments at the site, which included an anemometer to measure the wind speed, a wind vane to indicate the wind's direction, a sunshine recorder, and a microbarograph to measure the pressure of the atmosphere (Fig. 4.4). He said, "You can almost weigh a cloud with a microbarograph." Simpson also anticipated greenhouse gas studies to come when he said that a spectrometer would be installed the following year to measure the total water vapor above the summit. This would permit a determination of the effect of water vapor on the temperature of the summit at night when energy absorbed from sunlight during the day is radiated outward to space.

A similar instrument would be installed at Hilo to compare the temperature under a moist sky with that under the very dry sky at the summit (Lytle 1951).

The dedication speech was given by Emil J. Sady, chief of the Pacific Branch, Division of Territories, US Department of Interior. In his speech, Sady recalled the resolution of the Pan-Pacific Scientific Conference of 1920 that called for the establishment of a weather station atop Mauna Loa:

> It is seldom that a group of men have gathered together in awesome surroundings such as on this barren summit of Mauna Loa for a more significant purpose. . . . Through all the harrowing 31 years that have transpired since that resolution of the Pan Pacific Congress, the idea of a weather observatory has lived. It caused men of good will . . . to seize the opportunity to join together diverse federal, territorial, and civic agencies, the University of Hawaii, and universities on the mainland in a common effort to carry it out. A new instrument for understanding the earth and its atmosphere has here been founded. A new basis for faith in the enduring value of international cooperation has been established. (Pierce 1951)

Sady acknowledged the vital roles played by the key players, including Robert Simpson, Thomas Vance, the Kūlani prisoners, and the Hilo Lions Club, which had worked tirelessly to sponsor the road construction project undertaken with the US Navy's giant earth movers and prison labor. Alfred Bell of the Lions Club thanked the group for the opportunity to participate in the project with its sponsorship of the road.

Loring G. Hudson of the Pacific Science Congress said that the "value of the observatory was worldwide, since weather knows no international boundaries" (Lytle 1951). Hudson also observed that the group was the largest to ascend the summit of Mauna Loa since the US Exploring Expedition of 1840–1841.

After the group returned from the mountain, Emil Sady described for Tom Vance the awesome nature of his experience on the mountain:

> When we were seated at the dinner table at night, I asked Mr. Sady for a report on his trip. He replied saying: "It was the most awesome experience of my life. I was a witness to the continuing process of creation. I became aware of the fact that I was traversing an area where my Creator had been only yesterday. I was troubled by the thought that he might suddenly return. If so, what should I do? Should I introduce myself and what should I say? Washington has protocol for almost every circumstance but not for such as this. Admiration for our Creator grew as we realized that he had created this beautiful universe out of desolation such as this. We did get some relief from this horrible picture of utter desolation when we entered the snow-capped summit area but when we looked into the huge summit crater we get a picture of the violence which had created this great mountain. Unfortunately steam cracks were melting the beautiful snow cover and exposing the frightening picture of desolation. As we came back into the beautiful forest surrounding the prison camp and saw smoke rising from the chimneys of the camp, I breathed a sigh of relief, 'Home at last.' I never realized that a prison camp could look so good. I'm glad that I made the trip but I am even happier that I am not one of those who will be making this trip daily in the interest of scientific research. This awesome experience will doubtless become an interesting and pleasant memory but it will take quite some time to recover from the initial shock." (Vance 1969)

Maintaining the Observatory

Additional men were added to the Weather Bureau office in Hilo to allow the staff to manage the mountain site. Simpson recalled later how an observer was scheduled to visit the summit each week to recover the chart recordings, replace paper and ink, and check the status of the instruments (Fig. 4.5). Besides the summit observatory, the staff also serviced instruments at three additional sites: Kūlani, at 5,180 feet (1,579 meters); Kūlani Mauka, at 8,300 feet (2,531 meters); and Mauna Loa Slope at 11,000 feet

FIGURE 4.5. This five-channel chart recorder was installed inside the Mauna Loa Observatory. The gear assembly at right appears to be a weight-driven drive mechanism. *(MLO archive.)*

(3,353 meters) (Ellis 1988). The latter two sites were to play a significant role in the future of the observatory.

A first-person account of the summit servicing procedure was provided by James W. Steiner of the Hilo Weather Bureau Office in the *20th Anniversary Report of the Mauna Loa Observatory*. Echoing others who drove the old Kūlani Prison route to the summit, Steiner wrote about the "Strainback" Highway, the name first used by the Hilo Women's Club after their tour below Kūlani Mauka a few years earlier:

> How well I recall the weekly torture of driving to the summit site in the early fifties. As I recall, the distance to the summit from Hilo was about 50 miles. It took 4 hours for the drive: 2 hours to the 11,000-ft level, and then 2 hours for the last 10 miles to the summit. First it was in a reclaimed WWII ambulance-type vehicle, then in a new Dodge power wagon. (Steiner 1978)

Steiner also related an interesting anecdote about how the Dodge was acquired, and Ellis (1988) and the *Honolulu Advertiser* (1952) provided additional details. Steiner recalled, "The impetus for the purchase of the new wagon was undoubtedly the fact that the old vehicle gave out near the summit when I was there with [Peyton Harrison] the director of what was then the Hawaii Aeronautics Commission." The two men had driven to within a mile (1.6 km) of the observatory on April 15, 1952, when they were blocked by snowdrifts. They hiked through the snow to the observatory and returned to the truck at 12:30 p.m. to begin the descent. Unfortunately, the truck broke down at 12,000 feet (3,658 meters) about 1.3 miles (2.1 km) north of where the slope unit of the Mauna Loa Observatory would be constructed in 1956.

The two men walked some 10 miles (16 km) down to 8,300 feet (2,530 meters) at Kūlani Mauka, where a government aviation transmitter was located. Finding the building locked, they continued walking down the road. Steiner wrote that, in total, "We walked for some seven hours, until about eight in the evening, before we were 'rescued' by prisoners from Kulani prison at about the 6,000-ft level of the road." The prisoners had been sent to look for the men when they didn't return through the prison as expected. According to the *Honolulu Advertiser* (1952), Harrison, who returned to Honolulu tired from the experience, explained that while there was a relationship between the new observatory and the aeronautics commission, "I won't be going back up to check it." Ellis (1988) records that the incident led to concerns about safety and liability:

> Mr. Steiner made an attempt to stem the flood tide of suggestions for emergency procedures—radios, safety equipment, liability insurance coverage, workers' compensation, and on and on. . . . By May, emergency gear had been obtained, physical examinations at six month intervals for all employees making the trips were provided for, and a form was drawn up and needed to be signed by all non-employee passengers, freeing the WB [Weather Bureau] from liability.

The instrumentation at the summit was subject to none of these contingency plans, and the instruments did not always work properly. Only a month after the dedication ceremony, the wind mast blew down. Because the weekly trip (Fig. 4.6) was so time consuming and tiring, the Hilo office tried to modify the instrumentation to increase the time between summit trips. As Steiner (1978) recalled,

> Because of the arduous trip all the way to the summit, attempts were made to modify standard instruments for 30-day operation. Several Rube Goldberg contraptions were designed by me and my staff in Hilo, but none was effective. We returned to standard operation and made weekly trips. Even then there were problems. The diurnal lifting of the inversion layer and the influx of marine air caused the ink to blot, caused the paper to stretch, and in general, was a thorn in the side.

His work in Hawai'i completed, Robert Simpson departed on May 18, 1952, exactly four years after he had left Washington for Honolulu. Simpson's personal involvement and his creative ability to leapfrog bureaucratic hurdles, promote his dream through the media, and partner with the US Navy, Tom Vance, a prison, and a Lions Club had made possible the establishment of the first permanent observatory on the summit of Mauna Loa. Or, at least, permanence was Simpson's intention. As so often happens when a leader leaves the scene, enthusiasm for the project began to crumble along with the poorly maintained road to the summit. As Simpson (1978) recalls in the *20th Anniversary Report*,

> Unfortunately, after the road-building equipment was returned to the Navy, the road began to deteriorate rapidly, and getting to the summit site to service equipment and collect records became progressively more arduous. Failure to obtain funds to improve the road so that a standard passenger car could reach the summit prevented Tom Vance from establishing a ski lodge. Finally, in [1954] it was necessary to discontinue trips to the summit observatory, and observations started in [1951] were discontinued.

In the MLO archives are two photographs from November 16, 1952, that illustrate the poor condition of the road, one of which is Figure 4.7. Captions suggest that former Governor Stainback made a trip to the summit of Mauna Loa along the road that bore his name.

During the dedication of the little observatory on that cold December day in 1951, Simpson had voiced his prophetic dream for the facility:

> The weather bureau's newest, and perhaps most unique observatory, outwardly appears to be an unimposing installation. Indeed it is an autographic weather station of humble proportions—but nevertheless of signal importance. This is a humble beginning of an operation which, because of its strategic location, may ultimately provide data and serve as the advance base for some of the more important geophysical research of the century. (Pierce 1951)

Because of the continuing problems with the recording instruments and the difficulties of servicing them and retrieving the data charts, few records are available after the summit observatory opened for business in November 1951. The water vapor measurements proposed by Simpson were apparently never implemented. In June 1954, the little observatory on the summit of Mauna Loa was abandoned. Robert Simpson's dream was placed on hold.

A Virtual Visit to the Mauna Loa Summit Observatory Site

After the Mauna Loa Summit Observatory was abandoned, the Park Service removed the building. All that remains today is the concrete slab on which the building stood (Fig. 4.8) and the crude four-wheel drive access road. Thanks to aerial imagery and Google Earth™, anyone can visit the site from the comfort of a keyboard. The journey is begun by flying Google Earth (http://earth.google.com) to the Big Island of Hawai'i. Zooming in on the center of the island will take the visitor to Mauna Loa. Its summit is just below the center of the island, and dark streaks of lava radiate outward from the summit crater. Continued zooming and steering will fill the computer screen with the giant crater.

FIGURE 4.6. This four-wheel drive Dodge Power Wagon was used for weekly trips to the summit. The team shown here is checking a weather station near where the present slope unit of the Mauna Loa Observatory would be built in 1956. *(MLO archive.)*

FIGURE 4.7. This trip to the summit on November 16, 1952, illustrates the poor condition of the road. A caption suggests that former Gov. Ingram Stainback was on this trip. *(MLO archive.)*

FIGURE 4.8. This concrete slab, shown here in 1978, is all that remains of the Mauna Loa Summit Observatory. Mauna Kea is in the background. (John Miller. Used with permission.)

The concrete foundation of the abandoned Mauna Loa Summit Observatory is near the northwest side of the crater. The fastest way to find its approximate location is to steer toward it while referring to the geographic coordinates of the pointer given at the lower left corner of the Google Earth screen. The mouse or pointing pad should be tweaked until the pointer is at 19 degrees 29 minutes north latitude and 155 degrees and 36 minutes west longitude. Zooming in on this spot will reveal a white square on the lava at the end of a rough road from the northeast. This is the concrete foundation of the abandoned Mauna Loa Summit Observatory.

The date of the Google Earth imagery of Mauna Loa may not be specified, and the images may be several years old. A more immediate view of the summit and its giant crater is available from the Hawaiian Volcano Observatory of the US Geological Survey, which has installed an automated camera at the western side of the crater that rotates in five steps to provide a panoramic view of the entire crater. The image is updated at ten-minute intervals (http://hvo.wr.usgs.gov/cam2/). The view is sometimes blocked by ice, fog, and clouds, but generally a fine view of the crater can be had. The USGS camera is located about half a mile (805 meters) southwest of the concrete pad of the Mauna Loa Observatory site and is very close to the site of Henry Guppy's 1897 camp.

Salvaging the Dream

After the summit observatory was abandoned, the Weather Bureau did what it could to continue monitoring temperature and rainfall at three elevations along the accessible segment of the observatory road between the Kūlani Prison and about 11,200 feet (3,414 meters). Meteorologist Howard Tatum had first been assigned to the Weather Bureau station in Hilo in September 1951. When he returned in January 1955, the Weather Bureau was making weekly trips up the mountain to service the weather instruments along the road. Tatum (personal communication with author, July 3, 2007) describes some of his adventures while taking part in the difficult drive along the road:

> We had rain gauges (dip stick), weighing rain gauge, temp and humidity gauges and a barograph at three different sites: 1—Kulani Prison, 2—8,300 ft. level [Kūlani Mauka], 3—very near the site of [the present] MLO.
>
> The only road in those days took us through the Kulani Prison and was only a very rough trail on up the mountain. We drove an old Dodge Power Wagon that didn't have seat belts. [We] bumped our heads many a time on the roof. We also had no means of communication. So, as we went through the prison gate we told them we were on our way up the mountain. If we were delayed for any reason our wives would call the prison to find out if we had come back. If the answer was no, they would send up a rescue party.
>
> As we would change all the [recording] charts on our way up, we would check them on the way down. After having stopped at the 8,300 ft. site I drove on down and noticed that the Power Wagon WAS ON FIRE! I stopped and jumped clear. After waiting about 15 minutes I noticed no more smoke so decided to take a chance and drive on down. Got in the vehicle, took off the emergency brake and proceeded on down. No more smoke. You idiot, you were driving with emergency brake on since you stopped at 8,300, thus, the smoke.
>
> The boss and a new arrival had gone up on a very windy day. On their return, when they got a couple of miles from the prison there were several trees laying across the road. They walked down to the prison and got help.

The meteorological measurements made along the Mauna Loa road were important, but they did not begin to meet the promise of the little observatory on the top of the mountain. With the summit station closed, the tenacity of the Hilo Weather Bureau station staff and the Kūlani Prison would continue to play a role in salvaging what was left of Robert Simpson's dream for Mauna Loa.

References

Alexander, W. D. 1892. The ascent of Mauna Kea, Hawaii: Report of W. D. Alexander on the Mauna Kea trip of 1892. *Pacific Commercial Advertiser*, September 14. From Kepa Maly, MAUNA KEA—KA PIKO KAULANA O KA 'AINA, Kumu Pono Associates LLC. (Accessed at http://www.ifa.hawaii.edu/users/steiger/mk_survey.htm.)

Anon. 1897. "Scientific Notes and News." *Science* 6 (142): 443.

———. 1920. General resolutions of the Pan-Pacific Scientific Conference. *Science* 52 (1343): 286–287.

Apple, Russell A. 2000. Thomas A. Jaggar, Jr., and the Hawaiian Volcano Observatory. (Accessed at http://hvo.wr.usgs.gov/observatory/hvo_history.html.)

Associated Press. 1947. Plans scenic highway to Volcano. The *New York Times*, January 17: 25.

Bird, Isabella. 1875. The Hawaiian Archipelago: Six months among the palm groves, coral reefs, and volcanoes of the Sandwich Islands. (Accessed at http://www.gutenberg.org/dirs/etext04/hwap10h.htm.)

Dana, Dwight. 1852. Note on the eruption of Manna Loa. *American Journal of Science and Arts* 14 (14): 79–80.

Department of the Environment and Water Resources. 2005. History of Macquarie Island. Australian Government. (Accessed at http://www.environment.gov.au/coasts/mpa/macquarie/history.html.)

Duncan, George. 1961. Account of an automobile trip in 1909 around the Island of Hawaii. In R. A. Apple, 1965, *Trails: From steppingstones to kerbstones*. Hawaiian Archeology. B. P. Bishop Museum Special Publication 53. Honolulu: Bishop Museum Press.

Ellis, Howard. 1988. The observatory: On their own. (Accessed at http://www.mlo.noaa.gov/pdf/memorabilia/Ellis%20booklet%20edited.pdf.)

Friedman, S. Morgan. 2011. The inflation calculator. (Accessed at http://www.westegg.com/inflation/.)

Guppy, H. B. 1906. A naturalist in the Pacific between 1896 and 1899. Vol. 2, Appendix, Note 61, Meteorological observations on the summit of Mauna Loa, 586–587. London: Macmillan and Co. (Accessed at http://books.google.com.)

Honolulu Advertiser. 1952. Mauna Loa Summit Rd. cleared of snow. April 18: 1–2.

Honolulu Star-Bulletin. 1951. New apparatus atop Mauna Loa now in operation. November 27.

Lytle, Hugo. 1951. Unique isle weather post on Mauna Loa. *Honolulu Advertiser,* December 13: 1 and 10.

Miller, John, ed. 1978. *Mauna Loa Observatory: A 20th Anniversary Report.* Washington, DC: National Oceanic and Atmospheric Administration, US Department of Commerce.

Pierce, Howard. 1951. Unique weather observatory on Mauna Loa dedicated. *Hilo Tribune-Herald,* December 13: 1–2.

Shiramizu, Harry. 1951. New story of Kulani is in "the making." *Hawaii Times,* January 22.

Simpson, Robert H. 1978. Early days of the Mauna Loa Observatory. In John Miller, ed., *Mauna Loa Observatory: A 20th Anniversary Report.* Washington, DC: National Oceanic and Atmospheric Administration, US Department of Commerce.

———. 1989. Interview by Edward Zipser, September 6 and 9, 1989, for the American Meteorological Society, University Corporation for Atmospheric Research. (Accessed at http://www.ucar.edu/archives/publications/simpson-robert%20interview.pdf.)

———. 2009. *Lifetime Acquaintances, Adventures & Exploration That Made a Difference: A Memoir, Sketches from the Life of Robert H. Simpson.* July: 74–76. Unpublished.

Steiner, James W. 1978. Introductory letter in John Miller, ed., *Mauna Loa Observatory: A 20th Anniversary Report.* Washington, DC: National Oceanic and Atmospheric Administration, US Department of Commerce.

Stone, Peter, ed. 2006. Macquarie Island shipwrecks. In *Encyclopedia of Australian Shipwrecks.* Yarram, Victoria: Oceans Enterprises. (Accessed at http://oceans1.customer.netspace.net.au/macquarie-main.html.)

USGS. 2007. Wilkes Expedition revisits Kilauea. (Accessed at http://hvo.wr.usgs.gov/volcanowatch/2007/07_03_15.html.)

Van Rooy, M. P., ed. 1957. *Meteorology of the Antarctic.* Pretoria: Weather Bureau.

Vance, Thomas B. 1969. Unpublished memoir sent to Judith B. Pereira of MLO on December 9.

CHAPTER FIVE

Founding of the Slope Unit of the Mauna Loa Observatory (1955–1956)

Years of planning and hard work could not meet the icy challenges thrown at the tiny Mauna Loa Observatory atop Earth's biggest mountain. Repairing and servicing the weather instruments was made increasingly difficult by the crude and rapidly deteriorating access road. Like the explorers Archibald Menzies, Charles Wilkes, and Henry Guppy who preceded him, Robert H. Simpson was determined to persevere over the challenges posed by Mauna Loa. But his departure from Hawai'i in 1952 ended the leadership that had launched the fledgling observatory. Simpson's plans for the Mauna Loa Summit Observatory came to an end when the station was closed and abandoned in June 1954. Simpson later recalled that

> it got progressively tougher to maintain the facility at the summit site, and finally it was decided, much to my distress, to close it altogether. To me, we had established a foothold at a site with enormous potential for scientific observations of many kinds, and I felt we had been less than perceptive in tossing in the towel so soon. (Simpson 1989)

Among the observations that Simpson had hoped to pursue on the mountain was the measurement of the ozone layer over Hawai'i. Because 90 percent of the atmosphere's ozone lies in the stratosphere, Simpson believed that the ozone over Mauna Loa could serve as a tracer to study the "potential significance of the cold lows that continually paraded westward from the California area across Hawaii, many of them eventually reaching the western Pacific where they seemed to be associated in some way with the formation of typhoons" (Simpson 1989).

Ozone efficiently absorbs the ultraviolet wavelengths emitted by the sun. Simpson's interest in monitoring the ozone layer and another scientist's interest in measuring the sun's ultraviolet were soon to experience a serendipitous merger atop a mountain in southern New Mexico. Only a year after its closure, new hope for the Mauna Loa Observatory was on the horizon.

A Long-Deserved Vacation

After his four-year assignment in Hawai'i, Robert Simpson resumed his busy Weather Bureau career in Washington. One of his first assignments was to study the deadly Waco, Texas, tornado of 1952 and attend a contentious hearing called by the state's governor about why the Weather Bureau had not predicted the disaster. This was not to be

Simpson's last adventure with high-level officials and politicians, and this one led to his proposal to establish tornado warning radars in Texas and Oklahoma that were to eventually evolve into today's National Weather Service radar network.

Simpson spent his summers in Bermuda flying with US Air Force hurricane reconnaissance teams, became involved in serious hurricane research, and eventually became the first director of the National Hurricane Research Project (NHRP), a fitting position for a man whose family's Texas home was flooded by a major hurricane when he was a child.

Prior to Hawai'i, Simpson's budding meteorology career had become a personal whirlwind of moving from place to place that jeopardized his marriage. His family was so tired of moving that in 1948 they begged him not to accept the Hawai'i assignment. As he later recalled,

> my first marriage came to an end in 1948 principally because of my self-centered career interests when I finally accepted the assignment to Hawai'i. (Simpson 1989)

When the spring of 1955 arrived, it was finally time for Simpson to take a break from his arduous schedule, especially with a major new assignment on the way. During a telephone interview, he recalled that

> I had been transferred from pillar to post with the Weather Bureau, a lot of different jobs up to that time. There hadn't been a single year that I had taken any annual leave longer than about a week, because I just had too many things that I was interested in doing and was commanded to do with my various assignments. But this time, with the National Hurricane Research Project coming up, I could see it would be years before there would be another opportunity to have a real vacation, so I took off nearly a month of annual leave which I had accumulated in the years before and took my two daughters, Peggy and Lynn, who are in Washington with us now, and took them out to the canyon country out west. (Simpson, telephone interview with author, March 17, 2007)

Simpson picked up his daughters in San Antonio, Texas, in early June of 1955, and the three headed west in his 1950 two-door, forest green Buick sedan, "one of the early models with an automatic transmission," he recalled. It was to be a vacation none of them would forget.

Most of the first day of the trip was spent driving across West Texas. They spent the first night in a "tourist cabin" shortly after crossing into New Mexico. The next morning they headed across the desert for the village of Cloudcroft in the Sacramento Mountains of southern New Mexico. Harry Wexler, director of research for the Weather Bureau, had an interest in the sun's impact on the atmosphere, and he asked Simpson to visit the Sacramento Peak Observatory at Sunspot near Cloudcroft.

> I had been told in Washington to stop off at the observatory up there and see what's going on. "You might find something interesting, and we ought to at least know about it ourselves in Washington." So I did. Of course I had the girls, and I didn't want to leave them with nothing to do but suck on a lollypop. I put them in a dude ranch there for one overnight. They went horseback riding and had a lot of fun. (Simpson, telephone interview with author, March 17, 2007)

The Silver Lining in a Dusty Cloud

After taking care of his daughters, Simpson headed for Sunspot, home of the Upper Air Research Observatory and its solar telescopes. At 9,200 feet (2,804 meters) above sea level, Sunspot is high above the adjacent Tularosa Basin and its famous deposits of powdered gypsum known as White Sands, which can be seen from an overlook at Sunspot. Sunspot is much cooler than the adjacent desert, and Simpson arrived during a cooler than usual June. The last freeze had occurred on May 26, and the average high temperature that June was only 71 degrees Fahrenheit (22 degrees Celsius) (DeMastus 1976).

Unknown to Simpson, two scientists from the National Bureau of Standards were also visiting Sunspot that June. They were Ralph Stair and Russell G. Johnston, who had previously measured sunlight at Sunspot in July 1953 (Stair et al. 1954). A dry, dust-free atmosphere is essential for high-quality measurements of sunlight, and Stair had selected Sunspot because of its clean sky. Indeed, in November 2002 sun photometer calibrations made by the author at nearby Apache Point Observatory provided results nearly as good as those performed at the Mauna Loa Observatory.

The purpose of the repeat visit by Stair and Johnston was to refine their precise measurements of sunlight as part of a project to measure the solar constant, the sun's irradiance at the top of the atmosphere (Stair and Johnston 1956). Stair was not new to this highly specialized field of science, for he had previously measured sunlight from high elevations in Colorado and New Mexico. Ralph Stair was well on his way to becoming one of the world's leading experts on measuring the solar constant and the ozone layer during the pre-satellite era.

Stair had to know the amount of ozone overhead in order to measure the solar constant. In the 1920s, the British scientist Gordon Dobson had developed a method to measure the Earth's ozone layer by means of a double-prism spectrometer that compared the intensity of sunlight at two or more pairs of ultraviolet (Dobson 1930). In 1937 Stair had proposed a simplified method that requires only a UV detector and filter. He had tested the method during solar observations atop Mount Evans, Colorado, in 1936 and 1938 (Stair and Hand 1939). In June and July of 1947, Stair visited New Mexico to use his simple ozone photometer to measure the ozone layer over the Organ Mountains northeast of Las Cruces across the Tularosa Basin from Sunspot. The occasion was the launch of a V-2 rocket from the White Sands Proving Ground, later known as the White Sands Missile Range. Stair set up a table for his instrument at a parking area along the south side of Highway 70 at the crest of the San Augustine Pass. The site is still there, and it offers a spectacular prospect of the barren Tularosa Basin bordered by the Sacramento Mountains along the eastern horizon. Below is the White Sands Missile Range headquarters.

The V-2 launch was scheduled for July 3, and Stair began his ozone measurements on June 29. Summer is monsoon season in the Southwest, and while Stair was beginning his observations a cold front entered the area and slid under the tropical maritime air that had blanketed the region. Clouds and haze interfered with the measurements on the first few days, but the sky cleared noticeably on July 2. Tropical air returned on July 3 and clouds returned to the Organ Mountains. The US Army had successfully launched eight V-2s so far that year (V-2 Firing Record Summary 1947), but the V-2 launch scheduled for July 3 misfired. Stair continued with his measurements and stayed on until noon the next day, when clouds ended the project. While the V-2 misfire was a disappointment, Stair exhibited his expertise in measuring the ozone layer in a paper he wrote about the experience for the *Journal of the National Bureau of Standards* (Stair 1948). He also learned that clouds could ruin the most carefully prepared experiment.

On June 3, 1955, a week prior to Robert Simpson's 1955 visit, Stair and Johnston began using Stair's spectrometer to measure sunlight piercing through the crystal clear sky over Sunspot. The clear sky conditions held for the next several days, but then unexpected trouble blew in from the west and put a hold on sunlight measurements. Ordinarily the sky over Sunspot is bone dry and free of significant dust, but sometimes, especially during spring, the sky overhead is filled with dust originating from wind storms in the adjacent Tularosa Basin and Mexico. As shown in Figure 5.1(a), on dusty days at Sunspot the sun is surrounded by a huge aureole caused by the scattering of sunlight from particles of dust. This contrasts with Fig. 5.1(b), which shows the virtual absence of a solar aureole on a clear day at the Mauna Loa Observatory. Dry years have the potential of being followed by especially dusty springs, and a tree ring study for the lower Rio Grande Valley west of Sunspot showed that 1954 was the driest year since 1733 and the third-driest year from AD 622 to 1994 (Grissino-Mayer et al., 1997). Thus, the Southwest cycle of drought had predetermined that the spring of 1955 was likely to be much dustier than usual.

A dusty sky obscures direct sunlight and completely ruins delicate measurements designed to establish the solar constant. The paper Stair wrote about the results of the

FIGURE 5.1. The inset shows *(a)* a prominent solar aureole caused by dust over Sunspot, New Mexico, on March 27, 2004 and *(b)* the absence of a solar aureole in a very clean sky at the Mauna Loa Observatory on May 25, 2008. On a dusty day in June 1955, Robert Simpson met Ralph Stair at the Big Dome of what is now the National Solar Observatory/Sacramento Peak at Sunspot, shown here *(c)* during a dust event on March 14, 2008.

Sunspot research reveals that dusty conditions prevailed over Sunspot from June 8 to 15, 1955 (Stair and Johnston 1956). He had selected the spring following the driest year in more than two centuries to make the most precise sunlight measurements ever made. Ralph Stair was not a happy man.

A Chance Encounter

An unexpected dusty sky, a rare vacation, and an even rarer scientific experiment merged together amidst the ponderosa pines, Douglas firs, aspens, and telescopes at a remote New Mexico mountaintop in June of 1955. The stage was set for the reestablishment of Hawai'i's Mauna Loa Observatory when Bob Simpson arrived at Sunspot's new Big Dome Observatory, shown on a dusty day in Fig. 5.1(c), and almost immediately met Ralph Stair. When asked how he found Stair, Simpson replied,

> He found me. I walked in, and he was there working on something, and somebody said somebody from the Weather Bureau was here, and he came out, and immediately we became engaged in an interesting conversation. First time I had ever met him. I didn't know anything about him or about his work, either. (Simpson, telephone interview with author, March 17, 2007)

Simpson, in the same interview, chuckled as he recalled that Stair

> was in a bad mood. . . . He had been out there quite a while and had only gotten two or three days of solar . . . measurements.

Stair explained why he was unhappy in a memoir published in the *Mauna Loa Observatory 20th Anniversary Report:*

> It was at Sunspot, in June 1955, that we had the good fortune to receive a prominent visitor from the Florida Hurricane Center, none other than Dr. Robert H. Simpson himself. As there had been a dust storm of considerable magnitude over the adjacent New Mexico desert, the sky was very bright with the sun coming through weakly.

> Dr. Simpson was very sympathetic to our need for a clear sky with direct sunlight coming through at high intensity. . . . We wanted a better location but knew not where to turn, since we had considered all possible locations within the continental United States and found no place superior to Sunspot. (Stair 1978)

There would have been no reason for Stair to complain had the dust storm not occurred or if Simpson had arrived a few days earlier or later when pristine sky conditions prevailed. Indeed, if the sky had been clear, Stair would have been so preoccupied with his spectrometer that there might have been very little time to converse with a visitor, even a distinguished one. But the dust did its work in forcing Stair out of work, and Simpson arrived with the perfect solution. Fifty-two years later, Simpson recalled, as if it were only yesterday, that he asked Stair,

> Why don't you go some place where you can expect to have 340 or 50 days a year of observations without any interference from clouds? (Simpson, telephone interview with author, March 17, 2007)

Simpson described Stair's response and the discussion that ensued in his NCAR interview (Simpson 1989):

> "Is there such a place?" he asked.
>
> "Oh, yes," I replied. "At Mauna Loa summit. We had an observatory there for a while which could be reopened with a little help."
>
> "Tell me more," he said.

Simpson then told Stair about the ultraclear sky over the abandoned observatory at the end of the long lava road leading to the edge of the gigantic Moku'āweoweo caldera at the summit of Mauna Loa.

The Ozone Connection

Stair must have been especially intrigued by Simpson's interest in measuring total path ozone from atop Mauna Loa, for, as noted earlier, in June 1947 he had measured the ozone layer over the Organ Mountains just across the Tularosa Basin from a site within binocular range of Sunspot. Measuring the ozone layer was also a crucial part of Stair's measurements of the solar constant.

Ozone absorbs the invisible UV wavelengths of sunlight, and the ozone layer is best measured from the surface by comparing the magnitude of two closely spaced wavelengths of UV between about 300 and 320 nanometers. Because the shorter wavelengths are absorbed more than the longer ones, the ratio of the intensity of sunlight at the two wavelengths is directly related to the amount of ozone between the measuring instrument and the sun. This provided a vital missing ingredient in Stair's research, for measuring the solar constant from the surface of Earth requires that the data be corrected by adding the ultraviolet sunlight blocked by the ozone layer.

In June 1955, the ozone layer was a nuisance that interfered with Stair's delicate sunlight measurements. But for Simpson, the ozone layer provided a possible means for tracking a missing link in the origin of Pacific typhoons. The ozone layer had become the key to reestablishing the Mauna Loa Observatory.

Achieving a difficult scientific goal usually requires a good deal of planning. Implementing the goals Simpson and Stair eagerly discussed would be no different, and the conversation between Simpson and Stair turned to practical considerations. Simpson explained that "The biggest problem is building a permanent access road and a building suitable for continuous occupancy at the site" (Simpson 1989). As he recalled later,

> I told him it would take a little doing to restore [the observatory] and get it in operation, but it shouldn't be a great big deal provided we could find a little bit of money. (Simpson, telephone interview with author, March 17, 2007)

Simpson knew that the National Bureau of Standards—which, like the Weather Bureau, was under the Department of Commerce—was unlikely to provide the money required to achieve what he was proposing. But in the same interview, he indicated that he knew where the money could be found:

> I had an idea where we might find it, because we had been given more money than we could spend the first year on hurricane research. I also knew that the way the government works that if you have a big appropriation and you don't spend what you have appropriated the first year, it disappears, not only for the next year but succeeding years.

Years of Weather Bureau assignments had taught Simpson how to bypass bureaucratic obstacles to get work done economically and efficiently. He presented to Stair a creative way to avoid the red tape that would strangle a more conventional approach to rebuilding the observatory on Mauna Loa:

> The Weather Bureau is hamstrung in contracting for such tasks, which must be done through the GSA [General Services Administration] and usually turn out to be prohibitively expensive. However, I happen to know that the Bureau of Standards has enabling legislation which allows it more freedom to do such things. If we can work together, we may both be able to accomplish our objectives more easily than we could alone. If you, on behalf of the [Bureau of] Standards, would become prime contractor for the project, build the road and observatory building suitable for continuous occupancy, then purchase and install a Dobson ozone instrument, and measure total path ozone at the site for a minimum of two years, I think I can justify diverting funds from the NHRP [National Hurricane Research Project] to cover costs of the road and the building. (Simpson 1989)

Stair, who cared more for his solar constant measurements than government bureaucracy, liked Simpson's creative approach to achieving their goal, as Simpson noted in the same interview:

> We both agreed that at first blush it looked feasible and that we would pursue it with our respective bosses. (Simpson 1989)

Simpson (telephone interview with author, March 17, 2007) recalls more details about his historic meeting with Stair:

> We had a great chat and had a little lunch together, and then we went on from there. And I became very fascinated in his work, and he became very fascinated in the prospect of going someplace where he could work a large number of days. It went on from there.

The next day, Simpson retrieved his daughters from the dude ranch, and the trio headed for canyon country. They soon experienced an adventure nearly as memorable as the prospect of reestablishing the Mauna Loa Observatory.

> When we resumed our journey westward the next day . . . we found ourselves descending a rough gravel road (several miles of it) when an enormous beast resembling a mountain lion leaped onto the road in front of the car, stopped and snarled at us, then disappeared into a thicket on the other side of the road—a thrilling, memorable incident for the girls. . . . We went on down to Arizona, went to the Grand Canyon. . . . I think we spent about five nights in the immediate canyon area there in Arizona and southern Utah, and we went up to Salt Lake City and then came back down through Boulder and over the Rockies . . . Estes Park and so forth, and then came directly on back into Texas. (Simpson, personal communication with author, March 25, 2007)

When asked if his daughters remembered their Canyonlands vacation in June of 1955, Simpson enthusiastically replied, "Oh, yes, one of their brightest memories." The same response could be given by the entire atmospheric sciences and meteorological communities, for the chance meeting of Robert H. Simpson and Ralph Stair on a day made dull by a thick layer of dust had sparked what was to become an even brighter memory. For in only a few hours, a serendipitous meeting of two determined, kindred spirits had paved the way for the renewal of Robert Simpson's dream: a usable road and a permanent observatory on Hawai'i's Mauna Loa.

In my communication with Dr. Simpson in 2007, he reflected back on his historic encounter with Ralph Stair, which occurred almost exactly a year after the Mauna Loa Summit Observatory had been closed:

> I don't believe I ever had the opportunity of sitting across the table again with Ralph Stair after our meeting at Sunspot, although I did talk to him by telephone a number of times to check on the progress made or roadblocks encountered. . . . I don't believe he had returned to DC by that time, and I was soon "off and running," fully absorbed in completing arrangements for the inaugural year of National Hurricane Research operations from West Palm Beach, Florida. I didn't even manage to return to DC for nearly a year.

Reichelderfer's Conundrum

Simpson had a good relationship with his boss, Francis Reichelderfer, director of the Weather Bureau, and was optimistic that he would support transferring funds to the National Bureau of Standards (NBS) for reestablishing the Mauna Loa Observatory. But Reichelderfer's negative reaction surprised Simpson.

> When I returned to Washington, I had a rougher time with Reichelderfer on this matter than I expected. He was concerned about "congressional intent" for use of NHRP funds, and doubted the connection between ozone measurements and cold lows could be credibly related to tropical cyclone research. (Simpson 1989)

Before Simpson's dream could move forward, Reichelderfer had to be satisfied that the diversion of Weather Bureau funds to the National Bureau of Standards could be fully justified and above reproach. Would the diversion violate the intent of the Congress? Was the ozone study a legitimate scientific rationale for reestablishing the abandoned observatory? Simpson (telephone interview with author, March 17, 2007) explains the political realities behind Reichelderfer's concern:

> We had the three hurricanes, Carol, Edna and Hazel in 1954, which damaged so much property in the New England area and the upper Atlantic area that many senators were affected. In fact, more senators were affected, and their constituencies were affected, during that year than has ever happened in any hurricane before or since, though the hurricanes were not as strong as some have been. But it impressed the senators, and they immediately demanded that we submit a budget that would accommodate a continuing, concerted research program on hurricanes [the National Hurricane Research Project]. I was made the first director.

Simpson stood by his ozone point, but the vote for the transfer of funds remained tied at one to one. Interviewed in 2007, he remembered what happened next: "That was a hard sell to my boss, but it was an easy sell to the head of research, Dr. Wexler."

Wexler's Endorsement

"At this juncture," Simpson recalled during his NCAR interview, "unexpected support came from Harry Wexler, Director of Research." Wexler was highly respected for his scientific expertise, which extended well beyond meteorology, and Reichelderfer would seriously consider Wexler's recommendation. According to Simpson, Wexler said,

> This sounds great. And we could measure carbon dioxide there, too. It is a site completely free of upstream pollution and at elevation, which frees it from local pollutants. This could become truly a baseline station for observing and measuring increases of CO_2 in the atmosphere. (Simpson 1989)

Wexler was very aware of studies about the increase of carbon dioxide in the atmosphere and its potential impact on climate. His support sold Reichelderfer on the potentially risky plan, and Simpson was elated, as indicated during his NCAR interview:

> And with that, the Weather Bureau Central Office was off and running, the money was drawn from NHRP funds, the Bureau of Standards did its thing, and the present observatory and its important record of CO_2 measurements became a reality.

Wexler was so intrigued by the prospect of a revived observatory on Mauna Loa that he didn't bother to ask Simpson for the report he had earlier requested about the science being conducted at Sunspot. The report wasn't on Simpson's mind, either.

> Looking back on it, I was probably too excited at the prospects of reestablishing the Mauna Loa Summit Observatory to carry out the modest mission I had in visiting Sunspot—I don't recall ever writing up the report [about what] I was sent there to investigate! (Simpson, personal communication with author, March 25, 2007)

While Simpson was primarily interested in measuring the ozone layer from the revived observatory, Wexler's endorsement added carbon dioxide to the list while also proposing that Mauna Loa could become a baseline observatory in pristine alpine air largely unaffected by pollution. Harry Wexler had proposed an expanded version of Simpson's dream, and in our telephone interview of March 17, 2007, Simpson reflected on Wexler's role in persuading Francis Reichelderfer to approve the transfer of Weather Bureau funds to the National Bureau of Standards:

> [Wexler] was the one who was really thinking far ahead of the potentialities of variation in carbon dioxide up there, which no meteorologist really paid much attention to. . . . And he was a man of great vision, a man who didn't earn nearly the plaudits that he should have won for the scientific changes and the things he made different in science far beyond meteorology that he was involved in. As his deputy for a number of years, I've really come to have an enormous respect for the man, his ingenuity, his creativity, his untiring efforts to get things done. He never knew of [the phrase] "It can't be done." That wasn't in his vocabulary.

Simpson, ever the meteorologist, remained more interested in using the ozone layer to study the origin of Pacific hurricanes called typhoons. In the same interview, he recalled that

> What I wanted was total path ozone, . . . which we could use as a tracer in studying the cold lows that moved southwest from California over Hawaii, causing great floods there at times, or heavier rains, anyway, and passed on into the Western Pacific and associated themselves with typhoons there. That was the main [reason] that I had in mind using the observatory for, and the understanding I had with Ralph Stair if he could convince his people to use their authority to build the road and to build the observatory there without having to go through the GSA, or other government regulations, to get a low bid from somebody but go ahead and do it with their own design and with their own funds, wherever they got the funds. They could do it. The Weather Bureau could not. That's how Ralph Stair and I managed to get together to both of our advantages.

After Reichelderfer approved the diversion of funds, Simpson in this interview recalled that

> I got reports from time to time, very gratified by what was happening. Wexler turned the thing over to his research staff at the Weather Bureau, and Stair carried the ball for the Bureau of Standards, the details of which I don't have except for fragments.

From this point on, the implementation of Robert Simpson's dream was left to the National Bureau of Standards physicist he met on a dusty day at Sacramento Peak. Simpson himself was soon working full time establishing and directing the National Hurricane Research Project in West Palm Beach, Florida. Hawai'i's Mauna Loa was about to host a new station for the purpose of measuring Stair's solar constant, Simpson's ozone layer, and Wexler's carbon dioxide, none of which had been measured from the abandoned summit observatory.

Ralph Stair Takes Over

After Robert Simpson and his daughters continued on their once in a lifetime vacation in canyon country, Ralph Stair and Russell Johnston waited out the dusty skies at Sunspot. They had managed to collect good ultraviolet and visible sunlight data on June 3, 4, 6, and 7 before the dust storm and Simpson arrived. They had to wait until June 16 for the sky to clear. They then measured the infrared portions of the solar spectrum from June 16 to 19, packed their sunlight spectrometer gear and returned to Washington. The next mountain Stair's spectrometer would visit would be in the middle of the Pacific Ocean, thousands of miles away from New Mexico dust storms.

Stair had two projects on his mind as the research at Sunspot was completed. He knew that the solar constant measurements made with the instrumentation he had designed would be of high interest to the scientific community and especially to the Geophysics Research Directorate of the Air Force Cambridge Research Center, which provided the funds for the Sunspot research. Thus, analyzing his data to determine the solar constant and writing a report about his findings was a high priority project that would require considerable time. Simpson's proposal to provide the money to reestablish an observatory on Hawai'i's Mauna Loa also deserved high priority. Mauna Loa would provide a dust-free location for Stair's solar constant measurements. Moreover, water vapor interfered with his solar constant measurements even more than ozone, and Mauna Loa would provide a drier sky than Sunspot (Stair 1958).

Simpson's dream became Stair's goal as he pondered the obligation to complete the Sunspot report and the new opportunity to find a better site for future measurements of the solar constant. Stair's memoir about what happened next was published in the *Mauna Loa Observatory 20th Anniversary Report* (Stair 1978):

> Upon our return to Washington a number of conferences at the NBS and at the Washington office of the U.S. Weather Bureau resulted in an agreement for NBS to use some $25,000, which would be transferred from the U.S. Weather Bureau to the Radiometry Laboratory of NBS, in the establishment of a laboratory for the joint use of the Weather Bureau and NBS. The amount of money available was insufficient for the construction of a permanent building through the normal construction channels, but for use in establishing a durable "shelter" for practical use in all kinds of weather expected on one of the Hawaiian mountains it was considered adequate if strict economies were followed throughout.

While Stair was working on the observatory project from Washington, Robert Simpson was busy setting up the National Hurricane Research project in Florida. He was completely unaware that a crucial decision had been made to build the new observatory high on the north slope of Mauna Loa instead of its summit. Stair, who first visited the site in December 1955, described the reason behind the decision in the *20th Anniversary Report* (1978):

> Following preliminary discussions and evaluations in Washington, it was decided that the best location for a solar radiation laboratory in the Hawaiian Islands would be on the north slope of Mauna Loa, where as a result of the shielding effect of Mauna Kea the atmosphere would be least turbulent and most free of clouds. Furthermore, a mountain cinder road was already in existence in this area. . . . In picking the site for the observatory a number of considerations were taken into account. The site chosen was near the mountain cinder road at the upper terminus of its better condition, that is, at the highest elevation possible to be reached with two-wheel-drive vehicles. The elevation, approximately 3400 m, was considered near the limit for extended living and working for most individuals. The area was on a recent volcanic lava flow which, because of its elevation above adjacent areas to the right and left, offered promise of some protection from future flows on this side of the mountain.

The original summit observatory included a bunk bed for overnight stays. But altitude sickness can pose a serious problem at the 13,459-foot (4,102-meter) elevation of the summit, as the US Exploring Expedition learned during their three-week stay on the opposite side of the summit crater in 1840–1841, and as hikers who spend the night at the nearby Park Service cabin sometimes learn even today. A site at a lower elevation would reduce, but not entirely eliminate, the altitude sickness problem. Simpson experienced only shortness of breath and a rapid pulse during his hikes to the summit to select the original observatory site and plan the cinder road. While he is well aware of and highly supportive of the achievements of today's Mauna Loa Observatory, he always preferred the summit site. More than half a century later, he said,

> if given an opportunity, I would have reopened it again and probably at the summit rather than down where it is. I wouldn't have compromised the altitude there. (Simpson, telephone interview with author, March 17, 2007)

At 11,141 feet (3,396 meters), the proposed new slope site was 1,941 feet (592 meters) above the elevation of Sunspot and curiously close to the 11,190 feet (3,411 meters) elevation of the site at Climax, Colorado, where in 1952 Stair had made high-altitude ultraviolet observations (Stair 1952). Could it be that the altitude of the Climax site played a role in selecting the new slope site for the Mauna Loa Observatory? This would provide a way to compare solar constant measurements at the two locations without having to correct for the difference in altitude if the new observatory was constructed on the summit of Mauna Loa. In the end, Stair and others involved in planning the new observatory held out hope for the reestablishment of an observatory at the summit, for they later proposed that the new structure they planned for the north slope of the mountain would someday serve as a staging area and dormitory for scientists and technicians who would work on the summit (NBS and US Weather Bureau 1956). But after the observatory was built and Stair had made his first round of solar constant measurements there, he wrote a few lines about a key advantage of the new MLO slope location that greatly reduced the priority of returning to the summit:

> This altitude has a further advantage over that of the summit location in affording conditions of atmospheric pressure (and hence oxygen concentration) more in keeping with the normal requirements of the average individual. Many observers would be unable to live and work at a much higher level. (Stair and Johnston 1958)

As for placing the new observatory on a "recent volcanic lava flow," John P. Lockwood of the Hawaiian Volcano Observatory provided the geological details in the *20th Anniversary Report*.

> When the written accounts of witnesses to nineteenth century Mauna Loa eruptions are reviewed, one eruption is notable because its site has not been previously recognized. At 3:30 a.m. on February 17, 1852, citizens of Hilo were awakened by a Mauna Loa summit eruption. Reverend Titus Coan (1852, 219–220), Hilo's precise chronicler of Hawaii's volcanic eruptions from 1840 to 1881, wrote that "immense columns of burning lava shot up heavenward to the height of 300 or 400 feet, flooding the summit of the mountain with light, and gilding the firmament with its radiance. Streams of light came pouring down the mountain, flashing through our windows, and lighting up our apartments so that we could see to read large print. . . . In two hours the molten stream had rolled, as we judged, about fifteen miles down the side of the mountain. The eruption was one of terrible activity and surpassing splendor. But it was short. In about twenty-four hours all traces of it seemed to be extinguished."
>
> It thus seems likely, all considered, that the *a'a* lavas underlying MLO may be the products of the eruption viewed by Reverend Coan on the early morning of February 17, 1852. For those interested, a dramatic summary description of the 1852 eruptions was published in the *Hilo Tribune Herald* (Allan 1929). (Lockwood 1978)

Governor Samuel Wilder King

In 1951 Robert Simpson had experienced major opposition from the Park Service staff in establishing the original summit observatory and road. Stair was aware of this, and he cleverly bypassed the Park Service by going directly to Hawai'i's new territorial governor.

> Accordingly, an arrangement was made for a conference (in December 1955) with the Honorable Samuel Wilder King, then Governor of Hawaii, for the purpose of transferring title to a suitable area on the north slope of Mauna Loa (within the Mauna Loa forest and game reserve) from the Territory of Hawaii to the Department of Commerce of the federal government. Mr. King agreed to the transfer, and it was accomplished through the proper Hawaiian channels, with survey by C. L. and D. J. Murray for an area of 4.05 acres (10 hectares) in the shape of a square 420 by 420 ft (tax map 4-4-16; C.S.F. No. 12333). The exact area was chosen previous to the survey by Ralph Stair, Roy L. Fox, and James W. Steiner, of NBS; Pacific Area, U.S. Weather Bureau; and Hilo office, U.S. Weather Bureau, respectively. (Stair 1978)

The Hawaiian territorial government gave the Department of Commerce a grant to the site in the form of an Executive Order (King 1956), a document that in coming years would demonstrate the wisdom of Stair's foresight in planning the new observatory. The order reads as follows:

> **Executive Order No. 1720**
> **Setting Aside Land for Public Purposes**
>
> By this Executive Order, the undersigned, Governor of the Territory of Hawaii, by virtue of the authority in me vested by Section 91 of the Hawaiian Organic Act, and every other authority me hereunto enabling, do hereby order that the public land hereinafter described be, and same is, hereby set aside for the uses and purposes of the United States of America.
>
> FOR a United States Weather Bureau Station Site, to be under the control and management of the United States Department of Commerce.
>
> U. S. Weather Bureau Station Site
> Kaohe 5, Hamakua, Hawaii, T. H.
> Being portion of the Government Land of Kaohe
>
> Beginning at the reference station "STAIR 1955" (Spike in concrete marked "STAIR 1955") at the northeast corner of this parcel of land.

The order continues with a listing of the geographical circumstances at each of the four corners, which are separated by "420.00 feet" (128 meters) and which enclose "AN AREA OF 4.05 ACRES." The order concludes with a signature block:

> In Witness Whereof, I have hereunto set my hand and caused the Great Seal of the Territory of Hawaii to be affixed. Done at the Capitol at Honolulu this 28th day of January, Nineteen Hundred and fifty-five.
>
> [signed] Samuel Wilder King
> Governor of the Territory of Hawaii
>
> Approved as to form:
> [signed] Edward N. Sylva
> Deputy Attorney General

Stair managed to finalize the site location, have it surveyed, and secure Governor King's Executive Order only seven months after his meeting with Robert Simpson at Sunspot the previous summer. In his memoir, Stair (1978) describes the survey markers placed at each corner of the site:

> The 4.05-acre area was marked at the northeast corner by a spike set in concrete which was labeled "Stair, 1955." The other three corners and the midpoints of the north and south sides were defined by 3/4-in iron pipes driven into the lava.

Today three of these markers are indicated by blue reflectors on metal posts and can be easily found. The northeast marker with Stair's faintly outlined name and date is under a small pile of lava rocks that supports its blue reflector. The area is now in a pit, the surrounding lava having been raised and leveled in 1996 with rubble from the site of the NDSC Building (now known as the NDACC Building).

The Observatory Building

Stair quickly moved ahead with the design and construction of the new observatory's building as efficiently as he measured the solar constant. He describes in his memoir (1978) details about the design and construction of the building.

> The transfer of this parcel of Mauna Loa to the Department of Commerce (of which NBS and the U.S. Weather Bureau were parts) having been arranged, numerous interviews were conducted with other government and business people within the Hilo area concerning the best procedures to follow in getting the most desirable building on the site at the least cost. A first thought was to construct the building of lava block, of which there was an abundance within the area. However, investigation showed this to be more expensive than the use of cast cinder block manufactured in Hilo. (The possibility of a lava block building was left open, however, for a time when more money might be available, through the incorporation of a concrete ledge on the building foundation.) Other interviews regarding available material resulted in the location of local overstocks of certain items and sizes (for example, aluminum sash and roofing sheets). The final specifications for the building included only materials available locally so as to permit construction without delay or extra cost. All possible contractors within the area were interviewed and made acquainted with the local supplies in stock.

Like Thomas A. Jaggar's Hawaiian Volcano Observatory in 1912 and Robert Simpson's Mauna Loa Summit Observatory in 1951, Stair was the recipient of valuable time and labor provided by prisoners from Kūlani, a relationship that was to continue for another decade. Stair described in his memoir (1978) how the assistance from the Kūlani Prison and other creative strategies kept the project within its budget:

> The Hawaiian Kulani prison authorities were extremely helpful in the construction of the new observatory. They not only continued to keep the roadway in repair but volunteered to "level off" the 4-acre area ahead of the contractor and to deliver to him the water required during construction. The latter was a most important item as it relieved the contractor of all expenses that might be involved in obtaining and operating water tanks.
>
> All developments in the specifications for the new building were handled by NBS, with consultations and interviews with U.S. Weather Bureau personnel. This work was performed primarily on regular Bureau schedule with no cost to the construction fund. As a matter of fact, to keep all costs to a minimum no drawings or blueprints were prepared, all work being performed from detailed specifications.
>
> With the building specifications in order, the requests for bids on construction were handled through the offices of NBS in the usual official manner.

Simpson's concern about the added expense and bureaucratic maneuvers required to build a facility under the supervision of the GSA had led to his proposal to provide funds directly to the Bureau of Standards. Stair proved the wisdom of this move as he planned the 20 x 40-foot (6 x 12-meter) concrete block building, which was partitioned into working, cooking, and sleeping sections. A concrete slab measuring 15 x 45 feet (4.6 x 14 meters) was installed along the south side of the structure (Fig. 5.2) to provide a stable place for outdoor instruments, such as the planned Dobson spectrophotometer for measuring the ozone layer (Price and Pales 1963) and, of course, Stair's solar constant spectrometer. An elevated outdoor instrument platform was provided in the form of a wood tower that rose just above the roofline at the east end of the building. A water tank installed at the west end of the building stored water captured by the roof. A small diesel electrical generator was installed under the water tank.

The contractor for the project was Jas. W. Glover, Ltd. Many years later, Hilo native and chemist Alan Yoshinaga, who would spend a career working at the observatory, recalled that his uncle had worked for Glover during the construction of the observatory building.

FIGURE 5.2. Workers complete the concrete slab along the south side of the new MLO building on May 7, 1956. *(MLO archive.)*

> He used to tell me, "Whew, we had a hard job on the hill today! We were building a little observatory at the mountain." (Yoshinaga, video interview, Hilo, Hawai'i, December 4, 2006)

The remoteness of the construction project posed problems for the contractor, for extra time was required to transport workers, supplies, and equipment to the building site at 11,200 feet (3,414 meters). Based on the performance of construction workers at the site today, the workers probably performed their tasks more slowly than at sea level, and some may have experienced symptoms of altitude sickness. Fortunately for all concerned, the construction of the new observatory building proceeded on schedule, and Stair later recalled some of the details.

> During construction, in order to keep costs down, all inspections and consultations with the contractor were handled by Hilo U.S. Weather Bureau personnel through occasional visits to the site followed by simple reports to NBS.
>
> By June 1956 the building was essentially complete [Fig. 5.3] as described and illustrated in the September 1956 issue of *Weather Bureau Topics,* the October 1956 issue of *National Bureau of Standards Technical News Bulletin,* and the November 1956 issue of *Discovery.*
>
> The construction of the observatory proceeded smoothly throughout. There were helping hands in all quarters with no dissidents anywhere. Donations came from several organizations. One, in the form of a radio transceiver, was supplied by the Radio Section of NBS. A small library [which can be found today at the observatory's Hilo office] was furnished by the Smithsonian Institution and the Department of Agriculture. Only minor problems developed, and those were primarily the result of "corner cutting" to get the job done at low cost.

FIGURE 5.3. The concrete-block MLO building was completed in June 1956. *(Stair family archive. Used with permission.)*

> For example, the placing of the diesel electric generator near the building under the water tank required moving it later to a more distant location because of vibrations and noise that interfered with experiments inside or near the building. Although the construction of the observatory building was carefully specified to contain steel rods to hold the structure securely attached to the foundation in case of a hurricane, no such forethought was included in the case of the "outhouse"; it went the ways of the winds upon the arrival of the first 100 mph hurricane. (Stair 1978)

During the considerable time he devoted to the construction of the new observatory, Stair somehow managed to complete the paper he and Johnston wrote about their solar constant measurements at Sunspot the previous June that led directly to the new observatory project. The paper was submitted to the *Journal of Research of the National Bureau of Standards* on April 20, 1956. Meanwhile, the first research project for the new observatory was being planned. Even before the building's official dedication, Carl C. Kiess and Charles C. Corliss of the National Bureau of Standards and their wives, Harriet K. Kiess and Edith L. R. Corliss, moved into the new building and began making high-resolution spectral observations of Mars during June 1956 in a project sponsored by the National Geographic Society (Kiess et al. 1957; Kiess 1968). Solar constant and ozone expert Ralph Stair prepared for this inaugural project using skills outside his scientific expertise, two examples of which he describes in his memoir (1978):

> In getting the "most possible for the least cost," one can run into real trouble. This happened in the case of the NBS project leader, who found it necessary to become plumber and janitor for a day to get the water running and the floors clean for the first scientific observing team to use the observatory (Dr. C. C. Kiess and associates of NBS for the National Geographic Society).

Dedication Day

Dedication day for what was called the slope unit of the Mauna Loa Observatory soon approached. Back in 1951, Robert Simpson had arranged for a few dozen distinguished visitors and newspaper reporters to be present at the dedication of the original summit observatory. Stair and the Weather Bureau followed his lead by planning an on-site celebration for the dedication of the new slope building. Two decades later, Stair (1978) recalled the final steps before dedication day:

> By dedication time, the building was completely furnished (down to bed sheets, kitchen utensils, dishes, and flatware) for comfortable living for six observers. Government inspection (by Ralph Stair) confirmed fulfillment of the contractor's obligation, although the observatory faced "magnetic south" rather than geographic south, which shows that one can never be too careful in writing a specification—the contractor used a compass. The building was duly accepted by NBS, the contractor paid, and the Mauna Loa Observatory dedicated on June 28, 1956, the introductory remarks being made by the Honorable Samuel Wilder King, Governor of Hawaii.

That morning, some seventy-six officials and visitors accompanied Governor King in cars and trucks on the arduous trip from Hilo to the 11,140-foot elevation of Mauna Loa. The contingent included a dozen or so reporters and Roy L. Fox, Saul Price, Ray Busniewski, Dominic Conte, and Dean Parry of the Weather Bureau. As they drove along the rough, bouncy road, they passed through a sequence of every vegetation and climate zone found on the Big Island, from luxurious tropical forests of tree ferns decorated with orchids to fields of black lava as far as the eye could see. Above the tree line, they drove along a rough road paved with chunky, sharp-edged lava clinkers.

Dr. Walter R. Steiger, professor emeritus of physics and astronomy at the University of Hawai'i, recalled in a telephone interview (July 11, 2007) how the caravan passed through the Kūlani Prison and that it included ordinary passenger vehicles. When he and his colleagues from the University of Hawai'i struck a particularly sharp rock, the gas tank of their car was punctured. Someone asked if anyone had a bar of soap, and Steiger just happened to have one with his travel gear. When the soap was pressed into the hole, the leak was plugged, and they were able to continue driving.

Finally, well above the clouds that blanketed the lower slopes of Mauna Loa, they reached their destination, a concrete block building with a corrugated metal roof. Steiger was no stranger to the site, for he had visited two weeks earlier, after which he wrote a memo in his notebook that describes the building's facilities:

> **Trip to Mauna Loa 6/13/56 with Dr. Edward Chupp**
> Purpose of trip: to investigate the possibility of the new 11,000 ft Weather Bureau station as a possible site for a cosmic ray monitor for the IGY. The building, just completed, has three small rooms (bedrooms) 10' x 10', a large general workroom 20' x 20', and a kitchen 10' x 10' equipped with butane stove, refrigerator & water heater. Large room has a propane space heater. Water is collected in 1000 gal tank from rain drained off roof. Power to be supplied by a 7.5 KW generator under water tank. (Steiger 1956)

While people were inspecting the site, Steiger walked some 130 feet (40 meters) upslope, turned around, and took a photograph of the new scientific outpost at one of the most desolate places on Earth with Mauna Kea looming in the background (Steiger, personal interview with author, Hilo, Hawai'i, May 7, 2009; NBS and US Weather Bureau 1956). "It was a beautiful day," Steiger recalled. "It was a nice warm, sunny day." Prior to the dedication ceremony (Fig. 5.4) , the guests toured the facility, took whiffs of oxygen provided for the occasion, and feasted on fried chicken provided by none other than the Kūlani Prison kitchen staff. The temperature was a pleasant 58 degrees F (14.4 degrees C), and the clear sky and altitude left most of the participants with a sunburn (Allen 1956).

Joel B. Cox, president of the Geophysical Society of Hawai'i, served as the master of ceremonies, which began Hawaiian style with an invocation in the island's native language by two women from Hilo's historic Haili Church. As the *Honolulu Star-Bulletin* reported,

> Gracious and lovely, silver haired and sweet, they invoked God's blessings on this study of God's handiwork in the heavens. They are Mrs. Eliza Leialoha Kamakawiwoole Osorio and Mrs. Anna Mauna Loa Kamau Hoopii. (Allen 1956)

FIGURE 5.4. The Mauna Loa Slope Observatory was dedicated on June 28, 1956. Roy L. Fox is standing at right in front of the row of chairs. To his right are Ralph Stair and the two women from Hilo's Haili Church who blessed the ceremony. Ray Busniewski is standing before the right pane of the window at left, and Dominic Conte is third from left in the front row behind the wood bench. *(MLO archive.)*

FIGURE 5.5. Roy L. Fox spoke at the dedication of the Mauna Loa Slope Observatory on June 28, 1956. Shown here *(from left to right)* are Dr. Walter Steiger, University of Hawai'i; James Kealoha, chairman, County of Hawai'i; Ralph Stair, National Bureau of Standards; Roy L. Fox, meteorologist in charge, Pacific Region, US Weather Bureau; Samuel Wilder King, governor of Hawai'i; and unknown. *(Stair family archive. Used with permission.)*

In view of the latter's middle name, the *Hilo Tribune-Herald* reported that the invocation was "given most appropriately by Mrs. Anna Mauna Loa Kamau Hoopii" (Pierce 1956). Equally significant, but unreported, was the historical link between the Haili Church and the Rev. Titus Coan, the Congregationalist missionary whose flock assisted with the US Exploring Expedition in 1840–1841 and who himself became the most noted observer of Hawaiian volcanic eruptions throughout his long service in the ministry. As noted earlier, it was the Rev. Coan who in 1852 observed and reported the brief eruption that formed the *'a'ā* lava flow over which the slope observatory was constructed a century later (Coan 1852, 219–220).

Following the invocation, Cox introduced Governor King, who congratulated those responsible for the new observatory and promised his future support and cooperation (Allen 1956; Pierce 1956). James W. Steiner, the meteorologist in charge of the US Weather Bureau office in Hilo, spoke next. Steiner had assisted Robert Simpson in finding the site for the original summit observatory, and he had made many trips to service the facility. Steiner spoke briefly about the history of observations at the summit of Mauna Loa.

Ralph Stair of the US Bureau of Standards then spoke on "Some Research Potentials of the Mauna Loa Observatory."

> Mr. Stair, who designed the "slope unit," declared the research potential of Mauna Loa Observatory is almost unlimited. Among possibilities he listed:
>
> Routine weather observations, solar radiation measurements, determination of ozone, atmospheric measurements of radio transmission, study of radiation of cosmic rays from outer space and study of atomic fallout, snow crystals, meteors, the sun's corona, the green flash of stars and the air glow of night light in the skies. He asserted, "There is no place in the United States with comparable clearness of atmosphere." (Pierce 1956)

Stair also suggested that a telescope might someday be installed at the observatory. An artificial Earth satellite would not be launched for more than a year, but Governor King was clearly aware of the budding space program. During Stair's presentation, as reported by Pierce (1956), the governor added to Stair's list of scientific applications for the new observatory:

> When Governor King suggested space ships, Mr. Stair's prompt rejoinder was "Yes. We can learn many things to help in this development."

Carl C. Kiess of the National Bureau of Standards represented the National Geographic Society when he spoke on "A Study of Water Vapor in the Atmosphere of the Planet Mars," an especially appropriate topic since it was the object of the ongoing study at the observatory begun by him, Charles C. Corliss, and their wives even before the building was dedicated.

Dr. Walter R. Steiger, representing the University of Hawai'i, spoke next on "A Milestone in Geophysics." It was fitting that Steiger's talk followed Kiess, for he was especially impressed by the pioneering astronomical work being conducted by Kiess and Corliss on Mauna Loa. "I was very pleased to see this kind of research taking place at this new facility," he recalled during a 2007 interview at the Caltech Submillimeter Observatory headquarters in Hilo, where, though retired, he still maintained an office.

Finally, Roy Fox, who with James Steiner had helped Stair select the site, spoke on "The Mauna Loa Observatory: Past, Present and Future" (Fig. 5.5).

> Abandoning his prepared talk, Roy L. Fox, Meteorologist in Charge, Pacific Area, U.S. Weather Bureau, raised the curtain on future plans that are directed toward bringing a full fledged summit observatory to reality. (Pierce 1956)

Fox explained that the Mauna Loa Observatory project was in the second of three phases. The first phase was Operation Summit, Robert Simpson's original observatory, and the second "retrenchment phase" was Operation Slope. The third phase would merge the summit and slope observatories with what he called

> Operation Summit-Slope, the long range aim, with the main observatory back at the 13,453 feet level and with

> the "slope unit" at the 11,134 feet level serving as a base of operations for housing personnel and handling supplies. Declaring, "We already are thinking and planning for the third stage," Mr. Fox commented that prospects appear encouraging after conferring with Superintendent John Wosky of Hawaii National Park for relocating the main observatory within park boundaries [at the summit]. (Pierce 1956)

Fox observed that the road to the summit would have to be significantly improved to achieve Operation Summit-Slope. He also suggested the importance of an emergency escape road in the event of an eruption of the volcano.

The speakers (Fig. 5.6) and their topics were introduced in a three-page brochure that briefly described the history and purpose of the observatory, including the fact that

FIGURE 5.6. Roy L. Fox (pointing toward the summit of Mauna Loa), Governor Samuel Wilder King, Carl C. Kiess, and Ralph Stair during the dedication of the Mauna Loa Slope Observatory on June 28, 1956. *(Stair family archive. Used with permission.)*

> Recent increased interest in such problems as atmospheric circulation, the ozone content of the middle atmosphere and solar radiation in relation to intense tropical storms has focused the attention of meteorologists and radiation physicists on observations from mountain locations in the Pacific. . . .
>
> The dedication of this unit does not mark the end of efforts to establish an adequate facility near the summit of Mauna Loa. Rather, it is another step toward the goal of developing a complete functional unit of the Observatory near the highest point on this enormous volcanic mountain. The realization of this aim is dependent upon many things, including approval for future use of the existing site near the summit, the extension of the improved Mauna Loa Highway to that point and the erection and equipping of an appropriate building there.
>
> When these are accomplished, it is envisioned that most of the research programs of the Observatory will be carried on at the "summit" unit. The building being dedicated today will then serve primarily as a staging point and living quarters for the scientific personnel. (NBS and US Weather Bureau 1956)

While the author of the brochure is not given, Ralph Stair himself is the most likely candidate. This is suggested by the listing of the Bureau of Standards ahead of the Weather Bureau on the brochure's title page and the recitation of Simpson's goals for studying ozone and eventually returning to the summit. While Harry Wexler was a key supporter of the new observatory, he was not present. Although Wexler's interest in measuring carbon dioxide was unmentioned in the brochure, it was very much on his mind.

The Press Approves

The dedication of the revived Mauna Loa Observatory was covered favorably in both the popular press and the scientific literature. The *Honolulu Advertiser, Honolulu Star-Bulletin,* and *Hilo Tribune-Herald* covered the event. Articles appeared in *Nature* (December 1956), *Bulletin of the American Meteorological Society* (November 1956), *Journal of the Washington Academy of Sciences* (October 1956), *Weather Bureau Topics* (September 1956), and *Technical News Bulletin* of the National Bureau of Standards (October 1956).

The new Mauna Loa Observatory building was dedicated almost exactly one year after the chance meeting of Robert Simpson and Ralph Stair at Sunspot in June 1955. Scientists and reporters alike knew that important meteorological and atmospheric observations would be made there. After all, no other observatory on Earth was located at such a pristine, alpine site. But they could have no idea how important some of those observations would become.

The scientist behind all this excitement was nowhere to be seen and unaware of the dedication of the new observatory. Robert Simpson was in Florida busily preparing for the 1956 hurricane season while directing the fledgling National Hurricane Research Project, the organization whose budget provided the $25,000 for the construction of the second incarnation of Simpson's dream, the new "slope building" of the Mauna Loa Observatory.

References

Allan, V. M. 1929. Eruption of 1852 one of splendor. *Hilo Tribune-Herald,* October 6.

Allen, Riley H. 1956. Scientist's believe location is ideal. *Honolulu Star-Bulletin,* June 29.

Coan, Titus. 1852. On the eruption of Mauna Loa, Hawaii, February, 1852. *American Journal of Science* 2nd ser., 14 (40–52), 219–224.

DeMastus, Howard L. 1976. A twenty-year summary of Sacramento Peak weather: August 1954 through July 1974. Sloan Digital Sky Survey Telescope Technical Note 19970822, US Air Force Geophysics Laboratory. Report Number AFGL-TR-76-0096. (Accessed at http://www.apo.nmsu.edu/Telescopes/SDSS/eng.papers/19970822_Weather/19970822.html.)

Dobson, G. M. B. 1930. *Proceedings of the Royal Society* 129: 411.

Ellis, Howard. 1988. The observatory: On their own. (Accessed at http://www.mlo.noaa.gov/pdf/memorabilia/Ellis%20booklet%20edited.pdf.)

Grissino-Mayer, Henri D., Christopher H. Baisan, and Thomas W. Swetnam. 1997. A 1,373 year reconstruction of annual precipitation for the southern Rio Grande Basin. Legacy Program, Directorate of Environment, Natural Resources Division, Fort Bliss, Texas. Appendix 1: Reconstructed precipitation for water year rainfall (in inches) for the central Rio Grande Valley in south-central New Mexico, AD 622–1994: 58–63. (Accessed at http://web.utk.edu/~grissino/downloads/Legacy%20Final%20Report.pdf.)

Kiess, C. C. 1968. Spectrographic studies of the atmosphere of Mars and Jupiter. Research Reports 1955–1960, National Geographic Society.

Kiess, C. C., C. H. Corliss, H. K. Kiess, and E. L. R. Corliss. 1957. High dispersion spectra of Mars. *Astrophysical Journal* 126: 579–584.

King, Samuel Wilder. 1956. Executive Order No. 1720: Setting aside land for public purposes.

Lockwood, John P. 1978. The volcanic environment of Mauna Loa Observatory: Hawaii history and prospects. In John Miller, ed., *Mauna Loa Observatory: A 20th Anniversary Report*. Washington, DC: National Oceanic and Atmospheric Administration, US Department of Commerce.

Miller, John, ed. 1978. *Mauna Loa Observatory: A 20th Anniversary Report*. Washington, DC: National Oceanic and Atmospheric Administration, US Department of Commerce.

National Solar Observatory. 2002. Virtual tour. (Accessed at http://nsosp.nso.edu/pr/virtual_tour.html.)

NBS and US Weather Bureau. 1956. Dedication of the "slope" building, Mauna Loa Observatory, Thursday, June 28. Jointly sponsored by the National Bureau of Standards and the US Weather Bureau.

Pierce, Howard. 1956. Greater role seen for Mauna Loa Observatory. *Hilo Tribune-Herald,* June 29.

Price, Saul, and Jack C. Pales. 1963. Mauna Loa Observatory: The first five years. *Monthly Weather Review* 91: 665–679.

Simpson, Robert H. 1989. Interview by Edward Zipser on September 6 and 9, for the American Meteorological Society, University Corporation for Atmospheric Research. (Accessed at http://www.ucar.edu/archives/publications/simpson-robert%20interview.pdf.)

Stair, Ralph. 1948. Measurement of ozone over the Organ Mountains, New Mexico. *Journal of Research of the National Bureau of Standards* 40: 9–19.

———. 1952. Ultraviolet radiant energy from the sun observed at 11,190 feet. *Journal of Research of the National Bureau of Standards* 49: 227–234.

———. 1958. Some studies of atmospheric transmittance on Mauna Loa. *Journal of Research of the National Bureau of Standards* 61: 419–425.

———. 1978. The construction of the present observatory. In John Miller, ed., *Mauna Loa Observatory: A 20th Anniversary Report*. Washington, DC: National Oceanic and Atmospheric Administration, US Department of Commerce.

Stair, Ralph, and I. F. Hand. 1939. *Monthly Weather Review* 67: 331.

Stair, Ralph, and Russell G. Johnston. 1956. Preliminary spectroradiometric measurements of the solar constant. *Journal of Research of the National Bureau of Standards* 57: 205–211.

Stair, Ralph, R. G. Johnston, and Thomas C. Bagg. 1954. Spectral distribution of energy from the sun. *Journal of Research of the National Bureau of Standards* 53: 113–119.

Steiger, Walter R. 1956. Personal notebook, p. 41 (unpublished).

V-2 Firing Record Summary. 1947. White Sands Missile Range Museum, Archival Holdings. (Accessed at http://www.wsmr-history.org/FiringRecords1947.asp.)

CHAPTER SIX

Startup

The First Science at MLO (1956–1957)

Serious science began at the new Slope Facility of the Mauna Loa Observatory with the observations of the Martian atmosphere by the Kiess-Corliss team (Kiess et al. 1957) even before the dedication of the building on June 28, 1956. While the facility was specifically developed for measuring and monitoring solar, atmospheric, and meteorological parameters in the free atmosphere over Hawai'i, the Weather Bureau was not quite ready to begin work at the observatory. Instead, the bureau was making plans for a variety of investigations during the International Geophysical Year (IGY), which was scheduled to run from July 1957 through 1958.

The IGY was planned to investigate both Earth and space, and it eventually involved some thirty thousand scientists in seventy countries who conducted a wide range of studies involving meteorology, space flight, plate tectonics, and polar research. The program would lead to the launch of the first artificial satellites, the discovery of the Van Allen radiation belt around Earth, support for the theory of plate tectonics, and increased research in the Arctic and Antarctic. It would also lead to major new discoveries about changes in the composition of Earth's atmosphere.

The United States was a major participant in the IGY. When President Dwight D. Eisenhower received a letter about the IGY from Chester I. Barnard, chairman of the National Science Board of the National Science Foundation, he replied on June 24, 1954, as follows:

> I am glad to support this undertaking. It is a striking example of the opportunities which exist for cooperative action among the peoples of the world. . . . The findings of this research will be widely disseminated throughout the world, aiding in the further development of telecommunications, aviation, navigation, and weather forecasting. . . . I am sure that our participation in this far-reaching effort will very materially strengthen our bonds with many cooperating nations and make a constructive contribution toward the solution of mutual problems. (Eisenhower 1954)

As the IGY and the completion of the Mauna Loa Observatory approached, Ralph Stair arranged for various scientific studies to be conducted at the new observatory. Water vapor is the gas most responsible for weather and for keeping the average temperature of the earth well above the freezing point of water, and three of the earliest projects at MLO involved the measurement of the water vapor in a column through the atmosphere, much like a project Robert Simpson had proposed for the Summit Observatory at its 1951 dedication. Column water vapor is expressed as the depth of liquid water that would result if all the water vapor in a vertical column through the atmosphere could be brought down to the surface at standard temperature and pressure. Thus, column water vapor is the liquid equivalent of the water vapor in the atmosphere over a given point on the surface.

Stair's solar constant measurements had required that he become an expert in measuring column water vapor, for, as with ozone, it was necessary to correct his measurements by adding back the power in the sunlight that was absorbed by water vapor. This played a key role in the need for Stair to make solar constant measurements from alpine sites, for the column water vapor can reach 70 or more millimeters on a sweltering summer day near sea level but can be less than a millimeter atop a high mountain on a dry day, as it is on some clear mornings at MLO. While it was visible dust that most strongly impacted Stair's solar constant measurements at Sunspot, New Mexico, in June 1955, invisible water vapor was the principal interference during days free of dust.

> As a matter of fact the large fluctuations [in column water vapor] noted at Sunspot were a primary reason for seeking the new location on Mauna Loa where a lower water-vapor concentration would interfere less with the measurement of the spectral intensities in the infrared. But water vapor is such a highly absorbing constituent of the atmosphere that even at the very low concentrations ordinarily present above Mauna Loa there is still a marked absorption within the principal water-vapor bands, even near noon on days having a relatively low total water-vapor concentration. Then small changes in water vapor produce large changes in total optical absorption within the regions of the principal absorption bands. (Stair and Johnston 1958)

Stair and a handful of other scientists inferred the column water vapor using a method pioneered by Frederick E. Fowle (1912) of the Astrophysical Observatory of the Smithsonian Institution. The method involved measuring the intensity of a narrow band of infrared sunlight that is strongly absorbed by water vapor and a nearby infrared band unaffected by water vapor. When corrected for the angle of the sun through the atmosphere during the measurement, the ratio of the intensity of the two bands of sunlight is directly related to the vertical column water vapor.

The first of the science projects Stair approved for the new observatory was a study to measure not just the water vapor over MLO but that in the atmosphere of Mars. The project was very likely the first serious astronomical project conducted on the Big Island, and it may be the only study conducted at MLO by a pair of married couples.

Kiess and Corliss Study the Atmosphere of Mars (June–July 1956)

With support from the National Geographic Society and the full cooperation of the Weather Bureau, National Bureau of Standards' scientists Carl C. Kiess and Charles H. Corliss and their wives Harriet and Edith moved to the Mauna Loa Observatory during June 1956. As evidenced by their signatures in the first guestbook, which was entitled *Visitors*, all four of them were present during the dedication ceremony on June 28. During the ceremony, Kiess spoke about their planned spectrographic measurements of water vapor in the atmosphere of Mars. The lofty elevation of the Mauna Loa Observatory meant that on many days there was very little water vapor over the site, perhaps as little as the liquid equivalent of a few millimeters or less between the observatory and the top of the atmosphere. Thus, they hoped measurements at MLO would experience relatively little interference from water vapor in the atmosphere over the mountain.

Following the June 28 dedication, the Kiess and Corliss team set up in front of the new building a crude shed, the first of many temporary buildings and shelters to be installed at MLO. They installed in the shed a laboratory-style spectrograph, a cumbersome instrument that divided light into its respective wavelengths. Ralph Stair's spectroradiometer that led to the establishment of the observatory when he met Robert Simpson at Sunspot, New Mexico, the previous year used a pair of quartz prisms to break sunlight into its constituent wavelengths so that it could be measured electronically. The ultraviolet through the visible wavelengths of the spectrum were swept across a phototube, and the infrared spectrum was measured by replacing the phototube with a solid-state photoconductive cell composed of lead sulfide.

The Kiess and Corliss team used a National Bureau of Standards spectrograph that employed a diffraction grating to divide the light from Mars into its spectrum. Two gratings were available for use with the instrument, one ruled with 15,000 and the other with 30,000 parallel lines per inch (2.5 cm) in the speculum metal deposited on the surface of a concave mirror. Instead of electronically sampling the spectrum, the spectrograph projected the rainbow of light reflected from the grating onto a rectangular photographic plate measuring 2 x 10 inches (5 x 25.4 cm) that was sensitized prior to exposure by a solution of ammonia and water. The telescope that projected the image of Mars onto the spectrograph's grating had a lens with a diameter of 12 inches (30.5 cm).

The Kiess-Corliss team began observations of Mars the night of July 12, 1956, during windy conditions and freezing temperatures. Observations were also made of the Moon to provide a reference for comparison with Mars. Since the Moon has no atmosphere and thus no water vapor, comparing the infrared-absorbing water vapor spectra reflected by both bodies would compensate for the water vapor in the Earth's atmosphere over Mauna Loa and allow the detection of any water vapor in the Martian atmosphere. The best spectrograms were obtained the nights of July 21 and 22.

The Kiess-Corliss team was unable to detect water vapor in the atmosphere of Mars. They concluded that a future study should employ a spectrograph better able to detect the infrared portion of the Martian spectrum, where water vapor is indicated by a series of significant and easily recognizable dips in the solar spectrum that are known as absorption bands.

The Kiess-Corliss team published their MLO research in the November 1957 issue of the *Astrophysical Journal*. The paper included a few findings of fact that, in hindsight, seem puzzling. For example, the paper noted that

> For the meteorological conditions that prevailed during these observations the water vapor content of the air in the optical path was equivalent to a film of water approximately 0.15 mm thick. (Kiess and Corliss 1957)

As we shall shortly see, the measurements made by Ralph Stair shortly thereafter indicated a minimum amount of column water vapor of about 1.0 mm, which is more than six times greater than that measured by Kiess-Corliss and very much in line with measurements made at MLO today. Stair's spectrometer featured an electronic detector rather than a photographic plate, and this may explain the difference. Support for this possibility is provided by the curious caption for Figure 2 in the paper: "The O_2 bands apparently are unaffected by the difference in altitude" (Kiess and Corliss 1957). This refers to the absorption by oxygen of near-infrared bands of sunlight. Because the pressure at MLO is typically 680 hPa, or two-thirds the pressure at sea level, there is less oxygen over MLO than at sea level, and the difference in the reduced absorption by oxygen is easily measured using a simple spectrometer that connects to a laptop computer or a basic sun photometer that can fit in a coat pocket. Perhaps the Kiess-Corliss finding that there was no oxygen effect was due to the inability of the photographic emulsion used with their spectrograph to adequately respond to the oxygen absorption bands.

Taken together, these measurements illustrate the difference in results obtained using the old photographic method of spectroscopy and the ultrasensitive electronic detectors employed by Stair. In spite of possible instrumental shortcomings, the Kiess-Corliss paper, the first to emerge from the fledgling observatory, was a remarkably significant publication that proved the value of MLO as a highly suitable site for a major study of Mars long before astronomical observatories were installed atop the neighboring Mauna Kea. The Kiess-Corliss team also demonstrated that four investigators, who also happened to be a pair of married couples, could transport a complex experiment from Washington, DC, to Hawai'i and then perform highly precise observations during freezing and windy conditions while living and dining at MLO. In fact,

> The couples seemed to have had a grand time and in particular they tried out all kinds of canned gourmet food they had purchased since the National Geographic Society footed the bill. (Ellis 1988)

Harold Yates Measures Water Vapor (July 1956 and July–September 1957)

Harold W. Yates and Jack H. Taylor of the Office of Naval Research signed the MLO *Visitors* book on July 14, 1956, only two weeks after the dedication of the slope observatory. Their plan was to measure atmospheric transmission in the infrared and the water vapor, which absorbs infrared, between Mauna Kea and MLO. According to Ellis (1988), it would have been impractical for Yates and Taylor to install their equipment inside the main building at an open window facing Mauna Kea, for this would have made the building quite chilly for the Kiess-Corliss couples, who were still staying there. Therefore, Yates improvised an alternative site for his instrument.

> The platform tower had been built to the east of the building so that it was possible to have a free flow of air beneath it. Dr. Yates, in need to get his study started, enclosed the space beneath the tower with a roof, floor and sides. He then had his view of Mauna Kea through a square cut in the north side of this structure that he had added. (Ellis 1988)

While full details of the July 1956 visit could not be found, some details about a subsequent visit during 1957 have provided information about the project (Yates 1957; Price and Pales 1959). Yates and a colleague, whose first name was Joseph but whose last name is illegible, signed the *Visitors* book on August 9, 1957. The light source for this study was a pair of carbon arc searchlights 5 feet (1.5 meters) in diameter installed at about 10,000 feet (3,048 meters) on the south slope of Mauna Kea, 17 miles (27 km) from the new slope unit of the Mauna Loa Observatory. As noted previously, water vapor efficiently absorbs certain infrared wavelengths, and an infrared spectrometer installed at an opening cut in the newly enclosed room under the tower at MLO permitted the water vapor between the two mountains to be measured. After Yates' experiment was completed, the enclosed space under the tower was used as a work space. Today the same space is used for storage.

Meteorological Parameters (December 1956)

The first regular observations by the Weather Bureau at MLO were of temperature, relative humidity, and pressure, the same suite of meteorological parameters first measured at the summit by the US Exploring Expedition during 1840–1841 and automatically at the Summit Observatory from 1952 to 1954. Wind conditions that were estimated by the US Exploring Expedition were recorded automatically at both the summit and the new slope observatory.

The *Visitors* book records that Dominic Conte of the Weather Bureau in Honolulu, who was present for the dedication of MLO on June 28, 1956, returned for "Installation of equipment" from December 4 to 8. The meteorological measurements at MLO were viewed as routine and not part of the IGY program (Price and Pales 1959). Nevertheless, they were important for understanding the mountain's circulation that had first been studied in detail by Henry Guppy in 1897 (Guppy 1906) and that directly affected the IGY surface ozone and carbon dioxide projects.

Harry Wexler and the Snowflake People (1956–1957)

In September 1955, Dr. Ukichiro Nakaya of Hokkaido University in Sapporo, Japan, visited the Weather Bureau office in Honolulu to make plans for a study of snow at the summit of Mauna Loa to be financed by the US Geophysical Research Directorate. Walter Steiger (personal interview, Hilo, Hawai'i, December 4, 2007) recalled the visit and how Nakaya and a colleague became known as the "snowflake people." Howard Ellis (1988) provides more details in his history of MLO:

> He became enthusiastic about living in the summit observatory building with two others for the entire winter. When Mr. Busniewski [of the Hilo Weather Bureau] heard of this, he pointed out that the 8' x 10' shack there was not at all adequate. He outlined the array of arrangements that needed to be made before a project like this could be undertaken. He proposed the use of the larger more sturdy Park Service cabin on the east-southeast side of the crater.

The snow study was postponed until the opening of the MLO slope unit, and the *Visitors* book provides its dates. Ukichiro Nakaya and Mikio Shoda of the Snow Experiment Station of Japan's Railway Technical Institute were at MLO from December 7, 1956 to January 21, 1957. According to Ellis (1988),

> They made an extensive study of condensation nuclei and snow crystals at the summit of Mauna Loa. Many physical [tasks] that were thought by many, before and after, to be impossible were done by these hard working happy scientists. They lived that winter in the new structure at 11,150 feet, but walked, and some say ran to the summit as frequently as days on end. Dr. Wexler visited during their stay and worked some with them. An illustrated report of this work became contribution No. 1, of the Hawaii Institute of Geophysics of the University.

Dr. Harry Wexler's visit to MLO was likely his first. His signature in the *Visitors* book was dated December 30, 1956, and was followed by "En route to Antarctica." As demonstrated by Robert Simpson, Ralph Stair, Saul Price, and a long string of other scientists, the big mountain had a special attraction for those who had visited and, especially, done science at MLO. This would not be the last time Wexler would pay a visit to his favorite atmospheric science observatory.

Walter Steiger at MLO (1956–1957)

Dr. Walter R. Steiger, the University of Hawaii physicist who spoke at the dedication of the slope unit of the Mauna Loa Observatory on June 28, 1956, helped pioneer astronomy in Hawai'i. Since 1953 he had as a goal the establishment of a solar observatory on one of the mountains of Hawai'i. A solar observatory was eventually constructed at MLO in 1965, and, in his history of astronomy in Hawai'i, Steiger (2001) describes why he didn't select Mauna Loa a decade earlier:

> There were three possible sites: Mauna Loa (13,680 ft.), and Mauna Kea (13,784 ft.), both on the island of Hawai'i, and Haleakala (10,025 ft.) on the island of Maui. Mauna Loa was an active volcano with very difficult access and was deemed unsuitable. Mauna Kea, though volcanic in origin, as is all of Hawai'i, was considered dormant or extinct. But like Mauna Loa, it was very remote and without vehicular access.

Later MLO was a possible candidate for a cosmic ray monitoring site for the International Geophysical Year, but the Makapu'u Point Solar Observatory on Oahu was selected due to its easier access.

Steiger was also involved in selecting an alpine site to track artificial Earth satellites planned to be launched during the IGY, and he and J. Allen Hynek of the Smithsonian Astrophysical Observatory visited MLO on June 13, 1957, along with Roy L. Fox, meteorologist in charge of the Pacific Supervisory Office of the Weather Bureau in Honolulu. Hynek signed the *Visitors* book and added "Chasing the Satellite." A reporter for the *Honolulu Star-Bulletin* and Norris W. Rakestraw of the Scripps Institution of Oceanography were also present. Rakestraw was apparently there to inspect the premises prior to the installation of the carbon dioxide monitoring apparatus in March of 1958. Howard Ellis (1988) describes why MLO was not selected for the new satellite tracking station:

> On October 19 the headline in the *Honolulu Star-Bulletin* read: "Island Station to Track Earth Satellite." The paper went on to say how the University of Hawaii was to direct the project on Mauna Loa. Dr. Steiger saw the news as embarrassingly premature and when he finished

> his study, his decision was that the weather conditions were good enough on Maui and that the logistics were so much easier there than on Mauna Loa.

Steiger was right, for at the time the road to the summit of Haleakalā on Maui was in much better condition than the road to MLO. In early 1957 Steiger finally became involved in an MLO science study, an intriguing effort to capture meteor dust described by Ellis (1988):

> During this academic year, an internationally known oceanographer from Sweden [Hans Pettersson] was visiting scientists at the University of Hawaii [as guest professor of geophysics]. His interest was in tiny magnetic spherules of a unique iron-nickel alloy. These were thought to have originated from meteoric dust settling to earth. He and Dr. Steiger made several trips to Mauna Loa, thinking to collect dust samples there to help to solve the problem of the accretion rate.
>
> Dr. Steiger said that even with the cleanliness of Mauna Loa's air, there still was enough dust to mask the meteoric portion. These trips were made during the time when the station was unmanned and a locked chain at the 8,300 foot level supposedly secured the area. Dr. Steiger expressed the need for a mirror in the kitchen for shaving and an electric light in the outhouse.

During telephone (July 11, 2007) and personal (Hilo, Hawai'i, December 4, 2007) interviews, Dr. Steiger recalled the meteor dust project in exquisite detail. After he drove Pettersson to MLO in a rented Jeep, they were greeted by "huge flakes of snow" and very cold temperatures. Unfortunately, neither the generator's starter motor nor its hand crank could be made to get the generator running, so they drove back to Hilo and returned the next day with a fresh battery and some ether capsules that enabled them to start the generator. Pettersson signed the *Visitors* book on January 10–11 and 17–19 and both he and Steiger signed on February 4–6, 1957.

Pettersson's experiment collected cosmic dust by filtering large volumes of air through filters. The filters were analyzed for particles of meteoric origin and radioactive fallout (Pettersson 1958), with the latter observations anticipating a major program conducted during the 1960s at MLO by the US Atomic Energy Commission. Steiger recalls that cosmic particles were also collected by means of a magnet placed in a rainwater downspout of the building.

Harry Wexler's Carbon Dioxide Plan for MLO (1956)

While Robert Simpson's goal for the Mauna Loa Observatory was the measurement of the ozone layer during the International Geophysical Year, the Weather Bureau's Harry Wexler was more interested in the impact of atmospheric aerosols and infrared-absorbing gases on Earth's radiation balance (Wexler 1953). He had advocated the measurement of carbon dioxide on Mauna Loa in 1955 when Robert Simpson first proposed to provide the funds for Ralph Stair to build the main building. Plans were already in place to make daily measurements of the ozone layer at various stations in Europe and the United States during the IGY, but no one was making regular measurements of the concentration of carbon dioxide (CO_2). Moreover, the published measurements of CO_2 at the time were often inconsistent.

Because the concentration of CO_2 is so small—slightly less than four molecules for every thousand molecules of air—it is considered a trace gas. Yet CO_2 is a key precursor ingredient for the construction of plants and fungi. Plants extract the carbon from the gas to build their stems, roots, fibers, leaves, flowers, and seeds. The carbon in food products and in every sheet of paper, block of wood, and strand of cotton is recycled from the carbon dioxide taken into the leaves of plants through countless numbers of tiny pores called stomata. Too small to be seen through a hand lens, stomata can be observed through a microscope at a magnification of one hundred times. When a fresh dandelion leaf is illuminated from above and viewed through a microscope, the stomata appear as tiny ovals sprinkled across the surface of the leaf. They can be seen both in living leaves still attached to a plant and in well-preserved leaves encapsulated in ancient amber. Through these tiny apertures pass the carbon that builds the plant. In short, carbon dioxide is essential to plants. But can there be too much of a good thing?

The Greenhouse Effect

Carbon dioxide, like water vapor, strongly absorbs infrared radiation. During the day, sunlight warms exposed soil, rocks, vegetation, and water, which then emit some of the energy that has been absorbed as infrared radiation, which is perceived as heat. Water vapor, carbon dioxide, methane, and other trace gases in the atmosphere absorb specific bands of infrared radiation, thus retaining some of the heat and causing a net warming of the planet.

Calculations made well over a century ago showed that the oceans would be frozen were it not for the warming effect of the atmosphere. John Tyndall (1863), who discovered that water vapor and carbon dioxide absorb infrared radiation, summed it up best when he wrote that water vapor

> is a blanket more necessary to the vegetable life of England than clothing is to man. Remove for a single summer-night the aqueous vapour from the air . . . and the sun would rise upon an island held fast in the iron grip of frost.

The warming effect of Tyndall's blanket of aqueous vapor came to be compared with how air trapped in a greenhouse becomes heated by sunlight because the glass absorbs the infrared emitted by the materials heated within. Thus, infrared-absorbing gases such as water vapor, carbon dioxide, and methane are popularly known as *greenhouse gases*. The analogy is flawed by the fact that a greenhouse covered with infrared-transmitting polyethylene film performs about as well as one in which the glazing is infrared-absorbing glass (Bowling 1987). Indeed, commercial growers in the United States prefer polyethylene over glass. The heating effect of a greenhouse is due primarily to the absence of convection; that is, the absence of the exchange of hot air inside the structure with cooler outside air, not the infrared absorption of its glazing. The net result has been succinctly summarized by the Intergovernmental Panel on Climate Change (IPCC 2007):

> The glass walls in a greenhouse reduce airflow and increase the temperature of the air inside. Analogously, but through a different physical process, the Earth's greenhouse effect warms the surface of the planet.

While the greenhouse analogy is flawed, *greenhouse effect* and *greenhouses gases* have become widely adopted phrases that describe the warming effect of infrared-absorbing gases. In these pages, these phrases are used with the understanding that they simply describe the ability of infrared-absorbing gases to reemit the infrared they absorb, thereby warming the atmosphere. The effect is routinely demonstrated by water vapor, the most potent greenhouse gas, for even the casual observer knows that the outdoor temperature remains warm overnight when the air is humid but becomes cool when the air is dry. The importance of the natural greenhouse effect to life on Earth has been concisely summarized as follows:

> Without a natural greenhouse effect, the temperature of the Earth would be about zero degrees F (–8°C) instead of its present 57°F (14°C). So, the concern is not with the fact that we have a greenhouse effect, but whether human activities are leading to an enhancement of the greenhouse effect. (NOAA 2007)

Human activity can add small quantities of water vapor to the atmosphere, as when large agricultural regions are artificially watered by means of irrigation. But the human contribution to the atmosphere's vast pool of water vapor is dwarfed by natural processes like ocean evaporation. The addition to the natural background of additional carbon dioxide from the combustion of fossil fuels, however, is more significant, for the lifetime of a molecule of CO_2 in the atmosphere is decades or centuries while that of a molecule of water vapor is only days or weeks.

The Carbon Conundrum

The fact that human activity has added carbon dioxide to the atmosphere is well established. Spencer Weart (2003) has provided a detailed history of the subject in *The Discovery of Global Warming.* Many others have also covered the topic. For example, more than a century ago Arvid Högbom (1894) proposed that factories were adding CO_2 to the atmosphere. While water vapor is the gas most responsible for warming the atmosphere, Svante Arrhenius (1896) calculated that cutting the concentration of CO_2 in the atmosphere in half would lead to ice age temperatures across Europe. He also calculated that doubling the CO_2 would raise the temperature of Earth by 9 to 11° F (5 to 6° C).

Guy Stewart Callendar (1938) calculated that doubling CO_2 would raise global temperature by 3.6° F (2° C), an increase that he believed would be beneficial for those who inhabit northern climes. He succinctly summarized his hypothesis in a 1939 paper:

> From the best laboratory observations, it appears that the principal result of increasing atmospheric carbon dioxide would be a gradual increase in the mean temperature of the colder regions of the earth.

Because he was an engineer, Callendar's carbon dioxide papers were not taken seriously by some of his professional contemporaries. Interest in the potential warming impact of CO_2 waned after water vapor was found to completely dominate the greenhouse effect, and an official US government publication explained why:

> The theory received a fatal blow when it was realized that carbon dioxide is very selective as to the wavelengths of radiant energy it will absorb, filtering out only such waves an even very minute quantities of water vapor dispose of anyway. (Russell 1941, 94)

This became the official view of the American Meteorological Society, which also dismissed the role of CO_2 in climate change in its 1951 *Compendium of Meteorology:*

> The [greenhouse] theory of climate change was never widely accepted and was abandoned when it was found that all the long-wave radiation absorbed by CO_2 is also absorbed by water vapor. . . . This theory is not considered further. (Brooks 1951)

Callendar became a contrarian as he persisted with his hypothesis and protested that the methods then available for measuring the infrared absorption of carbon dioxide were inadequate. New analytical methods eventually supported his contention, and during the mid-1950s there was renewed interest in the potential influence of carbon dioxide on climate. The year that the slope unit of the Mauna Loa Observatory was dedicated, Gilbert Plass (1956) wrote that

> The results of more accurate calculations of the radiation flux have recently become available. Thus it seems worthwhile to reappraise the CO_2 theory of climatic change. . . . The present evidence suggests that variations in the atmospheric CO_2 concentration are one of the most important reasons for changes in the climate.

The carbon dioxide pendulum had begun to swing back to the warming side, and Callendar's writings eventually came to be viewed as among the most important in establishing the hypothesis that the addition of carbon dioxide to the atmosphere by human activity could cause an increase in Earth's temperature. Scientists began to ask if plants and oceans could soak up the excess carbon dioxide released by the burning of fossil fuels to produce electricity and to power cars, trucks, trains, and airplanes. They also wondered if there could be unintended consequences from the transfer of carbon to the atmosphere and oceans at a much more rapid pace than in the past.

Roger Revelle and Hans Suess (1957) of the Scripps Institution of Oceanography wrote about the potential role of carbon dioxide in climate change and made a significant proposal:

> Human beings are now carrying out a large scale geophysical experiment of a kind that could not have happened in the past nor be reproduced in the future. . . . An

> opportunity exists during the International Geophysical Year to obtain much of the necessary information.

Revelle's influence would soon enable him and Scripps to play an active role in fulfilling Harry Wexler's goal for a CO_2 monitoring program at the Mauna Loa Observatory. Neither Wexler nor Revelle could have had any idea that the CO_2 findings to come would eventually trigger a contentious scientific debate, scary predictions of impending climatic disaster on a global scale, and a vast array of research programs whose collective budgets would dwarf the funds required to maintain MLO and its tiny staff. Nor could they have anticipated that the measurements planned for MLO would twice play a key role in saving the observatory from being closed.

These and related developments will be covered in chapters to come. But first we must meet a determined young scientist who would not allow institutional and governmental bureaucracies to interfere with his goal of tracking carbon dioxide better than anyone else to answer the question uppermost on his mind: Was the concentration of carbon dioxide in the atmosphere really increasing?

Charles David Keeling (1956)

Charles David Keeling (1998) developed an interest in science at the age of five when his father demonstrated how the earth revolves around the sun. Known as Dave by his family and friends, Keeling had planned to study chemistry at the University of Illinois, but he switched to general liberal arts and earned a bachelor's degree in 1948. He returned to chemistry at Northwestern University and in 1953 earned a doctorate following completion of a dissertation on polymer chemistry. While at Northwestern, Keeling acquired a serious interest in geology, which he actively pursued through course work. He then merged his two specialties by accepting a fellowship under Professor Harrison Brown of the Geochemistry Department at the California Institute of Technology (Caltech) in Pasadena, California.

At Caltech, Dr. Keeling began studying the concentration of CO_2 in air and in bodies of water under an Atomic Energy Commission grant. He enjoyed the opportunity to do science in the scenic outdoor settings along the California coast, where he collected his air samples. During this work he found that the CO_2 in air was relatively constant in remote areas, a finding that differed with earlier measurements in Europe that showed much more variability. Keeling was not surprised by their findings, for he had found high variability in the measurements he made in heavily polluted Pasadena. That's why he sought out pristine air uncontaminated by automobile engines and power plants. In his memoir, Keeling (1998) describes the significance of this early research:

> Why did I devise such an elaborate sampling strategy when my experiment didn't really require it? The reason was simply that I was having fun. I liked designing and assembling equipment. I didn't feel under any pressure to produce a final result in a short time. It didn't occur to me that my activities and progress might soon have to be justified to the sponsoring Atomic Energy Commission. At the age of 27, the prospect of spending more time at Big Sur State Park to take suites of air and water samples instead of just a few didn't seem objectionable, even if I had to get out of a sleeping bag several times in the night. I saw myself carving out a new career in geochemistry. I did not anticipate that the procedures established in this first experiment would be the basis for much of the research that I would pursue over the next forty-odd years.

Keeling's first CO_2 samples were taken during his Caltech postdoc in 1955 from areas remote from human influences, in the western United States and off the coast of Central America. There he found CO_2 concentrations that, "in contrast to previous studies," were nearly constant in well-mixed air on sunny days (Keeling 1978, 37).

Harry Wexler had anticipated Keeling's concern about the need to accurately measure the CO_2 concentration at a pristine site in the free atmosphere well above and away from sources of pollution. Both men were well acquainted with the fact that CO_2 is a greenhouse gas that might someday have an impact on climate if its concentration was rising as a result of the combustion of fossil fuels. At issue was the best way to measure the gas.

The concentration of carbon dioxide in air can be measured by chemical and optical methods. The chemical methods of the time suffered from reliability and accuracy limitations until Keeling developed an apparently novel approach using liquid nitrogen extraction. This highly accurate method was to become Keeling's key to calibrating his future measurements of CO_2 using an optical method that could be deployed in the field (Drew Keeling, personal communication, July 31, 2008). But the optical method was new and untested outside the laboratory. The method relied on the fact that CO_2, like water vapor, strongly absorbs specific bands of infrared radiation, the characteristic that makes it a greenhouse gas. By measuring changes in the transmission of infrared through a sample of air from which the water vapor has been removed, the concentration of carbon dioxide in the air can be measured with high accuracy, at least in a laboratory environment.

Robert Simpson's plan to measure the ozone layer over the Mauna Loa Observatory during the International Geophysical Year was much more of a sure thing than Wexler's goal to measure the concentration of carbon dioxide. The ozone layer could be measured using a Dobson spectrophotometer, a time-tested, reliable instrument with a well-established protocol. Plans were already in place to use Dobsons for daily measurements of the ozone layer during the IGY at various stations in Europe and the United States. But there was then no tested and proven method for accurately measuring CO_2 in the field. No one was making regular, high-quality measurements of the concentration of the gas, and the published measurements of the time were sometimes contradictory. When Wexler learned about Keeling's CO_2 work, he invited him to fly to Washington in early 1956. It was Keeling's first airplane trip, and he described what happened in a memoir written expressly for the 20th anniversary of the MLO:

> I talked to [Wexler] about the possibility of setting up a continuous recording CO_2 gas analyzer on Mauna Loa since it would be possible to live there and tend the analyzer as necessary. As far as I knew, no one had ever before suggested measuring atmospheric CO_2 continuously. Wexler asked a number of questions in rapid-fire, covering both the scientific and the practical. He was especially interested in costs. We went so far as to discuss setting up a second continuous CO_2 analyzer in Antarctica. Then the interview was over. Altogether it took almost exactly 15 minutes, as scheduled. Wexler had made up his mind to press for CO_2 measurements at Mauna Loa using monies which he hoped would be made available by the participation of the United States in the IGY. (Keeling 1978)

Many years later, Keeling reviewed his notes and found that the meeting had actually lasted about an hour. More important than the length of the meeting was that Wexler offered him a position at the Weather Bureau in Washington to establish a carbon dioxide measurement program. Meanwhile, Roger Revelle, director of the Scripps Institution of Oceanography in San Diego, California, was also keenly interested in carbon dioxide. During a pleasant luncheon at Scripps, Revelle offered Keeling a position and a small office. In his memoir, Keeling (1998) notes, "After briefly considering the alternative of a basement room in Washington, I agreed to come." Wexler endorsed this arrangement. He offered to help fund it, and Revelle established a new lab at Scripps for the study of CO_2, which Keeling was invited to direct. Since Scripps promised much more than a small office space, Keeling accepted the position with the approval of Wexler, who worked closely with Keeling and Scripps to establish a CO_2 monitoring program for MLO. As Keeling (1978, 39) observed twenty-two years later,

> Wexler's support of continuous measurements of atmospheric CO_2 at MLO was a bold decision not widely accepted at the time. Wexler knew that I had located a manufacturer of nondispersive infrared CO_2 gas analyzers, but he also knew that I had not yet been able to test such an analyzer. Even the firm itself did not claim that its infrared analyzer was accurate enough for the task. It had been designed principally for industrial uses which did not demand high accuracy. I was relying on the judgment of one of the firm's engineers that the device was

> inherently very sensitive and stable. The firm couldn't even lend me one to test. The basic instrument was expensive and required costly additional equipment to operate as an air monitor at a remote field station. Reference gases to calibrate the instrument did not exist.

The Model 70 Infrared Analyzer (1956)

The expensive, untested infrared gas analyzer method (Smith 1953) that Keeling wanted to use at MLO to monitor carbon dioxide was greeted with skepticism by some of the IGY planners. Had they read the instruction manual for the analyzer eventually selected by Keeling, an original copy of which remains at the Mauna Loa Observatory, they might have been even more skeptical, for on page 1, the manual described the typical application for the apparatus:

> The Model 70 analyzer has been successfully applied to a number of analytical problems typical of those arising in *petroleum refinery operation* [emphasis added]. These problems have involved mixtures of C1 through C5 hydrocarbons, both saturated and unsaturated, with as many as four absorbing components present in the sample in important amounts. . . . The list of applicable materials [that can be detected] includes water vapor, ammonia, carbon monoxide dioxide [*sic*] cyanogens and hydrogen cyanide, hydrogen chloride and bromide, stibine, and arsine. (APC 1954)

The one-of-a-kind spectrometers used by Kiess and Stair in their early measurements at MLO of Mars, Jupiter, and sunlight employed expensive prisms or diffraction gratings to disperse light into its spectrum of wavelengths. Keeling selected a simpler, nondispersive infrared analyzer that did not use prisms or gratings, the Applied Physics Corporation (APC) Model 70, which was manufactured in Pasadena near Caltech. The company was not sure the instrument would fulfill Keeling's requirement, and APC engineer Kenyon George admitted as much in response to Keeling's many questions. Keeling (1978) later expressed his appreciation for George's candor:

> I am indebted to Kenyon George, engineer of the Applied Physics Corporation, Pasadena, California. George patiently replied to my detailed questions during 1956 about the performance characteristics of his firm's nondispersive infrared gas analyzer. He was not himself convinced that atmospheric CO_2 could be determined by infrared analysis to the accuracy I sought, but his frank answers and total lack of bias provided sound arguments in favor of trying out the infrared method during the IGY.

Notwithstanding these limitations and the fact that the continuous measurement of the concentration of carbon dioxide in a stream of air was not yet a routine laboratory analysis—much less one that could be easily implemented high on a remote mountain in Hawai'i and at the South Pole—both Wexler and Revelle were impressed by Keeling's qualifications. Wexler dipped into his IGY budget and paid for four infrared analyzers and their associated equipment.

Because the Model 70 infrared analyzer that Wexler bought for MLO would be used for nearly half a century and provide the best-known set of data from the observatory, it's of interest to explore how it worked. A simple explanation was provided by Pales and Keeling (1965) in their first paper on progress at MLO:

> The radiation, emitted by a glowing nichrome filament, is mechanically chopped at 20 cps [cycles per second] and then directed through the gas sample into a detector cell which is permanently filled with CO_2 diluted with argon [see below]. Absorption of radiation by the CO_2 gas in the detector cell produces a cyclic pulsation in pressure which is transmitted to a tantalum diaphragm of a condenser microphone and thence converted to an alternating voltage, amplified and recorded.

The instrument included two of the systems described by Pales and Keeling. Outdoor air flowed through the first (the analyzer path), and calibration gas flowed through the second (the reference path). The mechanical and electronic components of the analyzer were elegantly designed, carefully crafted, and installed within a sturdy metal cabinet the size of a kitchen oven. Dave Moss (2008), who served as an engineer for the Scripps carbon dioxide program for more than thirty years, best described the quality of the instrument's design and construction when he said that it was a product of the "watchmaker generation." For all intents and purposes, the instrument appeared to be self-contained and ready to use. But this was far from the case, for the analyzer itself was quite complex, tricky to maintain, and only the key part of a larger system.

The pair of internal detector cells was the cleverest part of the analyzer. These chambers were based on the photoacoustic principle discovered by Alexander Graham Bell during his pioneering experiments in lightwave communications in Washington, DC, in the 1880s. Each stainless steel cell included an enclosed microphone, and both cells were filled with a mixture of CO_2 and argon (Pales and Keeling 1956) through an attached valve assembly. Infrared radiation that passed through the reference or analyzer cell entered the detector cell through an infrared-transmitting window made from rock salt. A threaded capsule filled with ascarite (granular asbestos) removed any stray water vapor and carbon dioxide from the detector cell (APC 1954).

The detector cells were designed for long-term use, and the CO_2 within them made them self-tuned to respond best to the same wavelengths of infrared that were absorbed by CO_2 in the sampled air. However, there is overlap in the absorption of some bands of infrared by CO_2, water vapor, methane, and carbon monoxide, an issue addressed by the analyzer's manual:

> In general, the Model 70 analyzer will detect traces of an absorbing gas of the order of one part per million in the presence of the non-absorbing gases. When two absorbing gases are present, it is theoretically possible to make the instrument to respond to the concentration of the selected gas and ignore the interfering gas over a wide range of composition. In practice, perfect discrimination against interference cannot always be obtained. (APC 1954)

Water vapor was by far the most important interfering gas, and it was imperative that it all be removed. This was easily achieved by passing the sampled air through a *freeze trap,* a device that chills the air to well below the freezing point of water so that any water vapor in the air is left behind as ice.

The freeze trap was only one of the extras required by the carbon dioxide analyzer. Also required were four air sampling towers, hundreds of meters of aluminum air sample lines, air pumps, regulator valves, equipment racks, cylinders of reference gas with a precisely known concentration of CO_2, purge gas, a chart recorder, chart paper, ink, stable electrical power, and a temperature-regulated room. All this required the careful attention of skilled technicians equipped with plenty of spare vacuum tubes, suitable tools, test equipment, manuals, circuit diagrams, a log book to document repairs, meals, and sufficient dedication to keep the apparatus properly running twenty-four hours a day while spending six days and nights a week at the end of an unpaved lava road on a desolate mountain 11,200 feet (3,414 meters) above the Pacific Ocean.

The Applied Physics Corporation Model 70 Infrared Analyzer cost $6,000 in 1956 dollars (Keeling 1998), which is $47,583 in 2010 dollars (Friedman 2011). This was a considerable investment for an instrument intended to serve only one purpose. By way of comparison, consider a modern instrument such as the Li-COR 7000 that NOAA uses today at MLO to measure CO_2. Both instruments detect carbon dioxide using similar methods. The modern instrument cost $15,500 in 2008 dollars, which is only a third of the inflation-adjusted price of the Model 70. Yet the Li-COR is far more reliable and much more compact, it stores data digitally, and it includes a digital readout. (Not included in these costs are the reference gas cylinder, power supply, sampling pump, and air supply lines, all of which are required for both instruments.)

Of course, this comparison is unfair, for many kinds of modern scientific instruments cost less and work better than their predecessors; it is included here because no one then knew that the APC Model 70 analyzer would remain in operation for more than four decades after the IGY. The Model 70 was an expensive, temperature-sensitive, temperamental vacuum tube device housed in a large and heavy metal cabinet. But it was the state of the art in 1957, and it was to remain a key part of Keeling's program for the remainder of his career.

Dave Keeling made plans to move to Scripps, where he arrived in the summer of 1956 to begin the US IGY CO_2 program. The job involved much more than the proposed MLO program, for Scripps was planning to measure CO_2 in Antarctica and to collect air samples from aircraft and at ground sites. Keeling soon found himself in a tiny office immersed in paperwork and correspondence. The IGY began in July 1957, and the Weather Bureau people in Washington and Hawai'i began to wonder about the delay in beginning the CO_2 program at MLO. The traditional view among some Mauna Loa Observatory old-timers was that Keeling was responsible for the delay. But the reason for the delay was more complex, and Keeling's son Drew (in a personal communication with author, July 31, 2008) summarized the key issues:

> In his article for the 20th anniversary of the MLO observatory, Keeling attributed the delay to Revelle's insistence that aircraft and shipboard sampling take priority. Underlying this prioritization was the lingering concern that CO_2 concentrations might naturally vary from place to place, and widespread sampling during the IGY would be needed to establish a "benchmark" global average that might be updated every 10 or 20 years. Widespread sampling on ships and aircraft during the IGY, together with the establishment of measuring stations at Mauna Loa and the South Pole, amounted to a considerable burden of tasks to accomplish in a short period of time. Correspondence in the Scripps archives provides considerable further detail and suggests a still more complex version of events during the IGY.

Ralph Stair Validates MLO (May–June 1957)

Back at the Mauna Loa Observatory, on April 30, 1957, National Bureau of Standards' scientists Carl and Harriet Kiess and Charles H. Corliss returned with the spectrometer they had used to search for water vapor in the Martian atmosphere during June of the previous year (Kiess et al. 1957). As during their previous visit, they were sponsored by the National Geographic Society. They were accompanied by Ralph Stair and Russell Johnston, who arrived at MLO with the spectroradiometer shown in Figure 6.1, the same one they had used to measure the solar constant at Sunspot, New Mexico, in June 1955, when they were visited by Robert Simpson.

For the first time, MLO was being used around the clock by two teams of scientists, both of which were equipped with sophisticated spectroscopy equipment. Stair and Johnston did their observing during daylight hours, while Kiess and his team did their observing at night. Kiess and his team photographed the spectra of Jupiter in a search for the gases responsible for the colored bands around the giant planet. They soon found evidence for a layer of nitrogen tetroxide that "may be responsible for the variable color patterns" in the Jovian atmosphere (Kiess et al. 1960). They completed their observations in good time and departed MLO on May 13, Kiess signing out with "Good results" in the *Visitors* book. Stair and Johnston stayed on to continue their solar constant measurements, which included simultaneous measurements of the column abundance of water vapor and the first measurements of the ozone layer over MLO (Stair and Johnston 1958). The water vapor and ozone measurements anticipated the future, for MLO would soon begin measuring both the ozone layer and column water vapor on a regular basis.

As discussed in chapter 4, dust, ozone, and water vapor interfere with measurements of the solar constant from Earth's surface. Today Asian dust is often present at MLO during April and May (as during extended stays at MLO by the author during preparation of this book). It is significant that there was no mention of dust in the papers that the two teams published about their findings.

Chapter 5 briefly explained how the ozone layer is measured from the surface by pointing an ultraviolet measuring instrument at the sun. Two or more thin slices of the sun's ultraviolet wavelengths are measured, the ratio of which is related to the amount of ozone between the instrument and the top of the atmosphere. The total ozone

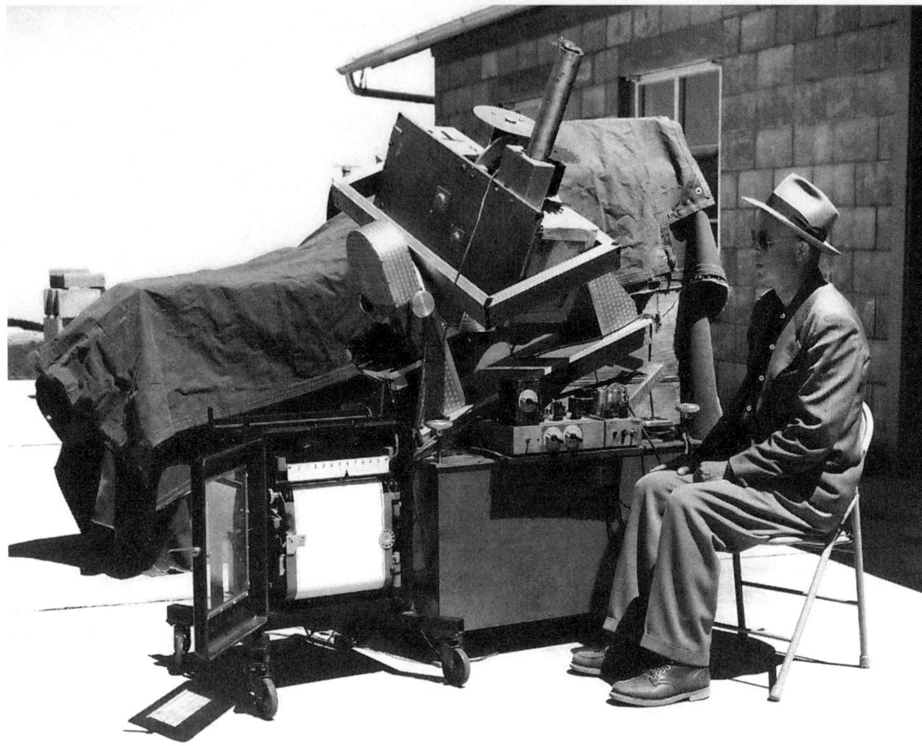

FIGURE 6.1. Ralph Stair and his solar constant spectrometer on the concrete pad in front of the main building during May 1957. This is the same instrumentation that Stair used at Sunspot, New Mexico, during the visit by Robert Simpson in June 1955. *(MLO archive.)*

in a column between the instrument and the sun is then computed by subtracting the measured ratio of the two wavelengths at the surface from that which would be found at the top of the atmosphere and by invoking the absorption coefficients for ozone at the measured wavelengths of UV. Finally, the data are adjusted to represent the ozone in a vertical column overhead by correcting for the angle of the sun.

Stair and Johnston found that the 1913 ozone absorption coefficients of Fabry and Buisson gave an ozone amount over MLO of about 240 Dobson units, while the 1953 coefficients of Vigroux gave an ozone amount of 320 Dobson units, an unacceptably large difference.

> Since the calculated value of atmospheric ozone depends upon the laboratory measurements of the absorption coefficients of ozone, and since there is considerable spread between the values obtained through using the two sets of data, it seems advisable to report both values until further experiment definitely determines the absorption coefficients of ozone. (Stair and Johnston 1958)

Based on the high quality of measurements by Stair and Johnston, it is likely that their ozone measurements could be brought up to date by applying today's refined ozone absorption coefficients. Stair and Johnston's measurements of total column water vapor over MLO were at least as important and much more certain than their ozone measurements. As noted earlier, the measurements of column water vapor by Kiess and Corliss were much too low. But those by Stair and Johnston were well within the range of measurements by modern instruments at MLO, and this deserves elaboration.

Stair and Johnston converted their measurements of near-infrared into the column water vapor over MLO by using the time-tested absorption coefficients experimentally determined by Fowle (1915) forty years earlier. Figures 8–11 of their report (Stair and Johnston 1958) show the column water vapor at frequent intervals over four days (June 7, 8, 12, and 13). The minima are of particular interest, for they better represent the free atmosphere over the observatory than when moist air arrives as the inversion layer rises toward the observatory. The column water vapor minima during the four days that can be estimated from the plots in their paper were about 1.1, 1.2, 1.5, and 1.8 mm. While these amounts were significantly higher than the minima at MLO for June 2006, they are slightly below the minima of June 2007 (NOAA 2006, 2007). Thus, in light of the variability of the column water vapor over MLO and the care with which Stair measured the near infrared spectrum, his results were reasonable.

A group of seven Hilo residents visited MLO on June 9, including Angie Nobriga, Stair's future wife. She wrote in the *Visitors* book, "I dearly love the observatory and the people in it are wonderful." Stair and Johnston completed their measurements on June 20, 1957, and signed the *Visitors* book before departing. Johnston wrote, "Big, barren but beautiful. We 'love' it and will miss our many friends." Stair left behind various tangible improvements, including technical books and a rare set of hardbound annals of the Smithsonian Institution that included papers detailing solar constant research by the Astrophysical Observatory. These annals, which can be seen in various photographs made inside the main building, are now in the Mauna Loa Observatory museum in the observatory's office in Hilo.

While the paper by Stair and Johnston was not the first to come from MLO, it is the most significant of the earliest papers, for it describes the first measurements from the station of the solar constant, the ozone layer, and columnar water vapor. The paper validated MLO, and it was fitting that Stair and Johnston used the same spectroradiometer that they employed to measure the solar constant from Sunspot, New Mexico, in June 1955 when Robert Simpson stopped by during his vacation and suggested the measurements might be better made from Mauna Loa.

The Weather Bureau Prepares for the IGY (1957)

The initial studies at MLO by teams of visiting scientists were completed in only a matter of weeks. The Weather Bureau planned continuous observations during the International Geophysical Year, and this required considerable planning and a formal budget. Dr. Harry Wexler, the research director of the Weather Bureau, understood the potential of MLO. He had written about the climatic effects of major volcanic eruptions (1952), the sun's effects on the atmosphere (1956), and the impact of the Earth's radiation balance on "climatic change" (1953). The radiation balance is the ratio of the energy in sunlight absorbed by the Earth and that emitted by the Earth as infrared. Aerosols in the atmosphere block or scatter sunlight, which can generally cause cooling, while increases in the gases that absorb the infrared radiation emitted by the Earth can cause warming. Depending on their physical nature and the time of day, clouds can cause warming or cooling. Wexler knew that Mauna Loa was an ideal site for studying and quantifying the radiation balance, especially as it is influenced by the free atmosphere.

Wexler and Roy L. Fox coordinated the IGY planning for MLO. Their plans depended on the availability of an enthusiastic staff willing to spend six-day weeks working, cooking, and sleeping on the mountain.

Wexler and Fox would also need a bright Weather Bureau scientist to coordinate the science to be conducted at MLO. A year and a half before Robert Simpson returned to the mainland, he had persuaded Wexler to allow his friend Saul Price to move to Hawai'i (Simpson, personal communication with author, April 23, 2008), where Price worked under Simpson as research meteorologist at the Weather Bureau's Pacific Supervisory Office (PSO) in Honolulu, the central office for all of the Weather Bureau's stations scattered around the Pacific. Price, who had spoken at the dedication of the Mauna Loa Summit Observatory in 1951, was assigned the task of coordinating the science to be performed at MLO. His leadership was to set the scientific standard for the many hundreds of scientists who would eventually produce research reports and papers based on their research at the observatory. Walter

Steiger (personal interview with author, Hilo, Hawai'i, December 4, 2007) came to know Price from their joint participation in IGY planning meetings and reflected back on their acquaintance of half a century ago:

> Saul Price was a gem. He was a well spoken, very knowledgeable person. I always admired him.

Saul Price's role at the MLO slope unit can best be understood by returning briefly to the days of the original Summit Observatory that was closed in 1954. James W. Steiner, who had directed the Hilo Weather Bureau since 1949, also directed the Mauna Loa Summit Observatory. By 1954 the road to the summit had deteriorated to a point that it was only a matter of time before observations at the summit would end. Meanwhile, the US Air Force asked Roy Fox if the Weather Bureau could assist in an air sampling project on Mauna Loa. Howard Ellis (1988) relates what happened next:

> Mr. Fox came up with a seven-page letter evaluating the proposal.
>
> He presented suggestions for worthwhile additional observations to be made simultaneously "on our own." He felt that the job would be difficult but should be tackled, stating: "that headaches usually go away in time, and if the anticipated gains are continually kept in mind, annoyances and inconveniences are more easily coped with."
>
> One man was to be left at the 9,000 foot level while two others journeyed onto the summit each day that the air sample was to be taken. Mr. Fox made some cautions about providing a shelter for the man left alone, as well as a Coleman burner for heating food and that radios should be used. Costs for setting up the project were estimated at $18,000.
>
> He recommended that Mr. Raymond G. Busniewski, of the Hilo staff, be in charge with two assistants hired locally and that Mr. Busniewski's slot on the Hilo staff be filled with a man detailed from elsewhere within the WB [Weather Bureau].
>
> One compressor was to be operated at 8,000 feet and another at 11,150 feet, filling six liter steel cylinder tanks with air. The Navy diverted to this project 240 empty tanks that had held helium used in their weather balloon work in the Pacific in exchange for eventually getting them back to their Lakehurst, New Jersey, air station.
>
> Mr. Saul Price, WB research forecaster at the Honolulu airport, came over to Hilo to coordinate the meteorological measurements that they needed to carry out during the Air Force project. The program started June 1 [1954] and was to last for about four months. The Air Force gave the WB $10,000 to do it.
>
> Mr. Price kept up the pressure to get more data. It would be "very interesting," he said. "It will take more than a little effort; however, we would be seriously remiss not to take the fullest advantage of this rare opportunity." The Oahu Research Center was going to make similar observations on the slopes of Haleakala on Maui and this "tremendously increases the interest and value of the data," he said.

Clearly, Price recognized the value of Mauna Loa for serious science, and he was not alone. Back in Washington, Harry Wexler continued to keep a close watch on events at Mauna Loa. Meanwhile, the Air Force project was about to meet the same fate as the Summit Observatory, as Ellis (1988) notes:

> One trip had an observer sitting on foam rubber on top of the truck holding in his hands an aerometeorograph, a weather recording instrument normally flown on aircraft. The truck went over a rough place in the road, dumped him into the lava, and gave him cuts and bruises. However, he was able to save the instrument with only splashed ink destroying the record.
>
> Hawaiian Airline personnel were talked into the free use of a forklift and an operator to load the cylinders onto the C-54 plane which came over for the pickups. Once a stewardess hand carried a barometer over from Honolulu.
>
> There were stations at the following elevations in feet: 1,000, 3,700, 5,181, 7,030, 8,300, 9,300, 10,080, 10,958, and 13,456. By September, Hilo reported that the Air Force project could not be continued if the weather stations had to be tended.

> On October 8 [1954], the project ended with the compressors, manifolds, and the three flying suits being shipped back to the Air Force.

Although both the Summit Observatory and the Air Force project had ended, Saul Price had only begun. He had spoken at the dedication of the Summit Observatory on December 12, 1951, and at the dedication of the slope unit on June 28, 1956. Like Simpson and Wexler, he fully understood the value of an observatory on Mauna Loa.

After the dedication of the MLO slope unit, Fox continued planning the Weather Bureau's observational program that was scheduled to begin in early 1957. According to Ellis (1988), he developed a budget that

> included half the salary of a clerk in his Honolulu office. He favored a staff of three people at the observatory, one GS-9 and two GS-7s. Each would spend two weeks continually at the mountain, with each week having six days paid and one day as compensatory time off.
>
> His yearly costs were estimated at $35,000. A program of synoptic observations was to be the core of the work. The men would work out their own arrangements for buying and cooking their food, "on their own."

In January 1957, Steiner left Hawaiʻi for a new assignment, and Raymond Busniewski was placed in charge of MLO. Busniewski, who was present when the slope unit was dedicated on June 28, 1956, was already very much involved with the observatory.

> His work on the mountain continued with the building of a generator shed, the installation of two generators and a diesel fuel tank. He replaced the gas space heater with an electric one. All this was done mostly "on his own."
>
> There was frequent trouble with the Dodge power wagon. Sometimes the trip was made in a vehicle Mr. Busniewski borrowed from someone in the community willing to help the observatory. To most of the [Weather Bureau] airport staff the trip was a nice break. It was not so often as to be a hardship. Some used it as a picnic outing for their family. (Ellis 1988)

The Weather Bureau Arrives Full Time (July 1957)

Back in Washington, Harry Wexler sent a memo to his boss, Weather Bureau chief Francis Riechelderfer. Wexler was MLO's biggest booster, and his note listed equipment and personnel necessary for the IGY work at the new observatory. His request of $20,000 for the necessary expenses included a prescient observation:

> During the IGY one of the Weather Bureau's most important research programs will be the observational programs at Mauna Loa. The Observatory has attracted considerable interest here and abroad and already noted scientists have taken advantage of the unique location and facilities. (Ellis 1988)

Wexler was correct about the attention MLO was receiving. The World Meteorological Organization (WMO) was beginning a formal program to develop standardized procedures for measurements of the ozone layer through its Global Ozone Observing System program (WMO 2008). The WMO took note of the fledgling observatory on the island of Hawaiʻi, an interest that was to lead to many international collaborations over the years.

Riechelderfer apparently approved at least some of the requests, for on June 12 Wexler informed Roy L. Fox

> that Jack C. Pales of the Weather Bureau's Physical Research Division in Washington, D.C., had left the Central Office to assume the duties as Physicist in charge of the Mauna Loa Observatory for the duration of the IGY. Mr. Robert Williams of the WB [Weather Bureau] Research station at Las Vegas, Nevada and Mr. Clifford Kutaka of the Honolulu WB office had been asked to serve as assistants as soon as their transfers could be arranged. (Ellis 1988)

For the first time since the original Summit Observatory was placed in operation, the Hilo Weather Bureau office would no longer be directly involved in operating and maintaining MLO on a part-time basis. Instead, the facility would be transferred to a permanent crew whose sole responsibility was to staff the observatory. Howard Ellis (1988) describes the transfer of responsibilities in his account:

> Mr. Fox wrote Mr. Busniewski that he and his staff could discontinue their weekly trips "up the mountain" until the end of the IGY. The Dodge power wagon was to be turned over to Mr. Pales and furnished office space was to be made available to him and his staff at their airport office in Hilo.
>
> Mr. Pales, yet another tall slim southerner, arrived [in June]. The airport weather people made their last trip on July 11.
>
> Honolulu proclaimed the Mauna Loa Observatory a first order station which should take surface observations exactly as prescribed in Circular N, the observer's "bible," and all the standard forms should be completed and sent to them. The climate stations were transferred over to Mr. Pales.

Jack Pales signed the MLO *Visitors* book for the first time on June 20, 1957 (Fig. 6.2), the same day that Ralph Stair and Russell Johnston completed their solar constant measurements and moved out of MLO after their eight-week visit. It was just eight days before the first anniversary of the dedication of the facility.

Robert Williams and Clifford Kutaka signed the *Visitors* book on July 9. With his team in place, Pales began working around the clock seven days a week making plans for operating the observatory on a continuous basis. He kept Wexler and Price advised with progress reports, the first sent soon after his arrival, and "Progress Report No. 2: June 28–July 27, 1957," sent a month later (Price 1957).

Bumps in the Administrative Road (1957)

After Pales' arrival, the Weather Bureau bureaucracy began to sort out the management and funding of the new observatory. Harry Wexler's Office of Meteorological Research (OMR) had provided limited startup funds, and, as suggested in Ellis' account (1988),

> [OMR] hoped that the Honolulu people might get excited enough about the observatory to divert, from some place in the Pacific, funds for running the place.

Even though OMR did not fully fund the observatory, Wexler maintained full control through Price and Pales over the science conducted there. Meanwhile, the Weather Bureau Central Office designated the installation on the slope of the giant mountain as the "Weather Bureau Mauna

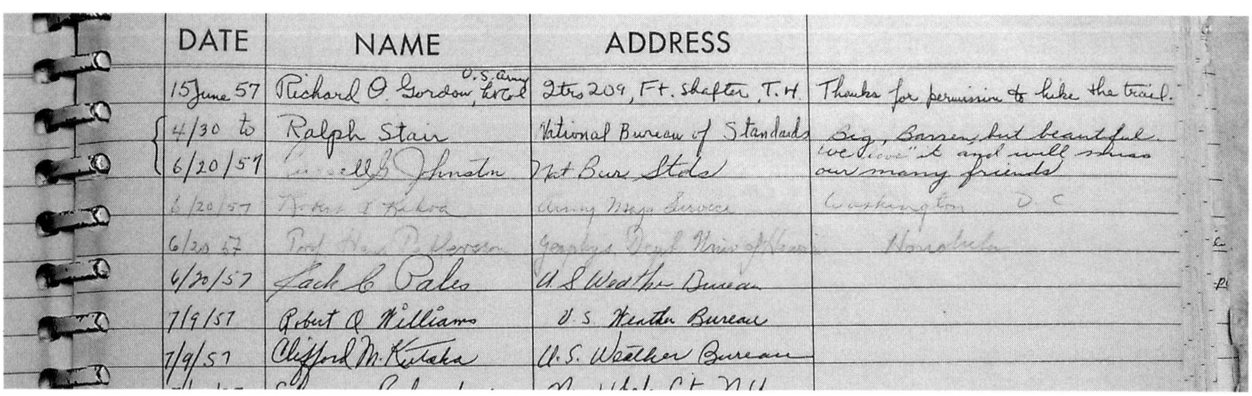

FIGURE 6.2. The MLO *Visitors* book records when Jack Pales, Robert Williams, and Clifford Kutaka first signed in for duty on the mountain. Pales signed in the same day that Ralph Stair and Russell Johnston departed.

FIGURE 6.3. Jack Pales *(left)* strides toward the main building at MLO between July and November 1957 while visitors enjoy the view of a snowcapped Mauna Kea from the instrument platform. The structure in front of the main building is the temporary shed erected by the Kiess-Corliss team to shield them and their spectrometer from wind and chilly air when they made spectroscopic observations of Mars. *(MLO archive.)*

Loa Observatory" (Fig. 6.3). Because there was no airport or general public to serve, the Central Office declared that the facility was not a "first order station" but should be considered as such for administrative purposes (Ellis 1988).

While MLO was not considered a "first order station," it was continuously manned twenty-four hours a day by Pales, Williams, and Kutaka, with each man having only one day off each week. As Ellis observes, "If there had been a big busy airport there in the lava fields, it would have been well served."

Pales and his men soon learned that the long, rough road to the observatory was easier to negotiate than some of the Weather Bureau's red tape imposed by administrative staff in Honolulu and in Washington, DC. The Weather Bureau's field stations had traditionally been operated with minimal budgets using supplies, instruments, and equipment sent at regular intervals. But MLO was not a traditional station.

> Mr. Pales got chided by the Honolulu office for each purchase he made, no matter how small. . . . Mr. Pales, coming from a research laboratory, was accustomed to freer spending. He wrote that there were enough dishes, linen, and chairs for twenty people, but he said, "I wouldn't want to hurt anyone's feelings, but the observatory was built with absolutely no planning for anything." (Ellis 1988)

FIGURE 6.4. The southwest corner of the main room, shown here in 1957, became an important work station for Jack Pales and his team at the new observatory. *(MLO archive.)*

Pales left Washington for Hawai'i with the impression that he and his men would receive overtime pay for the many extra hours they worked. But when he submitted the time sheets, the overtime pay was denied. The administrators in Honolulu and Washington were enthusiastic about the role MLO was playing during the International Geophysical Year but they didn't fully appreciate the difficulties involved in living at a high-altitude research station, the frequent need for supplies and spare parts, and the wear and tear on the vehicles caused by the increasingly rough road maintained only when the staff took time to stop and toss chunks of lava out of the way. Nor did they fully appreciate the challenge of requiring that a team of only three observers man the observatory twenty-four hours a day for seven days a week.

Pales' persistence paid off, and he was eventually able to make needed improvements. The observatory had two four-wheel drive vehicles, one of which was always at the mountain site for safety purposes, and the observatory was linked to both vehicles and the Hilo Weather Bureau office with fifty-watt radio transceivers. By the end of 1958, the small generator originally used to power MLO was replaced by two thirty-five-kilowatt diesel generators installed in a shed west of the building. Price and Pales (1959) describe how the two generators provided a major power upgrade to the station:

> These are ordinarily used alternately to permit preventive maintenance and minimize power interruptions, but they can be synchronized to operate in unison. Frequency regulated current is supplied to the more critical circuits.

Although the generators were noisy and required frequent attention, when they were working they provided the observatory with a stable source of power twenty-four hours a day.

Initiating the Legacy Observations at MLO (1957)

During the summer of 1957, the Soviet Union was quietly preparing to launch the first Earth satellite, and a sophisticated set of observations was being readied for MLO's participation in the International Geophysical Year, including measurements of the ozone layer, sunlight, and the concentration of surface ozone and carbon dioxide.

Installation of the instruments was not simple. Power and data cables had to be laid across the lava. Air sampling towers had to be built and sample return lines had to be installed. Once installed, instruments had to be calibrated and carefully checked for accuracy. Some instruments needed frequent repair. Vacuum tubes had only begun to be replaced by transistors in 1957 and then only in a few consumer products, such as portable radios and hearing aids. The integrated circuit had not yet been invented. Personal computers were unheard of, and data collected by electronic sensors was recorded by servo-driven pens that drew undulating lines of ink on rolls of graph paper. Data was manually extracted from the charts at one of the desks in the observatory's central living room or even on its red floor. The desk shown in Figure 6.4 eventually became the key work station for the MLO staff.

Surface Ozone Measurements (August 1957)

Because of its adverse effects on human health, crops, trees, and exposed materials, surface ozone was considered among the most important of the IGY measurements. Continuous surface ozone measurements were begun in August 1957 using an instrument made by Regener based on the classic method of detecting ozone by means of its reaction with potassium iodide. The instrument, which was about the size of a file cabinet, was installed in the northwest corner of the main room in the building. Outside air was collected through a glass tube inserted through the roof of the main building where today several air collection tubes and pipes protrude. A glass collection tube was used since rubber and plastic react with and decompose ozone. A photograph of the complex-appearing Regener apparatus is in Figure 6 of Price and Pales (1959).

Surface ozone had been measured at various locations around the world, and there were no major surprises in the concentrations measured at MLO early in the program. The measurements soon showed a pronounced diurnal variation, with the peak ozone occurring in the early morning. Ozone reacts with organic matter, including vegetation, and the reduced ozone in the afternoon was attributed to the arrival of air from well below the station that had circulated within the plants and trees across the lower slopes of Mauna Loa. Thus, the surface ozone served as a tracer to study the mountain's circulation. By the end of the first year, an obvious seasonal change was evident, with a spring maximum and a fall minimum that nicely matched the seasonality of the total ozone column (Price and Pales 1959).

The ozone monitoring system proved to be very difficult to operate, and many days of data had to be scrapped as unreliable. The surface ozone measurements begun in 1957 continue today using a pair of modern, briefcase-size ozone analyzers installed only a few feet away from where the old, bulky Regener once stood. Their air intake line passes through the roof very near where the original glass tube once passed.

Saul Price Visits the Mauna Loa Observatory (September 1957)

Prior to Pales' arrival, the early Weather Bureau observations at the Mauna Loa Observatory were mainly routine measurements of meteorological parameters, including temperature, relative humidity, and wind speed and direction. Saul Price was keen to advance the science at MLO by installing an all-sky camera system to determine the impact of the giant mountain on cloud types and their sky coverage.

Price made a working visit to the observatory in September 1957, and the first night he and Jack Pales shared the facility with visitors. A 1957 photograph shows the

Visitors book in the second shelf of a cabinet beside a worktable on the east wall of the main room between the two bedroom doors. But the first night of Price's visit, the book was misplaced or neglected, and none of the visitors signed in. This oversight was more than made up for by Price, who wrote an enthusiastic report about his visit, which was copied to Jack Pales and Harry Wexler. A battered copy of Price's seven-page, single-spaced report is in a file cabinet at the MLO Hilo office and is likely the copy sent to Pales. The report, unpublished until now, discusses in detail some of the research at MLO, addresses important questions about the motion of clouds and air around and up and down the slopes of the mountain, and provides a rare account of life at the new observatory. In view of Price's significant role in directing the early science at MLO, his "Report on a Visit to the Mauna Loa Slope Observatory, September 27 to October 2, 1957" is especially well worth reading, and most of it is reproduced here.

> [Saul Price] Research Forecaster, WBAS
> [Weather Bureau Airport Station], Honolulu
> October 14, 1957
>
> **ITINERARY**
>
> I left Honolulu at 4:40 p.m. on Friday, September 27, 1957 via Trans-Pacific Airlines and at 5:50 p.m. arrived at Hilo Airport, where I was met by Mr. Jack C. Pales, Physicist-in-Charge, Mauna Loa Slope Observatory. On Saturday morning I accompanied Mr. Pales to the Observatory, which we reached shortly after noon. I remained there until 11:30 a.m. on Tuesday, October 2, returning to Hilo with Mr. Robert Williams of the Observatory staff. I departed Hilo via TPA at 2:30 p.m. and arrived at Honolulu at 4:15 p.m.
>
> My original intention had been to ascend the mountain and return to Honolulu the preceding day, but the temporary disability of one of the Observatory's vehicles necessitated the postponement.
>
> **PURPOSES AND ACTIVITIES**
>
> I had three specific reasons for visiting the Observatory at this time.
>
> 1. To deliver to its staff the time-lapse camera developed by the electronic technicians at PSO, and to instruct them in its use. Present plans are to leave this at the Observatory for several months or longer, where it will be used to obtain for Honolulu a record under various meteorological circumstances of the visual aspect of that very large area of the island of Hawaii visible from the Observatory.
>
>   The equipment consists of four units: the camera itself, enclosed in a transparent plastic waterproof case; the battery power pack; the intervalometer; and the modified tripod mounting. All have been weatherproofed and the whole designed for easy portability and operation. The equipment's complete independence of external power sources and its wide range of single-frame exposure-intervals—one second to one hour—make it eminently suitable for field use and any anticipated time-lapse purpose, from ordinary long-interval recording of sky cover and cloud-type to short-interval movies of special meteorological events.
> 2. To investigate at first hand certain aspects of the weather and climate of Mauna Loa, in part for a paper presently being prepared on this subject for presentation at the Ninth Pacific Science Congress.
>
>   Prior to the permanent staffing of the Observatory in July 1957, no manned stations had been operated on the upper slopes of Mauna Loa, except for several brief stays by scientific parties outside the Weather Bureau: Nakaya, Stair and Corliss-Kiess. Hence, our knowledge of the climate and weather at its higher elevations was largely inferential, being derived chiefly from the fragmentary and incomplete autographic records of hygrothermographs, barographs, and rain gages placed at intervals along the slope.
>
>   My few days at the Observatory gave me the opportunity to check some of these inferences against the current autographic records, the visual scene and other observations, and proved helpful in suggesting possible solutions to a number of questions which the past incomplete data had evoked

but been inadequate to answer. This is not to say that no mysteries remain!

A less anticipated outcome of the visit was that the distribution of clouds observed over the central portion of Hawaii may suggest the need for certain revisions in the mean isohyetal maps for that area. As close a study of cloud patterns in other regions of that and of the other islands might well lead to similar conclusions concerning them.

3. To discuss with Mr. Pales, on the interest he had previously expressed in doing so . . . , additional projects which he and his staff might undertake, as time and their primary programs permit. Some of these had previously been outlined in [Notes on Discussion with Mr. Jack C. Pales Concerning Certain Projects to be Undertaken at the Mauna Loa Observatory: June 19, 1957].

[The report continues with 1.5 pages of discussion about time-lapse photography of clouds, Clifford Kutaka's helpful tabulations of data begun when he was Price's assistant in Honolulu and resumed at MLO, possible continuation of a failed 1954 experiment to make continuous records of temperature, humidity, and wind from an apparatus mounted on an observatory vehicle, influence of Mauna Loa on the local air flow, the possibility of obtaining cloud drop size, and a note that many other ideas and projects were discussed with Pales that "might be pursued by the Observatory staff within their time limitations."]

IMPRESSIONS

Although I have been greatly interested ever since my arrival in the Hawaiian Islands in the possibilities offered by Mauna Loa as a site for meteorological and other observations, and was present at the dedication of the summit unit in December 1951 and the Slope Observatory in June 1956, this was the first time I had seen the latter staffed and in full operation and my first opportunity to spend more time on the mountain than the very few hours of my other two visits.

It turned out to be one of the most memorable experiences of my life. The Observatory itself, reached after nearly three hours on tortuous roads which seem at times to be leading nowhere, is perched pink and tidy in the midst of tumbled fields of brown and black lava which stretch in every direction like enormous slag-heaps. The other great mountains of the island—Hualalai, Kohala and Mauna Kea—and in the greater distance Haleakala and the peaks of west Maui, were visible with a clarity which belied the distances printed on the map, while the clouds in their development and motion provided an ever-changing landscape of light and shadow. Several small and far-distant villages and an occasional solitary building could be seen if looked for carefully through binoculars, and the saddle below was spotted with greenery, but otherwise life seemed as remote as the moon. There was an overpowering feeling of detachment from the ordinary cares of the world, and I found myself with a first appreciation of the monastic life or of the retreats maintained by some faiths.

I had expected to be troubled, as some others have been, with headaches and quick exhaustion brought on by the relative lack of oxygen (the Observatory's pressure averages about 680 mb). Quite the contrary. Not only had I no awareness of this, but I experienced instead a feeling of well-being and exhilaration as I had not had in many years. For this the bright springlike days and cool nights may have been responsible, with temperatures oscillating between the high fifties and low thirties. We slept under four army blankets, and on one early sunny morning the shallow pools of water left on the Observatory's concrete apron by late afternoon showers of the day before were crisped over with ice.

The sheer physical beauty and immensity of the landscape, the invigorating weather, were only part of what I found attractive about the Observatory. At least as impressive was the efficient tidiness of its interior, the camaraderie among its staff, their interest in their work and their shared understanding of its purpose, and their readiness to take on additional responsibilities. I was glad to see that Cliff Kutaka [who had worked for Price in Honolulu] was pleased with his work and associates and they with his. When men live as closely together as these do, sleeping in one room of the Observatory, cooking and eating in another (and what superb cooking and eating, and how many dishes to be washed afterward!), working in a third, sharing the housekeeping chores, hostilities can develop as readily as friendships. Yet morale was excellent, and the visit on the first night of my stay of six scientists—two from the Army Meo [Meteorological] Service and the four University of Hawaii people who are taking over their lunar occultation program—demonstrated that these good and easy relations extended outside the Bureau personnel. For this, it appeared to me, one must credit first Mr. Pales, himself, for his unassuming, yet firm and knowledgeable, leadership; his assistants, Mr. Kutaka and Mr. Williams, and for their willingness and pleasant temperaments; Mr. Fox and the others at PSO for their farsightedness in providing the Observatory with the requisites for comfortable living, and their continuing helpfulness; and Mr. Busniewski and his staff at WBAS, Hilo, for many assistances.

CONCLUSIONS

After only three days at the Observatory, I felt as Jack Pales must have when in one of his first progress reports he half jestingly, but yet with sincere enthusiasm, called for a five-year extension of its program.

The excellence of the site for meteorological and other geophysical observations seems self-evident. On the other hand, certain considerations will have to enter into the use of the data obtained there. These have to do with the nature of mountain climates. It is evident, and generally recognized, that even small mountains, to say nothing of one as massive as Mauna Loa, not only create their own weather and climate, but also affect those of surrounding regions and, to some degree, modify the "free" atmosphere into which they thrust themselves. The motion of the air, for example, may be mechanically disturbed to windward, to leeward and above such an obstacle for distances equal to many times its own height.

In addition, the daily cycle of heating and cooling produces up- and down-slope thermal currents, local cloudiness, horizontal pressure gradients and other locally anomalous characteristics.

These mountain influences must affect not only the most commonly observed meteorological elements, but also the atmosphere's content of ozone, carbon dioxide, particulate matter and its other properties and constituents. It might be a serious error, therefore, to regard measurements of these or other variables made at a mountain observatory as representative of or directly comparable with those obtained elsewhere at equivalent altitudes in the free air. . . .

These topographically-induced circulations bring to the Observatory, at times in rapid succession, air from diverse sources and of diverse properties. The corresponding fluctuations in the elements under observation may be of the same order of magnitude as the synoptic or other factors being sought, or larger. An analogy that comes to mind is that of an accurate thermometer carefully mounted in an instrument shelter near the concealed outlet stack of a restaurant kitchen. One's intention may have been to determine the temperature differences between, say, the land- and sea-breeze regimes, but what he actually measures is whether the instrument is up- or downwind of the stack.

At the Mauna Loa Slope Observatory, the ordinary and supplementary observations, with whose interpretation we have so much more experience, may become of prime importance not only for themselves but in facilitating the evaluation of the other and less familiar data.

Saul Price

Price's enthusiasm about his stay was typical of those who are unaffected by the effects of breathing ultradry air containing only two-thirds the oxygen of air at sea level. But not everyone who was to work at or visit MLO would be so tolerant of the conditions, especially when driving up and down the mountain on a daily basis. Even some of the best-adapted workers were occasionally affected by an aching head, a stuffy nose, or both.

In the end, Price's visit to MLO paved the way for a close collaboration with Jack Pales and the MLO staff that was to lead to the first scientific papers by Weather Bureau staff based on science conducted at the observatory. His concern about whether or not the air at MLO, which rises upslope during the day and sinks downslope at night, could be considered representative of air elsewhere in the free troposphere was eventually answered by the carbon dioxide program, which had not yet begun. In the years to come, measurements of the concentration of carbon dioxide made at MLO and elsewhere agreed both in terms of their upward trend and their amplitude. It was likely that Price's concerns influenced the protocols soon to be developed for the sampling of carbon dioxide.

The First Weather Log (October 1957)

While Saul Price was visiting MLO, the staff, perhaps under his direction, began entering daily weather summaries into a small book labeled "*Weather Log*" (Pales et al. 1957–1958), which in 2006 was discovered by Leslie Pajo in the files of the Hilo office. The first entries describe the incredibly clear days for which MLO is so famous:

> Oct. 1, 1957
> A beautiful day. Clear to about 1300 H, then to broken to about 1600 H, then to clear again. Min temp 34 degrees. Max temp 57 degrees. No rain. Winds generally from NE. Sunshine 640 min.
>
> Oct. 2, 1957
> Another nice day, a clear day. . . .
>
> Oct. 3, 1957
> Pretty much clear up to noon.

On October 4, 1957, the Soviet Union amazed the world by launching Sputnik I, the first artificial Earth satellite and among the most significant achievements of the International Geophysical Year. During twilight, sunlight glints from its tumbling booster rocket were seen by people around the world, including the author. The effect was that of a slightly flickering star moving across the sky. If the crew at MLO was aware of this historic development, they didn't record their impressions in the *Weather Log*.

> Oct. 4, 1957
> Clear day. Cu [cumulus] and Sc [stratocumulus] W thru E with bases below station all day. Temp range from 800 H to 1700 H 49° to 58°. Dew point 14 to 33. Relative humidity 19 to 46%. Had 706 minutes of sunshine. Wind mostly Nly and NEly about 8 kts. No rain. Looks like a clear night. Max Temp 58°: Min Temp 37°. Sunshine 706 minutes.
>
> Oct. 5, 1957
> A truly beautiful morning with visibility over 100 miles. Molokai and Lanai were both visible. Clear all day.

The weather wasn't always pleasant, as exemplified by this entry:

> November 30, 1957
> A very nasty day to end another month! It rained all day with 0.97 falling by 0800 H. Visibility was also obstructed by fog in afternoon. Ceilings were very low then and most of the day so were the pressures obtained by the impending hurricane. Pressures ranged from a high of 19.965 to a very low 19.865. Temps ranged from a low of 38.5 to a max of 48.9. Winds were S'ly increasing in velocity to 28 kts by late afternoon. RH's naturally were very high, ranging from 86 to 98%. Total sunshine = 0 minutes.

Holidays were rarely mentioned in the *Weather Log*, but their days were, which confirms that the observatory was manned seven days a week. Here is the entry for Thanksgiving Day, 1957:

> Nov. 28, 1957.
> Clear morning. Clear skies until 1000, then broken up to 1700.

The US Exploring Expedition worked at the Mauna Loa summit during Christmas of 1840, and Pales and his team continued the tradition.

> Dec. 25, 1957
> This Christmas was a cold day. Temp ranged from 33 to

> 42. Dew points 12 to 36. Winds S'ly averaging 17 to 20 kts during the day. All instruments were out by 0930. Generator troubles. Had some rain. Had snow 1715. Wind now (2200) blowing S'ly and about 40 miles per hour; looks like a cold night. . . . Sunshine recorder out. Have a gasoline generator going outside the door.

These entries in an obscure weather log exhibit the dedication of Jack Pales, Clifford Kutaka, and Robert Williams, who worked six-day shifts at 11,200 feet (3,414 meters) to implement the dream that had begun with Robert Simpson and was now shared and promoted by Ralph Stair, Harry Wexler, Saul Price, and a growing number of leading scientists. Pales, Kutaka, and Williams worked far longer schedules than most other civil servants of their time and their successors at MLO today. Yet while their work schedule was already full, it would grow even more as new instruments were placed in service, instruments that would eventually provide data more important than any of them could have anticipated at the time.

When the MLO staff was not maintaining and repairing their instruments, they were kept busy extracting data from the chart recordings, sometimes on their knees while the paper rolls were laid out across the red tile floor. This tedious task required a skilled eye and careful checking to find errors. Consistency was also required, for the ink lines on chart recordings are two-dimensional. This raises various questions, as those who have worked with chart recordings know so well. Should the data be extracted from the center of the trace or one of the edges? How are data extracted when the trace is slightly thinner or wider than usual? How does one analyze the occasional ink blob or gap in the trace?

Carbon Dioxide Delays (1957)

While progress at MLO was rapid, work on the carbon dioxide program at the Scripps Institution of Oceanography seemed to advance at a snail's pace, especially when compared with Saul Price's enthusiastic progress report. Paul A. Humphrey was a senior Weather Bureau meteorologist under Harry Wexler at the Weather Bureau headquarters in Washington, DC. He was also the project leader of the Weather Bureau's IGY Antarctic Meteorological Observing Program. On July 26, 1957, Humphrey (1957a) advised Charles Keeling that Wexler would be visiting the Mauna Loa Observatory about the first of November and that "He would like to see the [carbon dioxide] analyzer in operation during his visit." The Scripps CO_2 program workload was so busy that a November date was impossible. Bad news arrived seven weeks later when it appeared that the CO_2 project would be given up by Scripps. On Saturday, September 14, Dave Keeling (1957) prepared a letter for Humphrey with the details:

> Since your telephone call on Thursday [September 12] the status of the whole Scripps CO_2 program looks uncertain. Next week I will have an opportunity to discuss the situation with Dr. Rakestraw and Dr. Revelle, but I do not expect any radical new developments. I should like to be considered for a job in Washington to continue the atmospheric CO_2 program if the Weather Bureau makes definite plans to transfer the program there. I am not sure I would be able to accept such a job but I certainly would like to investigate it further.

The carbon dioxide program was apparently at a crossroads, but something happened after the weekend to keep Scripps from abandoning the program. Someone printed "NOT SENT" across the top of the first page of Keeling's two-page letter to Humphrey. Exactly what happened in this case, as in various other interactions within and between Scripps and the Weather Bureau during the early days of the carbon dioxide program, remains unknown and may be revealed when the Scripps Library completes its processing of the extensive Charles David Keeling archives.

On September 30, while Saul Price was staying at MLO, Humphrey (1957b) wrote Keeling with new information that may provide a clue for why Scripps had entertained thoughts about abandoning the CO_2 program:

> No doubt, both of us have given much thought to the best way of re-organizing the CO_2 program and carrying

> on. As I interpret Dr. Wexler, he envisions your carrying on with the infrared analyzer and using it to the best scientific advantage. . . . Also, we will independently set up and operate a carbon dioxide sampling network. We would co-ordinate the entire program closely with you and make any of the data available as soon as possible.

Humphrey's letter also included "urgent" suggestions about his planned visit to MLO in late October and Wexler's visit in early November:

> I am looking forward to seeing you on Mauna Loa, and am hoping that you, Dr. Wexler and I can talk together at the Observatory to develop a carbon dioxide program which will bring maximum scientific results.

The "urgent" suggestions about visiting MLO soon changed, and on October 9 Humphrey (1957c) wrote Keeling that the first priority was for Keeling to get the carbon dioxide aircraft sampling program in operation:

> In view of the urgency for getting the sampling program in operation, it is suggested that you do not attempt to go to Mauna Loa for a while. . . . I believe that you might consider the trip around the first of the year.

Meanwhile, Humphrey planned to stop over at Scripps on October 28 for a four-day visit prior to his trip to Hawai'i. He instructed Keeling to "ship the analyzer to Mauna Loa as soon as possible." Humphrey (1957c) closed by expressing appreciation for "even the short handwritten notes from you."

In his autobiography, Keeling (1998) confirmed the rationale for the delay in his travel to Hawai'i:

> In the fall of 1957, with the IGY under way, I became concerned about the delay in establishing measurements at Mauna Loa Observatory. The equipment was ready for installation, but Revelle was reluctant to sign travel orders for me to go to Hawaii because the aircraft program was not yet operational.

Revelle's failure to sign the travel orders was not unilateral, for, as reported above, Humphrey had already made his recommendation about delaying the trip, which would be paid for by the Weather Bureau. (As noted in chapter 7, Harry Wexler also recommended delaying or even canceling Keeling's first trip to Mauna Loa.)

Charles Keeling had been assigned responsibility for the entire land, air, and sea carbon dioxide sampling program, and the workload was impossible for one person to fulfill. Harry Wexler was becoming increasingly concerned by the delays at Scripps and decided to help by sending Weather Bureau research meteorologist Eugene M. Wilkins to California to assist Keeling. The plan was for Wilkins, who worked for Dr. Lester Machta doing research on the distribution of radioactive fallout from nuclear tests, to arrive with Humphrey on October 28 with Dr. E. Eichers, chief of the Chemistry Division of the National Bureau of Standards. Eichers would remain at Scripps for a week, while Wilkins would remain much longer and be responsible for getting "the field CO_2 sampling program underway."

Eichers' role grew out of an October 10, 1957, meeting at the National Bureau of Standards at which the IGY carbon dioxide program was discussed. On October 14, Harry Wexler (1957a) sent Roger Revelle at Scripps a report by air mail about this and related matters. The letter expresses concern about the reliability of the planned carbon dioxide measurements:

> It seems there are more technical problems to be encountered in obtaining reliable measurements than originally anticipated. Also, this meeting brought out the need for checking certain results by an organization such as the Bureau of Standards, and perhaps by a world recognized laboratory of a foreign country, if the results are to be generally accepted for IGY purposes.

Wexler's concerns about the accuracy of the measurements and $5,000 requested by Keeling (through Revelle) to complete his laboratory were addressed in the letter from Humphrey (1957c) to Keeling dated the day prior to the meeting at the NBS:

> Dr. Wexler questions the need for elaborate calibration facilities at Scripps, providing the Bureau of Standards will insure the accuracy of the technique. He would appreciate knowing the various purposes for which your laboratory may be used in the future. Put into other words, now that the Bureau of Standards has consented to give us assistance, Dr. Wexler would like to know what you need for laboratory facilities, and your plans for using such facilities.

In the end, Wexler was anxious to move forward, and in his letter to Revelle, he concluded that "Our aim is definitely to have the IGY carbon dioxide program in operation by January 1, 1958" (Wexler 1957a).

Paul Humphrey Visits MLO (November 1957)

On November 2, 1957, one month after Saul Price ended his stay on the mountain, Paul Humphrey signed in at Mauna Loa Observatory. It was a clear, dry day, and the temperature reached 56 degrees that afternoon. Early the next morning, the temperature fell to 32.1 degrees at the beginning of another dry, nearly clear day with a few scattered cumulus clouds and an afternoon high of 57.8 degrees. As he enjoyed the view from the main room inside the MLO building, Humphrey (1957d) sat at a desk and typed a letter to Charles Keeling:

> Dear Dave,
>
> I am writing from the Mauna Loa Observatory; and I am sitting by a window from which I have a fine view of 13,792 foot Mauna Kea, the top of Haleakala about 93 miles distant [on the island of Maui], cloud sheets on either side of the island of Hawaii, and the distinct atmospheric horizon which is the base of the trade wind inversion, two or three thousand feet below.

Humphrey continued with a list of supplies at MLO that would assist the carbon dioxide program. He also discussed plans to ship two steel towers from Honolulu to provide air sampling points for the project. He closed by noting that "Ordinary living arrangements at the Observatory are very comfortable. I am sure that you will enjoy your visit here."

Later that day, Humphrey (1957e) typed a second letter to Keeling in which he addressed concerns that Keeling had expressed during Humphrey's visit to Scripps:

> Mauna Loa Observatory
>
> Dear Dave,
>
> As you might suppose, I have been mulling over our recent discussions; and am concerned that there is a possibility that you really do consider that recent Weather Bureau action has been arbitrary, disregarding your recommendations.
>
> Actually, the recommendations which we agreed upon, and which were in my report, were considered; but there were some of these recommendations which could not be put into effect. Also, further consideration of some of the recommendations showed that they did not satisfy the logistic requirements of the program.

Humphrey addressed various aspects of the carbon dioxide program and the temporary assignment of Eugene Wilkins to Scripps. Apparently Keeling would have preferred the help of a technician and not a research meteorologist, and Humphrey wrote that he would recommend that a technician be assigned to Scripps. He would also endorse Keeling's request to travel to Hawai'i on a commercial flight rather than on an Air Force plane. Humphrey (1957e) concluded that

> I left Scripps with the impression that scientifically the program is sound; and that when organized, Scripps has the resources and potential to fulfill all IGY requirements. With continued effort I feel that the program will develop as it should. . . .
>
> Because of the great interest in the CO_2 program you are in a very fortunate position.
>
> This will probably be my last contact with you for a month or two [Humphrey was en route to Antarctica]. Until I see you again I hope that you will see to it that Wilkins helps the program as much as possible.

Even before Keeling received Humphrey's letter, on November 5 Wilkins demonstrated his worth in the form of a letter to E. A. Volbrecht of the Weather Bureau in which he requested that $6,000 be sent to Scripps (Wilkins 1957). A two-page report, which Wilkins also sent to Volbrecht, summarized significant progress with the CO_2 program and the satisfactory working relationship Wilkins had worked out with Keeling.

Meanwhile, Harry Wexler was preparing to leave for Antarctica. On November 13, the day he left, Wexler (1957b) sent Keeling by air mail a two-page letter about Keeling's suggestion that an aircraft sampling program would provide better quality carbon dioxide data than a surface network:

> Dear Dave:
>
> As a result of our telephone conversation on November 7, and at conferences the next day with Dr. Friedman and Dr. Creitz, I now have a better feel for the situation with regard to the CO_2 sampling program. As I understand it, you desire to place major emphasis on the aircraft sampling program because you feel you'll get away from surface contamination effects by obtaining samples from the more representative upper air.

Wexler then reminded Keeling that "almost one third of the IGY has already passed," and suggested that Keeling quickly initiate aircraft sampling "from the West Coast, under different weather conditions." Wexler (1957b) was concerned about Keeling's reduced interest in ground sampling, and he addressed its importance:

> I do not think that we should dismiss [the] surface network program because of "contamination." We may find recognizable large-scale patterns based on the difference between land and ocean, foliated areas and desert areas, etc., which are facts of nature and would exist whether or not any artificial contaminants were present. However, we will never know this until we actually try to plot such large-scale charts.

Wexler (1957b) also addressed the frequency of air sampling planned for what remained of the International Geophysical Year:

> Our original intention, as you know, was to take observations three times each month for the remainder of the IGY. In an attempt to cut down the shipping costs and analysis labors, you recommended only one observation every three months. For budgetary reasons, I was tempted to go along, but in the discussion next day with Dr. Friedman and Dr. Creitz the consensus was that once every three months would be far too small a sample to obtain anything of value. It was agreed that it would be best to adopt a compromise—three times a month every three months.

Finally, Wexler (1957b) addressed Keeling's concern about flying to Honolulu aboard a Military Air Transport Service (MATS) aircraft:

> I can well understand objections you might have against MATS, but if these are not too serious I would hope that you would take advantage of free travel—again to conserve our dwindling dollar supply.

MLO Founder Robert Simpson Visits MLO (November 1957)

On November 10, 1957, only a week after Humphrey's MLO visit, MLO founder Robert Simpson took time from directing the National Hurricane Research Project to visit—for the first time—the observatory that sprang from his chance visit with Ralph Stair at Sunspot, New Mexico, in June 1955. During his visit, the visibility was good and the temperature reached a pleasant 54.2 degrees Fahrenheit. But it was a cloudy day, with thick cirrus during the morning and thick altocumulus and stratocumulus during the afternoon.

Besides standard meteorological equipment, Simpson saw an array of much more sophisticated instruments that had been installed by Pales and his team. The instruments included the Wegener surface ozone recorder, the direct incidence solar pyrheliometer and recorder, the weighing rain gage, the Naval Research Laboratory radioactivity filter apparatus, and a micrometeor collector with paper filters.

Total hemispherical and net exchange radiometers were installed, along with horizontal incidence pyrheliometers, but these were awaiting chart recorders. A theodolite was installed for making balloon observations for wind speed and direction aloft. The Dobson ozone spectrophotometer needed a better power supply before it could be used (Ellis 1988).

Although the Dobson was not yet in operation, Simpson was no doubt delighted at the prospect of measuring changes in the ozone layer from MLO. As he later remarked, "One of many objectives had been that these observations could also contribute significantly to the understanding of tropical cyclone formation and intensification" (Simpson, personal communication with author, July 22, 2008).

Enormous progress had been made since Simpson spoke at the dedication of the Summit Observatory in 1951, and he succinctly summarized the status of MLO when he signed the *Visitors* book (Fig. 6.5). After his signature, Simpson, who had been devastated by the closure of the original Summit Observatory in 1954, wrote, "A long step ahead. We must preserve this progress."

Sunlight Measurements (November 1957)

While the carbon dioxide program slowly advanced at Scripps, at MLO measurements of the total irradiance in a direct beam of sunlight and the sunlight scattered by the sky were begun in November 1957. As explained by Price and Pales (1959), the solar observations were begun as part of an international program designed to increase the knowledge of the radiation fluxes to and from the earth's surface.

The direct sun measurements were made with an Eppley normal incidence (or direct sun) pyrheliometer, an instrument that is continually pointed at the sun by means of a motor-driven tracker. In operation, the pyrheliometer, which is shown in Figure 6.6, generates a small voltage that is proportional to the intensity of the sunlight entering the instrument, which is sensitive to wavelengths across the sun's optical spectrum from the ultraviolet to the infrared. For many years, the voltage was continuously recorded by pen and ink on a strip chart recorder.

With the exception of occasional power outages and the presence of clouds that blocked the view of the sun, the ongoing direct sun measurements at MLO have exceeded fifty years and have become the longest of the IGY time series measured at the station. Direct sunlight is significantly reduced when major volcanic eruptions propel

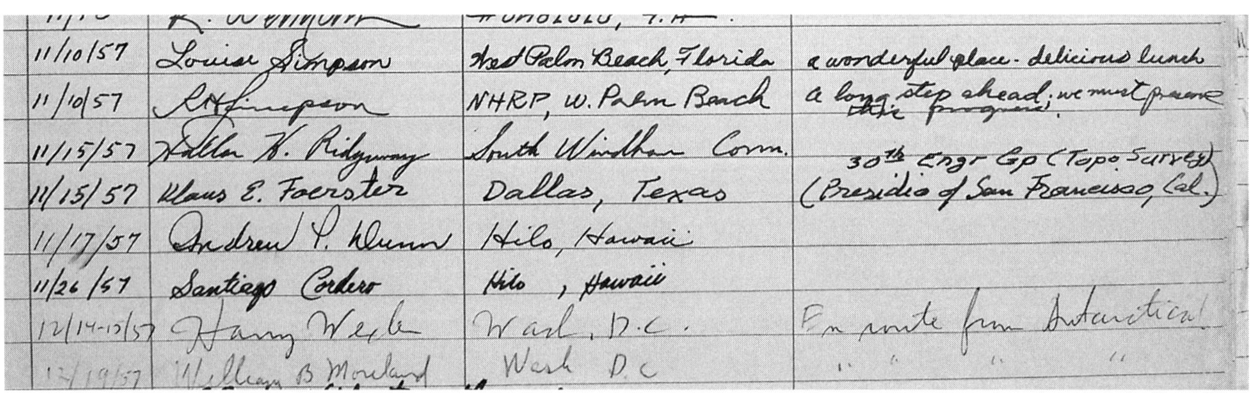

FIGURE 6.5. Robert Simpson visited MLO for the first time on November 10, 1957. Harry Wexler signed in a month later and spent a night at MLO while en route from Antarctica.

FIGURE 6.6. Instruments that measure direct and total sunlight were installed at MLO in the fall of 1957. The most important was the Eppley normal incidence pyrheliometer, the instrument on the second post from the right. The instrument was automatically pointed at the sun by a tracking motor. *(MLO archive.)*

particles and gases high into the stratosphere, and the direct sun measurements have permitted a series of major volcanic events to be observed at MLO over the years. The direct sun measurements also provided data for a series of key papers by MLO scientists that refuted suggestions that a global blanket of anthropogenic aerosols was responsible for the cooling of Earth that occurred during the 1960s and 1970s. (The series of papers is introduced in chapter 10.)

The aerosols from major volcanic events reduce direct sunlight while increasing the diffuse sunlight scattered from the sky. The sum of the scattered sunlight and the light directly from the sun, which is known as global or full-sky irradiance, was measured by an Eppley horizontal incidence pyranometer, a detector that looks straight up from a fixed position. (Later instruments measured the scattered light from the sky by means of a disk that tracks the sun so that a shadow of the disk falls directly on the light-sensitive portion of the pyranometer.) As with the direct sun instrument, for many years the data were recorded on a strip chart recorder. Several other solar instruments were also installed in 1957.

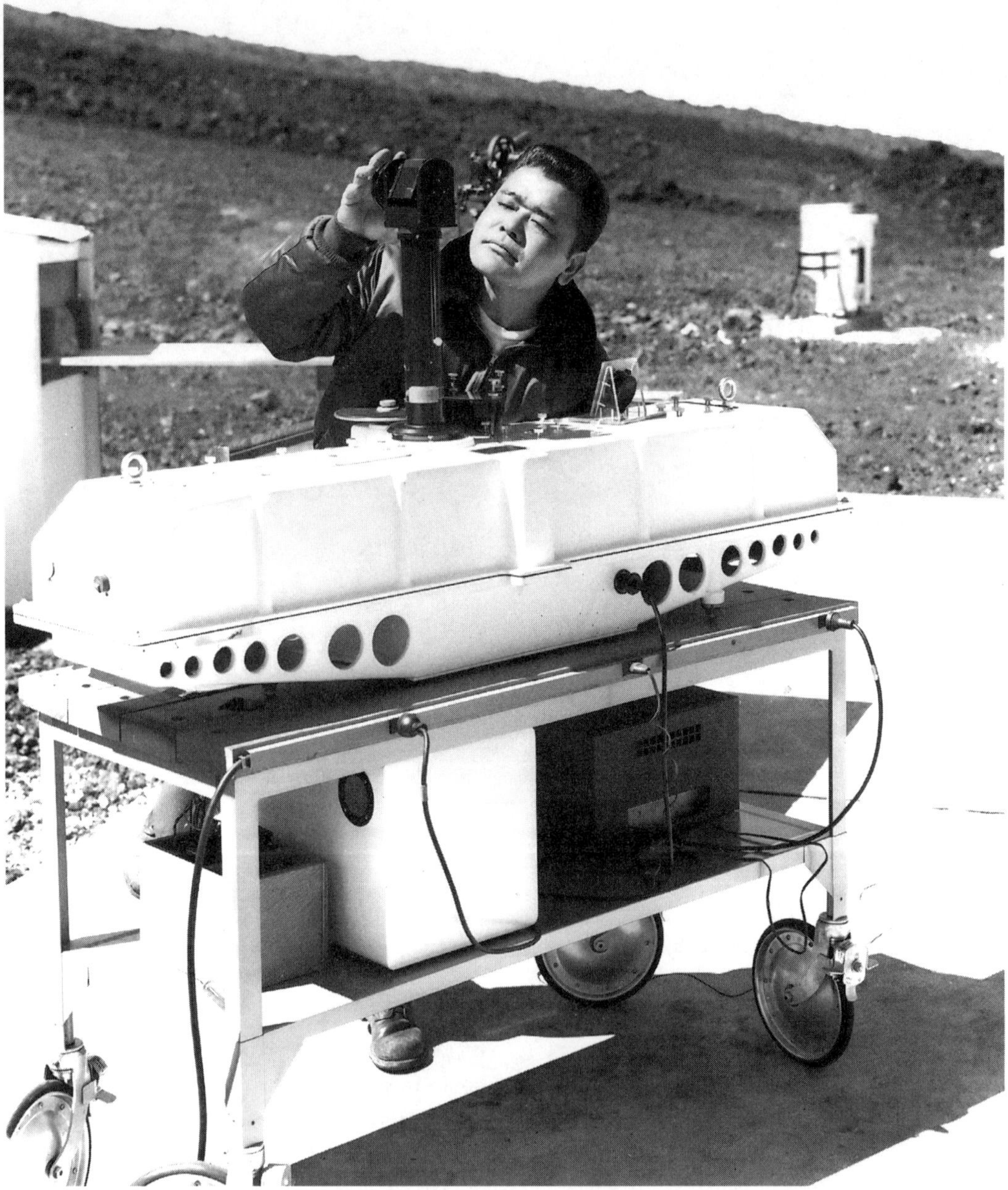

FIGURE 6.7. Clifford Kutaka adjusts the periscope of the MLO Dobson spectrophotometer prior to making a measurement of the ozone layer. This photograph was probably made by Jack Pales during a morning observing session in late November or early December 1957. *(MLO archive.)*

Ozone Layer Measurements (November 1957)

Mauna Loa Observatory founder Robert Simpson had always stressed the importance of measurements of the ozone layer as a justification for MLO, for ozone changes could indicate the presence of Pacific cyclonic disturbances. Ralph Stair needed to measure the ozone layer in order to remove its effect from his measurements of the solar constant, which made the 1955 meeting of Stair and Simpson that led to the founding of the observatory all the more remarkable. When Stair and Johnston (1958) conducted their solar constant experiments at MLO during June and July of 1957, they became the first to measure the ozone layer there.

The Dobson instrument arrived at MLO in October or early November 1957, and the instrument was placed in operation November 27 after a power supply problem was resolved (Ellis 1988). As shown in Figure 6.7, the Dobson was initially operated from the concrete pad in front of the building. A shed with a sliding roof that was opened by pulling on a lanyard was soon built to house the heavy instrument and shield the operator from the wind. In their first paper on the scientific progress at MLO, Saul Price and Jack Pales (1959) discussed the significance of the new Dobson instrument:

> Its elevation (it is the highest permanent installation of its kind), the clarity of the overlying atmosphere, and the relatively small ozone variations typical of the latitude are expected to make its observations valuable not only for their contribution to the synoptic ozone program, but also for calibration purposes—principally the determination of the extraterrestrial constants. The Mauna Loa spectrophotometer is to be used as a standard for the United States Dobson network.

As noted in chapter 5, the ozone layer is measured by comparing the magnitude of closely spaced wavelengths of solar ultraviolet. The "extraterrestrial constants" referred to by Price and Pales represented the intensity of UV at various wavelengths that the instrument would measure if it were located at the top of the atmosphere, parameters that are inferred by taking measurements against the direct sun over the course of a morning as the sun rises in the sky. The extraterrestrial constant is also of use when measuring the optical thickness of the atmosphere caused by haze and other aerosols between an observer and the sun. Because the ozone layer and the optical depth have long been monitored at MLO, a simplified account of how the extraterrestrial constant is determined will be of interest to some readers.

The method is based on the Beer-Lambert-Bouguer law and is known as the Langley method after Samuel Pierpoint Langley, who refined the technique for his program to measure the solar constant at the Smithsonian Institution in the early 1900s. Briefly, the natural logarithm of the measured intensity of sunlight at a specific wavelength is plotted against the thickness of the atmosphere (the air mass) through which the instrument observes the sun during each measurement. Connecting the points yields a perfectly straight line if during the measurement period the atmosphere between the sun and the observer remain stable and free of pollutants, dust, haze, and clouds. The extraterrestrial constant of a particular wavelength is inferred simply by extending the straight line until the air mass is zero. The difference in irradiance inferred at the top of the atmosphere and measured at the surface is directly related to the total amount of aerosols or ozone in between. For much of the twentieth century, this procedure was accomplished by hand using graph paper, a ruler, and a sharp pencil. Today it's done much more quickly and accurately using a computer and simple software.

The thickness of the atmosphere directly overhead at the zenith for a location at sea level is defined as one air mass. The air mass (m) with respect to the sun is therefore always higher than 1 when the sun is away from the zenith, and m is approximately the reciprocal of the sine of the sun's angle over the horizon. At MLO, the extraterrestrial constant is best measured by making measurements from early morning when m is about 6 to midmorning when m falls to about 2 or 1.5. Stopping the measurements near $m = 1$ avoids the noise caused by turbulence in the sky as the sun reaches its highest point in the sky. Morning is almost always better than afternoon, for the downslope air is usually free of pollution, and the moist inversion layer is far down the mountain.

Harry Wexler Returns to MLO (December 1957)

The United States attempted to launch its first satellite on December 6, 1957. The launch failed spectacularly when the untried Vanguard launch vehicle fell to the ground in a huge ball of flame just after lifting off its pad. Eight days later Harry Wexler made his second visit to MLO "En route from Antarctica!" (see Fig. 6.5). The day was December 14, and that afternoon the temperature reached a pleasant 51 degrees. The MLO crew and their distinguished guest were treated to a view of some unusual altocumulus castellanus, cirrocumulus, and lenticular clouds in a nearly clear sky. Wexler spent the night at MLO and awoke to a crystal clear sky. The temperature rose from 28 degrees before sunrise to a balmy 52 degrees. As one of the MLO crew wrote in the *Weather Log*, it was "A beautiful day in which to say goodbye to Dr. Wexler."

Wexler's visit provided an opportunity for Pales to inform his Washington boss about pressing needs at the observatory, and Howard Ellis (1988) records some of the details in his MLO history:

> Mr. Pales was able to get [Dr. Wexler] to okay twenty-four hours of overtime pay each week for each man. It was as though all the questions about ozone would soon be answered. Dr. Wexler was very pleased. The carbon dioxide equipment had not yet arrived and there was trouble with the freezing nuclei counter. There were two Fairbanks-Morse diesel generators of six kilowatts each and a ten-kilowatt gas [generator]. Besides the Dodge power wagon, a new all-wheel drive International Carryall had been purchased. All caused quite a bit of trouble and the road needed repairs.

In addition to the ozone measurements, a month of measurements of the transmission of direct sunlight through the atmosphere had been collected in what was to become the observatory's longest measurement series. But Charles Keeling's Model 70 infrared analyzer, which Wexler had hoped to see, had not yet arrived. The International Geophysical Year was well underway, and measurements of the concentration of carbon dioxide at MLO, which was Wexler's highest priority, had yet to begin.

References

APC. 1954. Model 70 infrared analyzer (modified shell design). Applied Physics Corporation.

Arrhenius, Svante. 1896. On the influence of carbonic acid in the air and upon the temperature of the ground. *Philosophical Magazine* 41: 237–276.

Bowling, Sue Ann. 1987. How do greenhouses work? Article #817, Alaska Science Forum, Geophysical Institute, University of Alaska at Fairbanks. (Accessed at http://www.gi.alaska.edu/ScienceForum/ASF8/817.html.)

Brooks, C. E. P. 1951. Geological and historical aspects of climatic change. In Thomas F. Malone, ed., *Compendium of Meteorology*, American Meteorological Society: 1004–1018.

Callendar, Guy Stewart. 1938. The artificial production of carbon dioxide and its influence on temperature. *Quarterly Journal of the Royal Meteorological Society* 64: 223–237.

———. 1939. The composition of the atmosphere through the ages. *Meteorological Magazine* 74: 33–39.

———. 1958. On the amount of carbon dioxide in the atmosphere. *Tellus* 10: 243–248.

Eisenhower, Dwight. 1954. (Accessed at http://www.eisenhower.utexas.edu/dl/IGY/LetterDDEtoBarnardJune241954.pdf.)

Ellis, Howard. 1988. The observatory: On their own. (Accessed at http://www.mlo.noaa.gov/webmuseum/publications.html.)

Fowle, F. E. 1912. The spectroscopic determination of aqueous vapor. *Astrophysical Journal* 35: 149–162.

———. 1915. The transparency of aqueous vapor. *Astrophysical Journal* 42: 394.

Friedman, S. Morgan. 2011. The inflation calculator. (Accessed at http://www.westegg.com/inflation/.)

Guppy, H. B. 1906. A Naturalist in the Pacific between 1896 and 1899. Vol. 2, Appendix, Note 61, Meteorological observations on the summit of Mauna Loa, 586–587. London: Macmillan and Co. (Accessed at http://books.google.com.)

Högbom, Arvid. 1894. Om sannolikheten för sekulära förändringar i atmosfärens kolsyrehalt." *Svensk kemisk Tidskrift* 6: 169–177. See Arrhenius (1896) for an account of this work in English.

Humphrey, Paul A. 1957a. Letter to Charles D. Keeling (July, 26, 1957). Keeling Papers, 2003–18. Box 3, Folder 8, Scripps Oceanographic Institute Library.

———. 1957b. Letter to Charles D. Keeling (September 30, 1957). Keeling Papers, 2003–18. Box 3, Folder 8, Scripps Oceanographic Institute Library.

———. 1957c. Letter to Charles D. Keeling (October 9, 1957). Keeling Papers, 2003–18. Box 3, Folder 8, Scripps Oceanographic Institute Library.

———. 1957d. Letter to Charles D. Keeling (November 3, 1957). Keeling Papers, 2003–18. Box 3, Folder 8, Scripps Oceanographic Institute Library.

———. 1957e. Letter to Charles D. Keeling (November 3, 1957). Keeling Papers, 2003–18. Box 3, Folder 8, Scripps Oceanographic Institute Library.

IPCC. 2007. Frequently asked questions. Based on *Climate Change 2007: The Physical Science Basis*. Contribution of Working Group I to the Fourth Assessment Report of the Intergovernmental Panel on Climate Change. Cambridge and New York: Cambridge University Press, 996 pages. (Accessed at http://ipcc-wg1.ucar.edu/wg1/Report/AR4WG1_Pub_FAQs.pdf.)

Keeling, Charles D. 1957. Unsent letter to Paul A. Humphrey (September 14, 1957). Keeling Papers, 2003–18. Box 3, Folder 8, Scripps Oceanographic Institute Library.

———. 1978. The influence of Mauna Loa Observatory on the development of atmospheric CO_2 research. In John Miller, ed., *Mauna Loa Observatory: A 20th Anniversary Report*. National Oceanic and Atmospheric Administration Special Report, September: 36–54. Boulder, CO: NOAA Environmental Research Laboratories.

———. 1998. Rewards and penalties of monitoring the Earth. *Annual Review of Energy and the Environment* 23: 25–82.

Kiess, C. C., C. H. Corliss, Harriet K. Kiess, and Edith L. R. Corliss. 1957. High dispersion spectra of Mars. *Astrophysical Journal* 126: 579–584.

NOAA. 2006, 2007. NOAA/GSD ground-based GPS meteorology (GPS-MET): Current IPW and MET observations. (MLO data for June 2006 and 2007, accessed at http://gpsmet_test.fsl.noaa.gov/cgi-bin/gnuplots/rti.cgi and plotted for comparison with data from Stair and Johnston 1957.)

NOAA. 2007. What is the greenhouse effect, and is it affecting our climate? Global warming. (Accessed at http://www.ncdc.noaa.gov/oa/climate/globalwarming.html.)

Pales, Jack, Clifford Kutaka, and Robert Williams. 1957–1958. *Weather Log*. (Unpublished log in MLO files.)

Pettersson, Hans. 1958. Rate of accretion of cosmic dust on the Earth. *Nature* 181: 330.

Plass, Gilbert N. 1956. The carbon dioxide theory of climate change. *Tellus* 8: 140–154.

Price, Saul. 1957. Report on a visit to the Mauna Loa Slope Observatory, September 27 to October 2, 1957. (Unpublished report in MLO files.)

Price, Saul, and Jack C. Pales. 1959. The Mauna Loa high-altitude observatory. *Monthly Weather Review* 87 (1): 1–14.

Revelle, Roger, and Hans E. Suess. 1957. Carbon dioxide exchange between atmosphere and ocean and the question of an increase of atmospheric CO_2 during the past decades. *Tellus* 9: 18–27.

Russell, Richard J. 1941. Climatic change through the ages. In US Department of Agriculture, ed., *Climate and Man: Yearbook of Agriculture*. Washington, DC: US Government Printing Office.

Smith, V. N. 1953. A recording infrared analyzer. *Instruments* 26: 421–427.

Stair, Ralph, and Russell G. Johnston. 1958. Some studies of atmospheric transmittance on Mauna Loa. *Journal of Research of the National Bureau of Standards* 61: 419–425.

Steiger, Walter R. 2001. Origins of astronomy in Hawai'i. (Accessed at http://www.ifa.hawaii.edu/haleakala/History.html.)

Tyndall, John. 1863. On radiation through the Earth's atmosphere. *Philosophical Magazine* 4 (25): 200–206.

Weart, Spencer R. 2003. *The Discovery of Global Warming.* Cambridge, MA: Harvard University Press. (A comprehensive overview is at http://www.aip.org/history/climate/index.html.)

Wexler, Harry. 1952. Volcanoes and world climate. *Scientific American* (April): 74–80.

———. 1953. Radiation balance of the Earth as a factor in climatic change. In Harlow Shapley, ed. *Climatic Change,* 73–105. Cambridge, MA: Harvard University Press.

———. 1956. A look at some suggested solar-weather relationships. In *The Sun's Effects on the Earth's Atmosphere,* Tech. Report No. 2, Institute for Solar-Terrestrial Research, High Altitude Observatory of the University of Colorado, October.

———. 1957a. Letter to Roger Revelle (October 14, 1957). Keeling Papers, 2003–18. Box 3, Folder 8, Scripps Oceanographic Institute Library.

———. 1957b. Letter to Charles D. Keeling (November 13, 1957). Keeling Papers, 2003–18. Box 3, Folder 8, Scripps Oceanographic Institute Library.

Wilkins, E. M. 1957. Letter to E. A. Volbrecht (November 5, 1957). Keeling Papers, 2003–18. Box 3, Folder 8, Scripps Oceanographic Institute Library.

WMO. 2008. Background and history. World Meteorological Organization. (Accessed at http://www.wmo.int/pages/prog/arep/gaw/history.html.)

Yates, Harold W. 1957. Recent atmospheric transmission studies. 42nd Annual Meeting, Optical Society of America, Cleveland, Ohio, October 17–19. Abstract in *Journal of the Optical Society of America* 47 (November): 1054.

CHAPTER SEVEN

The International Geophysical Year (1958–1959)

On New Year's Day of 1958, the eighteen-month International Geophysical Year was one-third complete. The team at the Mauna Loa Observatory recorded in the *Weather Log* that

> This first day of the year was cold. The min temp was 22°. The 0700 obs temp was 23°. Sky was clear. Visibility over 80 miles all day. Temp ranged from 23 to 48° (0700 to 1700). Winds SW'ly for 3 obs in the morning and NE'ly the rest of the day. Speeds 1 to 11 kts. Dew point –1 to 27. Pressure lower than normal (19.960 about). Cu [cumulus] and Sc [stratocumulus] N thru E below station all day. R.H. 19 to 46%. Max Temp 49°; Min 22°. No rain. Sunshine 623 minutes. (Pales et al. 1957–1958)

This entry nicely illustrates the regularity of the routine meteorological observations at MLO, which were made on an hourly basis just as the US Exploring Expedition did during 1840–1841 at its Observatory 1 in Hilo, aboard the USS *Vincennes*, and at the summit of Mauna Loa. These basic observations at MLO have continued until the present, with the major difference being the lack of human intervention. Today the data are automatically collected minute by minute, stored electronically, and then transmitted to the Web and to NOAA in Boulder, Colorado. While these basic meteorological data have played an important role in understanding the climatology at MLO, they have received far less attention than more specialized measurements, especially those begun during the IGY. Various other MLO projects have received even less attention, including a botanical survey begun during the IGY.

Lava Life (January 1958)

The road to the Mauna Loa Observatory provided a means for botanists and biologists to more easily explore life on the upper slopes of Mauna Loa. F. R. Fosberg was a leading expert on the ecology and botany of the Pacific islands, and on January 5, 1958, he and M.-H. Sachet visited MLO with the assistance of Chester K. Wentworth. They were the first to sign the *Visitors* book in 1958 and among the few botanists to visit Mauna Loa since Henry Guppy spent twenty-three days alone on the summit in 1897. The *Weather Log* reported that

> Beautiful weather prevailed once again. Very warm and dry and visibility unrestricted. Winds slow 6 kts from a Northerly direction. All clouds remained below the

> station during entire day. Temps ranged from a min of 33.2° to a balmy 54.1°.

The weather was ideal for the purpose of the visit, a botanical survey along the remote road to the observatory. The fields of jagged, sharp-edged *'a'ā* lava that cover most of the mountain and all of the observatory site were found to be devoid of vegetation above about 10,000 feet (3,048 meters). But sheltered in cracks and crevices in old, exposed sections of ropy *pāhoehoe* lava, Fosberg (1959) found a gray alpine moss (*Rhacomitrium lanuginosum* var. *pruinosum* Hook. & Wils). He also found two species of plants:

> The fern, *Pellaea ternifolia* . . . and the grass, *Trisetum glomeratum* . . . are occasional here in very small tufts in crevices at any angle of exposure, but mostly exposed upward. These crevices are either quite deep or are covered by an accumulation of small rock debris which protects the root crowns and rhizomes from undue desiccation.

Fosberg found no plants in the vicinity of the observatory, which reinforces the notion that the twenty or so specimens of some half a dozen species found there today are successful because of the warmth provided by the pavement and buildings. Rain runoff from the pavement and structures and, as suggested by MLO veteran Steve Ryan, construction sand may also play a role. On his return to Washington, Fosberg (1959) wrote a report on his findings. It was published in the journal *Ecology,* where it joined the earliest of the scholarly papers to emerge from the new observatory.

Growing Pains (January 1958)

The beginning of the IGY experiments at MLO greatly increased the workload and highlighted the need for more work and storage space than provided by the main building. The observatory had limited workshop space in the uninsulated, enclosed space under the observing tower and in the noisy generator shed. But more space was needed, as was office assistance.

Meanwhile, Pales and his men were working far more hours than typical government employees as they maintained the IGY experiments they had installed. Pales expected much from Williams and Kutaka, and he shared in the work duty on the mountain. The three had no idea how much more work would be required after the carbon dioxide project arrived, and in coming years some of the MLO staff would joke about the hard-working Pales being "a slave driver."

> Jack Pales was a real hard worker, and he always made sure everybody was doing something. (Chin, video interview, Hilo, Hawai'i, December 4, 2006)

The MLO *Visitors* book records that six months before Pales and his team arrived, Dr. Harry Wexler first visited MLO on December 30, 1956, while on the way to Antarctica. As noted in chapter 6, Wexler stopped by MLO again on December 14–15, 1957, and this provided an opportunity for Pales to inform his Washington boss about pressing needs at the observatory. Pales reported directly to the influential Wexler, who had played a key role in supporting Robert Simpson's dream for the observatory. After Wexler returned to Washington following his two-day stay at MLO in December 1957, Pales sent his boss a letter on January 7, 1958.

> He told [Wexler] that the carbon dioxide program would require an additional hour each day to "work up" the data and because of this he felt that he must hire another person. He recommended that his wife be hired as a GS-5 to run errands in Hilo, do secretarial work and reduce data. He said that this would greatly free him to do more work, save the government travel money and most importantly; "tie us to the observatory." He explained this further by saying, "The advantages to both me as Physicist-in-charge and to the Bureau are overwhelming . . ." for continuing the project past the IGY. He asked for $700 to equip space in his home for an office and said that he would move out of the airport office, radio link and all. (Ellis 1988)

Pales' letter crossed with one sent the next day by Wexler (1958a), who enthusiastically wrote

> how impressed I was with the ship shape appearance of the Mauna Loa Observatory and the high enthusiasm of yourself and observers. You have been doing an excellent job with slender resources at the end of a long pipeline.

Wexler approved $250 for minimal road repairs and authorized the darkroom Pales had requested for making photographic copies of printed items. Pales was a skilled photographer, and the early black and white photographs in the MLO archive were apparently produced by him in this darkroom. More importantly, Wexler addressed the concern that he and Pales shared about the future of MLO after the end of the International Geophysical Year, which had been extended through 1959. Wexler wrote Pales that significant research results would help keep MLO open, and he suggested that Pales collaborate with Saul Price in Honolulu to conduct science studies using instrumentation and data from MLO. This suggestion was to culminate in two of the most important early papers from MLO, and it was helped along by Wexler, who advised Price to be patient with the harried and overworked Pales as the two men developed a working relationship (Wexler 1958a).

The Scripps Report Card (January 1958)

Following his visit to Antarctica, Paul Humphrey visited Charles Keeling at Scripps for the second time on December 30, 1957. In a four-page report to Harry Wexler dated January 10, Humphrey candidly described progress and problems with the carbon dioxide program. In spite of plans to ship the CO_2 analyzer to MLO by January 1, Humphrey reported that the instrument would not be shipped until January 16 and would not arrive at Hilo until the first of February. Keeling planned to travel to Hawaiʻi by either MATS or commercial aircraft as soon as the analyzer arrived at Hilo. Humphrey's report also listed several tasks for which Scripps was responsible but which had not yet been completed or which lacked a status report. Most of a page was devoted to the problem of office and laboratory space for Keeling's program:

> The construction of Dr. Keeling's laboratory facility is not as yet started. . . . The actual work to make Dr. Keeling's laboratory operational is considerable. . . . The conservative estimate of the time which may elapse from the beginning of the construction until samples are being processed routinely will be four to six months. (Humphrey 1958a)

Humphrey's report was not optimistic about the space problem, to which it devoted nearly half a page:

> The outlook for a suitable space for the program is very dim. The possibility of using laboratory space across the hall seems to have disappeared.

The report illustrated the problems faced by Keeling by mentioning that the only telephone in the Scripps Chemistry Division was in Keeling's office. "It is noted that Keeling stopped several times in the course of the afternoon to answer calls which were not his own," Humphrey (1958a) wrote.

Humphrey's report didn't spare his colleagues in the Department of Commerce, for it noted that the Bureau of Standards had not yet prepared the standard tanks of carbon dioxide it had promised to send Scripps.

On the same day that Wexler read Humphrey's report, he sent a terse note to Keeling. The letter mainly addressed issues over which Keeling had only partial control, which may explain why Wexler copied it to S. Ruttenberg of the IGY and Roger Revelle of Scripps.

> As you already know from my letter of 13 November 1957, I am greatly concerned that a third of the IGY has passed without the establishment of the U.S. sponsored air sampling program at a network of surface stations. Not having received a reply to that letter, I am further disturbed by the impressions received by Paul Humphrey during his recent visit to Scripps. His report, copy of which is attached, indicates little if any progress has

> been made in attaining this goal in the intervening two months.
>
> It would appear that emergency action is called for and I would hope that the suggestions made by Humphrey will assist in speeding up early dispatch of the air sampling containers [flasks] to the designated stations here and abroad.
>
> For the sake of the over-all sampling program it would appear best for you to postpone your trip to the Mauna Loa Observatory until Mr. Pales has had a chance to set up the carbon dioxide measuring equipment and observations. If no problems arise and the program proceeds smoothly a visit might be unnecessary, thus permitting you to devote the full time to other aspects of the sampling program.
>
> I would be interested in receiving your reaction to Humphrey's suggestions and your estimate of the date of inauguration of the surface network air sampling program. (Wexler 1958b)

In a draft of a letter to Humphrey dated January 13, 1958, Keeling explained that flying to Hawai'i on MATS aircraft was not a good idea:

> The fact that life insurance policies are often voided for MATS travel is evidence to me that there is some statistical basis for believing that such travel carries a degree of risk. Having avoided that risk I decided to avoid the risk of air travel altogether (except between Honolulu and Hilo) by patronizing the Matson Steamship Company. (Keeling 1958a)

The letter was not sent, for Wexler's letter halted Keeling's plans to travel to Hawai'i.

Humphrey's prior dealings with Keeling had been cordial, but Wexler's letter and Humphrey's report raised a red flag at Scripps. The already overworked Keeling interrupted his schedule to prepare a four-page memorandum to the Scripps Carbon Dioxide Committee to be sent to Revelle after "any necessary changes."

The Carbon Dioxide Committee was apparently the "CO_2 Guidance Committee" discussed in an attachment to a November 5, 1957, letter to the Weather Bureau's E. A. Volbrecht by Eugene Wilkins, the one-man "meteorologist in charge" of the WBO (Weather Bureau Office) assigned to Scripps by Wexler. Wilkins, who had become an invaluable aide to Keeling, described the committee in the attachment:

> The CO_2 Guidance Committee: Drs. Revelle, Craig, Suess, Wheelock and Keeling, and committee secretary E. M. Wilkins, convened last Friday to discuss the problem of laboratory and office space for the CO_2 program. (Wilkins 1957)

Norris Rakestraw was also on the committee, but he did not attend the meeting since he was measuring carbon dioxide during a Scripps IGY cruise to Tahiti. While the Carbon Dioxide Committee was well aware of the space problem two months prior to Humphrey's January 10, 2008, report to Wexler, the problem had not been solved.

Keeling's memo, which was written in the third person, rebuts most of Humphrey's report:

> His observations except where they are qualified by personal opinion are accurate but often trivial. Those remarks, affected by personal judgment, which suggest that Humphrey may view parts of the program darkly need not, however, be taken in this sense, since at the time of the visit, in terms so distinct that Keeling even now cannot fail to recall them, Humphrey remarked:
>
> (1) He and those whom he represents have unbounded faith (or something approaching this) in Keeling's ability to prosecute the carbon dioxide program.
> (2) It is recognized that certain problems such as suitable space for the program are entirely internal matters of Scripps and therefore are not of concern to the U.S. Weather Bureau.
> (3) As a result of spending several weeks in the Antarctic amid ice and snow, he now has a "detached" attitude toward the IGY carbon dioxide program in general. (Keeling 1958b)

Unless otherwise credited, all photographs taken by the author.

PLATE 25. The MLO building was remodeled during May–June 1977. *(John Miller, MLO archive.)*

PLATE 26. An air inlet tower was installed over the main building in 1977. *(John Miller, MLO archive.)*

PLATE 27. Workers completed the installation of the new air intake tower on June 24, 1977. Construction of a much higher solar radiation tower was begun on July 5. The dome that housed the Dobson spectrophotometer *(left)* is still in use. *(John Miller, MLO archive.)*

PLATE 28. The remodeled observatory in 1978. The original wood tower was replaced by a taller tower for the solar monitoring instruments. The white structure at left is the Butler Building that was used by the atmospheric electricity project in the early 1960s. It was moved to the position shown here by Howard Ellis ca. 1964, where it is used for storage. *(John Miller, MLO archive.)*

PLATE 29. The Rev. Edward Kealanahele, husband of former MLO secretary Marge Kealanahele, blesses MLO director John Miller, visitors, and the observatory during the dedication of the renovated MLO on January 28, 1978. Dr. Lester Machta, director, Air Resources Laboratory, is standing to the right of Dr. Miller. The Honorable Daniel Akaka, US representative from Hawai'i, is to the right of the Rev. Kealanahele. *(MLO archive.)*

PLATE 30. Aerosols from the 1982 eruption of Mexico's El Chichón volcano created spectacular twilight glows at MLO, including this one before sunrise in May 1982. *(Kinsell L. Coulson, MLO archive.)*

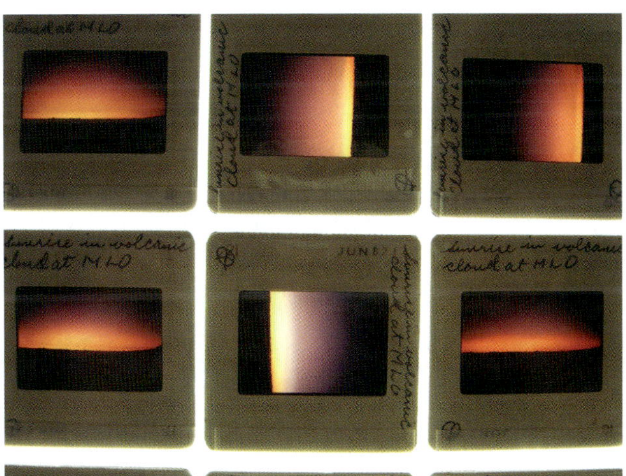

PLATE 31. Some of the many El Chichón twilight glows photographed by Kinsell Coulson in 1982.

PLATE 32. Darryl Kuniyuki *(left)* and Preston Sato review El Chichón twilight photos on December 7, 2006.

PLATE 33. The 126-foot (38.4-meter) meteorological and atmospheric sampling walk-up tower was erected in only four days in December 1986. *(MLO archive.)*

PLATE 34. The top of the walk-up tower at sunset on May 3, 2009. Visible are air sample lines, meteorological instruments, lightning arresters, and a red warning light.

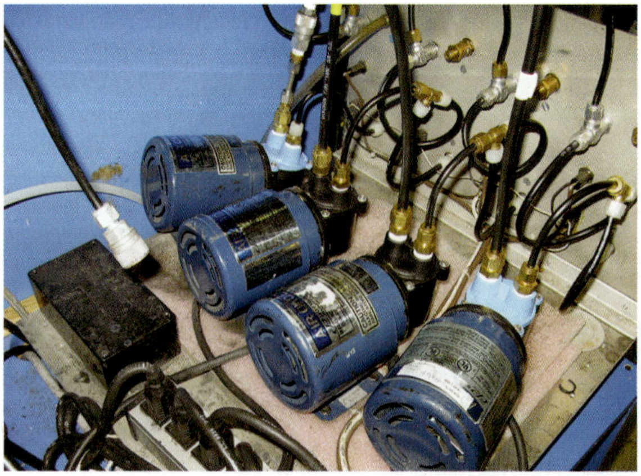

PLATE 35. This array of noisy pumps in the main building continuously sucks in air from inlets on the walk-up tower.

PLATE 36. Clean Rest Rooms, one of the two original buildings, has escaped remodeling during the many changes at MLO over the years.

PLATE 37. The USSR research ship *Akademik Korolev* visited Hilo from April 28 to May 1, 1987, during SAGA II, the Soviet/American Gas and Aerosol expedition. *(Darryl Kuniyuki, MLO archive.)*

PLATE 38. MLO's Judy Pereira helped host a welcome feast for the visitors from the *Akademik Korolev*. *(Darryl Kuniyuki, MLO archive.)*

PLATE 39. The USSR research ship *Akademik Korolev* departed Hilo on May 1, 1987. *(Darryl Kuniyuki, MLO archive.)*

PLATE 40. The NDSC Building neared completion in 1997. The solar radiation deck was not yet installed. Note the brown water tank at the back (south side) of the building. *(Russell Schnell, MLO archive.)*

PLATE 41. MLO director Russell Schnell and the newly completed NDSC Building. *(MLO archive.)*

PLATE 42. Group photograph of the fortieth anniversary field trip to MLO on November 19, 1997. *(Darryl Kuniyuki, MLO archive.)*

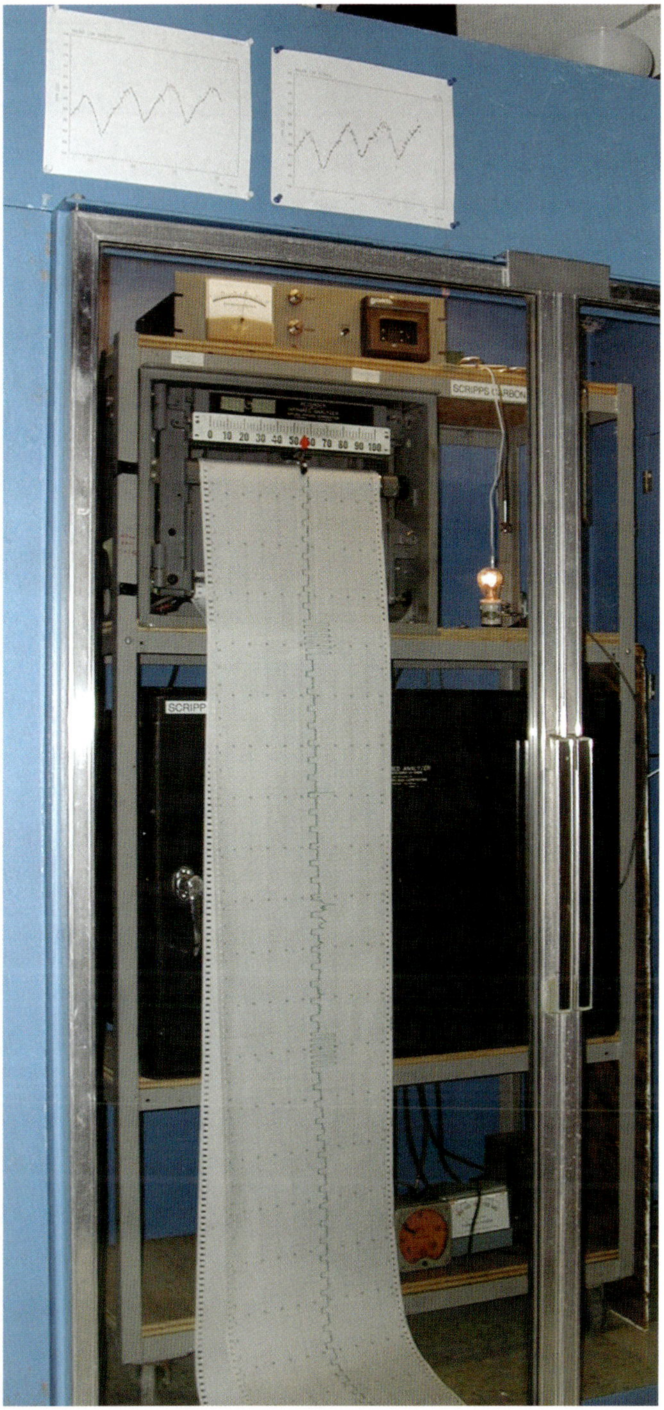

PLATE 43. The APC gas analyzer *(behind paper chart)* was sensitive to temperature changes. In this June 22, 2001, photograph, the analyzer is installed in an enclosure in which a lamp *(upper right)* regulated the temperature.

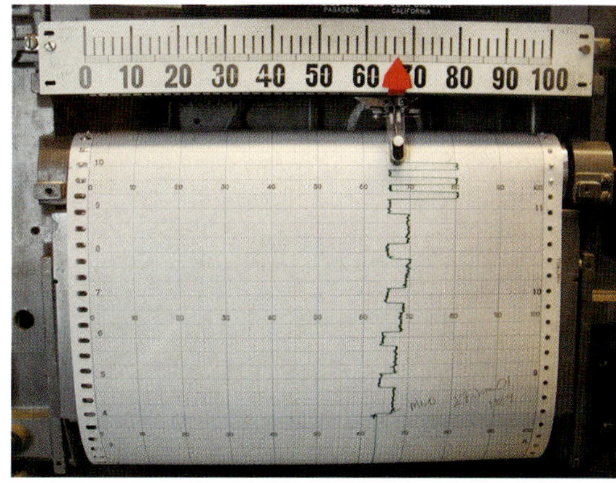

PLATE 44. The APC gas analyzer at work on June 27, 2001.

PLATE 45. One of the vacuum tubes from the APC gas analyzer.

PLATE 46. Dr. Charles Keeling *(left)* and John Chin on November 30, 1997, when the MLO building was named the Keeling Building. *(John Miller, MLO archive.)*

PLATE 47. A snowstorm blocked access to MLO in March 1990. Darryl Kuniyuki and Steve Ryan hiked the snow-covered road to reach the observatory on March 6. *(Darryl Kuniyuki, MLO archive.)*

PLATE 48. Bob Uchida at a power line pole damaged by lightning. *(MLO archive.)*

PLATE 49. The MLO staff wisely left the mountain when a thunderstorm arrived on June 27, 2001.

PLATE 50. A summer hailstorm transformed the mountain station into a winter wonderland on July 1, 2001.

PLATE 51. Storms provide water at MLO, as when this tank overflowed *(lower left)* during the hailstorm of July 1, 2001.

The report continues with a detailed explanation of how Keeling planned to continue the previously agreed to plans for beginning the aircraft sampling program and to also visit the planned CO_2 project at Mauna Loa Observatory to conduct scientific investigations and to train the staff in the operation of the equipment.

> The primary purpose of Keeling's visit is to make a study of the influence of volcanic activity on the measurements of carbon dioxide at the station, a study which ought to be in the hands of someone thoroughly acquainted with the equipment and with previous experience measuring atmospheric carbon dioxide in the field. (Keeling 1958b)

Carbon Dioxide Committee members Harmon Craig, Norris Rakestraw, and H. E. Suess (1958) sent their opinion to Dr. Revelle the same day Keeling sent the committee his rebuttal:

> The CO_2 Committee feels that the IGY program on carbon dioxide abundances is being handled in a satisfactory manner. . . . The space problem still exists, but we are actively at work on this and believe it will be settled very shortly.
>
> This committee believes that Keeling should proceed to Mauna Loa on schedule to set up the analyzer station. . . . We agree that most of the points raised by Humphrey are trivial.

Scripps quickly found proper working space for Keeling, and Humphrey (1958b) extended an olive branch to Keeling in a letter dated January 29, 1958, that was copied to Wexler. Following two pages of program updates and a request to receive replies from Keeling, he wrote,

> I was pleased to note that excellent office space has been assigned in Ritter Hall. . . . Along with your title of "Senior Investigator" I hope you also have the feeling that you are "carrying the ball" as far as the program is concerned. We are alert to help, but you have the initiative so far as I am concerned. I believe that you should be the one to feel the responsibility for unanswered correspondence, or for starting correspondence, rather than our office. . . . I continue to hope that you will show sufficient progress for the program to be regarded as successful. If you can suggest any further way help can be given let us know.

Two days after Humphrey wrote Keeling, the United States finally joined the Soviet Union in the space race when a modified Jupiter rocket called the Juno launched Explorer I into orbit. The routine IGY measurements on Mauna Loa received none of the immense publicity devoted to the first U.S. satellite. No one then knew that Pales and his colleagues were helping establish the foundation for the ground-based calibration of future satellites that would measure the ozone layer and aerosols from space.

The Model 70 Analyzer Finally Arrives at MLO (March 1958)

Significant observational science was underway at MLO, and on February 25, 1958, Roy Fox and Ray Busniewski visited the station to see the progress. Jack Pales showed them the details, including a strip chart recording of a full day's solar irradiance measured by the instruments installed less than four months before (Fig. 7.1). Much had been accomplished since Fox helped Ralph Stair select the site for the observatory and since he and Busniewski attended the 1956 dedication (see Fig. 5.5).

Meanwhile, the carbon dioxide analyzer purchased for MLO was finally ready to go to Hawai'i, and Harry Wexler himself took charge, as described by Keeling (1998) in his autobiography:

> Then, in November, Mr. Ben Harl[i]n, who had operated the CO_2 equipment at Little America [Antarctica], conveniently returned to the United States. To prevent further delay Wexler enlisted him to install the APC analyzer system on Mauna Loa. This he did in March, 1958, with the help of Jack Pales, a recently arrived physicist-in-charge of the observatory. In advance, I gave advice including a suggestion of what CO_2 concentration

FIGURE 7.1. On February 25, 1958, Jack Pales *(standing)* showed Roy Fox *(seated, right)* and Ray Busniewski a chart recording of a full day of solar irradiance at MLO at the desk first shown in Figure 6.4. *(MLO archive.)*

> to expect if the equipment worked correctly. By assuming that the concentration there would match closely with test measurements of presumably unpolluted air made shortly before at the Scripps pier, I told Harl[i]n to expect 313 ppm [parts per million].

On February 28, Paul Humphrey notified Keeling by teletype that Harlin, who was a Weather Bureau meteorologist, would arrive at La Jolla to visit Scripps on March 3. On March 7, Humphrey (1958c) explained why it was not yet time for Keeling to visit MLO:

> With regards to your request for travel to Hawaii, Dr. Wexler is hesitant to take action to have you away from your laboratory while you are so busy inaugurating the flask sampling program. We hope that when the time comes for your trip you will travel by MATS and thus help preserve funds.

Humphrey's letter is annotated, presumably by the recipient, with several marginal markings that highlight some of the residual tension between Scripps and the Weather Bureau. "This is not new" is added next to the paragraph quoted above. "NO!" is printed adjacent to a suggestion by Wexler for flask sampling dates and a second "NO!" is added adjacent to a request that Keeling notify the Weather Bureau if he approved the dates.

Meanwhile, back in Hawai'i, the Mauna Loa Observatory was about to experience a major storm, an event that is described in the *Weather Log:*

> Mar. 6, 1958
> The day dawned with overcast skies and hail and freezing rain and snow. Winds were from SWly direction with a speed of 47 kts. All day the wind blew with speeds from 27 to 47 kts, occasionally 60 kts. Blew down the outhouse. Temp 36 to 45°. Sky threatening and overcast all day. Wet day! Pressure 19.865 to 19.930. Pressure very unsteady and was rising and falling rapidly all day. Worst weather so far this year. Max Temp 46°; Min Temp 36°. Rain 0.12. Sunshine 326 minutes. (Pales et al. 1957–1958)

Two days later, Ben Harlin signed the MLO *Visitors* book. The *Weather Log* described the day as very windy with a towering cumulus cloud and frequent lightning visible in the afternoon. The next day was windy with thunder, rain, sleet, and the first snow of the year. It was not an auspicious beginning for what was to become the observatory's best-known project.

The carbon dioxide monitoring system was not simple, as can be seen in Figure 7.2, which shows the system as it appeared some eighteen months following Harlin's installation. Figure 7.3 is an aerial photograph of MLO from this time that shows the air sampling lines laid along the lava to sampling points well away from the main building and its generators. The time required to install the carbon dioxide equipment and arrange for external air sampling lines is unknown. What is known is that everyone was surprised when the first measurements on the mountain were exactly what Keeling had predicted. Later he modestly wrote that the exact agreement was merely an "accident" and that subsequent measurements were 1 ppm different. The clear fact is that Keeling had done his homework. The background level of carbon dioxide around the world at that time really was very close to what he was measuring in

FIGURE 7.2. The Scripps APC carbon dioxide analyzer *(top center)* and its chart recorder *(lower right)* were installed in March 1958. This photograph is from September 1959. *(MLO archive.)*

FIGURE 7.3. Aerial view of MLO looking north-northeast (downslope) in early 1958. The outhouse and generator shed are at left. The Kiess-Corliss instrument shed is in front of the main building. The Dobson ozone building with sliding roof is at the right end of the main building. The row of solar radiation instruments is at center foreground. The white lines at left and right are air lines that lead from sampling points to the Scripps APC carbon dioxide analyzer in the main building. *(MLO archive.)*

rural California away from local contamination. Though it was largely unrecognized at the time, Charles David Keeling had become the world's leading experimental CO_2 scientist.

On March 25, Keeling (1958c) sent a four-page letter to Wexler thanking him for the updates and apologizing for not replying to Wexler's letters of November 13, 1957, and January 10, 1958. He diplomatically points out that the Weather Bureau was slow in informing him about recent developments in the air sampling program, and he then addresses the status of the Mauna Loa Observatory project:

> You probably know that I have taken your suggestion, seconded by Dr. Revelle, to postpone my trip to Mauna Loa. I feel that the Weather Bureau acted wisely in sending Ben Harlin out to help Jack Pales start up the equipment. However, I do not regard Harlin's visit as more than a temporary expedient. Last year, although he was conscientious and worked very hard to keep the equipment at Little America running well, the data he obtained suggest that he did not understand the chemical aspects of the problem well enough to insure their reliability.

Keeling's letter to Wexler goes on to affirm the good working relationship he had established with Eugene Wilkins, the Weather Bureau meteorologist whom Wexler had assigned to Scripps to assist Keeling. He closes with more good news:

> Now that the space and secretary problems are solved, and the major development work is nearly over, the carbon dioxide programs, Gene Wilkins assisting, ought to proceed reasonably fast. However the IGY is rapidly drawing to a close. Paul Humphrey told me that there are plans to keep Mauna Loa running next year. I hope very much that this can be done since 18 months of carbon dioxide data are needed to establish any sort of secular trend. I believe that Scripps will be able to continue other aspects of the program for another year if this can be arranged with the cooperating agencies. It has been a pleasure to be able to work with your office this past year. I hope that soon we will be able to carry out enough measurements to justify some of the hopes of you and Dr. Revelle who had so much to do with getting the program started.

On April 4, Keeling (1958d) sent the Weather Bureau's Paul Humphrey a detailed six-page status report on the carbon dioxide program at Scripps that includes an enthusiastic report on the installation of the Model 70 analyzer at MLO:

> Ben Harlin stopped at Scripps on Monday [March 31, 1958]. He brought back a written report of his activities which is one of the best compiled documents I've ever had the privilege of reading. I feel that Ben fulfilled his trust with distinction. From his report, which was accompanied by the original chart records, I have nearly as good an impression of the status of the project at Mauna Loa as though I had been at Mauna Loa myself. If Ben were to return to the station to take charge of the routine operations, and if the erratic high carbon dioxide concentrations which he found during the last day of his stay there could be expected to be infrequent, I would feel that the station could run successfully without my making a visit.

Keeling's letter continues with some concerns about the fledgling CO_2 program:

> However, Ben left some doubt in my mind that the operation after he left would continue in exactly the same vein. I am disturbed that Jack Pales was absent from the station most of the last week Harlin was there, during the time the equipment was running at its best. . . .
>
> In order to keep in close touch with operations at Mauna Loa until these uncertainties are cleared up I would like to have Pales compile daily reports similar to those Ben compiled. . . . A copy of the proposed form is enclosed for your consideration. Unless you object , I will instruct Pales to send these reports . . . to us each week until further notice.

The next portion of Keeling's letter to Humphrey deals with the Little America Program and particularly Ben Harlin's role. Two weeks earlier, Keeling's letter to Wexler of March 25 had questioned the reliability of the CO_2 data obtained by Harlin in Antarctica, suggesting the problem was owing to Harlin's failure to understand "the chemical aspect of the problem." But in his letter to Humphrey, Keeling (1958d) takes the blame for six months of bad data:

> Looking over my correspondence I don't find any letter about Ben Harlin's work last year in the Antarctic. He worked against adverse conditions, and it was a distinct accomplishment that he succeeded in operating the equipment at all. There is no doubt that he worked very hard to follow the instructions which he was furnished and that the shortcomings of last year's operation were a result of inadequate instructions and equipment proceeding from Scripps. Specifically, Scripps set up an inadequate program of inter-checking the [calibration] tanks . . . the Gast pumps were a poor choice for circulating air, and my original idea of long intakes [air sampling lines] appears to have been poor. The latter led to Ben's depending heavily on a 900 ft NE line which probably gave contaminated values nearly all the time from April through September 1957. I want to make it clear that these results should not reflect on Ben's work and that he did a superb job under the circumstances.

Keeling (1958d) closes his report to Humphrey with a friendly note about the progress that had occurred:

> Last night I read over the complete file of correspondence between Scripps and the Weather Bureau looking for unfinished business. This reading gave me the impression of the great quantity of business which has passed between these two organizations two thousand miles apart in order to bring the carbon dioxide program to its present status. It left me with a conviction that Scripps owes the Bureau a great debt of gratitude for their assistance.

Humphrey's critical report to Wexler about the Scripps carbon dioxide program was in the correspondence file that Keeling read, but he didn't mention it. Instead, Keeling (1958d) suggested that he and his friend Humphrey visit Cape Hatteras during Keeling's next visit to Washington:

> A visit there would be highly justified because it would allow an inspection of what ought to be one of the best of the land station program sampling sites if only a couple of experts like you and myself have a chance to perform a preliminary reconnaissance. If you are interested, I will bring my sleeping bag.

Humphrey was "very pleased" by Keeling's suggestion of a visit to Cape Hatteras. In a letter dated April 18, he suggested that they could drive to the cape and suggested that Keeling ship his sleeping bag and supplies.

Perhaps motivated by the successful installation of the Model 70 analyzer at MLO, on April 29, 1958, Wexler visited Scripps to discuss the CO_2 program. But the good news from MLO was soon followed by bad news when generator problems shut down the CO_2 analyzer for much of June 1958 (Keeling 1978; also see Keeling et al. 2005). Nevertheless, by July enough data had been collected from MLO, Little America, various other ground sites, and from flights over the Pacific for Keeling to conclude in a preliminary statement for an IGY meeting in Moscow that Wexler planned to attend that "the [global] concentration of carbon dioxide appears to be nearly constant" (Keeling 1958e).

Work, Tension, and More Work at MLO (1958)

From the outset, Jack Pales was much more than an administrator, for he played a key role in the day-to-day operation of the observatory. Pales was especially interested in the sun measurements made by the sun-tracking Eppley pyrheliometer, the earliest of the advanced experiments at MLO. Figure 7.4 shows Pales adjusting the position of the instrument. The growing workload increased the need to

FIGURE 7.4. MLO director Jack Pales adjusts the Eppley normal incidence pyrheliometer in 1959. This instrument and several identical successors have provided the world's longest series of direct sunlight measurements. *(MLO archive.)*

find additional staff, and in April 1958 Wexler assigned Ronald Taylor to MLO for four months. The *Visitors* book records that R. C. Taylor signed in on May 20, 1958, with the comment, "Entering on Duty." He was accompanied by Dr. Lester Machta, who also signed the log. In 1948 Wexler had selected Machta to direct the Special Projects Section of the Weather Bureau. Like Wexler, Machta supported the measurement of carbon dioxide in collaboration with Charles Keeling (NOAA 2001).

Howard Ellis (1988) describes in his history of the observatory the busy work schedule at MLO in those days and some of the interactions of its staff:

> All surface observations were called down to the [Weather Bureau station at the Hilo airport] within seconds after each observation at hourly and at shorter intervals as required by their interpretation of the airways manual.

A staff memo by Mr. Pales on May 23 [1958] outlining the duties gives some flavor of these times.

Williams:
1. Maintenance of vehicles and execution of truck logs.
2. Solar radiation program. (Data reduction, forms and calibration)
3. Air sampling program. (Micrometeor and radioactivity)
4. Freezing nuclei program.

Kutaka:
1. Water and fuel oil supply.
2. Surface observations, operations record.
3. Surface ozone. (Records, calibration)
4. Cloud photography.

Taylor:
1. Laundry, distilled water, helium.
2. Atmospheric ozone (forms, calibration).
3. Carbon dioxide program (forms, calibration).

Mr. Williams [shown at work in Fig. 7.5] had a hi-fi system that Mr. Kutaka didn't always enjoy hearing. By July, Mr. Williams wanted out at the end of the year. Mr. Kutaka wanted to stay through the fiscal year. Mr. Pales didn't want Mr. Taylor replaced after his stay was over.

Mr. Pales defended a request he had been denied to visit Washington, D.C., saying that it was for technical help and not for personal reasons.

Clearly the long work periods on the mountain were beginning to take their toll on the MLO staff. In a sign of things to come, the scientists on the mainland didn't always appreciate the rigors of life at MLO, the difficulties and hazards posed by the trips to and from the station, and the need for flexibility.

In a July 28 letter, Harry Wexler (1958c) informs Charles Keeling that both Saul Price and Jack Pales had made "strong recommendations" that he should visit MLO. He states that the Weather Bureau would pay for the trip and asks Keeling to advise him when he could leave. Wexler also writes about a new development at MLO:

> In a memorandum, dated July 24, Jack Pales states that "Doc" Foss, the [new] electronics technician at MLO, has made some improvements on the CO_2 analyzer, which Mr. Pales believes has increased its accuracy and reliability considerably. He suggests that these be inspected and approved by you before MLO gets too much farther along in the program.

While the "improvements" are not described, based on problems listed in the CO_2 *Log,* possible candidates include changes to the sample and reference gas lines and pumps to reduce leaks and modifications to the freeze trap that froze water vapor in the air sampling line so that it could not reach the analyzer and contaminate the measurements.

Two days later, presumably before Wexler's letter arrived from Washington, Keeling (1958f), who was already aware of Pales' suggestions about improving the carbon dioxide system, expressed concerns about Pales in a letter to Humphrey:

> I am enclosing a copy of a recent letter to Jack Pales which, I hope, conveys no unnecessary irritation. It was difficult for me to decide how to approach Pales regarding some of his activities which I consider to be un-satisfactory. I would like your opinion of my final choice of rather direct criticism. You doubtless remember my worry about Ben Harlin during his stay at Little America. I know now that this was unjustified; that Ben did an exemplary job. It is possible that my worry about Pales is equally unfounded—a mere skittishness. Nevertheless, although I have tried to embrace the art of positive thinking about Pales, I have not been able to shake off the feeling that the carbon dioxide program at Mauna Loa will not run properly until Pales adopts a fiercer interest in the operation. Pales appears to have two attributes which, taken in conjunction, considerably restrict his value as a technician operating other people's equipment: first, his independent spirit regarding

FIGURE 7.5. Robert Williams, one of the three original MLO staff members, checks the chart recorder for the Eppley normal and horizontal pyrheliometers at the observatory's south wall work area in mid-1958. *(MLO archive.)*

suggestions of a technical nature; second, his lack of initiative regarding activities which involve the consumption of time. We do not yet have enough information from Pales to even begin to properly assess the contamination situation at Mauna Loa [caused by CO_2 from the diesel generators and the mountain itself]. Unless Pales can be convinced of the utility of carefully adhering to our instructions, and transmitting information promptly and adequately, I believe that Scripps ought not to be considered responsible for the quality of the data produced at Mauna Loa.

Keeling also wrote that the MLO data "may not be of a quality to justify its inclusion in the IGY annals" until the contamination problem was resolved. He then pointed out that a visit by him to MLO would come near the end of the IGY and asked if the program would be continued at MLO through 1959. Keeling (1958f) then brought up a new idea that some in the Weather Bureau would later consider:

> On the other hand if Mauna Loa turns out to be unsuitable do you know how we might be able to find personnel to operate on Haleakala [Maui] during 1959? When I talked with Saul Price last month we discussed this possibility. Haleakala should not have any orogenic [volcanic] carbon dioxide. In other respects Mr. Price regards Haleakala to be as good a location as Mauna Loa.

Humphrey shared Keeling's letter about Pales with others before sending Keeling his response on August 7:

> It seems that you are correct in writing to Jack Pales and explaining in detail all of the necessities for the Carbon Dioxide Program. Your letter was read and appreciated by Ronald Taylor and Saul Price. The consensus is that Pales would have made a better showing if it had been possible for him to devote more time to carbon dioxide. The generators have given him very much trouble; but at present, the original generators are performing well. We are hoping that the two new 25 KW generators, already ordered, will perform even better. With the generator problem solved, and with more experience, we are hopeful that Jack's work on the observational program will become much more satisfactory. (Humphrey 1958d)

As is apparent from the MLO record, Pales hardly lacked initiative. Indeed, he had joined his crew of two in living at MLO for days at a time since the summer of 1957. Humphrey's letter explained that generator maintenance occupied much of Pales' time, but there was much more to his responsibilities at the observatory. Besides sharing multiple-day, twenty-four-hour shifts on the mountain, including holidays and weekends, with Clifford Kutaka and Robert Williams, Pales had to attend to vehicle repairs, the purchase and transport of generator fuel, generator maintenance, frequent maintenance of the surface ozone instrument, preparation of reports (without a secretary), paying bills, the shipping and receiving of gas cylinders, supplies and instruments, and numerous other chores involved in running a remote mountain observatory.

Amidst all this, Pales had to find time to maintain and repair the CO_2 analyzer and its air intake lines, pumps, water vapor freeze trap, leaks, vacuum tubes, and gas cylinders. Added to this was the time required to extract the data from the chart recordings, preparing and carefully folding the charts for Keeling, and his family responsibilities down in Hilo. In short, Pales was stretched much too thin.

Keeling was well aware of the competing responsibilities of observers at Little America, for he had written Wexler about the difficulties in Antarctica faced by Ben Harlin and Charles Williams:

> Williams gives the impression of having too many responsibilities. He has found that maintaining the equipment takes nearly all his waking hours, seven days a week. I don't see how he can maintain his present high spirits and careful attention to details unless the situation improves or he receives some assistance. It might benefit the carbon dioxide program considerably if there were some way of relieving him of some of his other duties. (Keeling 1958c)

Eventually Keeling understood that Pales faced even more responsibilities than Harlin and Williams had in

Antarctica, and he and Pales developed a good working relationship. Meanwhile, Humphrey, who had been copied on the letter to Wexler, realized that Keeling was hoping for a full-time observer to attend the carbon dioxide project at MLO. He introduced the subject in his defense of Pales:

> Comparing the work of Harlin, Williams and Pales, I would say that desirable performance has been directly proportional to time available. Conditions will improve further with the addition of Doc Foss to the staff; but to actually please you, we need to have a competent individual at the Observatory who will have the CO_2 program as his entire responsibility, or at least his primary responsibility. With our limited funds, I hardly see how we can afford to assign another man to either the Observatory or to the CO_2 program; but a special scientific technician specifically for the carbon dioxide program continues to be worth our consideration. (Humphrey 1958d)

As for the future of the CO_2 program, Humphrey replied that

> It is expected that the carbon dioxide observations at the MLO will extend through FY-59 [Fiscal Year 1959]. The future of the Observatory, and the future of most Weather Bureau operations, cannot be foreseen beyond June 30, 1959.

Humphrey also addressed the sensitive topic of moving the program to Haleakalā, which had a much better access road but much more traffic. He suggested that Keeling reserve judgment until visiting MLO, and he pointed out that there are "probably many artificial sources of CO_2 on Haleakala."

Wexler had reported in his letter to Keeling on July 28 that both Price and Pales had strongly recommended that Keeling visit MLO. Price (1958) followed up on this theme in a letter to Keeling on August 12:

> As the Office of Meteorological Research may have indicated, I strongly recommended to them on the basis of our telephone conversations that you visit the Mauna Loa Observatory as soon as convenient to inspect the station and its environs, especially the suspected [volcanic] vents, and to review completely the CO_2 observations and instrumentation there. Only thus, I felt, could the heavy investment of time and money in that program at MLO be made to pay off in data acceptable to you and compatible with similar observations made elsewhere.

Meanwhile, Keeling was on an extended vacation, apparently a Sierra backpack trip that Humphrey had originally hoped to join. On August 22, Eugene Wilkins (1958) wrote Humphrey in response to Humphrey's August 7 letter to Keeling. The letter provides new insight into the work routine at Scripps and illustrates the wisdom of Wexler's decision to assign Wilkins to temporary duty at Scripps:

> For some time now it has been my job to perform all of the analyses while Dave has been tied down completely with administrative work—no small part of which is just plain cost accounting. While this may seem like an odd arrangement it has been found to be a necessary one. It would not be feasible for me, without Scripps employee status, to do the administrative work. Unfortunately we have not yet succeeded in making the flask analysis technique sufficiently simple that sub-professional help can be used. I do not mind doing the analysis work—in fact prefer it, as I would find it difficult to accept results obtained by someone else (I am getting just like Keeling). From experience I can say that 8 hours at a stretch is about all a human being can endure working on the vacuum line.

Charles Keeling (1958g) confirmed Wilkins' account of their division of labor in a September letter to Humphrey that emphasizes the important role that Wilkins had assumed in the program:

> At present he [Wilkins] is operating the flask analysis program with very little help from me and managing completely the Land Station program. This allows me to put my time on managing details of the Antarctic and Mauna Loa programs, field aspects of the Aircraft program, compiling data and working on the construction

> of the absolute calibration equipment. . . . After the calibration equipment is running, my time will be even more taken up so that the loss of Gene will mean the virtual cessation of flask analyses until I receive some other form of assistance.

An Unexpected Plan to Close the Observatory (1958)

MLO founder Robert Simpson had headed the Weather Bureau operations in Hawai'i, and his successors continued their support of the observatory. This changed in 1958 when Nels E. Johnson became head of the Pacific Supervisory Office of the Weather Bureau. On September 17, 1958, Johnson signed the MLO *Visitors* book for the first time. As noted by Ellis (1988) in his MLO history, Johnson

> seemed a little cooler than previous Heads toward the observatory concept and mainly commented that the walk to the outhouse was pretty rough going. Wooden walkways were built and laid over the lava along the main routes of travel and that helped a lot. By September, Mr. Johnson advised that the observatory function would be moved to [the summit of] Haleakala on [the island of] Maui.

Johnson's plan to close the observatory was as bold as his signature in the *Visitors* book and a total surprise to the staff, particularly in view of the strong support for MLO from Robert Simpson, Harry Wexler, and Lester Machta. But the idea was not new to Charles Keeling, who, as noted earlier, had introduced the possibility in his July 30, 1958, letter to Paul Humphrey should Mauna Loa not prove a satisfactory observing site due to carbon dioxide contamination from the volcano. Keeling was also concerned about the prospect of the carbon dioxide measurements being contaminated by the minimal automobile traffic at MLO. But the Hawai'i part of his program would probably be ended if the observatory was moved to the summit of Haleakalā, a site with a much better road and plenty of tourist traffic.

> Mr. Pales fought against this move and won out, using as his main argument that the 8,000 cars per year reported driving up Haleakala would preclude the carbon dioxide benchmark measurements from being made there. (Ellis 1988)

Charles David Keeling Visits MLO (1958)

Revelle finally approved the requests by Wexler, Price, and Pales that Keeling make his long-delayed initial visit to Mauna Loa Observatory. It was one year and twenty days after Paul Humphrey had written to him during his visit to MLO that Keeling signed the *Visitors* book for the first time on November 22, 1958, as "C. D. Keeling" from "Del Mar, Calif." Keeling arrived just two months after Johnson announced his plan to close MLO. During his visit, Keeling restarted the carbon dioxide analyzer after a power shutdown had occurred. Had Johnson prevailed with his plan to move MLO to Haleakalā, it is likely that Keeling's visit would have been to shut down the carbon dioxide program.

When Keeling visited MLO, the CO_2 analyzer had or would soon have an elaborate air sampling network. Because of significant local CO_2 contamination caused by the MLO generator, multiple sampling points for the CO_2 analyzer were required. Each sent sampled air to the main building via flexible aluminum tubing with an inside diameter of 8 millimeters.

> Air intakes are mounted on four 20-foot towers 450 to 500 feet distant and in mutually perpendicular directions from the main observatory buildings. Two upwind flows are sampled concurrently, and difference between them being interpreted as being due to local contamination. (Price and Pales 1959)

After only a year of operation, the CO_2 measurements provided significant findings. The most important concerned the similarity of the concentrations measured at MLO with those measured elsewhere. Again referring to Price and Pales (1959),

> Data secured at Mauna Loa since sampling began in March 1958 are sufficiently similar to those obtained elsewhere to support previous indications that in the absence of local pollution the carbon dioxide concentration in the lower atmosphere is substantially uniform over the earth.

This uniformity was consistent with near-constant "characteristic background concentration" Keeling had discovered in his early West Coast CO_2 sample-gathering of 1955 and 1956. After less than a year, the MLO CO_2 data also showed interesting diurnal and seasonal variations, which are remarked upon by Price and Pales:

> The Observatory data also suggest a diurnal variation, perhaps related again to the mountain circulation, and a seasonal trend, with mean monthly values decreasing gradually from approximately 314 p.p.m. (parts per million) in May to 312 p.p.m. in August.

The findings of daily and seasonal variation were important discoveries, but a much bigger discovery was yet to be made. Wexler had provided the Weather Bureau's money to buy the Model 70, and the investment was about to pay off in ways neither he nor Keeling could have then foreseen.

Extending the International Geophysical Year (1958)

While the Weather Bureau pondered the future of MLO after the International Geophysical Year, back in Washington the space race was in full gear. On October 1, 1958, the National Aeronautics and Space Administration (NASA) was officially born, and NASA began work with a team of some eight thousand scientists, engineers, and technicians inherited from the National Advisory Committee for Aeronautics (Launius et al. 2006).

Meanwhile, the International Geophysical Year was so successful that at the end of 1958 it was extended until the end of 1959 under the official name International Geophysical Cooperation. The timing couldn't have been better, for Saul Price's collaboration with Jack Pales, which had been suggested by Harry Wexler, had culminated in the first scientific paper describing the IGY research at MLO. "The Mauna Loa High-altitude Observatory" was published on the cover of the Weather Bureau's *Monthly Weather Review* dated January 1959 and issued April 15, 1959. The abstract (Price and Pales 1959) is only one sentence:

> A description is given of the physical setting, facilities, and current program of Mauna Loa Observatory in Hawaii, with emphasis on its suitability as a site for field studies on a wide variety of phenomena.

The paper describes the history and climate of MLO and its desolate setting on Mauna Loa. The access road, main building, work areas, power source, and living accommodations are described in detail. Photographs with numbered markers, probably prepared by Pales in his new darkroom, show the observatory site and the instrumentation section of the main building. The station photograph shows the row of four solar radiation instruments whose descendants are still in use at MLO today. The earliest photographs inside the building show very few instruments, but the paper includes an updated photograph from late 1957 or 1958 (an annotated version of Fig. 7.6) that shows many instruments along the west end of the south wall. Included are the sunshine panel; wind, sunshine, and rainfall recorders; frequency regulators; 50-watt radio transceiver; microbarograph; and chart recorders for measurements of ozone, horizontal (full sky) and normal (direct sun) incidence pyrheliometers, radiometer temperatures, and net exchange and total hemispherical radiometers.

The paper by Price and Pales devotes a special section to "the IGY observations" of surface ozone, total ozone, carbon dioxide, and solar radiation. Also covered was the detection of radioactive particles, micrometeorites, and freezing nuclei in the air. The new paper provided an important boost to MLO's reputation. While the IGY activities at MLO were known to some within the meteorological community, no significant results had been released prior to the paper by Price and Pales.

FIGURE 7.6. The south wall work area at MLO about September 1958. The chart recorders to the right of the door recorded sunshine, wind, and rainfall. The recorders at right *(from the top)* were for the horizontal and normal incidence pyrheliometers, net exchange and total hemispherical radiometers, and surface ozone. *(MLO archive.)*

Several months before publication of the Price and Pales paper, the latest paper by CO_2 scientist G. S. Callendar (1958) had been published. While the status of the CO_2 program at MLO was apparently unknown to Callendar at the time, his paper discusses the CO_2 measurements that were planned for the IGY. The paper somehow found its way to MLO, where a photostatic copy was found stashed in an old file. The paper's discussion of the status of the measurements of CO_2 in the tropics must have been of high interest to Jack Pales as he gained experience with the APC Model 70 Infrared Analyzer.

The First MLO Office Is Opened in Hilo (1959)

The increasing responsibilities and the heavy workload at MLO led to a need for more office space in Hilo than available at the Weather Bureau office. Electronics technician Colby Foss, who first signed in at MLO on March 17, 1957, had been reassigned from the Honolulu Weather Bureau to MLO in 1958. While Foss' chief legacy was his ability to keep the MLO instruments working properly, he was a cross-disciplinarian who was well able to assist in other ways. According to Howard Ellis (1988), it was "Doc" Foss who found space for an MLO office near the Hilo Weather Bureau office in the southwest end of the Air Cargo Terminal at the General Lyman Field. A large grass field adjacent to the end of the building was available for outdoor instrumentation. Foss added partitions inside the warehouse-like facility to provide offices and a centrally located work space that included a long worktable. A loft along one side provided storage space.

The Hilo office of the Mauna Loa Observatory opened on April 15, 1959 (Ellis 1988). The new office was less than 0.8 mile (1.25 km) from the 1840–1841 site of Observatory 1 of the US Exploring Expedition (chapter 3). The Air Cargo Terminal is still in use today, and the space that once served as the office for MLO is now occupied by FedEx. Delivery trucks are serviced where the worktable once stood.

The new Hilo office was acquired shortly after Price and Pales submitted the final version of their first paper about the observatory to *Monthly Weather Review*. The facility finally provided the space and shop facilities the MLO staff had long needed. Meanwhile, up on the mountain all the IGY projects were up and running and the observatory was being powered by a pair of new 35-kilowatt diesel generators. Howard Ellis (1988) summarizes the status of the observational program as of 1959 in his MLO history:

> Surface weather observation (0700–1700)
> Atmospheric ozone observations
> Surface ozone measurements
> Freezing nuclei counts
> Carbon dioxide measurements
> Radioactivity by air filter collection
> Time lapse photography
> Solar radiation observations (normal incidence, total sun and sky, and total hemispherical and net exchange)

The expected programs were the following:

> Atmospheric electricity measurements
> Condensation nuclei counts
> Turbidity measurements
> Cloud droplet size distributions
> Infrared hygrometer development
> Expanded freezing nuclei counts
> Expanded surface and atmospheric ozone measurements and Umkehr observations
> Expanded carbon dioxide measurements
> All sky cloud photography

Looking for Work (1959)

The spring of 1959 was especially busy for Jack Pales. The previous year he had been denied permission to visit Washington, DC. But he was allowed to attend a science conference in Washington in April 1959. Aware of the impending end of the extended IGY, Pales took the opportunity to look for new research projects for MLO.

> During a two-week period in April, scientists at many locations around the nation were surprised, when upon answering their telephones, they heard the sound of coins being placed into a pay telephone. It was Mr. Pales, who had been granted a trip back to Washington, D.C., calling with his own coins from various places to drum up projects for the observatory, following some leads he came across, just as would have a hungry salesman.
>
> He scored best at the U.S. Army Electronics Proving Ground, Fort Huachuca, Arizona where he soon after put together "on his own" a program to measure solar radiation, dust, ozone and surface weather at both Hilo and the observatory for eighteen months. This was justified by the general promise of defining climatic extremes so that the Army could better write specifications for contractors who might furnish any sort of material for them. (Ellis 1988)

The US Army program, which is described in chapter 8, required two years to plan and place in operation.

While Pales was in Washington, the acting chief of the Weather Bureau's Meteorological Satellite Section visited MLO and on May 16, 1959, sent a four-page report to Wexler questioning who was really in charge of the observatory. Howard Ellis (1988) quotes from the report in his MLO history:

> He [the visiting official] advocated expanding the building so as to have a more pleasant place for off duty hours. He thought that the "large number of partly finished installations at the observatory" was "indicative of too many tasks being undertaken by the limited staff."
>
> The people there, he said, "were there to make a living and couldn't be expected to have the motivation of research scientists," and anyway, "the three-hour ride up the road was enough to shake out what little enthusiasm they might have had."

Wexler's response to these criticisms is unknown, but a hint of how he might have responded is provided in the letter that he sent Jack Pales on January 8, 1958, following his visit to MLO in December 1957. A portion of this letter was quoted earlier and, in view of the above, it bears repeating here. Wexler (1958a) wrote

FIGURE 7.7. Shown here in June 1959, with the Air Cargo Terminal in the background, are *(left to right)* Howard Tatum, Robert Williams, Harry Wexler, Jack Pales, and Colby "Doc" Foss. *(MLO archive.)*

> how impressed I was with the ship shape appearance of the Mauna Loa Observatory and the high enthusiasm of yourself and observers. You have been doing an excellent job with slender resources at the end of a long pipeline.

The IGY Winds Down

Wexler visited MLO in June 1959 (Fig. 7.7), bringing $15,000 of $100,000 requested by the Kūlani Prison for proposed work on the deteriorating road. Ellis (1988) wrote that

> [Wexler] said that projects should be undertaken that could only be done at this unique site. He cautioned about accepting any programs from either within or without the [Weather Bureau] which might destroy the uniqueness of the location. He had the radiometer observations, which had not given data that anyone wanted to use, stopped and suggested a simple temporary wind

observing network about the mountain. He asked for a once-a-year report detailing the status of each program and what had been done with the data.

Tom Tyrrell and Bill Cobb Report for Duty (1959)

In November 1959, Nels Johnson and Ray Busniewski of the Weather Bureau joined some of the MLO staff for a rare group photograph (Fig. 7.8). The photograph included two newcomers, Thomas "Tom" Tyrrell and William "Bill" Cobb. Tyrrell first logged in at the station on July 19, 1959. The MLO assignment was very difficult for both staff and their families, and the *Visitors* book records various visits by family members. On January 27, 1960, Dorothy and Patrick Tyrrell "came to visit Daddy for a few days," and two months later, Tyrrell decided to leave. On April 5, 1960, he signed out with "I have enjoyed my stay but am very happy to be going home."

Hawai'i became a state on August 21, 1959. That month, four Weather Bureau men arrived at MLO for a major study of atmospheric electricity. The team included Bill Cobb, Byron Phillips, Paul Allee, and summer student worker G. T. Severynse. They worked from a small room that they added to the back of the shed that housed the Dobson spectrophotometer that measured the ozone layer. Soon they assembled a metal Butler Building east of the main building to house their equipment and provide work and office space. A large metal ground plane was installed behind the building to detect signals for their instruments. The Butler Building was about the size of a one-car garage, which made it the second-largest structure at the observatory.

Cobb, who signed in at MLO on August 17, was a US Navy veteran who saw duty in the Pacific during

FIGURE 7.8. Weather Bureau (WB) and MLO staff at MLO on October 18, 1959. *From left to right,* Bill Cobb (MLO), Ray Busniewski (WB, Hilo), Nels Johnson (WB, Honolulu), Clifford Kutaka (MLO), Russell Grubb (WB, Washington), Jack Pales (MLO), and (kneeling) Thomas Tyrrell (MLO). *(MLO archive.)*

FIGURE 7.9. Bill Cobb *(left)* and Byron Phillips established the atmospheric electricity project at MLO during the summer of 1959. *(MLO archive.)*

World War II. Like Jack Pales, he came from the Weather Bureau's Washington headquarters, where he had worked on atmospheric electricity projects under Harry Wexler. The team that accompanied him to MLO stayed only a few months, and he remained to operate the atmospheric electricity project (Fig. 7.9). His research at the observatory would continue well into the 1980s.

The atmospheric electricity team operated independently of the MLO staff and drove its own vehicle along the rough road between MLO and Hilo. They were among the first of many government agencies, universities, and independent scientists that would conduct various science investigations on the mountain during the years to come. After they completed their research and left the mountain, Howard Ellis moved their Butler Building across from and south of the main building, where it has served as a storage building with workbench space ever since.

Some of those assigned to work on the mountain didn't stay long, as MLO veteran Robert Williams learned when he was preparing to leave for a new assignment with the Pacific Weather Patrol. As recorded by Ellis (1988),

> At lunch, the first day [Williams'] replacement was on the mountain, it was revealed that he did not drive and was not willing to learn. So, Mr. Pales would not keep him and Mr. Williams had to wait until June when another replacement came from the mainland. . . .
>
> Mr. Williams' second replacement, young, newly married, and valiantly acclaiming great good fortune in being assigned to Hawaii but without either him or his wife finding local links back to their familiar culture, began to "act up" on the mountain. The other two observers feared for him, but Mr. Pales said that he never observed anything wrong. . . . When, upon returning to the mountain after being allowed to stay in Hilo for a while, he had to be talked out of shooting himself in the foot to get off the mountain. He was transferred.

FIGURE 7.10. Bill Cobb at work with atmospheric electricity data during late 1959. Each morning the shortwave radio at upper right was used to receive time signals broadcast by WWV. *(MLO archive.)*

Closing Out the IGY (1959)

The MLO staff stayed busy as the clock ran out on the extended IGY. Each morning the staff staying at MLO tuned the shortwave radio to the National Bureau of Standards time signal so the observatory clocks could be reset. The generators were serviced, instruments were checked, rolls of chart paper were replaced, and data analysis was done (Fig. 7.10). Then there were the routine chores of cooking and cleaning up afterward.

Several notable visitors signed the *Visitors* book as the extended IGY drew to an end. Electronics technician Floyd Malcolm "Mike" Keyes visited MLO from the Honolulu Weather Bureau office and signed in on May 21, 1959, as F. M. Keyes. His next visit would be when he was assigned to the MLO staff. Canadian scientist and ozone expert Walter Komhyr, who was eventually to move to NOAA in Boulder, Colorado, signed in on August 25. Nels Johnson of the Pacific Supervisory Office signed in on August 16 and October 18.

The year 1959 drew to a close routinely with Jack Pales conducting calibration measurements with the Dobson ozone spectrometer. Four visitors from the University of Hawai‘i signed the *Visitors* book on December 31, 1959, and the historic International Geophysical Year and its one-year extension ended at midnight.

References

Callendar, Guy Stewart. 1958. On the amount of carbon dioxide in the atmosphere. *Tellus* 10: 243–248.

Craig, Harmon, Norris Rakestraw, and H. E. Suess. 1958. Letter to Roger Revelle of Scripps from the Carbon

Dioxide Committee (January 15, 1958). Keeling Papers, 2003-18. Box 3, Folder 7, Scripps Oceanographic Institute Library.

Ellis, Howard. 1988. The observatory: On their own. Ed. Russell Schnell and Leslie Pajo. (Accessed at http://www.mlo.noaa.gov/webmuseum/publications.html.)

Ellis, Howard T., and Rudolf F. Pueschel. 1971. Solar radiation: Absence of air pollution trends at Mauna Loa. *Science* 172: 845–846.

Fosberg, F. R. 1959. Upper limits of vegetation on Mauna Loa, Hawaii. *Ecology* 40 (1): 144–146.

Humphrey, Paul A. 1958a. Report to Harry Wexler (January 10, 1958). Keeling Papers, 2003-18. Box 3, Folder 7, Scripps Oceanographic Institute Library.

———. 1958b. Letter to Charles D. Keeling (January 29, 1958). Keeling Papers, 2003-18. Box 3, Folder 7, Scripps Oceanographic Institute Library.

———. 1958c. Letter to Charles D. Keeling (March 7, 1958). Keeling Papers, 2003-18. Box 3, Folder 7, Scripps Oceanographic Institute Library.

———. 1958d. Letter to Charles D. Keeling (August 7, 1958). Keeling Papers, 2003-18. Box 3, Folder 7, Scripps Oceanographic Institute Library

Keeling, Charles David. 1958a. Unsent letter to Paul A. Humphrey (January 13, 1958). Keeling Papers, 2003-18. Box 3, Folder 7, Scripps Oceanographic Institute Library.

———. 1958b. Memorandum to SIO Carbon Dioxide Committee for transmission to Dr. R. Revelle (January 15, 1958). Keeling Papers, 2003-18. Box 3, Folder 7, Scripps Oceanographic Institute Library.

———. 1958c. Letter to Harry Wexler (March 25, 1958). Keeling Papers, 2003-18. Box 3, Folder 7, Scripps Oceanographic Institute Library.

———. 1958d. Letter to Paul A. Humphrey (April 4, 1958). Keeling Papers, 2003-18. Box 3, Folder 7, Scripps Oceanographic Institute Library.

———. 1958e. *Preliminary Results of the Atmospheric Carbon Dioxide Abundance Program Prepared for Presentation at the IGY Meeting in Moscow, USSR* (July 3, 1958). Keeling Papers, 2003-18. Box 3, Folder 7, Scripps Oceanographic Institute Library.

———. 1958f. Letter to Paul A. Humphrey (July 30, 1958). Keeling Papers, 2003-18. Box 3, Folder 7, Scripps Oceanographic Institute Library.

———. 1958g. Letter to Paul A. Humphrey (September 22, 1958). Keeling Papers, 2003-18. Box 3, Folder 7, Scripps Oceanographic Institute Library.

———. 1978. The influence of Mauna Loa Observatory on the development of atmospheric CO_2 research. In John Miller, ed., *Mauna Loa Observatory: A 20th Anniversary Report. National Oceanic and Atmospheric Administration Special Report,* September: 36–54. Boulder, CO: NOAA Environmental Research Laboratories.

———. 1998. Rewards and penalties of monitoring the Earth. *Annual Review of Energy and the Environment* 23: 25–82.

Keeling, C. D., T. P. Whorf, and the Carbon Dioxide Research Group. 2005. Atmospheric CO_2 concentrations (ppmv) derived from in situ air samples collected at Mauna Loa Observatory, Hawaii. La Jolla, CA: Scripps Institution of Oceanography. (Accessed at http://cdiac.esd.ornl.gov/ftp/trends/co2/maunaloa.co2.)

Kiess, C. C., C. H. Corliss, and Harriet K. Kiess. 1960. High-dispersion spectra of Jupiter. *Astrophysical Journal* 132: 221–233.

Kiess, C. C., C. H. Corliss, Harriet K. Kiess, and Edith L. R. Corliss. 1957. High-dispersion spectra of Mars. *Astrophysical Journal* 126: 579–584.

Launius, Roger, Colin Fries, and Abe Gibson. 2006. A selective chronology of defining events in NASA history. NASA History Division. (Accessed at http://history.nasa.gov/Defining-chron.htm.)

NASA. 1964–1972. Appendix E. Geology field exercises: Early training, field training schedule for the first 3 groups of astronauts (29). (Accessed at http://www.hq.nasa.gov/alsj/ap-geotrips.pdf.)

NOAA. 2001. Dr. Lester Machta, founding director of ARL (1919–2001), Air Resources Laboratory. (Accessed at http://www.arl.noaa.gov/ss/machta.html.)

Pales, Jack C., Clifford Kutaka, Robert Williams, et al. 1957–1958. *MLO Weather Log* (unpublished).

Price, Saul. 1958. Letter to Charles D. Keeling (August 12, 1958). Keeling Papers, 2003-18. Box 3, Folder 7, Scripps Oceanographic Institute Library.

Price, Saul, and Jack C. Pales. 1959. The Mauna Loa high-altitude observatory. *Monthly Weather Review* 87: 1–14.

Stair, Ralph, and Russell G. Johnston. 1958. Some studies of atmospheric transmittance on Mauna Loa. *Journal of Research of the National Bureau of Standards* 61: 419–425.

Wexler, Harry. 1958a. Letter to Jack Pales (January 8, 1958). Keeling Papers, 2003-18. Box 3, Folder 7, Scripps Oceanographic Institute Library.

———. 1958b. Letter to Charles D. Keeling (January 10, 1958). Keeling Papers, 2003-18. Box 3, Folder 7, Scripps Oceanographic Institute Library.

———. 1958c. Letter to Charles D. Keeling (July 28, 1958). Keeling Papers, 2003-18. Box 3, Folder 7, Scripps Oceanographic Institute Library.

Wilkins, E. M. 1957. Letter to E. A. Volbrecht (November 5, 1957). Keeling Papers, 2003-18. Box 3, Folder 8, Scripps Oceanographic Institute Library.

———. 1958. Letter to Paul A. Humphrey (August 22, 1958). Keeling Papers, 2003-18. Box 3, Folder 7, Scripps Oceanographic Institute Library.

CHAPTER EIGHT

Boom and Bust (1960–1963)

The extended International Geophysical Year ended on a quiet note at the Mauna Loa Observatory at midnight on December 31, 1959. But 1960 began with a disastrous thud.

Dobson ozone spectrophotometer 63 had recently been shipped from Boulder to Hawai'i, and its calibration was being laboriously checked by measuring the extraterrestrial constants of the ultraviolet wavelengths of sunlight it observed in order to measure the ozone layer. All this was of high interest to Canadian ozone expert Walter Komhyr, who had visited MLO on August 25, 1959, and who was to retain connections with the observatory even as this book was written. As explained in chapter 7, the extraterrestrial constant is the intensity of sunlight that would be measured if the instrument could be pointed at the sun from above the atmosphere. Determining this signal involved a series of measurements over the course of a morning as the sun rose in the sky, and Howard Ellis (1988) described the details of the procedure:

> About Thanksgiving [1959], Mr. Pales began with his usual high resolve to perform a calibration on the Dobson spectrophotometer No. 63. This was done by making as many observations on the sun as possible, particularly at low sun. He worked fast and used an assistant to write down the readings as he called them out. At night he would plot the data on large sections of millimeter graph paper laid out on the spotless red tile floor of the main room.

William Cobb (telephone interview with author, April 10, 2008) recalls watching Pales at work on the graphs, which were presumably Langley calibration plots, in what was then the living room of MLO's main building:

> I have seen Jack with a graph of Dobson data on the floor and a number six pencil trying to draw a line through the graph as accurately as he could.

Disaster occurred on New Year's Day of 1960, and the details were later recorded by Howard Ellis (1988):

> New Year's day [1960] dawned bright and clear with Mr. Pales and one of his observers making the Dobson measurements. When they were moving the instrument about on the original cart designed by the inventor of the instrument, two diagonally opposite wheels came off. The heavy instrument fell to the concrete and was damaged.

> All the hard work of the past weeks was now of no avail. The instrument was shipped back to Washington for repair and it was about one year before it was returned.

William Cobb was in the main building when the accident occurred and recalls what happened. The heavy, bulky Dobson instrument was stored on its cart in the living room of the building at night, and Jack Pales and Tom Tyrrell were rolling it out the door for the first observations of the day.

> I heard the noise and then total silence. The Dobson fell going through the doorway. (Cobb, telephone interview with author, April 10, 2008)

All the calibration data were now worthless, and the instrument Robert Simpson used to justify the construction of the slope unit of MLO was out of commission. The international ozone data archived by the World Ozone and Ultraviolet Radiation Data Centre shows a four-year gap between the end of 1959 and December 1963. The serial number of the original ozone instrument is not given, but it is known that Dobson No. 63 (or 163) was the one placed in service in December 1963.

The four-year absence of ozone data was a major blow to the MLO ozone observation program. While data were obtained during 1958–1959, the MLO ozone time series presented in publications usually begins no earlier than 1963 because of the lengthy data gap. Regular total ozone measurements have continued without major interruptions since December 1963.

Fire and Ice: Harry Arashiro Joins the MLO Team (1960–1964)

Harry Arashiro, who had worked as an engineering draftsman for the Hawai'i Highway Division, joined the MLO team as a scientific observation specialist in 1960. On Sunday morning, February 7, he witnessed lava fountains spewing 1,000 feet (305 meters) above the Kapoho volcanic eruption. That afternoon he arrived at the observatory for the first time and experienced a chilly Mauna Loa hailstorm. He signed the *Visitors* book, wrote "I have arrived," and began the rigorous six-day work schedule on the mountain. Arashiro (2006) describes what it was like to work on the mountain:

> Pluses were the job and the working environment, the climate with low rainfall and humidity, the clouds below you stretching to Haleakala on Maui and the stars so near at night. You were isolated from civilization five days a week, and the drive up and down was slow and treacherous.

On May 23, 1960, Arashiro experienced still another side of Hawai'i when a major tsunami devastated much of downtown Hilo, killed sixty-one people (Cooper 2005), and nearly reached the house where he lived.

Back on the mainland, physicist Theodore Maiman demonstrated the world's first laser on May 16, 1960, an event that would eventually lead to a new way to probe the stratosphere over MLO. The device was disarmingly simple. A ruby rod is inserted inside a helical flashlamp, which emits a bright flash of white light when a large capacitor charged to a high voltage is discharged through the xenon gas in the tube. Ruby is aluminum oxide doped with a small fraction of chromium. The green wavelengths emitted by the xenon lamp stimulate the chromium atoms to higher than normal energy levels. When some of these atoms resume their normal state, the excited electrons emit photons of red light, some of which are reflected back and forth within the ruby by a pair of facing mirrors on each end of the rod. During each pass through the rod, the red photons stimulate excited atoms to release additional photons of red light. In less than a microsecond, much of the red light bouncing back and forth within the ruby is emitted through one of the mirrors in a powerful burst of brilliant red light that forms a tightly collimated beam.

A few weeks before Maiman successfully operated the first laser, a U-2 spy plane piloted by Francis Gary Powers was shot down over the Soviet Union, instigating an intelligence crisis that hastened the development of the first spy satellites. As for the laser, after a dozen years Maiman's invention would become part of a major new addition to the array of MLO instrumentation in the form of a laser radar that probed the stratosphere looking for particles of dust from distant deserts and tiny droplets of sulfuric acid from volcanic eruptions.

Adeline Fujisawa Arrives at MLO (1960)

During the fall of 1960, Jack Pales was allowed to hire an assistant to help him with MLO's growing office and administrative responsibilities. Adeline T. Fujisawa, a twenty-six-year-old mother of two, came aboard on November 15 as a GS-5 scientific services assistant (Fujisawa 2006a). Fujisawa had previously worked as a legal secretary for a judge and had considerable bookkeeping, shorthand, and typing experience. Her MLO duties were to greatly expand her resume, and she recalls the details:

> So as a scientific service assistant at Mauna Loa, I was [Jack Pales'] assistant; I was procurement agent, I did all the purchase orders. I was in charge of the recorded charts, reduction of data. I assisted in the preparation of all the technical publications and communication. I was secretary to him. I calibrated scientific data, and the data that I specifically remember was solar radiation and carbon dioxide and . . . nuclear; but we also did surface ozone, atmospheric ozone and rainfall. (Fujisawa, video interview with author, Hilo, Hawai'i, December 2006)

Mrs. Fujisawa has also written about some of the work routine at MLO.

> We had an office at Gen. Lyman Field, and FedEx presently occupies the site. We had most of our files and some equipment collecting data to compare with data collected on Mauna Loa and a darkroom to develop our pictures. We used two pickup trucks, one used by two workers who manned the MLO site (two workers per night taking shifts). Usually the PIC [Physicist in Charge Pales] and I manned the Hilo office daily. Our only means of communication was a 2-way radio to communicate with staff at MLO or the mobile unit. . . . I can remember having to go up Kulani Road and [driving] to the observatory on an unpaved road. (Fujisawa 2006b.)

On May 9, 1961, Fujisawa signed in at the observatory for the first time. After her signature, she wrote, "My first day of work up here. I just love the place." During her trips to MLO to do inventories and take part in some of the observing program, Fujisawa didn't need to pack a lunch.

> They cooked up there; I did not have to pack anything. That is why the boys stayed there two or three days; they did not come down every day. They would leave their homes, their wives, children, whatever, and they would stay there about two or three days and then come down and the next shift would go up. While they were down, they would work with me down at the station, whatever they have to do with the equipment or calibration and stuff. (Fujisawa, video interview with author, Hilo, Hawai'i, December 2006)

Fujisawa has written about how she helped process the chart recordings brought down to the Hilo office by the mountain teams:

> Our equipment was not as sophisticated as that today. After calibrating data, for example, the solar radiation, I would type the hourly data for a 24-hour period on a 24" manual typewriter for an entire month! (Fujisawa 2006b)

The "calibrating data" step mentioned by Fujisawa required that she extract data from the chart recordings and type them onto spreadsheets. This and her monthly trips to MLO established her as the first woman to take part in Weather Bureau science conducted at MLO.

Fujisawa (video interview with author, Hilo, Hawai'i, December 2006) reflects on some of her work experiences and adventures on the mountain, including a major summer snowstorm and a scary airplane ride with Jack Pales:

> [The snow] must have come down about [10,000 feet] . . . I was in a heavy woolen coat, and it was still cold. But the snow was all the way down, and we made the

FIGURE 8.1. This aerial photograph of MLO is from early 1963. The atmospheric electricity building and ground plane array are at left, and the main building is at center. The generator shed and outhouse are at right. Note the air sampling lines leading toward the main building. *(MLO archive.)*

front page of the [newspaper], because they never saw Mauna Loa with that much snow before, especially in mid-July. It must have been at least two feet, I would think so. . . .

Halemaumau was erupting then, so we go to rent a Piper Cub, I think it was. And he and I went on a reconnaissance flight and we took pictures from the airplane, and we flew maybe 500 feet over Halemaumau erupting at its best. That was frightening for me, but Jack was really the adventurous one. He was gung-ho about getting this photo. . . . I was holding on to my seat for dear life for fear that I'm going to fall out.

Pales' interest in aerial photography extended to MLO, and he either photographed or arranged for photographs of the observatory from the air on several occasions (Figs. 7.3 and 8.1). These aerial photographs, some of which were published, document the expansion of MLO since the main building was constructed in 1956. Fujisawa assisted with and typed Jack Pales' annual end-of-the-year report requested by Harry Wexler. As Howard Ellis (1988) notes in his account,

> Early in January 1961, Harry Wexler responded to Mr. Pales' lengthy year end report with a short note saying that the most interesting thing about it was the lack of mention of the simple wind observing network that he had suggested.

Like so many others associated with MLO over the years, former staffer Robert Williams made a return visit to the observatory where he had once spent many nights. On March 15, 1961, he signed the *Visitors* book and wrote, "Hasn't changed too much during the last two years—a treat to see the site again."

Troubles with the Surface Ozone Measurements

Measuring the concentration of ozone in the ambient air at MLO was among the most important IGY projects at the observatory. The data showed that the ozone concentration ranged from next to nothing to around 90 parts per billion (ppb). Or did it? Because of inconsistent results, the original Regener ozone analyzer installed at MLO in August 1957 was replaced by a newer model, which was later replaced by a Model 724-1 MAST ozone analyzer.

Even today surface ozone analyzers must be very carefully calibrated and maintained, for the concentration of ozone in the air we breathe is much harder to measure than the total amount of ozone in a column through the atmosphere. Two side-by-side instruments made by the same or by different manufacturers may provide different ozone readings.

While Price and Pales do not discuss in their two MLO papers the differences in ozone measurements provided by two different instruments, they do insert a revealing footnote in their second paper (1963, fn 2):

> The lack of calibration standards for surface ozone analyzers, the difficulty of getting any two or them to agree closely in ozone levels and variations, and the general uncertainty of measurements at such very low concentrations leave the authors with some reservations concerning their (and similar) data.

The measurement problem is well demonstrated by a laboriously typed page of entries for ozone measured every hour of each day at MLO during June 1961. Below the sums and averages across the bottom is the inscription, "These values questionable. (Not forwarded to USAEPG [United States Army Electronic Proving Ground])." The MLO surface ozone sheet for December 1961 is incomplete. Scrawled across it in large letters in what appears to be the handwriting of Jack Pales are the words, "Values No Good. Do Not Use."

An old folder of monthly surface ozone data sheets in the files of MLO's Hilo office shows that breakdowns of the instrument were as big a problem as the reliability of its data. For example, the files include data sheets for November 1961, when simultaneous ozone readings were made each hour at both the Hilo office and at MLO. The Hilo analyzer was "out" from midnight on November 1 until noon on November 9. It was again out from 8:00 a.m. on November 25 until the end of the month. The mountain instrument was out from 9:00 a.m. on November 21 until 8:00 p.m. on November 26. On days when both instruments were functioning, the ozone on the mountain was generally lower than in Hilo.

The MLO measurements of surface ozone in Hilo ended long ago, and today the surface ozone concentration at the mountain station is recorded automatically using a pair of instruments installed in the main building very near where the old Regener ozone analyzer was located back in 1957. One instrument is a Thermo Environment Instruments Model 49C ozone analyzer that belongs to the Environmental Protection Agency (EPA). The other is an older Thermo Electric Corporation Model 49 that belongs to the Global Monitoring Division (GMD) of NOAA's Earth System Research Laboratory.

While these instruments are far better designed than their predecessors, even they can sometimes provide significantly different readings. For example, at 8:36 a.m. on December 9, 2006, a visitor noticed that the EPA and GMD analyzers indicated 47.5 and 33 ppb, respectively. Significant differences like this occurred for days, and major differences were also noted the previous May. The GMD analyzer was later upgraded to a Thermo Electron Corporation Model 49i, and the two analyzers then exhibited much closer agreement. For example, during a random check at 8:36 a.m. on May 25, 2007, the analyzers indicated 61.2 and 61.5 ppb. Of course, these raw data were preliminary, whether in close agreement or otherwise. But they were typical data snapshots that illustrated the great improvement in instrument performance and reliability that Price and Pales would have appreciated fifty years before as they pioneered the measurement of ozone concentrations from a remote alpine site.

Problems with the Carbon Dioxide Project

The carbon dioxide project was also not without difficulties. Aside from concerns about possible contamination from the generators, vehicular traffic, and seepage from volcanic vents, there were issues with the analyzer itself, the reference gas, and the freeze trap that removes water vapor from the air being measured. Removing water vapor from the sampled air is especially important because, like carbon dioxide, water vapor absorbs infrared radiation, which is the operating principle of the analyzer. Thus, water vapor in the sampled air will cause falsely elevated levels of CO_2 unless it is completely removed before the sample air enters the analyzer.

A shopworn, green ledger at MLO preserves a record of the many problems with the CO_2 analyzer during the early years. The cover of the ledger is marked "CO_2 LOG." The first entry is dated June 10, 1960, and is in Jack Pales' characteristic printing. Entry after entry describes burned-out vacuum tubes, defective parts, and mysterious problems that defied solution for hours or even days.

This entry by Jack Pales describes a simple problem that he quickly isolated and fixed:

> June 18, 1961
> 0730
> Noted abrupt shift in trace, and apparent decrease in sensitivity. Checked [vacuum] tubes in recorder amplifier. One side of one 12AU7 out. Replaced this 12AU7. Operation returned to normal. JCP [Jack C. Pales]

Vacuum tube problems like this were especially common, and the MLO staff used everything from common sense to their oscilloscope to find and replace defective tubes. Jack Pales describes in blue and red ink, then in pencil, a frustrating tube problem under a chart recorder trace taped to a page in the ledger. He then describes the solution on the back of the preceding page:

> December 11, 1961
> After tests on facing page, the identical trouble appeared. Both 866-As were changed, with no results. It was noted that with output section on but voltage at zero that one PL6549 was glowing red, the other not. Both 6549's were changed. Now both plates appear cooler—and of identical color. It would appear that this difference of color with no output might indicate a faulty 6549. J [Jack]

In spite of its cantankerous personality, the APC infrared gas analyzer consistently provided far more trustworthy results than the surface ozone instrument, which was based on a chemical reaction and not the absorption of ultraviolet or infrared by ozone in the air. This was consistent with Charles Keeling's concerns about the questionable reliability of chemical methods for measuring carbon dioxide.

The CO_2 analyzer wasn't the only apparatus to break down in those days. The MLO vehicles were frequently victimized by the rough lava road that in places had deteriorated to a jeep trail. On April 9, 1962, Curt Shoemaker of the Hawaiian Telephone Company in Hilo spent an unplanned night at MLO and explained why in the *Visitors* book:

> Be sure you get your wheels on tight before you leave. My left front one is resting 2 miles down the road 100 feet in the a‘a [rough lava]. Mahalo [thanks] for the night.

John Chin Arrives (1960)

John Fee S. Chin arrived in Hawai‘i after receiving his bachelor's degree in physics and found a job as a laboratory technician with Hawaiian Cement in Honolulu. In the fall of 1960, Chin (video interview with author, Hilo, Hawai‘i, December 2006) learned about a job opening on the Big Island:

> I heard of this opening at Mauna Loa Observatory, and since I was a bachelor, and I only have one suitcase to carry with me, I just applied for the job. And I went for an interview in Honolulu, and I got the job. . . . I was . . . an observer like everybody else. We had to do all different chores—taking solar radiation, carbon dioxide measurements and ozone, and also weather observations. We used to start from 6 a.m. to 6 p.m. everyday and we worked four days up the mountain [and] stayed three nights. And the rest of the time we were working in the Hilo office. . . . Since I did not have a family, I thought it was very exciting up there. The scenery was very nice and, also, just like living—that is my second home. We have kitchen facilities and we sleep overnight, and we usually have two people on duty at the same time. So we have a lot of cooperation and knew other people in the group well.

John Chin signed in at MLO the first time on November 17, 1960, and wrote "On duty."

Howard T. Ellis Arrives (1961)

John Chin was followed a few months later by Howard Ellis, who, like his father, had long been intrigued by the weather. When he was a young boy in South Carolina, Ellis' father retrieved a "small bright orange parachute, a lot of string, and a small box" that a friend had noticed in one of their pine trees. Neither father nor son had any idea about the purpose of their find, and the next day at school Ellis found the answer in an encyclopedia. "It was a weather balloon with a device called a radiosonde that sent data back to the releasing station" (Ellis 2005a). So began Howard Ellis' interest in meteorology, an interest cultivated by his father's ability to make reliable weather forecasts but interrupted by military service during the Korean War. A postwar, undated US Army publication describes Ellis as follows:

> A permanent, undefeated squad leader, Ellis fulfilled his calling with a loyalty not seen since Senator Vest's dog. We could always depend on his friendliness and his good nature and nobody considered him as anything but a hell of a good guy.

After being released from the US Army in 1954, Ellis resumed working for the Weather Bureau in Atlanta, Georgia. When the weather station was moved to Athens, Ellis was able to resume his college studies, and in 1957 he earned a Bachelor of Science degree in physics from the University of Georgia. The launch of Sputnik in October 1957 motivated Ellis to pursue a research position at the Naval Observatory in Washington, D.C. After three years of cloud research, Ellis went to work for the atmospheric electricity unit of the Weather Bureau.

In late 1960, Harry Wexler transferred Ellis to Hawai‘i to take over the leadership of the Mauna Loa Observatory from Jack Pales, who was exhausted from the frenetic pace required to keep MLO running. Wexler had wanted Ellis to replace Pales earlier, but Ellis wasn't interested in moving to Hawai‘i. He changed his mind when given a promotion and advised that his Washington job was being eliminated. Ellis arrived in Hilo in January 1961 and was met at the airport by Pales, who was unhappy that Ellis had not arrived earlier so he and his wife could spend Christmas on the mainland. Ellis briefly shared an apartment with Bill Cobb before the two bachelors partnered with Harry Arashiro and John Chin to rent a house on Kino‘ole Street.

Soon after he arrived in Hilo, Cobb introduced Ellis to Lena De Silva, who recalls the occasion:

> I met Howard at a gathering when he first came over. His boss gave him a welcoming party, and that is where I met him. And six months after that, we were married. (Lena Ellis, video interview with author, Hilo, Hawai'i, December 2006)

De Silva, a Hilo native with three children, had lost her first husband in a car accident. She was no stranger to the observatory, for Cobb had taken her for a visit on January 2, 1961, a month before Ellis arrived in Hilo. She wrote in the *Visitors* book "Wow!! Venus covered with snow—something different." That summer she and Ellis were married, with Bill Cobb serving as best man. The Ellises took up residence at Lena's home on Kīlauea Avenue, the old Volcano Road along which the US Exploring Expedition marched on its way to the summit of Mauna Loa in 1840.

Lena Ellis, who worked in the business office of the Hawaiian Telephone Company, became a valuable asset to Howard during his career at MLO. She planned and prepared all his meals for his stays on the mountain and hosted visits to the Ellis home by some of the visitors to the observatory, including MLO cofounder Ralph Stair and his wife.

According to his Statement of Federal Service, Ellis began work on February 19, 1961. He signed the *Visitors* book at MLO just two days later. But because Pales was unable to find a suitable position elsewhere in the Weather Bureau, Ellis found himself working under Pales rather than replacing him. For a time, Ellis was disappointed by and unsure of his predicament. He was also uncomfortable working on the mountain. During a lengthy interview at his home in Hilo in December 2006, he readily admitted that he was not fond of working on the mountain during the traditional four-day, three-night shifts. But he soon adapted to the work schedule, including kitchen duty and other housekeeping chores associated with living at MLO. Besides becoming deeply involved in the carbon dioxide project and sunlight measurements, his interest in improving access to the observatory would eventually lead to major changes in both the road and the work schedule.

The Army Sunlight Project Arrives

The US Army Electronics Proving Ground project proposed by Jack Pales in 1959 was finally in place by March 1961. The project involved measurements of surface ozone, dust, and weather conditions. A cornerstone of the project was simultaneous measurements of sunlight at Hilo and the mountain station, just as the US Exploring Expedition had done in 1840–1841. The US Army's solar radiation sensors measured four distinct wavelength bands, from the ultraviolet to the red, of "total hemispherical radiation" from the sun and sky. The sensors were placed atop a row of four posts installed in the grassy field adjacent to the southwest end of the Air Cargo Terminal (Fig. 8.2). An identical array was deployed at MLO (Price and Pales 1963).

While the measurements were made for only sixteen months, the US Army program joined the Scripps carbon dioxide project in what was to eventually become many dozens of cooperative research projects at MLO. The program ended in July 1962.

Malcolm "Mike" Keyes Arrives (1961)

Malcolm "Mike" Keyes of the Weather Bureau's Pacific Supervisory Office in Honolulu first signed in at MLO on May 21, 1959, and was assigned to the observatory in the fall of 1961. While serving in the US Navy during World War II, Keyes attended an electronics school. After the war, he became a civil service worker and was assigned to the Philippines and then to Guam, where he worked for the Weather Bureau. There he was part of a team that managed to launch the first weather balloon into the eye of a Pacific typhoon. He was eventually assigned to Honolulu to replace Colby "Doc" Foss, and he later joined Foss as an electronics technician at the Mauna Loa Observatory. He was welcomed to Hilo by Jack Pales and Howard

FIGURE 8.2. Four solar radiometers outside the Air Cargo Terminal at the Hilo Airport and an identical array at MLO measured sunlight for the US Army Electronics Proving Ground from March 1961 to July 1962. The first two radiometers *(from right)* measured near-infrared and red light. The third measured visible and the fourth ultraviolet. The fifth instrument (closest to building) was an Eppley direct irradiance pyrheliometer on a sun tracker. *(MLO archive.)*

Ellis (Mike Keyes, video interview with author, Keauhou Bay, Hawai'i, November 29, 2007).

Keyes had a sinus condition that sometimes caused problems when he worked at the observatory. But this didn't stop him from sleeping on the mountain, as he related in the same interview:

> People who know me know that I didn't have any problems sleeping. I'm gone by the time my head hits the pillow, and I sleep at about the fourth level. And I really don't need to sleep very long until I finish it, and then I wake up, which was not unusual up there, because there's always plenty to do. There's either data to break, or mail to phone in or something to saw.

Like the wives of the others who worked on the mountain, Keyes' wife Henrietta "Hank," who was Hawaiian and Filipina, was not particularly happy about the times he had to stay on the mountain. Henrietta's first husband had been killed in an aircraft crash, and she and Keyes were married in the summer of 1962 shortly after he was assigned to MLO. While Keyes was working at MLO, Henrietta stayed busy taking care of her two daughters from her first marriage, Keala Jean ("Princess") and Moana Lynn ("Bitty"), and her third daughter Piʻikea Warne ("Missy"), who was born in December 1963.

The Mountain Routine

During their four-day shifts at the station, the two-man MLO crews worked twelve-hour days from 6:00 a.m. to 6:00 p.m. John Chin (video interview with author, Hilo, Hawaiʻi, December 2006) recalls that their responsibilities didn't end there:

> Then after work, you have to do your own cooking and do your own chores, wash dishes. And after that, you feel a little bit already tired. And also, [they have to] take a couple of courses called expanding courses. They have to do homework, to study. So they do not have much time to do house cleaning at all. Also, we did not care how the place looked. So, [after] four days stay up there, and the other two crew member come up . . . that is the time, that is the good chance to have the whole place cleaned up. So we used to put all the chairs on the tables and mop the floor and wax the floor and dust off all the cabinets. And then, after all got done, then you can say, "Okay, we can go home now."

The cleanup schedule was supplemented when necessary, as when a newcomer was sometimes initiated by being told to open a can of instant coffee in the kitchen. The high altitude of the observatory placed the contents of the container under pressure, and when the seal was opened there was a loud pop and powdered coffee was sprayed across the room. Harry Arashiro remembers that he once inadvertently played this trick on himself.

Because water boils at slightly below 190 degrees F (87.8° C) at the elevation of MLO, a pressure cooker was kept in the kitchen to assist with meal preparation. Harry Arashiro made stews with the pressure cooker, which once transformed the potatoes into mashed potatoes. Bill Cobb reported that ice cream taken to MLO would expand beyond its carton and beer cans would spew when opened. Jack Pales enjoyed coconut cream pie, which doubled in size in the rarified air (Cobb, telephone interview with author, April 10, 2008).

Standard housekeeping expenses were routinely approved by the Weather Bureau, but Bill Cobb's order of a large container of aspirin tablets raised a red flag (Cobb 2008). The tablets, he explained, weren't for relief of altitude symptoms but for the detection of rainfall. One of the atmospheric electricity instruments was very sensitive to the increase in electromagnetic potential that accompanied precipitation events, and one of the staff had devised a switch that turned off the instrument during rain. The switch consisted of a pair of spring-driven contacts separated by an aspirin tablet, which dissolved when wet and disconnected the instrument.

Living and working at 11,200 feet (3,414 meters) didn't bother some of the staff, but others had a variety of reactions during the first day or two. John Chin (2001) describes the most common reaction:

> So, on the first day when you go up the mountain, on the way up you feel pretty good. Oh, what a nice drive. By the time you get up there, you are feeling a bit woozy, and you do not want to run too fast and walk slowly, and you forget where you put your things. Then after the next day, things are okay again. You are back to your own self.

Then as now, sleeping was difficult for some, especially during the first night of their shift at the observatory, as Chin recalls:

> [The] first night when you get there, it is very difficult to fall asleep. I guess either the altitude, or the air was too dry. . . . Then, especially in the early morning, the air is very dry. That affects your breathing. So many times you have to clear your throat and drink some liquid.

Many of the staff needed some recovery time after completing their four-day mountain shifts. Howard Ellis was unable to speak for a day or so after working on the mountain. Bernard Mendonca, who joined MLO in 1964, remembers how tired he was after returning home from a shift on the mountain. He was not pleased when his wife Clara, who had been cooped up in the house with two children during his absence, suggested they go for a drive. John Chin (2001) describes similar effects:

> Every time in coming down, every one of us all felt fatigue . . . you were tired, and did not want to talk to anybody anymore, and sometimes [you were] very cranky; that is one of those effects. And sometimes when I went home and did not want to talk to the wife at all, then she probably thought I was mad. But then she understood that this is all from the altitude change, [and] you can see your face is still red.

First-time and infrequent visitors to the observatory are more likely to experience nausea, headache, and other unpleasant symptoms of altitude sickness. The fastest cure is to take the affected person down to a lower altitude as soon as possible. But that was not practical when only one vehicle was available, for it was necessary to keep it at the station in the event of a more serious emergency. Chin and his coworkers were reminded of this when they took Janet Abe, a Hilo science teacher, along as an observer during a four-day shift on the mountain. She signed the *Visitors* book on December 22, 1961.

> I was a bachelor when I first arrived to Hilo. Finally, I met a girl over here and she was a science teacher so she had an interest in [MLO]. We went up Mauna Loa Observatory for her to have a visit. Anyway, when she visited up the mountain, she also stayed three nights with us. The first couple of nights she got altitude sick. She could not eat at all, but she could not get off the mountain, because there is no other transportation to go down. So we had to stay until the third night; then she got better . . . she just hated the altitude. (Chin, video interview with author, Hilo, Hawai'i, December 2006)

Janet felt better after a few days at the station, and Chin took a photograph of her standing on the lava road with the observatory building behind her. She didn't hold the unpleasant side of her visit against Chin, and in December 1963 she became Mrs. Janet Chin. She visited the observatory again in 1964.

Bill Cobb and Harry Arashiro were unaffected by the altitude, which had an exhilarating effect on Arashiro, who recalls what it was like to live on the mountain:

> I thought I was in heaven. The air was dry, and it felt almost like God could walk over on the clouds below. . . . I would have stayed there forever; I thought I was in heaven. (Arashiro, personal communication with author, April 9, 2008)

Farewell to Harry Wexler (1962)

Harry Wexler signed the MLO *Visitors* book for the last time on June 3, 1960. After his name, he wrote "Good to be back." In 1955 Wexler had played the pivotal role in supporting Robert Simpson's scheme to divert Weather Bureau money to the National Bureau of Standards to pay for construction of the main building at the proposed Mauna Loa Slope Observatory. He specifically endorsed the site as being ideal for carbon dioxide monitoring, and he later established the collaboration with Charles Keeling that implemented their shared vision. He never regretted his decision to support the establishment of MLO, and on December 5, 1958, the *Hawaii Tribune-Herald* reported that Wexler had stated, "I am more confirmed today in my original belief that there is no observatory like it in the world."

While Wexler made a number of working visits to MLO and Hilo, his reach extended much farther than the lonely outpost on the mountain. In 1944, he became the first meteorologist to intentionally fly into a hurricane. As director of meteorological research for the Weather Bureau, he expanded sunlight monitoring across the United States, supported the development of weather satellites, and pushed for the use of computers in weather forecasting. When MLO was dedicated on June 28, 1956, Wexler was preparing to give a paper on "A Look at Some Suggested Solar-Weather Relationships" at the National

Meeting of the American Meteorological Society (Wexler 1956). He did not know that within a decade MLO would host a world-class solar telescope facility. While making plans for MLO's participation in the International Geophysical Year, he was making personal plans for his participation in an IGY expedition at the South Pole.

Wexler was personally popular and enjoyed an international reputation as an accomplished scientist, and the meteorological community was saddened to learn of his untimely death at the age of 51 on August 11, 1962.

To commemorate his memory, the Weather Bureau invited scientists to contribute papers in honor of Wexler for a special issue of *Monthly Weather Review.* So many papers were submitted that the memorial issue was expanded to include three combined issues and published in October 1963.

Included in the special issue honoring Wexler is an especially relevant paper by Saul Price and Jack Pales, "Mauna Loa Observatory: The First Five Years." The paper was received by *Monthly Weather Review* on July 5, 1963. After receiving comments from reviewers, Price and Pales revised the paper, and the final version was received by the journal on September 23, 1963. The paper acknowledges Wexler's strong interest in MLO.

> It seems especially appropriate that a collection of papers dedicated to Dr. Harry Wexler should include this account of an enterprise so intimately associated with him. . . .
>
> The Observatory staff reported directly to Dr. Wexler; and although MLO was but one of his many responsibilities as Director of Meteorological Research, he gave it always a special measure of his interest and attention, and helped see to its continued existence. He visited the Observatory often, for Honolulu was a way station between Washington and the many destinations to which his activities took him; and he seldom failed, despite the press of other business, to leave time for the flight to Hilo and the drive to MLO. He took intense pleasure in the ecology of the lush rain forest, and found in the bizarre landscape of the upper mountain, on which the Observatory was the only reminder of the world below, renewed confidence in the promise of so remote yet accessible a site.
>
> As he foresaw, the years have brought a wide and increasing use of MLO's facilities and data and a growing awareness of its potentialities. . . . (Price and Pales 1963)

In 2006, NOAA named Harry Wexler as one of its ten Giants of Science and posted on its Web site some of the more memorable tributes to Wexler from the special combined issue of *Monthly Weather Review* (NOAA 2006). F. W. Reichelderfer, chief of the US Weather Bureau and Wexler's friend and boss, included in his contribution the key principles that made Wexler a successful scientist and revealed him to be a kindred spirit of early Mauna Loa scientists Archibald Menzies, Charles Wilkes, and Henry Guppy.

> Dr. Wexler was too occupied with his scientific work to spend much time talking about his philosophy of science, but those of us who worked closely with him knew his lively curiosity for knowledge and his in-born desire to understand. Balancing his enthusiasm was a scientific maturity that steered him away from the pitfalls of pseudo-scientific fads. He believed that progress comes through attention to fundamentals. An admirer of Benjamin Franklin's scientific work, Wexler faithfully practiced what Franklin had advised: "He who would master nature must obey her laws. He must learn her laws and then obey them." (Reichelderfer 1963)

During a 2007 interview, MLO founder Dr. Robert Simpson was asked about Harry Wexler's legacy and his interest in monitoring carbon dioxide:

> That of course was the idea that was advanced by Harry Wexler, and I was trying to convince Reichelderfer that he should divert those funds to support the observatory where it is right now. [Wexler] was a man of great vision, a man who didn't earn nearly the plaudits that he should have won for the scientific changes and the things he made different in science far beyond meteorology that he was involved in. As his deputy for a number of years, I've

> really come to have an enormous respect for the man, his ingenuity, his creativity, his untiring efforts to get things done. He never knew of a word "it can't be done." That wasn't in his vocabulary. He rubbed some people the wrong way. He was a little bit pushy. I guess that's the reason why he didn't get the acclaim that another person with a little milder manner might have gotten had they done the same things as he did and made the same contributions that he made. (Simpson, telephone interview with author, March 18, 2007)

The Five-Year Status Report

On October 28, 1962, the world breathed a collective sigh of relief when Soviet premier Nikita Khrushchev announced that the USSR would remove its nuclear-tipped missiles from Cuba. Back in Hawai'i, in January 1963 MLO was formally named a US Weather Bureau Research Station with the designation "WBRS, MLO." The paper by Saul Price and Jack Pales about the first five years at MLO was published in December 1963, and it remains the most significant status report about the atmospheric research conducted at the young observatory. The landmark paper outlines the geographical features and climate of the island of Hawai'i and its giant Mauna Loa volcano. It then reviews the history of MLO and describes its facilities. Finally, the paper describes the principal measurement programs, with emphasis on those begun during the International Geophysical Year. The programs were still underway several years after the extended IGY came to a close at the end of 1959—most notably the ozone layer measurements promoted by Simpson and the carbon dioxide program arranged by Wexler.

Bernard Mendonca worked under Saul Price in Honolulu from 1958 to 1964 and prepared the figures in the paper. He, Harry Arashiro, and Clifford Kutaka are acknowledged by Price and Pales for "compiling portions of the data presented in this paper." Mendonca remembers Price as the thinker behind the paper, while Jack Pales was the doer. Since his 1957 stay at MLO, when he installed a sophisticated cloud camera, Price had been extracting serious science from observations made there. Pales and his mountain team had worked almost nonstop to install, maintain, and repair the instruments in Hilo and on the mountain and, along with Mendonca in Honolulu, to extract the data from the chart recordings.

Price and Pales anticipated the long-term significance of some of their findings at MLO, especially the upward trend in carbon dioxide they had measured in collaboration with Dave Keeling since March 1958. In 1960 Keeling had announced a slight upward trend in carbon dioxide measured by his infrared analyzer at the South Pole (Keeling 1960), but the first paper devoted exclusively to the MLO carbon dioxide project would not be published by Pales and Keeling until 1965 (the unpublished paper was cited by Price and Pales as being "in press"). Thus, the Price and Pales paper of 1963 was the first major announcement from MLO of the CO_2 findings.

> The implications of a secular increase in the atmospheric carbon dioxide constitute a major cross-disciplinary problem in geophysics, but one which far transcends the interests of scientists alone. Measurements of CO_2 began at MLO in March 1958 and are being carried on in closest collaboration with C. D. Keeling (Scripps Institution of Oceanography), who supplies the reference tanks of the gas and performs the final evaluation of the data. The past five years comprise the longest record of its kind and have made Mauna Loa a key station in the worldwide investigation of the carbon dioxide problem. . . .
>
> Perhaps of paramount importance is the emergence from the Mauna Loa record of a steady rise in carbon dioxide since observations began there five years ago (fig. 15). The increase averages 0.7 ppm annually, which, as Keeling points out, is about half the rate at which carbon dioxide is currently being released to the atmosphere by the consumption of fossil fuels. However, since the ascending trend is superimposed on a seasonal oscillation of much larger amplitude and derived from but a few years of record, it can not be regarded as definitive until supported by further years of data. (Price and Pales 1963.)

Pales and Keeling had proved that Keeling's prior measurements of a seasonal cycle in the carbon dioxide concentration at low elevations was also present in the free atmosphere more than 2 miles (3 km) above sea level. They also demonstrated that human activity—that is, the burning of fossil fuels such as coal, petroleum products, and natural gas—was causing a steady and measurable increase in the concentration of CO_2 in the atmosphere. Because of the climate warming potential of CO_2, the findings announced by Pales and Keeling had significant implications. These significant new findings were achieved despite ongoing problems with the carbon dioxide analyzer, such as one described in the CO_2 LOG by Howard Ellis:

> Oct. 17, 1962 (1200)
> Summary, with so many things going wrong it is impossible to see what component was or is at fault. Two things were changed and the result was better but only slightly. T1 was changed and the phasing was adjusted. . . .
> Further repairs need to be made.

While some problems were frustratingly difficult to identify and repair, others were easily corrected, such as this one identified by Ellis:

> Jan. 26 [1963]. Unstable trace noted during operation of Friden adding machine. Was plugged into different circuit and instability vanished.

When shown this notebook entry forty-five years after it was written by Ellis, Bernard Mendonca, who was assigned to MLO the following year, was reminded about "one of the vexing problems we had with collecting the [tiny] millivolt signals from the instruments at MLO":

> Electrical grounding at the observatory was extremely difficult, because the porous lava of the mountain presented a surface that was difficult to secure a good electrical ground. To our dismay, proper electrical grounding for the sensitive instruments at the observatory proved very difficult to achieve. The result of this was instrument malfunctions and even destruction during any atmospheric electrical storms in the vicinity of the observatory, not to mention signal contamination when unshielded lines were run too close to each other. No matter how many copper/brass or metal grounding rods we hammered into the lava, they would never provide proper grounding for the observatory. Various other methods were tried over the years, such as floating grounds, and many more, but nothing ever quite filled the bill. This problem was also a challenge at the South Pole station as well. (Mendonca, personal communication with author, March 14, 2008)

The paper by Price and Pales covers much more than the carbon dioxide findings. It presents results for measurements of surface ozone, atmospheric electricity, ice nuclei, and solar radiation. Over the years, the sunlight measurements, which Ellis processed using the Friden machine, proved to be among the most stable of all the data collected at MLO. They predated the carbon dioxide project by four months and would eventually provide the world's longest record of the transmission of sunlight through the atmosphere, a topic that will be taken up in subsequent chapters. The five-year paper closes with a prophetic summary of the future of the observatory:

> It is certain that the Observatory will continue to attract projects as numerous and diverse as those listed earlier in this report. The tremendous spurt in the space-oriented disciplines, and a concomitant interest in high-altitude stations for astronomical and other purposes, is directing increased attention to MLO itself and to its possible use as a base for work done at Mauna Loa summit, 2500 ft. higher.
>
> But perhaps more valuable than any of these endeavors, actual or potential, is the Observatory's role as a bench mark station. With the continuing industrialization of the planet and the increasing atmospheric load of a variety of gaseous and particulate contaminants, the existence within a stable local environment of such a vantage point from which to calibrate the free atmosphere is of inestimable value. Thus, observations during the International Quiet Sun Year will be compared

> with those made when MLO began, at the height of the solar cycle.
>
> But beyond that time Mauna Loa Observatory will continue to regard as its primary responsibility the careful and uninterrupted surveillance of carbon dioxide, solar radiation, total and tropospheric ozone and, at intervals, the atmospheric electric variables—the significant indices of global atmospheric change. (Price and Pales 1963)

There were other applications for some of the data, as when the director of the Hawaiian Sugar Planters Association (HSPA) developed an interest in measuring sunlight in sugarcane fields to better forecast the yield of the crop. He visited MLO multiple times and asked Howard Ellis to assist with calibrating their sunlight monitoring instruments. After the Indonesian volcano Agung experienced a major eruption on March 17, 1963, Ellis noticed that the veil of aerosols from the volcano caused a noticeable reduction in direct sunlight measured by the Eppley normal incidence (direct sun) pyrheliometer. This was of high interest to the HSPA director, even though the reduction in total sunlight was not as significant.

> That guy kind of played a trick on me. . . . He went back and told them that there is a young, smart scientist up there that said that there is not going to be enough sunshine this year [because of] the Agung Volcano . . . and they better use more fertilizer. . . . That is the only way he could get them to do it—to tell that story. (Ellis, video interview with author, Hilo, Hawaii, December 2006)

The Legendary White Dog and the MLO Dump

The illustrations in the paper by Price and Pales include a map of Hawai'i, charts, data, and various photographs. Among the latter is a notable aerial photograph of MLO that reveals considerable expansion since the original 20 x 40-foot cinder block building was dedicated in June 1956. The photo shows several small outbuildings, including the generator shed that housed the two diesel generators that delivered electrical power to the site. Also shown is the atmospheric electricity apparatus and its accompanying Butler building, which was later moved just south of the main building and is still in use. Various instruments can be seen scattered about the site, including those that measure sunlight, precipitation, and wind speed and direction. A curious feature of the photo is an array of white lines that converge on the main building. These are the aluminum tubes that carried air to the carbon dioxide analyzer from distant sampling points around the observatory site. The array of collecting lines was necessary to assure that air was always collected upwind from the generator shed west of the main building.

The aerial photo also shows a splotchy white area west of the generator shed and the "CLEAN REST ROOMS" outhouse. This area, which does not appear in earlier images, is the former MLO dump that attracted a mysterious visitor from time to time in a region where the only living creatures besides the MLO crew were a few resident spiders and the insects that occasionally blew in with the inversion layer. Bernard Mendonca relates the details in his 1978 MLO memoir:

> No reminiscence of MLO would be complete without mentioning the white dog of Mauna Loa. Much has been told about the mysterious phantom dog that would appear on the mountain to forewarn of a volcanic eruption. Hawaiian legend relates a tale of Pele, who is the fiery goddess of the volcanoes on Mauna Loa, and her companion dog, whom she would send as a messenger to alert the people whenever an eruption was imminent.
>
> The white dog was first noticed by the observatory staff during the latter part of 1959. At that time the staff was living on site for up to a week at a time on rotating shifts. Because of this housekeeping, a rubbish dump was soon developed to the west of the observatory. The contention of the staff was that a stray white dog had discovered the dump and foraged it for food. Attempts by the staff to befriend it and later to capture it, no matter how persistent or devious, failed. The dog for some reason would have nothing to do with the observatory staff.

> Soon the dog disappeared and was presumed to have found its way back to the populated regions of the island. In December 1959, Kilauea Iki erupted.
>
> To the amazement of the staff the dog reappeared at the observatory several months later and again was spotted intermittently for a month or so and then disappeared. This pattern of appearances and disappearances continued until 1966. . . . It would never have anything to do with anyone and whenever pursued would always easily outdistance its pursuers over the rough lava and run to the top of the mountain. (Mendonca 1978.)

The white dog became a mainly internal legend of sorts at MLO, and for good reason.

> I'll tell you what was hard about the white dog. We were trained scientists, meteorologists, physicists who were well schooled in the scientific method and proof and research and discipline and everything. And we're running a station on top of the mountain and all of a sudden we're confronted with this white dog. Well, this was a hard thing to swallow, because it was legend. (Mendonca, video interview with author, Boulder, Colorado, April 27, 2007)

Word about the white dog began to leak out, and one day Mendonca became involved in an unplanned validation of the animal's existence while trying to start a conversation with a NASA official he was driving to the summit of Mauna Loa. In the same interview, he related that he asked,

> ["Have] you heard about the white dog?" He [replied], with kind of a sneer, "Yeah, the white dog. What is this? Come on you guys, what's going on?" I tried to explain, and the more I explained you could tell he was [wondering] how I'm going to handle this. So I cut the conversation . . . and then all of a sudden he screams, "Look at the dog! Look at the dog!" And I turn, and there's the white dog. And he goes, "Look at the dog!" [and I say] "I don't see any white dog." And he just couldn't believe it. That's the best story I have about the dog.

In 1959 Jack Pales had tried to capture the dog but failed. Mendonca tried to capture the animal and suffered the consequences.

> After a typical long day at the observatory, Jim Simpson, who had a post doctoral assignment at the observatory, and I were driving down the mountain in the area between the observatory and Kulani Mauka heading back to Hilo, when we saw the dog laying in the lava about twenty to thirty yards toward the summit from the road. Jim was driving, and I quickly asked him to stop and slowly reverse as close as possible to where the dog was without spooking him. I then told Jim, "Hang on; I'm going to get the dog." I slowly opened the door, got out of the car and bounded for the dog. The dog immediately took off over the lava heading toward the summit. I was just about to catch him when I stumbled and did a face plant in the rough lava as the dog continued to run up the mountain beyond our reach. Because of the fall, I was cut to shreds. Jim took me home a bloody mess. (Mendonca, personal communication with author, March 21, 2008)

Mendonca's wife Clara recalls that "He came home with blood dripping everywhere." Years later, Mendonca's white dog story was included in an anthology entitled *Hawai'i's Spooky Tales* (Carroll 2003). He still bears scars from his encounter with the lava that bloodied his face and hands.

Jack Pales, Colby Foss, Harry Arashiro, Bill Cobb, Howard Ellis, John Chin, Mike Keyes, Charles Garcia, and others also saw the dog, especially in and around the dump. On one occasion, Arashiro, Chin, and Ellis saw the dog outside the kitchen window. Cobb remembers that Chin photographed the visitor—but that its image did not appear in the processed photo. Keyes (telephone interview with author, April 12, 2008) says that an EG&G contractor at the site to photograph a nuclear test used a professional camera to photograph the dog—but forgot to load the film. The men sometimes left table scraps for the dog outside the main building, and when it snowed they sometimes saw its paw prints. Garcia reported seeing the dog during trips to the summit of Mauna Loa. Ellis recalls seeing the dog around 30 feet (10 meters) away at the upper edge of the dump one day when he looked out the west window of the generator shed.

Only one photo of the white dog was published in the press, but more than one photo was taken. Malcolm "Mike" Keyes remembers the time a reporter photographed the white dog while it was standing by the road. He told the reporter to "Shoot the picture through the windshield or he'll be gone." Howard Ellis (video interview with author, Hilo, Hawai'i, December 2006) reports another incident:

> And we had so much junk left over—a lot of packing material—and he had a box that he slept in. So that is where he was that morning when Jack Bryan and I went up and he took the only picture that was ever taken of the white [dog]. Just before I left in 1980, the end of '80, the dog ran very close to the parking lot while I was there at night. Really close. Yes, I never told anybody that, because I did not think they would believe me, because they had not seen the dog for quite a few years.
>
> I don't understand why that dog appeared so white in our only picture. It was kind [of] dirty brown. (Ellis, personal communication with author, May 25, 2007)

Bernard Mendonca (personal communication with author, March 20, 2008) was also with a reporter who photographed the dog:

> As we approached the observatory, to our astonishment we spotted the white dog basking in the sun beside the road in the lava. We stopped the car, and he took a photo of the dog. And as far as I know, it is the only known photograph taken of the white dog, which was of poor quality due to the dog's distance from the car. After the photograph was taken, it soon appeared in the local newspaper and generated a lot of interest, especially among the local Hawaiian people who connected the dog to the legend of Pele's dog and its appearances before eruptions of the volcano.

The white dog story attracted the attention of the Honolulu Zoo:

> Once the guy in charge of the Honolulu zoo came over, and he thought maybe it was an aboriginal dog that was barkless. I never heard him bark. (Ellis, video interview with author, Hilo, Hawai'i, December 2006)

The zoo director was Jack Throp, and he was intrigued with the possibility that the dog was a "missing link" to the original poi dogs that accompanied the Polynesians who settled Hawai'i. The appearance was certainly similar, but Throp was unable to find what the *Honolulu Star-Bulletin* designated as "The non-barking living legend of Mauna Loa" and the "phantom dog" (Wharton 1966).

Perhaps the mysterious white dog might have been left by or hitched a ride with the staff or inmates from the Kūlani Prison, who periodically worked on the MLO road. Whatever its origin, the MLO staff enjoyed regaling visitors with tales about the old Hawaiian legend of Madame Pele and her mythical white dog. But they didn't explain why the pet of a deity powerful enough to initiate volcanic eruptions was reduced to dining on table scraps they left for it or that it found in the dump. Ellis and other old-timers were embarrassed by the garbage dump they inherited.

> The trash dump was a shame. I tried to clean it up later. They were burning it when I arrived. John Chin had a picture of the smoke and sent it to his family as a volcano. Shame on us for dirtying the mountain, but everybody dirties the mountain, because nobody wants to haul off the trash. So that is one reason I stayed up for 15 days, [because of] so much junk people leave over here. Oh, everybody would just say, "You could have it." (Ellis, video interview with author, Hilo, Hawai'i, December 2006.)

The dump included packing boxes and crates, standard household and kitchen refuse, and discarded mechanical, electrical, and plumbing items. Most of the dump was cleaned up decades ago, and today few people even know it existed, much less its location. Future archaeologists will not find the dump site on current aerial photographs, for what was left behind blends in with or has been covered by lava pushed over it. If they manage to find the small patch of lava where the dump was located, they will find a few broken dishes, rusty food and drink cans, and glassware and bottles from the kitchen trash during the time when men lived and worked at MLO seven days a week. They will also find oil filters from the diesel generators, radiator caps, wire, metal, batteries, broken light bulbs, pipe, and some rusty nails and bolts.

The Road to MLO

When it came to his work at MLO, Jack Pales was a chain-smoking, type A personality who slept only three hours a night (Ellis 1988). Some of the old-timers believe that Pales was working himself to death on the mountain, a thesis supported by his marathon work sessions and a near tragedy that occurred in January 1963. Adeline Fujisawa recalls what happened:

> During my tenure at MLO, Jack C. Pales, Physicist in Charge, while on duty at our Hilo Airport office, was stricken with a heart attack. (Fujisawa 2006a)

During Pales' recovery, MLO founder Robert H. Simpson arrived from Washington and signed in at the station on February 11, 1963, with "Great to return to Mauna Loa." Also during this time, Pales allowed a team from the US Geological Survey to use the MLO darkroom. Their objective was to process imagery from an aerial survey of the Kīlauea area of Hawai'i (January 26 to February 20, 1963) using infrared scanning equipment to find hot spots (Fischer et al. 1964). According to Ellis (personal communication with author, September 1, 2009), this team made aerial photographs of MLO at Pales' request (see Fig. 8.1).

Jack Pales survived the heart attack. But it was a major setback for his assignment as director of the observatory, for his doctors ordered him to stay off the mountain. Malcolm Keyes, who was known as "Mike" in his MLO days, reflects on the man and the transition everyone saw coming:

> Jack, I suppose, somewhat like myself, was a . . . Type A personality. He smoked. He was excitable. He was dedicated. He was thorough. He was a family man, and when his heart let him down, and [he] was prohibited by the doctors to go up again, it was sad, because . . . the station was Jack Pale's station, and he called the shots. And although, you know, we all thought he could have done some things differently, we had to respect him as really the father of the observatory. (Keyes, video interview with author, Keauhou Bay, Hawai'i, November 29, 2007)

Howard Ellis (telephone interview with author, April 8, 2008) expresses similar sentiments and notes, "Of all the bosses, Pales was the best." Because Pales was unable to work on the mountain, Ellis finally assumed many of the leadership responsibilities that had led to his assignment to MLO two years earlier. While Ellis was not officially promoted to the directorship of MLO until January 1964, the first edition of NOAA's *Geophysical Monitoring for Climatic Change* listed his tenure as beginning in 1963 (Miller 1972).

One of Ellis' major concerns was reducing the wear and tear on MLO's staff and their two four-wheel drive vehicles caused by the terribly rough mountain road above the Kūlani Prison Camp. Bernard Mendonca, who was working at the Honolulu office of MLO at the time, provides a colorful description of what these trips were like, for he experienced the road firsthand during a special two-week project at MLO in 1963, which, as he notes in the *Visitors* book on February 25, "was educational and an exhilarating experience."

> The first route used for access began at the Kulani honor camp turnoff on the belt highway. This preexisting road was called the Stainback Highway and very soon was corrupted to "strainback" highway by those who made the weekly trips to the observatory. It terminated at the Kulani honor camp (elevation [5],000 ft). The daily search and shakedowns of people and vehicles by the guards at the camp were an unbelievable experience for first-time travelers to the observatory. Later the guards got to know the staff and waved them through the camp unless something was amiss. Above the camp the road ran along the northeast ridge and required

> four-wheel-drive vehicles. A good trip with no breakdowns required 2-1/2 hours for the 40 miles. It took a hard head, resilient behind, and cast-iron stomach to make the journey and keep one's breakfast. Maintenance of the road was virtually nonexistent, and the gravel surface deteriorated rapidly with the heavy tropical rains. It was not unusual to come across wild boar rooting among wild orchids on the shoulder of the road in the early morning and later to see wild goats eating ohelo berries along the road above the camp. (Mendonca 1978)

Malcolm Keyes (video interview with author, Keauhou Bay, Hawai'i, November 29, 2007) also remembers the old road to MLO. A day or so before he was scheduled to make a trip up the mountain, his wife Henrietta would notice that he would become very grumpy. When the day arrived,

> you hopped in the truck before sun up, and if it had been gassed up by the last guy who used it, you were lucky. And if not, you had to stop [at] the last little gas stop on the way out of town before what was about a two or two-and-a-half hour trip at normal speed. And you would drive up and hope that your sinus didn't plug up, because if you did, you had the worst headache a man could imagine. And the only way to break it was to take a deep draw on the oxygen mask and listen to the whistle as the sea level pressure ripped through your nostril.

The road sometimes caused problems for visitors' vehicles, as Don, Evelyn, and Ed recorded in the *Visitors* book on November 13, 1960: "We got stuck at Twin Peaks. This is the way to get stuck on a mountain." They were rescued the following day by Gilbert P. Aguiar, who was the next person to sign the book: "Rescue[d] above 3 citizens."

Bouncing along the rough lava road also took a toll on the visitors themselves. Gerald Warnock, John Knoska Jr., and Howard Jackson made the trip on July 3, 1960, and Warnock added a note after their signatures in the *Visitors* book: "Drove to the top 'moderately rough road.' John turned green. Everyone back in fine shape."

Some visitors left notes thanking the MLO staff for coffee, cookies, or pie, which they provided from their personal stock. Malcolm Keyes remembers that they were sometimes willing to share, because visitors were uncommon and provided conversation. But some visitors could get in the way and a few posed problems, such as those with car breakdowns and stranded hikers. John Chin remembers incidents in which people in trouble demanded help "because we [the staff] belong to the government." One visitor with a stuck car who expected the MLO staff to stop everything to help was advised that the staff was at work. The visitor said she expected help because, "Hey, we pay taxes!" (Chin, video interview with author, Hilo, Hawai'i, December 2006).

Breakdowns of the MLO vehicles required expensive, time-consuming repairs. Bernard Mendonca, who did temporary work at MLO before being reassigned from the Honolulu Weather Bureau office to MLO in 1964, describes the wear and tear on the vehicles:

> I would venture to say we would break a vehicle, one vehicle a week, at least, I mean total it, I mean busted spring, busted oil pan, ball joints, steering mechanism, brakes. You name it, we'd go through it. (Mendonca, video interview with author, Boulder, Colorado, April 27, 2007)

The many breakdowns interfered with the work schedule on the mountain, and Howard Ellis decided the time had come to look for a solution to the problem. Thomas Vance, who had pioneered the Stainback Highway and its extension to the summit, was about to make another vital contribution to MLO's history.

"Doc" Foss, who was working at MLO when Ellis arrived in 1961, remained Ellis' friend after he left government service to begin Hawai'i Research, a company that developed a variety of solar energy projects (Ellis 2005b).

> "Doc" is a nickname for Colby Foss, who is just the greatest guy and should get the most credit for the observatory. Really, he was an all-around great guy. He could do anything, and the only thing that held him back was his heart. (Ellis, video interview, Hilo, Hawai'i, December 2006)

Foss introduced Ellis to Vance, who together with Robert Simpson and the Hilo Lions Club had worked with the Kūlani Prison to develop the road to the summit of Mauna Loa in 1951. As related in chapter 4, Vance and the Lions Club dreamed of a cross-island route between Hilo and Kealakekua that would link the two sides of the island with a highway that would be shorter and safer than the tortuous Saddle Road. The project was prematurely ended, leaving behind what some of today's MLO staff call "The Road to Nowhere."

The beginning of the abandoned road was connected to Kūlani Mauka by a curvy access road. Only around 4 miles (about 6.4 km) separated the beginning of the abandoned road from Saddle Road down below. Saddle Road connects Hilo on the eastern side of the island with Waimea and the Kona Coast on the west, and Vance and Ellis realized that cutting a road between the beginning of "The Road to Nowhere" and Saddle Road would provide a direct link to Kūlani Mauka and greatly improve access to the north slope of Mauna Loa and to MLO.

Vance arranged for Ellis to appear before the Hawai'i County Board of Supervisors to request the county's assistance in constructing a link between Saddle Road and the abandoned cross-island road project. The morning of the meeting, MLO secretary Adeline Fujisawa informed Ellis that Jack Pales, who was still recuperating from his heart attack, had instructed her to accompany Ellis to the meeting. Many years later, Ellis (video interview with author, Hilo, Hawai'i, December 2006) recalled what happened:

> Adeline Fujisawa was probably the one that really did it. She came along with me the day I went before the council. . . . And she gave a pretty good spiel, and I pretty much kept my mouth shut, and we got the road. . . . She was Mr. Pales' secretary at the time. I never had a secretary, and I thought he had sent her along to keep me in line, but she is just such a smart lady. And she knew all the people on the Board of Supervisors.

The MLO staff had developed a good working relationship with scientists and engineers from the Atomic Energy Commission (AEC), who had erected several buildings at MLO from which they monitored and photographed nuclear tests. Some of the leftover AEC buildings at MLO are still in use, and Ellis recalled in the same interview how one of the secret explosions was visible from Hilo:

> It was south of here. I was supposed to be at the mountain all this time that we were waiting for it, but it just so happened that I was down here at home, and we lived in the house where my wife had lived [on Kīlauea Avenue], and I held up our youngest son to see it out the window. It was a bright flash to the south of here. It was scary—reddish and yellow.

The AEC had agreed to provide road work in partial exchange for the use of their facilities on the mountain for monitoring nuclear tests. There may have been more to this bargain than meets the eye, for during the first half of the 1960s, dozens upon dozens of scientists and engineers from the Sandia Corporation and various companies with AEC contracts signed the *Visitors* book. They were driving up and down the MLO road every day during their stays, and they would benefit from the new road as much as the MLO staff.

Ellis worked with Henry Auwae of the Kūlani Prison to organize the new road cut. Auwae was well qualified for the project, for in 1951 he helped cut the original Mauna Loa road to the summit. The morning that the new road to Saddle Road was scheduled to be bulldozed, Ellis joined Auwae and the bulldozer operator at the intersection of the road that connected the Kūlani–Mauna Loa road and the abandoned cross-island road. Ellis had second thoughts as he watched the bulldozer move into position to begin its work:

> I was inexperienced with such, and, when the bulldozer went against the lava in the direction of Mauna Kea, I became convinced that it could not be done and decided to give up the venture. A man from the AEC who was to pay the contractor was flying in from Honolulu that morning. I asked a fellow staff member [Mike Keyes] to bring him up. On their drive through Kulani, the energy guy was given a pitch for the new road. When they

> arrived, the OK to move the bulldozer ahead was given before I expressed my doubts. The road was cut through to the Saddle Road by mid-afternoon. It was opened to public use on Saturday, April 27, 1963. (Ellis 2005b)

In years to come, the new road segment would save wear and tear on tires, vehicles, and drivers as well as countless hours of driving time to and from MLO. Thus, the new road segment is one of Ellis' most enduring legacies, yet he relegates it to only a footnote in his MLO history:

> This four-mile link to the Saddle Road was . . . opened on April 27, 1963. Thereafter, this was the route used to get to the observatory, with the road through Kulani not being used for this purpose anymore. (Ellis 1988)

The Hilo Lions Club arranged a ceremony for the opening of the new road where it joined the cross-island Saddle Road. Jack Pales gave a short speech commending Ellis for his role in the project, but Pales was unable to make the trip to the high altitude observatory because of his recent heart attack.

Nels Johnson, meteorologist in charge of the Weather Bureau Pacific Supervisory Office in Honolulu, and various local officials were present at the road opening. There was friction between Johnson and Ellis that was soon to play a role in the future of MLO, and, although Johnson visited MLO many times, he did not on this occasion. According to Ellis (personal communication with author, May 25, 2007),

> [Johnson] tried to fire me when the new road was opened up, but Jack Pales was still here, although not going to the mountain. . . . [Johnson] gave me one of his many severe dressings down. [On the day of the road opening] he said he didn't want to see me anymore that day and drove up the road to 8300 ft. [the junction with the original Kulani road] alone in his rented car.

Those who drove all the way to MLO to celebrate the new road filled nearly two pages in the *Visitors* book with their signatures. Howard Ellis' wife Lena signed the book and added, "Much better road." Several members of the Hilo Lions Club also signed the book.

On June 2, 1963, a family from Honolulu drove the new road on their second visit to MLO. They wrote in the *Visitors* book, "Repeat trip—lot better road this time."

MLO old-timers who drove the old road through Kūlani Prison praised Ellis for his role in providing the new road. While the new road was unpaved and drivers had to stop and remove chunks of lava on many of their trips, it provided a much better ride and reduced the need for repairs to the two MLO vehicles. For several months Ellis removed rocks during every trip, even when he was taking official visitors to MLO. Two visitors bought gloves for him to protect his hands, but he didn't wear them.

An unexpected effect of the new road was local newspaper publicity for both MLO and Ellis. News stories, an editorial, and letters to the editor commended Ellis for initiating the project that simplified access to the north slope of Mauna Loa.

The Yamanouchi Garage

While the new road was a valuable shortcut, it was unpaved lava and still rough on the MLO vehicles. After it was opened, the MLO staff began refueling their vehicles at the Yamanouchi Garage on Kaūmana Drive in the Upper Kaūmana Community near where Saddle Road begins. They also took their damaged vehicles there, where garage owner and mechanic Nobuichi "Nobu" Yamanouchi worked his magic and restored the vehicles, often working long into the night. Because the breakdowns were so frequent, Yamanouchi became a major asset to the observatory, and he was well respected by MLO veterans Howard Ellis, Bernard Mendonca, Barry Bodhaine, and others.

Ellis, who always addressed the garage's proprietor as Mr. Yamanouchi, recalls how the friendly mechanic sometimes shared Japanese food with him. He provided personal advice about the staff's personal cars and recommended that they use oversize tires on the MLO vehicles to reduce wear and tear. "He was excessively kind," Ellis (telephone interview with author, August 9, 2008) recalls.

Mendonca (personal communication with author, May 6, 2008) also held Yamanouchi in high regard:

> Anything you say about "Nobu" is not enough. He was really a special person and a mentor to us young kids driving up and down the mountain, almost on a daily basis. His cheery smile and sage advice were exactly what we needed after a bumpy ride down the mountain. He did many midnight repairs just to get the cars ready for the next trip. It was amazing.
>
> Not only was Yamanouchi a champion to us, but he was a pillar of the Upper Kaumana Community. Besides running the only gasoline/repair station on the mountain side, he was a jack of all trades. And when anyone in the Upper Kaumana Community was in need, they always went to Nobu.
>
> He was definitely one of the most unsung heroes of the early days of the observatory. I cannot imagine how many days the observatory would have been left unmanned and unattended if not for his dedication and service to our vehicles. I miss his friendly smile on our return from the mountain to this day. And how sweet it was when on Fridays he would sometimes ask us to stop and get out of the car and share a beer and sashimi with him while we lamented about the drive up and down the mountain on that awful road. He was a very special guy!

A Distinguished Visitor

The new road meant that, with reasonable care, ordinary passenger cars could drive to MLO. The newspaper publicity informed Hilo residents about the improved route to MLO, and the number of signatures in the *Visitors* book tripled after the shortcut was opened. Official visitors from the Sandia Corporation in Albuquerque, New Mexico, who visited MLO to monitor nuclear tests, also benefitted from the road.

The new road reduced the time needed for MLO staff to escort visitors to and from the observatory, as when Howard Ellis escorted world-famous astronomer Dr. Gerard P. Kuiper of the University of Arizona and famed telescope maker Alika K. Herring on an overnight visit. Saul Price called Ellis the afternoon of May 31, 1963, to tell him about Kuiper's impending arrival. Thanks to the new shortcut road, Ellis was able to drive down to Hilo, meet the two men at the General Lyman Field, and drive them up to the observatory the same evening (Ellis 2005b). They arrived before dark, and the two visitors signed the *Visitors* book.

Kuiper and Herring, who was born in Hawai'i, had arrived after a week on Maui, where they evaluated the clarity of the night sky over the summit of Haleakalā. Their objective was to find the best site for a proposed new NASA observatory. While they were impressed by the excellent seeing conditions from atop Haleakalā's 10,023-foot (3,055-meter) summit, fog from the crater and clouds from the inversion layer sometimes obscured the sky, while the much higher summits of Mauna Kea and Mauna Loa remained in the clear. Thus, Kuiper became highly interested in the two Big Island peaks.

In an unpublished memoir, Ellis (2005b) describes Kuiper as an enthusiastic character who was excited by the various life zones they drove through en route to MLO and who was interested in learning the temperature at the observatory. While some of the distinguished visitors Ellis had escorted to MLO displayed an air of superiority, he found Kuiper to be an agreeable and conversant travel companion:

> [He was different from] other visitors. He quickly determined my level of knowledge in science and mathematics by direct questioning. I never picked up on any condescension on his part toward me that I had learned to expect from most other such individuals.

After they arrived, Ellis experienced what Walter Steiger (personal interview with author, Hilo, Hawai'i, December 4, 2007) describes as the "very demanding" side of Kuiper's personality when Kuiper asked Ellis to tie a spring-driven temperature recorder to the top of one of the 50-foot (15-meter)-high air sampling towers. Ellis did so with some difficulty, using his belt to secure the apparatus to the top. It soon became dark, and Ellis (video interview with author, Hilo, Hawai'i, December 2006) recalls Kuiper's excitement over the clarity of the sky:

> He just about blew his mind with—you know, the dark space between the stars and the lack of twinkle and everything.

But Kuiper was disappointed in not being able to see the lowest star in the constellation known as the Southern Cross. This was because MLO is on the north slope of the volcano, and the summit blocks a portion of the southern horizon. Kuiper asked Ellis to climb the second tower and announce when the star was visible, and Ellis obediently did so. Later that night Herring made test observations from MLO and found the sky remarkably clear.

The following morning, Ellis retrieved the temperature recorder from the first tower. With his legs sore from the tower climbing, Ellis and his visitors climbed into a jeep for the rough drive to the summit of Mauna Loa. Because Mauna Loa is higher and thus drier and less vulnerable to the inversion layer than Haleakalā, Kuiper found it better suited for a proposed NASA observatory. But he was not through searching, for Mauna Loa is an active volcano and Mauna Kea is not. Kuiper and Herring would later return to Hawai'i, and Howard Ellis would again assist them in their search for the best possible site for the NASA observatory.

The Snow People

Long before the road to the summit of Mauna Loa was constructed, Tom Vance and the Hilo Lions Club planned a recreation facility above the snow line. Their plans never advanced beyond a few smoothed areas where people could employ sheets of metal to slide on the snow (Ellis 1988). In 1956 a team of Japanese scientists dubbed the "snowflake people" stayed at MLO nearly two months to study snow. After the new road to MLO was completed, a new generation of snow people arrived, and Bernard Mendonca (1978) describes them in his memoir:

> In the winter the snow line of the mountain often descended to about 2.8-km elevation. The stage was set to permit the people of the island to drive up to freshly fallen snow after a winter cold front passage and experience for the first time in their lives snowman building, snowball throwing, sledding, and all sorts of winter sports. Once the word got out that this could be done, the result was mayhem on the slopes of the mountain after each snowfall in the winter. A good weekend would bring more than 500 cars a day packed with people, hot Thermoses, picnic lunches, makeshift sleds, and every possible adaptation of Hawaiian wear into winter fashions. The upper reaches of the mountain turned into an instant winter carnival land. The effect on the observatory staff driving to work was bewilderment. The net effect was that the local community was accepting the establishment of a scientific observatory on Mauna Loa. To this day, to a lesser degree, the arrival of the snow people every winter is a Mauna Loa phenomenon.

Years later, the opening of an access road to the summit of Mauna Kea attracted skiers and many of the snow people who might otherwise have driven up Mauna Loa. After a winter snowfall on Mauna Kea, many Hilo residents drove pickup trucks up the mountain to collect snow to take back to their families and friends. They still do.

Problems in the Road Ahead

Because the new route between Saddle Road and the MLO saved time, a major change in the traditional observatory work schedule was on the horizon. From July 1957 to after completion of the extended IGY at the end of 1959, MLO was manned continuously by at least two of the three staff, each of whom lived and worked on the mountain for six-day shifts. Eventually more staff was hired, and the shifts were reduced to a much more tolerable four days and three nights a week.

The new road meant that staying overnight at MLO was no longer as necessary as before, for the staff could make the drive more rapidly and arrive at the observatory less tired. Thus, they could leave Hilo early in the morning and return by evening.

Malcolm Keyes, who had worked as an electronics technician at MLO since 1961, saw no reason for the overnight stays, particularly since he was always on call to repair broken instruments. Moreover, he was seeing a doctor due to sinus problems exacerbated by the altitude and the extremely dry mountain air.

Fatigue was Bernard Mendonca's major reaction to the altitude change:

> The altitude was a definite impact. It wiped you out, that [altitude] change every day would wipe you out. You'd get tired; you'd get a little dizzy. (Mendonca, video interview with author, Boulder, Colorado, April 27, 2007)

Keyes (telephone interview with author, July 2007) found a diplomatic way to challenge the tradition of staying overnight on the mountain when he discovered errors in what was called "breaking the data" from the strip chart recordings. Today scientific data are usually stored electronically. But in the old days at MLO, most of the instruments were connected to chart recorders that recorded their data as a line of ink on a slowly moving paper roll. As noted in chapter 7, since July 1957, extracting data from charts had been a regular part of working at MLO. Howard Ellis (1988) describes how Jack Pales, Clifford Kutaka, and Robert Williams (Pales et al. 1957–1958) processed the data while staying on the mountain:

> All three made tabulations, analysis, and various graphic presentations of their observations, "on their own." They were ever up to selling the usefulness of their effort. Mr. Price, for whom Mr. Kutaka had worked, was sent many of these products. Right away there was pressure to get them in as soon as possible. This work was kept up to date daily and all the previous months' observations were finished and copies in the mail no later than the second of the month.
>
> There was particular pride in and much use made of a continuous plot along a common time base of all hourly average values reduced from the records and of each and every observation made. Ordinate values were clearly indicated and both standard and creatively designed symbols were cleverly used. This record was maintained meticulously on a continuous roll of chart paper which was finally neatly and appropriately folded.

Keyes told Pales that the errors in breaking the data from the charts were likely due to the long work hours and oxygen deprivation on the mountain. The task, he explained, could be accomplished faster and with more accuracy on the long table in the Hilo MLO office. Pales was sympathetic, and after the new road was in place, Keyes (telephone interview with author, July 2007) said, "The magnus of that schedule went away."

But the traditional schedule wasn't completely abandoned, at least not yet. The carbon dioxide project worked best at night, when cool air from the free atmosphere over the big mountain flowed down its slopes. This air was uncontaminated by pollution from the generators and the occasional car on the MLO road, and this is the air that Charles Keeling wanted to sample for carbon dioxide. He was insistent that someone be available twenty-four hours a day to check on the temperamental CO_2 analyzer and make sure power was being provided by the generators. It was also necessary to manually control which of the four air sample towers was feeding air to the analyzer so that the exhaust from the generators did not interfere and the clean, downslope air could be sampled at night. Reliable electrical power and an automatic method of switching the CO_2 analyzer's air sampling lines would be needed before the all-night schedule could be ended.

These and many other improvements would simplify work on the mountain, but they were far from consideration as the end of 1963 approached. The future of the observatory itself hung in the balance. When the landmark paper by Price and Pales that announced the significant progress made at MLO was published in December, plans were being made to close the observatory.

References

Arashiro, Harry. 2006. Former MLO staff information sheet. In author's files.

Bureau of Labor Statistics. 2006. Consumer price index for 1800–2006. In *Handbook of Labor Statistics*, US Department of Labor. (Accessed through the Federal Reserve Bank of Minneapolis at http://www.minneapolisfed.org/Research/data/us/calc/hist1800.cfm.)

Carroll, Rick. 2003. *Hawai'i's Best Spooky Tales*. Honolulu: The Bess Press, 69–71.

Chin, John. 2001. Unpublished retirement scrapbook, prepared by Leslie Pajo. Copy in author's files.

Cooper, Jeanne. 2005. Memories last after waves recede. *San Francisco Chronicle,* January 9, 2005. (Accessed at http://www.sfgate.com/cgi-bin/article.cgi?f=/c/a/2005/01/09/TRGOQAMKN71.DTL&hw=flinn&sn=062&sc=726.)

Ellis, Howard. 1978. The observatory. In John Miller, ed., *Mauna Loa Observatory: A 20th Anniversary Report.* National Oceanic and Atmospheric Administration Special Report, September: 24. Boulder, CO: NOAA Environmental Research Laboratories.

———. 1988. The observatory: On their own. (Accessed at http://www.mlo.noaa.gov/webmuseum/publications.html.)

———. 2005a. When the rooster crows. In Veterans memoirs: The Korean War educator. (Accessed at http://www.koreanwar-educator.org/memoirs/ellis_howard/index.htm#Childhood.)

———. 2005b. *My Mauna Kea Connection.* Unpublished manuscript.

Ellis, Howard T., and Rudolf F. Pueschel. 1971. Solar radiation: Absence of air pollution trends at Mauna Loa. Science 172: 845–846.

Fischer, W. A., R. M. Moxham, F. Polcyn, and G. H. Landis. 1964. Infrared surveys of Hawaiian volcanoes. *Science* 146: 733–742.

Fowle, F. E. 1915. The transparency of aqueous vapor. *Astrophysical Journal* 42: 394.

Fujisawa, Adeline. 2006a. Former MLO staff information sheet. In author's files.

———. 2006b. Unpublished memorandum. In author's files.

Keeling, Charles D. 1960. The concentration and isotopic abundances of carbon dioxide in the atmosphere. *Tellus* 12 (2): 200–203.

———. 1978. The influence of Mauna Loa Observatory on the development of atmospheric CO_2 research. In John Miller, ed., *Mauna Loa Observatory: A 20th Anniversary Report.* National Oceanic and Atmospheric Administration Special Report, September, 36–54. Boulder, CO: NOAA Environmental Research Laboratories.

———. 1998. Rewards and penalties of monitoring the Earth. *Annual Review of Energy and the Environment* 23: 25–82.

Keeling, C. D., T. P. Whorf, and the Carbon Dioxide Research Group. 2005. Atmospheric CO_2 concentrations (ppmv) derived from in situ air samples collected at Mauna Loa Observatory, Hawaii. La Jolla, CA: Scripps Institution of Oceanography. (Accessed at http://cdiac.esd.ornl.gov/ftp/trends/co2/maunaloa.co2.)

Kiess, C. C., C. H. Corliss, Harriet K. Kiess, and Edith L. R. Corliss. 1957. High dispersion spectra of Mars. *Astrophysical Journal* 126: 579–584.

Mendonca, Bernard. 1978. The first twenty years: An unscientific remembrance. In John Miller, ed., *Mauna Loa Observatory: A 20th Anniversary Report.* National Oceanic and Atmospheric Administration Special Report, September, 17–23. Boulder, CO: NOAA Environmental Research Laboratories.

Miller, John M., ed. 1972. *Geophysical Monitoring for Climatic Change No. 1* (Summary Report, 1972), 69. Boulder, CO: NOAA.

NASA. 1964–1972. Appendix E: Geology field exercises: Early training, field training schedule for the first 3 groups of astronauts (29). (Accessed at http://www.hq.nasa.gov/alsj/ap-geotrips.pdf.)

NOAA. 2001. Dr. Lester Machta, founding director of ARL (1919–2001), Air Resources Laboratory. (Accessed at http://www.arl.noaa.gov/ss/machta.html.)

———. 2006. Dr. Harry Wexler: Giants of science. (Accessed at http://www.history.noaa.gov/giants/wexler.html.)

Pales, Jack C., and Charles D. Keeling. 1965. The concentration of atmospheric carbon dioxide in Hawaii. *Journal of Geophysical Research* 70: 6053.

Pales, Jack C., Clifford Kutaka, Robert Williams, et al. 1957–1958. *MLO Weather Log* (unpublished).

Price, Saul, and Jack C. Pales. 1959. The Mauna Loa high altitude observatory. *Monthly Weather review* 87: 1–14.

———. 1963. Mauna Loa Observatory: The first five years. *Monthly Weather Review* 91 (October–December): 665–680.

———. 1964. Ice nucleus counts and variations at 3.4 km and near sea level in Hawaii. *Monthly Weather Review* 92 (May): 207–221.

Reichelderfer, F. W. 1963. *Monthly Weather Review* 91 (October): 475–476.

Stair, Ralph, and Russell G. Johnston. 1958. Some studies of atmospheric transmittance on Mauna Loa. *Journal of Research of the National Bureau of Standards* 61: 419–425.

Steiger, Walter. 2001. Origins of astronomy in Hawai'i: Epilog. Institute for Astronomy, University of Hawai'i. (Accessed at http://www.ifa.hawaii.edu/users/steiger/epilog.htm.)

Wexler, Harry. 1952. Volcanoes and world climate. *Scientific American* (April): 74–80.

———. 1953. Radiation balance of the Earth as a factor in climatic change. In Harlow Shapley, ed., *Climatic Change,* 73–105. Cambridge, MA: Harvard University Press.

———. 1956. A look at some suggested solar-weather relationships. 146th National Meeting of the American Meteorological Society, July 25–27. In John W. Evans, Sydney Chapman, and Harry Wexler, *The Sun's Effects on the Earth's Atmosphere,* Tech. Report No. 2, High Altitude Observatory of the University of Colorado.

Wharton, Nadine. 1966. Phantom dog could be missing link. *Honolulu Star-Bulletin,* September 7, A-4.

Yates, Harold W. 1957. Recent atmospheric transmission studies. 42nd Annual Meeting, Optical Society of America, Cleveland, Ohio, Oct. 17–19. Abstract in *Journal of the Optical Society of America* 47 (November): 1054.

CHAPTER NINE

Survival Mode and Making History (1964–1965)

In less than a decade, the slope unit of the Mauna Loa Observatory had produced a remarkable record of scientific achievement. During the year before the Weather Bureau took over in the summer of 1957, the initial scientific studies proved the merit of reestablishing the high-altitude observatory, even at a lower elevation than the original summit site. The observatory conducted pioneering measurements and experiments during the International Geophysical Year, and by the end of 1963, more than six years of ground-level ozone and atmospheric transmission measurements had been acquired. The ozone layer and carbon dioxide had been measured for more than five years. The annual cycle of carbon dioxide and its yearly increase had been accurately measured for the first time. Column water vapor had been measured for four years.

The impressive record of observations at MLO had led to the publication of significant research reports and papers in the fields of planetary science, water vapor, and the atmospheric transmission of light. A paper by Saul Price and Jack Pales (1963) on the first five years at MLO was about to be published in the *Monthly Weather Review.* They had just completed work on a paper on concurrent measurements of ice nuclei at Hilo and at MLO (Price and Pales 1964). And the first paper by Pales and Charles Keeling (1965) devoted exclusively to the MLO carbon dioxide findings was slated for publication in the *Journal of Geophysical Research.*

The Mauna Loa Observatory was gaining a worldwide reputation in the scientific community, and its founder, Dr. Robert Simpson, visited on November 18, 1963, to see for himself the progress. The weather conditions were much better than during any of his prior trips up Mauna Loa, and he wrote in the *Visitors* book, "Best trip up the mountain I ever made." Accompanying Simpson was Dr. Joanne Malkus, who was the first woman in the United States to earn a doctorate in meteorology and whom Simpson married on January 6, 1965. They were driven to MLO in the Dodge Power Wagon by Malcolm Keyes, who knew about Simpson's history on the mountain but was concerned that Malkus might have a reaction to the high altitude. Therefore he brought along an oxygen bottle, and every few thousand feet in elevation, he announced that it was time for a regular oxygen stop in case Malkus needed any (Keyes 2008).

Four days after the visit to MLO by Simpson and Malkus, President John F. Kennedy was assassinated in Dallas, and Adeline Fujisawa (video interview with author, Hilo, Hawai'i, December 4, 2006) reflects on that day:

> That is when our office had to close. . . . He was assassinated when we were on the job. And, of course, Jack took it real, real badly, so he said, "Let's go home." So we all went home.

When Pales and his team returned to work the following Monday, he knew that the next closure at MLO might be its last.

Orders to Close MLO (1964)

Just as the Mauna Loa Observatory was acquiring world-class status, major across-the-board federal budget cutbacks led to a decision in the fall of 1963 to close the observatory. John Chin had left in 1962 after finding a job at the Marshall Space Flight Center in Alabama. The remaining staff began looking for work elsewhere.

Less than two months after MLO founder Robert Simpson visited the observatory, Nels Johnson and Saul Price of the Weather Bureau's Pacific Supervisory Office in Honolulu visited Hilo on a Friday in January 1964 (possibly January 3 or 10). Their purpose was to personally inform Jack Pales and his staff that the observatory was being closed. Howard Ellis and Malcolm Keyes were working at MLO that day, and Johnson and Price drove to the observatory to give them the bad news. They did not sign the *Visitors* book.

Ellis (telephone interview with author, June 10, 2008) vigorously objected to the news and told Johnson he would run the observatory himself if necessary to keep it open:

> In November 1963, word reached us from the U.S. Weather Bureau that the observatory would probably have to be closed down because of lack of funds and that everybody would be transferred. Despite this grim outlook I was determined to continue anyway and stated that I would find my own funds for running the place if it came to that extreme. (Ellis 1978)
>
> And when they came to tell us they were closing it down, well, I put up a big fuss about it, and I said I can do it myself. . . . And I argued real, real strong, and they said, "Well, you cannot do it yourself. We will give you these two guys [Malcolm Keyes and Bernard G. Mendonca] who do not have a place to go." (Ellis, video interview with author, Hilo, Hawai'i, December 6, 2006)

Ellis (personal communication with author, March 25, 2007) said that Johnson later told him, "Close that place down. Stop waving your flag about MLO." Ellis replied, "No, I'll stay with it." Notwithstanding his love-hate relationship with the mountain, Ellis was serious about keeping the observatory open, and forty-four years later his determination was still evident during several personal and telephone interviews and in correspondence, as noted in the letter of March 25: "My motivation for keeping the place open was something I never considered. I just knew I had to and would."

Lena Ellis (video interview with author, Hilo, Hawai'i, December 6, 2006) recalls Howard's determination to keep the observatory open:

> I was surprised when he came home one day and told me the story about [the plan to close MLO]. And he said he was going to run the observatory by himself, and I could not understand that. All by himself? You know, go up and handle everything? And no money at all? But he said he can do it, and I believed him, because he was strong in those days.

Bernard Mendonca (1978), who worked for Saul Price at the MLO office in Honolulu, writes about the crisis in his MLO memoir:

> There was serious talk and maneuvering to shut down the observatory permanently. Funds suddenly were not available, interest in long-term monitoring had lost its champions, and the expense and effort of running an observatory thousands of miles distant on a volcano in the middle of the Pacific was hard to justify.

Ellis (personal communication with author, July 19, 2008) recalls a letter sent by Saul Price to Lester Machta about a week after the encounter on the mountain. According to Ellis, Price wrote that all the staff members opposed the idea of closing MLO. Ellis was upset that the letter failed to mention that he "was the only one to really speak up with total rejection" of the proposed closure. He wanted to run the observatory by himself, but the letter proposed that three unnamed staff be left to staff the observatory. Price suggested a budget of slightly more than $100,000 per year.

While there had been disagreements between Howard Ellis and Charles David Keeling over the operation of the carbon dioxide program, the two men were allies when it came to saving the observatory. Keeling (1998) protested against the closing of MLO, and he describes the situation in his memoir:

> During 1963, the first signs appeared that unfettered support of my studies of atmospheric CO_2 wouldn't be sustained indefinitely by the United States government. The U.S. Congress mandated agency budget cuts for 1964 that seriously affected any Weather Bureau studies not directly related to weather forecasting. Mauna Loa Observatory was threatened with closure. Harry Wexler was not alive to defend it. Lester Machta [founding director of NOAA's Air Resources Laboratory] argued strenuously for its survival, using the CO_2 program as one of the arguments to prevent its closure. It was saved, but the staff was drastically reduced from eight to three, including the transfer, without replacement, of Jack Pales, the physicist-in-charge.

The hard-working Jack Pales, who was exhausted and had suffered a heart attack the previous January, was ready to leave for a less stressful assignment at the Weather Bureau Research Station at Las Vegas, Nevada, and among his final duties was to inform Adeline Fujisawa (2006b) that MLO was about to close:

> Jack was a very stoic person; he was more or less a jolly person. So he never would tell me about any of these things that he expected. But then toward the very end he said, "I think we are all going to leave here. Would you like to go to Washington, DC? We would have a job there for you. Would you like to go to Lockheed Aircraft?" He gave me a lot of avenues, but I told him, being that I had my two children, I would rather remain in Hilo.

Pales (1964) then wrote a letter of recommendation on behalf of Fujisawa. The letter, which was typed on Weather Bureau stationery, places a human angle on the administrative side of MLO that kept things humming and in tune during better days:

> January 17, 1964
>
> To Whom It May Concern:
>
> Mrs. Adeline Fujisawa has been employed by the U.S. Weather Bureau, Mauna Loa Observatory, Hawaii, as a GS-5 Scientific Services Assistant for a period of three years.
>
> We deeply regret that a reduction in force of the observatory staff has caused her position, as well as several others, to be abolished.
>
> In the position of Scientific Services Assistant, Mrs. Fujisawa has acted as stenographer, bookkeeper, procurement and supply agent, librarian, technical writer and data analyst.

Pales' letter goes on to praise Mrs. Fujisawa's attendance, punctuality, personality, and enthusiasm for her work, qualities that were very much in evidence during a visit with her forty-three years after the reduction in force that cost her position. Like the staff scientists and technicians, Fujisawa also possessed cross-disciplinary skills that were ideal for working at MLO:

> She is experienced in the reduction and preliminary analysis of scientific data and has assisted the scientists at the observatory in the preparation of technical papers for publication in scientific journals. She is not only a highly qualified secretary but also possesses a high degree of mechanical and technical ability not ordinarily found in a person of clerical background.

The letter is signed, "Jack C. Pales, Physicist in Charge."

Mrs. Fujisawa proudly displayed a copy of Pales' letter of recommendation when she visited the MLO office in Hilo in December 2006. She also provided a memorandum (2006b) that lists the MLO staff with whom she worked during her time at MLO:

> Jack C. Pales, Physicist in Charge (Lubbock, Texas)
> Howard Ellis, Meteorologist (North Carolina)
> William E. Cobb, Meteorologist (Washington, DC)
> John Chin, Meteorologist (New York)
> Clifford Kutaka, Meteorologist (Honolulu)
> Harry Arashiro, Meteorologist (Honolulu)
> Benjamin LeBlane, Meteorologist (Albuquerque, NM)
> Mr. Jackson, Meteorologist
> Colby "Doc" Foss, Electronics Technician
> Mike Keyes, Electronics Technician

While some of these staff were temporary, Pales and Kutaka had been on duty since the Weather Bureau took over MLO on a full-time basis in July 1957. As the staff members left one by one, there were no farewell parties, and sometimes no good-byes. Shortly after writing the letter on behalf of Fujisawa, Pales himself was preparing to depart. Ellis (personal communication with author, July 19, 2008) was deeply concerned about the fate of his coworkers on the mountain, who were also among his friends, especially Cliff Kutaka, Ben LeBlane, and Harry Arashiro. He recalls that "Ben and Cliff got very bad deals. I don't know what happened to Harry." Ellis had worked well with Pales and had managed the mountain station since Pales' heart attack. But as Pales was leaving, Ellis shouted, "I wasn't asked about where they would go!" Pales replied, "It can't be helped" (Ellis, telephone interview with author, July 29, 2008).

Only Howard Ellis and Malcolm "Mike" Keyes were left at the Hilo office. Bernard Mendonca remained at the MLO Honolulu office, where he had worked under Saul Price since 1958.

The versatile Adeline Fujisawa went on to pursue a series of careers on the Big Island, where she still works today. Did her brief career at MLO influence her future?

> Well, you will not believe this. I even told all my grandchildren if I had my life to live over, I would be a physicist or a meteorologist. Why are these kids not in these fields? They all tell me, "I am going to be an electronic technician." "I'm going into computer." "I'm going to be this and that." But nobody even tells me about weather. I told them this is the field to go into. (Fujisawa, video interview with author, Hilo, Hawai'i, December 4, 2006)

Bernard Mendonca (1964)

During an interview at the Global Monitoring Division of NOAA's Earth System Research Laboratory in Boulder, Colorado, in May 2007, Bernard G. Mendonca traced his interest in science to his fascination with caves when he was growing up on the island of Kaua'i. He recalls a picnic near a cave when he was twelve when he was ordered by a concerned adult, "Don't you go in!" Before entering the meteorology program at the University of Hawai'i, Mendonca worked for the Weather Bureau at Enewetak (formerly Eniwetok), the nerve center of the Atomic Energy Commission's Pacific Proving Ground for testing hydrogen bombs.

While he was a student at the University of Hawai'i in August 1958, Saul Price hired Mendonca on a half-time basis to process data for what was to become known as the Honolulu branch of the Mauna Loa Observatory. This was a major break for Mendonca (2006), who badly needed work:

> I was a starving college student and needed to eat. I had just been fired from my U. of H. [University of Hawai'i] cafeteria job for growing a beard.

Mendonca attributes Price with teaching him much about data analysis, statistics, and meteorology. Following graduation in 1961, Mendonca became a full-time employee of the Honolulu Weather Bureau. Like the MLO staff in Hilo, by then he had become an old hand at extracting data from squiggly lines on rolls of chart recorder paper.

He later made the illustrations for the second paper published by Price and Pales (1963).

This was long before modern electrostatic copy machines and desktop computers, as Mendonca and other old-timers remember well. Numerical data retrieved from chart recordings were either printed by hand onto forms or typed. Copies were made using carbon paper.

> It was carbon copy, baby! Carbon copy. Three layers thick. And even when we collected our data, we had to layer the carbon copy in and copy it. So we go back to carbon copy. Forget the Xeroxing copiers and stuff. And, yes, the change [in technology] has been remarkable and so relieving. . . . I can't believe that we did everything without computers. I mean, I just can't believe it. And we had a calculator on the station . . . that wasn't even electric. It was mechanical, and you'd crank the thing, and these gears would pop up and down with the numbers and everything. So, yes, we go a way back. (Mendonca, video interview with author, Boulder, Colorado, April 27, 2007)

Malcolm Keyes (video interview with author, Keauhou Bay, Hawai'i, November 29, 2007) remembers Mendonca "as just always pleasant. [He was] 'a local boy' who had basic human values that included, you know, punctuality, perseverance, honesty." His perseverance extended to MLO, and the day before Pales wrote his letter of recommendation on behalf of Adeline Fujisawa, Mendonca (1964) sent a four-page letter to Pales through Saul Price, his supervisor, and Nels Johnson, the meteorologist in charge of the Weather Bureau's Pacific Supervisory Office (PSO) who had proposed closing MLO in 1958. The letter is in the form of an official memorandum, and it outlines a strategy for salvaging the observatory by enhancing its scientific mission.

While the memo was addressed to Pales, the intended recipient—as Mendonca recently disclosed—was really Johnson. The memo was received by Johnson and Price at the same time final decisions were being made about the fate of MLO, and only days later Mendonca received notice of an unintended consequence of his memo. He was given two weeks' notice that he was being moved from Honolulu to MLO, and the opening paragraphs of his memo may explain why Johnson and Price selected him to assist Howard Ellis and Mike Keyes in keeping MLO alive:

> As a consequence of my assignment at PSO as the research assistant and subsequent affiliation with the Mauna Loa Observatory (MLO) and its staff, the pending reorganization of MLO, and because I feel strongly about its potentialities and its future, I present these suggestions and comments about the MLO and its operation for whatever value they may have in light of my experience in working with MLO data in this time of transition at the Observatory.
>
> Almost without exception, everyone who has visited the Observatory or who has been closely associated with it, herald[s] its uniqueness and its ideal location to observe atmospheric parameters to obtain baseline values and in conjunction with synoptic networks. They also proclaim MLO's natural potential to conduct scientific research. And yet, in my opinion, MLO has failed to fully live up to its reputation. The purpose of MLO at times has been hazy partly because the Observatory deals with basic research (it being fickle itself) and partly because there [have] been differences in opinion as to what its primary purpose is. The result has been unsteadiness of effort and purpose.
>
> A decision, even if a temporary one, should be made outlining a definite course for the observatory to take, whether it be a data gathering outfit or an active scientific circle not only for and in the Weather Bureau but with other organizations as well, or both. Then, yearly plans defining specific programs should be given to the staff as an incentive to accomplish. If this is not done the danger to the staff to lapse into a routine as in radiosonde stations is paramount. I believe that if MLO is to remain a scientific station, this is the biggest danger it must overcome. (Mendonca 1964)

Mendonca's memorandum continues by explaining the importance of an interdisciplinary approach to the acquisition of data at MLO:

> The data received from the Observatory for the most part has been good, but it has been naked. What I mean is, data can be collected in two ways—one, as data in itself and two, as data in response to its environment and other data.

Mendonca was well qualified to offer the three pages of suggestions that followed, for he began processing data from MLO in 1958 when he was hired by Saul Price. Thus, his involvement with the observatory went back to within a year of when Jack Pales, Robert Williams, and Clifford Kutaka began manning the station on a full-time basis.

Howard Ellis, Bernard Mendonca, and Malcolm Keyes Remain at MLO (1964)

The strong appeals by Howard Ellis from the Mauna Loa Observatory on the Big Island, Bernard Mendonca's memo, and Charles Keeling's appeal from Scripps did not go unnoticed. Lester Machta, who had succeeded the late Harry Wexler at the Weather Bureau in Washington, endorsed their support. It was decided that MLO would remain open with a sharply reduced staff and budget.

Ellis had already taken over many of Jack Pales' leadership roles after Pales' heart attack in January 1963, and he had received positive recognition when he arranged for the new shortcut to Saddle Road that greatly improved access to MLO. Ellis was the logical successor to Pales, whom he was supposed to have replaced in 1961, and it was decided that he would be assigned to run the observatory with the assistance of two others. Malcolm Keyes from the Hilo office would remain as the electronics technician. They would be joined by Bernard Mendonca of the Honolulu MLO office. Saul Price would remain at the Honolulu office and later become the state climatologist for Hawai'i.

Charles Keeling (1998) writes in his autobiography that Jack Pales was not replaced when he left Hilo in late January 1964, but he was almost immediately replaced by Ellis, who on February 2 was reassigned as supervisory physicist (GSA 1981). He was addressed in correspondence as "Physicist-in-Charge," the same title that had been applied to Jack Pales and the position he had been promised by Wexler when he left for Hawai'i in 1961. He was provided an annual budget of $100,000 for salaries, fuel, vehicle repairs, road maintenance, and all other expenses. Howard Ellis was finally in charge of the Mauna Loa Observatory.

Mendonca arrived in Hilo on February 4, 1964, with his wife Clara and their infant son Karl. He and Clara were given just two weeks' notice to move from Honolulu to Hilo.

> We packed in a hurry and had little time to prepare or to say goodbye to our family and friends before we left. As a matter of fact, some or our belongings were even sent by Hawaiian Airlines as airfreight instead of through a mover, including Clara's house broom with a name tag on it hurriedly put in a container and shipped as air cargo with some of our other last minute collection of household furnishings. . . . We really didn't have that much at the time. We (Howard Ellis, Floyd "Mike" Keyes and me), were soon on site and running the observatory, and everyone else was gone. (Mendonca, personal communications with author, March 14–19, 2008)

Mendonca was familiar with the MLO routine, for he had previously spent two weeks on the mountain during a special project and had amassed considerable experience working with MLO data in Honolulu. Like the Hilo team, he was an old hand at breaking out the data from chart recordings, and this made him well qualified for the data reduction and analysis aspect of his new assignment. While he found the trips to and from the mountain station fatiguing, he liked the new opportunities provided by the mountain assignment,

> because when we went to the observatory, and worked, we also had the chance . . . to do research as well. So, I could do what I was doing in Honolulu on the mountain next to the instruments. (Mendonca, telephone interview with author, July 3, 2007)

A few days before Mendonca arrived, Ellis closed the MLO office at the Air Cargo Terminal at Hilo's General Lyman Field to save the rent money. The Hilo office became a post office box, and the mountain station was the sole workplace.

> In February 1964, Jack Pales left for a new job in Las Vegas, and the administration of the observatory was turned over to me. To save money, I immediately closed the administrative office in Hilo and personally hauled all the equipment to the mountain. (Ellis 1988)

Ellis (video interview with author, Hilo, Hawai'i, December 6, 2006) "worked around the clock" and used a three-quarter-ton truck to move everything in the office. To borrow his phrase, for the next two years he, Mendonca, and Keyes would be "on their own." As Mendonca (1978) recalls,

> As a result of the cutback during this austere time the Honolulu branch was closed, the staff was reduced to 3, the Hilo office was closed, and a starvation budget was given for salaries, gasoline to get to the mountain, and chart paper for the recorders, with a few dollars extra for the light bill. Under Howard Ellis, the director at the time, the basic programs were kept going, and we waited. For 2 years, nothing much changed, and the observatory hung in limbo.

Prior to the budget crisis, the schedule at MLO for each employee was continuous duty on the mountain for four days and three nights. Ellis and Mendonca alternated the schedule so that at least one of them was always on the mountain. Like Pales before him, Ellis often worked an even longer schedule. His usual routine was to leave for the mountain station on Sunday afternoon and return the following Thursday, but often he worked a longer shift. "I tried to accomplish as much as I could by frequently staying on the mountain overnight and working 7-day weeks" (Ellis 1988).

The *Visitors* book lends support to Ellis' claim that the observatory, which was not locked in those days, was manned continuously or nearly so. The log reveals that in 1964 there were visitors on 84 days, representing every day of the week. There were 16 visits on Wednesdays, 15 visits on Tuesdays, Thursdays, and Saturdays, and 11 visits on Fridays. Sundays and Mondays had 6 visits each.

Keyes (video interview with author, Keauhou Bay, Hawai'i, November 29, 2007) made his own schedule and spent nights on the mountain usually when working on instruments. Because he was needed to attend to instruments whenever they broke down or needed routine servicing and because of a sinus condition that was aggravated by the mountain environment, he saw no reason for becoming part of the regular mountain schedule.

> So I was pretty much on my own, called my own schedule. But a lot of times it was just more convenient or socially acceptable to pull a shift, and so, you know, I know what it is to stay two, three, four nights in a row up there.

Mendonca (video interview with author, Boulder, Colorado, April 27, 2007) recalls his first night alone on the big mountain. "It was eerie," he says. Before going to sleep, it was necessary to walk to the generator shed to check the oil level and the voltage and make sure there was no fuel or oil leak. Mendonca left the main building at 10:00 p.m. and walked toward the generator shed under a jet black sky illuminated only by the Milky Way. When he reached the shed and opened the door, "the roar of the generator stopped me in my tracks. [I thought] what am I doing here?" In spite of the demands of the job and the lonely days and nights on the mountain, Mendonca says, "I'd do it again in a heartbeat."

Manning the observatory seven days a week with only two men made for a grueling, lonely work schedule. During an interview in Hilo on December 6, 2006, Ellis and his wife Lena remembered the difficulties of living on the mountain. Lena recalls how she prepared and packaged all of Howard's meals in plastic containers marked for each day of his five-day, four-night shifts:

> I was raising three children, so it kept me busy, although it was very difficult having him being away that long. . . .

> There were a lot of turkey sandwiches. That is real turkey, not the cold cut, the real turkey. And hamburger sandwiches. It was difficult, but I felt so sorry for him being up there with no access to anything nearby to buy, so I tried to make it easy and pleasant for him.

Lena also recalls how Howard was unable to speak for a day or so after returning from mountain duty. By the time he was back to normal, it was time to return to the mountain station.

Mendonca and his wife Clara recall how exhausted Mendonca was when he came down from the mountain. They also remember November 19, 1964, when Bernard was preparing to leave home to work his shift on the mountain that Clara, who did not drive, reminded him that their second child was due any day. Clara's concern was that there would be no possibility of contacting her husband on the mountain, because there was no telephone there and the radio used for communications could only reach the Hilo Weather Office. Bernard decided to forgo his mountain schedule for the first time and stayed in Hilo without notifying Howard Ellis until the last minute, thereby leaving the station temporarily unmanned. Later that day, Clara gave birth to their daughter Laureen (Mendonca, personal communications with author, March 14–19, 2008).

While the shortcut link to Saddle Road that Ellis had arranged in 1963 with the Atomic Energy Commission was a major improvement, the 18-mile (30-km) mountain route remained unpaved. Ellis managed to set aside $10,000 of his annual budget for repairing the badly deteriorated road by arranging for free and low-cost support from the community and by hauling the generator fuel up the mountain instead of paying extra for it to be delivered. But he lacked government administrative experience and didn't know that he couldn't hold the money over until the next fiscal year. Much to Ellis' chagrin, the saved money went back to the Weather Bureau instead of going toward repairs to the road.

Mendonca recalls that, in addition to maintaining the IGY projects, they had to continue the hourly weather observations at MLO. The official weather forms had provisions for the major cloud types—but not for clouds *below* the observatory. In hindsight, he suggests that MLO was likely designated as a Weather Bureau Research Station (WBRS MLO) in January 1963 to help keep the observatory alive.

Howard Ellis Helps Spark the Rush to Mauna Kea (1964)

While the Mauna Loa Observatory was struggling to survive, a growing number of astronomers had their eyes on the barren summit of Mauna Kea. As related in chapter 8, in May 1963 Howard Ellis escorted the famed astronomer Dr. Gerard P. Kuiper and telescope maker Alika Herring to MLO during their first trip to Hawai'i to scout potential sites for a proposed new NASA observatory. Ellis and his wife Lena had also taken Kuiper on a tour of the Kīlauea area at Volcanoes National Park. While Kuiper and Herring were impressed by the clarity of the air over MLO, Kuiper knew that an active volcano would be unacceptable to NASA. Mauna Kea was his goal, but there was no road to its summit.

After Kuiper and Herring returned to the mainland, Ellis visited Mitsuo Akiyama of the Hawai'i Island Chamber of Commerce. He told Akiyama about the distinguished visitors he had hosted and said, "So often I look across at Mauna Kea high above the clouds and think it would be an ideal place for an astronomical observatory" (Parker 2001: 25). The chamber had been searching for new business opportunities since the devastating tsunami of May 22, 1960, that destroyed much of downtown Hilo, and Akiyama immediately recognized the potential of attracting a major astronomical observatory to the Big Island. Ellis (2005) relates in an unpublished memoir about Mauna Kea how he gave Akiyama the names of several scientists who had made inquiries about the astronomical potential of Mauna Kea. Akiyama sent letters to these contacts and shortly thereafter excitedly informed Ellis that Dr. Kuiper had responded favorably.

Kuiper arrived at Hilo's General Lyman Field on Sunday, January 12, 1964, where he was met by Ellis and Akiyama. After introductions, Ellis returned to a birthday party for Jack Pales, who was about to leave Hawai'i for his new assignment in Nevada. Frustrated over the budget crunch at MLO, Pales expressed displeasure with Ellis over his Chamber of Commerce connections, "saying that they had no money" (Ellis 2005). Ellis fully understood Kuiper's vision for Mauna Kea, even though he and Tom Vance had hoped Kuiper would select Mauna Loa. Ellis was impressed by the visionary approach to the future of Hawai'i pursued by Vance ever since.

Vance and Robert Simpson worked together to build a road to the summit of Mauna Loa in 1951. While Ellis respected Pales as the best boss he ever had, his free spirit side led him straight back to Kuiper. MLO had some of the very few four-wheel-drive vehicles on the island, and Ellis offered to drive Kuiper and Herring to the slopes of Mauna Kea. When Kuiper expressed interest in flying above the mountain's summit, Ellis (2005) knew what to do:

> There was a crop duster pilot working here who owned a Cessna. His name was Bob Beals and he may still live in Honolulu. He flew Dr. Kuiper, Alika and me over the mountain. He charged $65 and I worried if my people would pay. In time, they did. Dr. Kuiper sat on the right in front with Alika behind him. I was in the other back seat. We made climbing turns out over Hilo Bay within range of the airport until we were at an elevation higher than the mountain, then quickly over flew, making several passes. Dr. Kuiper took photos. We flew over Mauna Loa as well and I was amazed at the eerie prospect of the spine of that mountain stretching to the southeast of its summit crater.

NASA would not pay for a road to the summit of Mauna Kea, but maybe Hawai'i would. Back in Hilo, Akiyama arranged for Kuiper to meet Governor John Burns. During a meeting in Honolulu, Kuiper persuaded Burns to begin the process of obtaining permission to build a road and to secure funds for its construction. Burns committed $25,000 and later added an additional $17,000. The money paid for the bulldozing of a crude road to the summit in April 1964, and Kuiper mapped the route. When the road was completed, NASA provided funds for Kuiper to build a small telescope dome for Alika Herring to evaluate the seeing conditions. The dome was built on Pu'u Poli'ahu, the second highest of the cluster of volcanic peaks atop Mauna Kea, the highest being avoided because of its special status to Native Hawaiians. The dome was completed in June, and Akiyama arranged for a special ceremony and invited Howard Ellis to participate.

> Dr. Kuiper gave a speech at Hale Pohaku [where support facilities and dormitories for summit workers were eventually constructed]. Governor John Burns attended. Over my protests, Mr. Akiyama got me to drive the lead vehicle to the summit. I wanted him along but he insisted. There I was driving, with Governor Burns seated to my right in an austere manner. In the cramped space in the back on benches that folded down from each side sat an exuberant Dr. Kuiper, Mr. Stephenson, the governor's adjutant general, and my coworker Mr. John Chin. (Ellis 2005)

Alika Herring soon began observations at the new dome, and Walter Steiger describes what happened next in his memoir:

> When completed, the study, to Dr. Kuiper's delight, showed that Mauna Kea was a truly superb site, the finest he had ever seen. With this established, he submitted a proposal to NASA for a new telescope to be placed on Mauna Kea.
>
> At this point, NASA decided to open the competition to other proposals and invited both the University of Hawaii and Harvard University to do so. The University of Hawaii had no astronomers experienced in telescope development, only the new team of solar astronomers. To Dr. John Jefferies, now in charge of the solar program, fell the responsibility of organizing an observatory development plan and submitting a proposal to NASA. The plan called for an 88-inch telescope. It was a well-conceived plan that ultimately won NASA's approval for support! This outcome was, understandably, a terrible blow to Dr. Kuiper, who felt "his mountain" was "stolen" from him. Regardless of the outcome, Dr. Gerard Kuiper must be

> acknowledged as the discoverer of Mauna Kea as a superb astronomical site. (Steiger 2001)

Jefferies' leadership of the program had a less than auspicious beginning, as Howard Ellis experienced earlier when Saul Price arrived from Honolulu with some astronomers from the University of Hawai'i.

> First they saw our facility on Mauna Loa, and then I took them to the summit of Mauna Kea. They appeared unenthusiastic over the need for astronomy work on either mountain. I thought they showed more pain than enjoyment, which was new for me. One of them was John Jefferies. It was his first visit to Mauna Kea. (Ellis 2005)

Parker (2001) provides more details about Jefferies' first trip to the summit of Mauna Kea:

> Strangely enough, Jefferies was not enthusiastic about the project at first; he later admitted that his first ride to the summit of Mauna Kea was not an enjoyable experience. The road was dusty and he didn't tolerate the altitude well. When he got back down he said he hoped he would never have to go up again.

But go he did, and John Jefferies was destined to join Walter Steiger and Gerard Kuiper as the pioneers of astronomy in Hawai'i. The slope unit of the Mauna Loa Observatory sprang from the chance meeting of Robert Simpson and Ralph Stair at New Mexico's Sunspot Observatory in June 1955. In a remarkable coincidence, Jefferies' decision to submit a proposal for the NASA telescope that Kuiper hoped to build also sprang from a meeting at Sunspot when he met with solar astronomer Dick Dunn. Jefferies (2001) sought advice on how best to collaborate with NASA's proposers when Dunn suggested that Jefferies submit his own proposal. He did, and what happened next is history.

The rush by the astronomy community to the summit of Mauna Kea had begun, and Howard Ellis had played a key role in its genesis. The summit so visible from MLO over the years was soon to sport the first of what was to become a dozen major astronomical observatories, each of which would cost many times more than the little atmospheric observatory across the Saddle on neighboring Mauna Loa. Meanwhile, back at Mauna Loa serious trouble was brewing for Ellis, Mendonca, and Keyes.

The Infamous Carbon Dioxide Data Gap (1964)

Bernard Mendonca (telephone interview with author, July 3, 2007) recalls the enthusiasm he, Ellis, and Keyes shared about keeping MLO open after they were left alone to run the observatory. "We had lots of drive, hope and determination," he says. Then disaster struck.

The story began before Howard Ellis arrived in Hilo in January 1961, when he spent time at Scripps with Charles Keeling learning about the operation of the carbon dioxide analyzer. He began working with Jack Pales on the Scripps project soon after he arrived at MLO. In an entry in the CO_2 *LOG*, Ellis describes how he used the CO_2 in his breath to check the operation of a newly installed set of air intake lines on a new tower:

> April 26, 1962
>
> A 50' tower is now in use with L3 midway and L5 at 50 feet and L6 at 2 feet. A check was made of L6 by breathing into the input after air from the line had been passing through the analyzer for about 4 minutes, making a straight line trace. About 4 minutes later the trace moved abruptly off scale. HE

Ellis' final entry in the CO_2 *LOG* discusses installation of a new air sampling line and mentions no problems:

> Oct. 22, 1963
>
> New line put to intake #2.
>
> This leaves us with all four lines new plus a test intake on the north wind mast.

The average carbon dioxide measured during December 1963 was 319.57 parts per million (ppm) (Keeling et al. 2005), and all was well with the CO_2 project unless Charles Keeling called MLO with new instructions. Because the MLO staff worked for the Weather Bureau, Ellis (video interview with author, Hilo, Hawai'i, December 6, 2006) wondered why Keeling was much more demanding than their employer.

> [Charles Keeling] was extremely demanding. I had a lot of suggestions I thought that would help him out, but he did not like any of them. And he just wanted people to be very careful about watching his machine all the time. And he even prescribed how we should fold the [chart] paper.

Ellis went on to recall that Keeling did not allow the MLO staff to make any modifications to the CO_2 equipment to try to improve its performance. Folding the chart recordings was made tedious by the altitude and was a distraction during busy times on the mountain. There was no such requirement for the government chart recordings, which were simply rolled before being mailed to Honolulu (as was eventually the protocol for the Scripps chart recordings).

> He just would not take it. He just wanted that data—and wanted [the chart recordings] folded to an eighth of inch. (Telephone interview with author, June 8, 2008)

Ellis applied southern hospitality during discussions with Keeling, but Keeling expected full compliance with his instructions and allowed no changes to his measurement protocols, which Bernard Mendonca (video interview with author, Boulder, Colorado, April 27, 2007) still remembers:

> Keeling was a hard taskmaster. He was tough. He demanded—I'm searching for words here; I want to be kind because he's such a great man—Keeling demanded rote obedience to his methods and instructions. You were never allowed to be innovative, adaptive, and, by all means, don't change anything, ever, without his signed, sealed approval. Because of this, there was a tension, because he was at Scripps in California, [and] we were on the field trying to make his instrument run. And there was always, "Why don't we do it this way instead of that way?" and it was always, "No, you can't do that." So this was hard. But, to give Keeling his due, he knew exactly what he was getting, because we couldn't change anything.

Malcolm Keyes (video interview with author, Keauhou Bay, Hawai'i, November 29, 2007) employed a more proactive approach, as when Keeling called MLO shortly before he was about to leave the station for Hilo:

> Most everybody tried to get off [the mountain] before dusk, because [it] wasn't everybody's ideal thing to drive in the dark down that road, because the headlights went up and down and around and created kind of spooky shadows in the lava. So most everybody left an hour or so before sundown, except when Dr. Keeling would call, and you sometimes wondered if he didn't do that intentionally. . . . We all know that the order of priority in one's life is your wife, the President of the United States, and the Director of the Weather Service. [When Keeling called and] was anxious to know how everything was . . . I tried to respect that, except when it got to pushing 7:00 and it was dark. So I finally hit on a plan or a technique of terminating the phone call by pressing the button down when I was talking, not when he was talking. That technique has served me well in subsequent life.

Then, shortly after 8:00 a.m. on Tuesday, January 21, 1964, only a few days after Jack Pales left Hilo, the carbon dioxide measurements suddenly came to a halt when a then unknown component of the analyzer ceased to function properly. Problems with the analyzer were common, but this one was different. A hand-lettered inscription on the manually plotted chart of the daily carbon dioxide data prepared at the Scripps Institution of Oceanography reads "ANALYZER CLOSED DOWN. NO TANKS." (Scripps 1964a.) The Scripps daily log sheet for January 21, 1964, reports that the analyzer system was shut down at 10:00 a.m. because of low pressure in the tanks and no tanks available (Dave Moss, personal communication and telephone interview with author, August 18, 2008).

The absence of tanks refers to the reference gas used to calibrate the CO_2 analyzer, and Malcolm Keyes recalls that they had indeed run out of gas. But the absence of reference gas did not cause the malfunction, and forty-three years later Keyes (video interview with author, Keauhou Bay, Hawai'i, November 29, 2007) relates what happened:

> Well, it sort of shut itself down. An integral part in the system broke, to the point where it was not replaceable by me or fixable by me. And what we needed was a spare chopper motor. . . . And so I think I went to another electronics technician or two and invoked their opinion. There was a consensus that we needed a new chopper motor.

The chopper motor was crucial, for it spun the wheel that divided the infrared beam projected through the two air sample tubes into twenty pulses per second. This sent pulsating bursts of infrared to the detector microphone at the opposite end of each sample tube. Without the chopper, the uninterrupted beams of infrared would elicit no usable signal from the detector microphones. But was the chopper motor really the problem? Howard Ellis (telephone interview with author, June 8, 2008) believes the culprit was one of the two detector microphones, but he and Keyes lacked the expertise and the test equipment to determine which part was defective. Had appropriate spare parts been available, perhaps they could have substituted a spare for a new part until the system was again working.

Meanwhile, as Mendonca (personal communications with author, March 14–19, 2008) recalls, "Howard was afraid to tell Keeling," who micromanaged the project from his office at Scripps. The Scripps project was only one of many at MLO, and Ellis had an observatory to run. He did not look forward to telephoning Keeling with bad news about the carbon dioxide analyzer. Instead, he telephoned Saul Price of the former MLO office in Honolulu.

A full review of Weather Bureau and Scripps records may shed more information on the details of the data gap. Meanwhile, enough has been gleaned from the manually plotted Scripps CO_2 chart for 1964, the Mauna Loa Observatory CO_2 *LOG*, and both personal and telephone interviews with each of the three retired staff members who were present at the time to provide a synopsis of their perspective of what happened during what came to be known as the "data gap."

"Mission Impossible" (1964)

Howard Ellis and Mike Keyes independently tried to repair the broken analyzer. The old CO_2 *LOG* records that Ellis had previously made many minor repairs to the CO_2 analyzer, and one weekend he and his wife Lena stayed at MLO while he worked on the instrument. Saul Price had suggested that he heat and cool the detector cell that Ellis thought might be defective, so he removed the suspect part and alternately refrigerated it and exposed it to sunlight. This procedure failed to correct the problem (Ellis, telephone interview with author, June 8, 2008).

Keyes disassembled much of the analyzer in an effort to find the problem. "Mike had that thing in pieces all over the floor," Mendonca recalls (telephone interview with author, June 9, 2008). No matter what he tried, Keyes was unable to repair the instrument. Meanwhile, Keeling eventually learned about the breakdown, perhaps from Saul Price or from the last chart recording Ellis mailed to him. According to Keyes (telephone interview with author, June 9, 2008), "[Keeling] was a micromanager. He called me about the breakdown more than once."

Though Ellis never telephoned Keeling about the breakdown, Keeling called Ellis at home some time thereafter. In what both Ellis and his wife Lena recall as an unpleasant conversation, Keeling "was kind of rough with me. [He] was demanding. He wanted a guarantee" that either Ellis or Mendonca would stay on the mountain twenty-four hours a day (Ellis, telephone interview with author, June 8, 2008).

Because of Ben Harlin's experience in Antarctica in 1957, Charles Keeling well understood the difficulties involved in operating a remote observatory and the need for a full-time observer to keep everything running. Several months later, after Scripps received funding to hire its own technician, John Chin, to work alongside Ellis and

Mendonca at MLO, the mountain responsibilities were outlined in detail on Chin's Job Description Card of May 1964, including this time breakdown:

> A breakdown into percentage of time spen[t] on each duty is meaningless in appraising this job since the technician on duty is in reality undertaking several duties simultaneously with no clear demarcation lines. When alone at the station 24 hours a day, he is operating all equipment 100% of the time. At the same time he also reduces data, services the instruments, and in case of breakdown repairs equipment. There will be times when the job demands only his presence at the station to cope with unforeseen interruption in any one of the instruments. (Scripps 1964b)

As for Keeling's demand that the station be manned continuously by Ellis and Mendonca, Ellis explained during the telephone call that government personnel rules did not allow the schedule that Keeling wanted. According to Ellis, Keeling replied, "Rules are made to be broken." Ellis didn't tell Keeling that he and Mendonca had continued the practice of manning the observatory on a nearly continuous basis, because he didn't want Keeling to control their schedule (Ellis, telephone interview with author, July 29, 2008). The fact that the instrument was broken was not mentioned, perhaps because Keeling had already placed calls to Mike Keyes about the problem (Keyes, telephone interview with author, June 9, 2008).

Ellis and Mendonca didn't see each other on a regular basis, for they were working alternating, multiple day and night shifts on the mountain. When the two men were next together, Ellis told Mendonca about Keeling's demand that they staff the observatory all day every day, and Mendonca (telephone interview with author, June 9, 2008) reflects on the matter: "We both kind of dismissed it as mission impossible." Yet, as noted earlier, signatures and dates in the *Visitors* book established that either Ellis or Mendonca was on the mountain most of the time.

Ellis had many other responsibilities on the mountain that went well beyond Keeling's expectations that Ellis or Mendonca should stay on the mountain away from their families for days at a time. Keyes, defending Ellis, recalls how Keeling sometimes called Ellis to express his frustration over an instrument problem that was Keyes' responsibility, for he, not Ellis, was the electronics technician.

As for the broken analyzer, even had Keyes repaired the instrument, they had run out of reference gas, and receiving a new cylinder from Scripps would have taken still more time. In separate interviews, Mendonca and Keyes both recalled how the matter exacerbated what was already a delicate situation:

> His reaction was, I think, not constructive, and the result was the [four] months of lost data. I think after that the working relationship was never 100 percent trusting from either side. (Mendonca, video interview with author, Boulder, Colorado, April 27, 2007)

While the carbon dioxide project at MLO was but one of many programs on the mountain, it was destined to become among the most important and by far the best known. Back in 1956, some had wanted only a "snapshot" of the carbon dioxide concentration during the IGY to provide a reference point that could be checked again after twenty years. Keeling had a better idea. His pushy perseverance and the long hours on the mountain spent by Jack Pales and his team had provided him with an unprecedented time series of nearly six years of CO_2 data before the 1964 data gap.

More than three decades later, Keeling (1998) describes in his autobiography how the breakdown of the Mauna Loa analyzer temporarily closed his entire carbon dioxide observing program:

> Our equipment, at the first glitch in performance, was shut down by the remaining station personnel. Even earlier, in August, 1963, my assistant from the Weather Bureau, Tom Harris, had been reassigned and left my program. I ordered the continuous analyzer from the South Pole returned to Scripps but, without Harris' help, I did not attempt to send flasks to the South Pole for the next field season. The ship and aircraft programs were also ended. Suddenly there were no precise measurements being made of atmospheric CO_2 anywhere.

Keeling describes the breakdown of the carbon dioxide analyzer as merely a "glitch" and not the latest in a long series of maintenance headaches described in the CO_2 *LOG*. He also writes that the analyzer had been intentionally shut down when in fact the instrument was inoperable and had broken down on its own. When asked about the "glitch," Malcolm Keyes (telephone interview, June 23, 2007, and personal interviews with author, Keauhou Bay, Hawai'i, November 28–29, 2007) used colorful language to express his displeasure with Keeling's account and recalled the many hours he spent on the mountain trying to repair the troublesome instrument without adequate spare parts to substitute for possibly defective ones.

Also unmentioned by Keeling was that the temperamental vacuum tube CO_2 analyzer was only one part of a complex system that was often monitored day and night by "the remaining station personnel," especially Howard Ellis, who voluntarily spent many nights on the mountain to watch over the instrument before and after the new shortcut road was completed. Even when the analyzer was working properly, problems might arise with the water vapor freeze-out trap, the chart recorder, the calibration gas supply, the air sampling tubes, the switching devices that selected which sampling tubes would feed air to the analyzer, and the diesel generators that powered the observatory. Without a man on the mountain at night when the pristine downslope air is flowing down over the observatory, Keeling's program would have produced considerably less reliable data.

In years to come, Keeling became much more generous in his recognition of the men on the mountain in Hawai'i who made possible what had become his most important research program, as in a paper he wrote in 1976:

> All members of the staff of the Mauna Loa Observatory contributed to maintaining this long term program. We especially thank Mr. Howard Ellis, Dr. Lothar Ruhnke and Dr. Rudolf Pueschel, successive directors of the observatory, and Dr. Lester Machta of the National Oceanic and Atmospheric Administration who has guided the destiny of the station through the years of this report. We gratefully dedicate this paper to the late Mr. Jack Pales who, by great personal effort, established the observatory as an environmental facility of worldwide significance. (Keeling et al. 1976)

Returning to 1964, there was a silver lining in the crisis with the broken analyzer, for the inability of the MLO staff to repair the instrument provided the opportunity for Charles Keeling to finally find support to hire and train his own technician to repair the analyzer and help maintain his program at MLO. Keeling (1998) describes in his memoir what happened next:

> I had seen the budget cut coming early in 1963 and had tried to prevent its terminating the CO_2 program at Mauna Loa Observatory. I even went to Washington to plead for supplemental funding. This had no tangible effect, however, until the cessation of measurements actually occurred. The National Science Foundation then found funds to pay for an additional technician at Mauna Loa. In May, 1964, with Mr. John Chin hired for the task, CO_2 measurements were restored there.

Some MLO veterans have suggested that the extended CO_2 data gap provided Keeling with leverage for keeping his project out of the control of government agencies that extended well beyond Chin's hiring. Ellis (telephone interview with author, June 8, 2008) recalls, "It was obvious to me that he had things up his sleeve." Mendonca (telephone interview with author, June 9, 2008) also comments on the data gap:

> Keeling blamed us for many years. From his point of view, he used [the data gap] very effectively to justify funding. I can understand that, for it gave him justification to get funding to continue his program. But it kind of stuck in our craw, because we knew what had happened.

Keeling's 1963 trip to Washington to plead for supplemental funding suggests that he was much less concerned about the ability of the remaining men on the mountain to maintain the CO_2 analyzer than about the congressionally mandated budget cut, which, as noted earlier, he viewed as a sign that

> unfettered support of my studies of atmospheric CO_2 wouldn't be sustained indefinitely by the United States government. The U.S. Congress mandated agency budget cuts for 1964 that seriously affected any Weather Bureau studies not directly related to weather forecasting. (Keeling 1998)

Thanks to the new funds that permitted the hiring of Chin, the first priority in 1964 was getting the carbon dioxide analyzer back in operation. Keeling did not send spare parts for Keyes to try after the analyzer broke down in January. Instead, he trained John Chin in the operation of the APC analyzer and in May sent him to Hawai'i with a handful of replacement parts.

John Chin Returns to MLO (1964)

John F. S. Chin was no stranger to the Mauna Loa Observatory, for he had worked there from 1960 to 1962. He and Ellis had roomed together for a time after Ellis arrived in Hilo in 1961, and Chin (2001) was very familiar with the established routine of four days and three nights on the mountain.

> There were no showers and electricity was [from] two generators that needed constant monitoring. We passed the time away by playing checkers, tossing horse shoes, or taking correspondence courses. In the winter we made snowmen.

When the problem with the carbon dioxide project occurred, Chin was working for Lockheed in Sunnyvale, California. Ellis told Chin's wife Janet that he had heard from Mike Keyes that Dave Keeling was looking for someone to work for Scripps and maintain the CO_2 project, and she notified her husband. This was welcome news for John Chin, for he wanted to return to Hawai'i, and he missed the scenery and remoteness of Mauna Loa.

> I heard that information, so I went down to San Diego, to La Jolla, to look for Keeling, and he was very surprised that I showed up. . . . I was an adventurer type. I liked to do all these unusual things and, also, the work was interesting. Then, after I left in '62, I went back to the mainland and worked on other projects, and in my mind still thinking back of Mauna Loa. It always comes back to me. What a beautiful place it was. I always tried to come back to this island. So I was hoping that I could get the job back. Not too many people can have that kind of luck and opportunity, because Mauna Loa is a very unusual place to work. Work also has a clean atmosphere, and, also, it is like untouchable scenery. A lot of people in Hilo [have not seen] this kind of scenery at all. . . . Every day when you drive up the mountain, it is different; you feel different, have a very good feeling about going to work. (Chin, video interview with author, Hilo, Hawai'i, December 2006)

Keeling offered Chin the job, and he accepted. According to his Scripps Job Description Card, Chin's work assignment matched the traditional MLO schedule that he had worked during his first tour at the observatory: "Continuous Duty at the Observatory for 4 days and 3 nights. Equivalent to 40 hours." The real meaning of the forty-hour equivalency was explained on the Job Description Card quoted earlier ("When alone at the station 24 hours a day, he is operating all equipment 100% of the time") (Scripps 1964b).

After two weeks at Scripps to learn how to operate and repair the CO_2 analyzer, Chin reported to MLO on May 20, 1964. His first entry of many to come in the green *CO_2 LOG* at the mountain station was dated only one day later, May 21, 1964. He signed the *Visitors* book on May 30 and wrote "Home again since 5-20-64."

John Chin Repairs the Broken CO_2 Analyzer (1964)

Chin (telephone interview with author, June 8, 2008) relates how Keeling gave him the spare parts needed to repair and maintain the CO_2 analyzer, including two microphone detector cells, an amplifier, and a chopper motor assembly. The duration of the carbon dioxide data gap is traditionally given as three months, but Chin's first entry in the old CO_2 *LOG* that Jack Pales and Howard Ellis kept before him was dated exactly four months after the breakdown on January 21. The log provides details about what was wrong with the CO_2 analyzer and how Chin managed to repair it.

According to the APC analyzer's operating manual (APC 1954), which is still at MLO, its two detectors were designated by the serial numbers 89 and 118. Detector 89 was mounted at the end of the stainless steel analyzer tube near the back of the cabinet, the one that sampled outdoor air. Detector 118 was mounted at the end of the reference tube near the front of the cabinet, the one that sampled the reference gas. Chin began troubleshooting the analyzer by zeroing in on Detector 118 closest to the front of the cabinet. This cell was the easiest to change, and Chin (telephone interview with author, August 12, 2008) decided "trial and error" would be the best approach.

Based on an inspection of the analyzer, the detector cell was removed by first using a screwdriver to loosen the two screws on the upper end of the cylindrical detector that secured the three electrical wires connected to the analyzer's amplifier. The four screws that held the collar that secured the upper end of the detector in place were removed next. The detector could then be pulled from the cylindrical aperture at the end of the steel tube through which the reference gas was passed. A replacement detector could then be inserted into the instrument, the collar secured in place, and the electrical leads reconnected. Chin recorded in the CO_2 *LOG* the substitution of the new detector:

> May 21, 1964
> Action: Detector V replaced Detector 118. Detector V from Scripps. Believed it is a good detector.

On Friday, May 22, Chin changed the ascarite capsules in the analyzer that absorbed water vapor and carbon dioxide inside the apparatus. No entries were made over the weekend. On Monday, May 25, Chin wrote, "Check T1 [vacuum tube 1]. Let analyzer run for day and night."

On May 26, Chin filled a page with notes about various adjustments and tests that he made, including switching the position of the two detector microphone assemblies: "Checked detectors. Detector V placed in analyzer path. Detector 89 on reference path."

Chin finally began collecting data on Wednesday, May 27, but the trace on the chart recorder was "too nois[y]." After the weekend the situation improved dramatically, and Chin recorded what happened on Monday in the CO_2 *LOG:* "June 1, 1964. Got a perfect phasing. Analyzer back to normal operation."

At 4:00 p.m. the system was working well enough to provide the first usable data in four months and eleven days. The daily chart plotted by hand at Scripps shows no data from January 21 through the last day of May. A hand-lettered inscription on the chart that was apparently made after June 1 reads "STATION CLOSED DOWN," but it was the analyzer, not the station, that was closed down. A second line below the inscription above reads "NO DATA UNTIL JUNE 1, 1964" (Scripps 1964a). Figure 4(a) in Keeling et al. (1976) also shows no measurements from the end of January until June 1. Actually, some of data Chin acquired while testing the instrument during the last days of May might have been valid, for the Scripps carbon dioxide record lists a concentration of 322.23 ppm for May 1964, the same value given for May 1963 (Scripps 2008).

Following gas leak repairs and another phasing adjustment on June 3, the analyzer continued to function. But trouble soon developed.

> June 9, 1964. Difficulty in obtaining phasing. Cleaned contact points [in the chopper motor assembly]. They were dirty. Failed to obtain phasing adjustment.
>
> June 10, 1964. Tried to adjust phasing. Trace [of chart recording] wandering. Decided to let analyzer run overnight and check phasing the following day.
>
> June 11, 1964. Analyzer back to normal operation. Perfect phasing.

But the data were questionable and were not plotted by Keeling (Keeling et al. 1976). Usable data resumed at the end of the month. Chin had managed to repair the analyzer, but "normal operation" still meant frequent trouble with vacuum tubes, leaks in the tubing that carried outside air and reference gas to the analyzer, and various other problems. Vacuum tube T1 (designated as 6AK5-W) was especially fidgety, for it was the amplifier vacuum tube connected directly to the microphones in the detector cells.

Earlier, on May 22, Chin entered in the log an inventory of supplies associated with the carbon dioxide project, including "two trunks containing 12 five liter flasks [each]." Two flasks contained air samples from December 17, 1963, and two contained air samples from January 15, 1964; none of these had been shipped to Keeling. Chin (telephone interview with author, June 11, 2008) and Ellis (telephone interview with author, June 11, 2008) explained why the flasks had not been shipped to Keeling prior to Chin's arrival. The wood shipping trunks were designed specifically to protect the fragile glass flasks, so the trunks were returned to Scripps only when all twelve flasks had been exposed. The flask samples were designed to complement the analyzer measurements of carbon dioxide, and both Chin and Ellis said that new air samples were probably not collected between January 15 and Chin's arrival since the analyzer with which they would be compared was out of commission.

By June 30, Chin was sufficiently confident in the analyzer's operation that the first flask sample since January 15 was collected. Prior to the data gap, the concentration of carbon dioxide in the flask samples and that measured by the analyzer usually agreed to within 1 ppm (Pales and Keeling 1965: 6067–6068). The CO_2 concentration measured by the analyzer on June 30, 1964, was 4.69 ppm less than that in one flask sample and 2.18 ppm less than that in a second (Keeling and Pales 1965: 179). The next comparison was on July 13, and the CO_2 concentration measured by the analyzer was 1.27 ppm less than that in the flask sample. Not until 1965 did the agreement approach that regularly achieved prior to the data gap.

Because MLO no longer had a Hilo office, Chin took the chart recordings from the analyzer back to Hilo and extracted the data from them at home before they were mailed to Keeling. Carbon dioxide measurements had finally resumed at MLO after a delay of four months.

Chin helped fill a void in the MLO work schedule, for now there would again be some nights when two men stayed on the mountain rather than only one. But Chin's presence did not completely eliminate the need for help with the CO_2 program, for he could not remain on the mountain twenty-four hours a day, seven days a week, and he did not like to remain alone. Despite Ellis' growing concern about Keeling's impact on his career, which he would later share in a letter to his boss on the mainland, Ellis remained involved in the CO_2 program.

Charles Keeling made his second visit to MLO June 28–30, 1965, and spent a night or two at the observatory with Ellis and Chin. The tension of earlier times was absent, and Ellis (personal communication with author, April 8, 2008) was impressed that Keeling exhibited no altitude symptoms and that he practiced Russian during the visit. The visit occurred during a time of relative tranquility for the CO_2 analyzer, for there are no entries in the CO_2 *LOG* between September 30, 1964, and August 16, 1965. Numerous entries were made thereafter.

More Carbon Dioxide Breakdowns

The carbon dioxide analyzer continued its old ways after John Chin was hired by Keeling in May 1964, and he filled page after page of the old CO_2 *LOG* with details about breakdowns, problems, and repairs much like those that had plagued Jack Pales and Howard Ellis. During one week in December 1965, the problems had accumulated to a point where the analyzer experienced a major breakdown akin to the one in 1964 that led to the famous data gap. Chin dutifully noted the problems in the CO_2 *LOG* page shown in Fig. 9.1 and transcribed here:

> 12/6–10/65 Major Breakdown.
> Line traces drift up scale continuously. Final zero setting was 1000 on Range 1. Great efforts have been made to overhaul the analyzer. We have replaced:
>
> 1) T-3 6AK5 6X4 0E3 [vacuum tubes]
> 2) Chopper Motor (original one hard to turn)

3) Commutator (original one spring too loose)
4) Replaced one 12AX7 [vacuum tube] in the thermal unit (one side very weak)
5) Reverse the two [infrared] detectors.
6) Replaced South Pole Amplifier with Amplifier 58

Based on a chart in one of Keeling's papers (Fig. 4b in Keeling et al. 1976), this episode cost only a few days of data. Yet the aging, vacuum tube–equipped carbon dioxide analyzer was more temperamental than usual, a hint that its moods might only worsen in the future. The CO_2 measurements were still being made, but the frustrated Chin made no more entries in the old CO_2 *LOG*.

Breakthrough: The First Major MLO Carbon Dioxide Paper (1965)

While John Chin was struggling with the latest CO_2 analyzer breakdown in December 1965, a major new development was about to affirm the thousands of hours and nearly eight years of days and nights on the lonely mountain that had been devoted to the CO_2 project by Jack Pales and his team prior to Chin's arrival. The temperamental CO_2 analyzer that was so frustrating to maintain had led Pales and Charles Keeling to a major scientific discovery.

Keeling's obsessively careful measurements in California and elsewhere had affirmed the seasonal cycle in CO_2

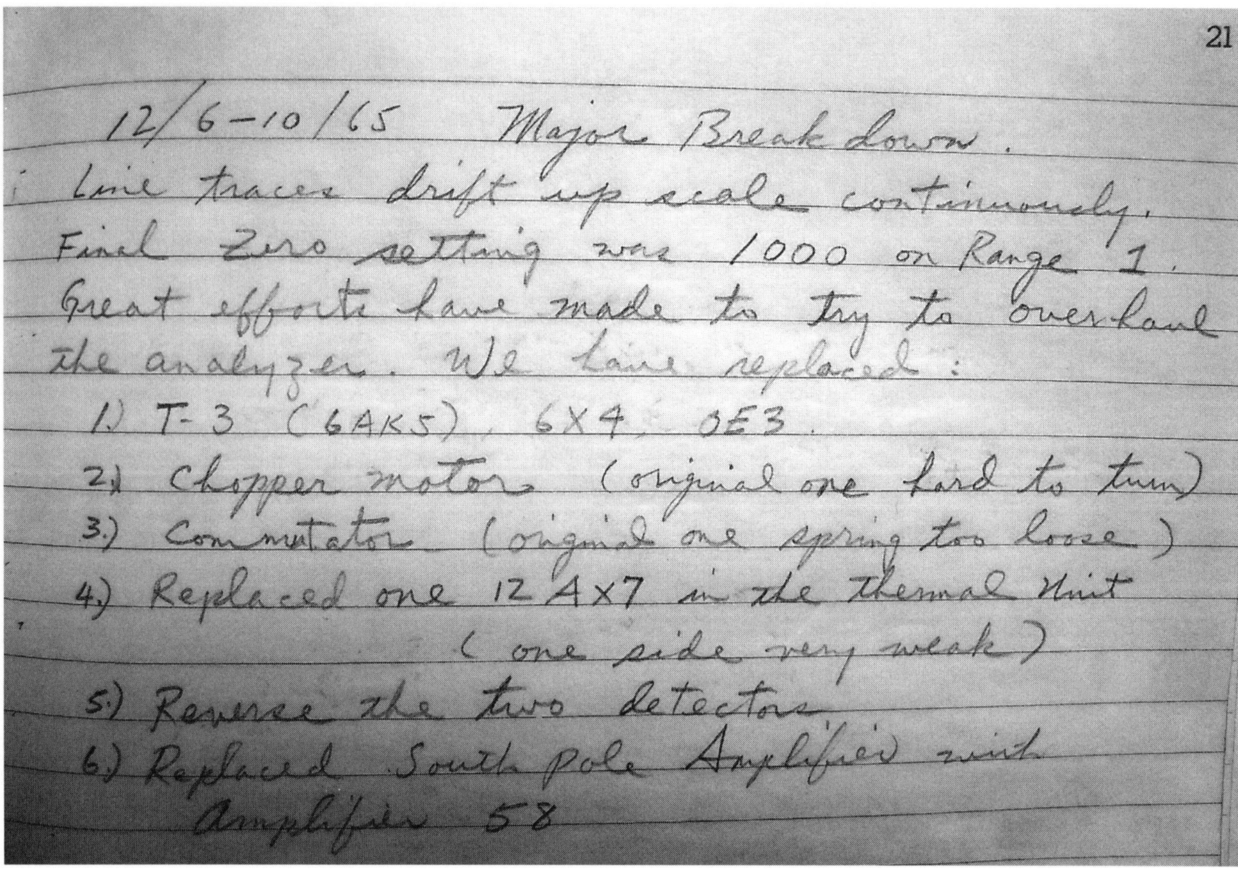

21

12/6–10/65 Major Break down.
Line traces drift up scale continuously.
Final Zero setting was 1000 on Range 1.
Great efforts have made to try to overhaul the analyzer. We have replaced:
1.) T-3 (6AK5), 6X4, OE3.
2.) Chopper motor (original one hard to turn)
3.) Commutator (original one spring too loose)
4.) Replaced one 12AX7 in the thermal Unit (one side very weak)
5.) Reverse the two detectors
6.) Replaced South pole Amplifier with Amplifier 58

FIGURE 9.1. John Chin listed a string of problems with the Scripps carbon dioxide analyzer in his final entry in the old CO_2 *LOG*.

measured at MLO "and possibly a worldwide rise in CO_2 from year to year" (Keeling 1998). Keeling had good reason to believe that CO_2 was uniformly distributed around the planet, and Saul Price and Jack Pales (1959) announced in their first MLO paper how the first year of MLO data supported this hypothesis. They also announced their finding of both a diurnal and a seasonal cycle in their first MLO paper in *Monthly Weather Review:*

> Data secured at Mauna Loa since sampling began in March 1958 are sufficiently similar to those obtained elsewhere to support previous indications that in the absence of local pollution the carbon dioxide concentration in the lower atmosphere is substantially uniform over the earth. The Observatory data also suggest a diurnal variation, perhaps related again to the mountain circulation, and a seasonal trend, with mean monthly values decreasing gradually from approximately 314 p.p.m. (parts per million) in May to 312 p.p.m. in August.

The following year, Keeling (1960) reported on his findings at Scripps in a paper of only 1,200 words that concisely summarizes his measurements from fall 1957 to spring 1960. The paper was published in *Tellus,* and Keeling (1998) later summarized its principle conclusions:

> I proposed that the activity of plants growing on land was the cause of the seasonal cycle. This activity explained why maximum CO_2 concentrations in both hemispheres were observed in the spring, when most plants begin to grow. The observed year by year rise in concentration was close to that expected if all of the industrial CO_2 from combustion of fossil fuels remained in the air.

Six days after Chin had spent most of a week trying to overhaul the malfunctioning analyzer in December 1965, the first scientific paper devoted exclusively to the CO_2 program at MLO was published. "The Concentration of Atmospheric Carbon Dioxide in Hawaii" by Jack Pales and Charles Keeling is the first of a section of four papers on carbon dioxide discoveries featured in the prestigious *Journal of Geophysical Research* on December 16, 1965. The paper confirms both the seasonal cycle and the yearly increase previously reported by Keeling in *Tellus* (1960) and by Price and Pales in *Monthly Weather Review* (1959 and 1964). The significance of the confirmation is that the absolute concentration of CO_2 in the free atmosphere at Mauna Loa and at sea level at Scripps were in close agreement and that the concentration was slowly rising at both locations. The paper immediately following the Hawai'i report, which Keeling also coauthored, announced very similar findings from the South Pole:

> [There is] a small seasonable variation superimposed on a small yearly increase. The increase is very nearly the same as observed at Mauna Loa Observatory, Hawaii, by Pales and Keeling in 1965. It is attributable to the burning of fossil fuels, which is causing the amount of CO_2 in the atmosphere to rise on a world-wide scale. (Brown and Keeling 1965)

Following the two papers on the concentration of CO_2 in the atmosphere of Hawai'i and Antarctica comes a paper on the distribution of CO_2 in the surface of the Pacific Ocean by Keeling and two of his Scripps colleagues (Keeling et al. 1965). The fourth paper in the group, which is by Keeling alone, is on the exchange of CO_2 in the Pacific Ocean with the atmosphere and was published in the same issue of the *Journal of Geophysical Research.*

The four papers are landmarks in the history of atmospheric science, for they conclusively demonstrate that a significant fraction of the carbon emitted during the combustion of fossil fuels is not absorbed by the oceans or transformed into wood and fiber by plants. Instead, the surplus CO_2 was remaining in the atmosphere, where calculations showed that its infrared absorption properties could contribute to a warming of the planet.

On December 15, 1965, the day before publication of the CO_2 paper, NASA's astronauts aboard Gemini 6A and Gemini 7 staged the first rendezvous of two manned space vehicles (Grinter 2000). Earlier in 1965, Arno Penzias and Robert Wilson of Bell Telephone Laboratories discovered the cosmic microwave background radiation, for which they shared the Nobel prize in physics in 1978 (Wilson

1978). Both of these historic achievements received considerable publicity in the scientific and popular press, and today a major international project at MLO is devoted to the latter discovery.

The four CO_2 papers received little attention outside the scientific community. Yet their confirmation of Keeling's earlier discovery was to lead to increasing concern that the annual increase in CO_2 would eventually have vastly more impact on the world's economy, transportation systems, agriculture, coastlines, and population than manned space flight or cosmological discoveries.

The significance of the paper about CO_2 measured at the Mauna Loa Observatory extends beyond the fundamental discoveries it announced. The paper showed that a large and clumsy vacuum tube instrument, when working properly, could reliably measure carbon dioxide with sufficient precision to detect a trend of a fraction of a part per million in the concentration of CO_2. The paper demonstrated that the arrangement of the instrument's four air sampling points and a knowledge of the local meteorology allowed occasional pulses of volcanic CO_2 to be reliably detected and removed from the data. Finally, the paper would never have been written without the combination of Keeling's expertise, determination, and obsession to detail and six years of effort at MLO by Jack Pales and his team, who maintained the instrument and who laboriously broke out the data from the chart recordings.

Dissecting the CO_2 Paper

As noted in the acknowledgments section of "The Concentration of Atmospheric Carbon Dioxide in Hawaii," the paper passed through "several manuscripts which led to the present version" (Pales and Keeling 1965). In view of the significance of the paper, its history is of interest, particularly since it did not include the latest year of CO_2 findings prior to its submission.

Some of the history can be gleaned from an early, partial draft of the paper found in the files of the MLO's office in Hilo. The nine-page draft, which is marked "copy 3," bears a longer title than the published paper: "The Concentration of Carbon Dioxide at Mauna Loa Observatory, Hawaii." The draft includes minor copyediting marks and several notes and sketches. At the bottom of page 3 is an outline sketch of how an aluminum air sampling tube was connected to a pump and a needle valve that could be adjusted to provide a flow of 0.5 liters per minute. The draft ends halfway down the ninth page, after a detailed description of the diurnal and seasonal variation in the concentration of CO_2. While much of this draft is preserved in the paper published in the *Journal of Geophysical Research* (*JGR*) in 1965, it lacks the introduction and it ends only halfway through the published version. From the context and the fact it was found in the MLO files, it appears that the draft is by Pales, and that Keeling added the major sections on flask sampling and the detailed analysis of the findings. Because the CO_2 data in the paper ends in December 1963, and Pales left Hilo for Las Vegas in late January 1964, it is likely that the draft found in the Hilo MLO office was completed by Pales in Hilo in January 1964 shortly before his departure.

A paper having the same title as the draft from the MLO files was referenced and described as "in press" in the second MLO paper by Price and Pales, which was received in final form by *Monthly Weather Review* (*MWR*) on September 23, 1963, and published in the combined issue dedicated to Harry Wexler dated October–December 1963 (see ref. 22 of Price and Pales 1963). However, no such paper can be found, which suggests that the "in press" status of the cited paper was incorrect or that the paper was sent back for revisions or retracted after acceptance.

The 1965 *JGR* paper by Pales and Keeling, which incorporates much of the draft described above but whose title omits "Mauna Loa Observatory," was not received by the *Journal of Geophysical Research* until December 14, 1964, and was not revised and accepted until August 3, 1965. Thus, there is a puzzling gap between the "in press" reference in *MWR* and the paper accepted by *JGR*. One possible explanation is that the original paper might have been pulled from *JGR* prior to publication so that all four CO_2 papers could be published together. This supposition, while speculative, is supported by the fact that Keeling coauthored three of the papers and was the sole author of the fourth, and that all four papers bear the same received and revised dates, suggesting they were intentionally submitted and accepted as a package. Charles Keeling's papers are in the process of being archived by Scripps, and they may explain the delay.

Of interest is that the CO_2 plots were not updated to reflect the delay in the submission of the final paper to *JGR*. Processing the data to Keeling's rigid standards was time consuming, and that may be the reason. Another possibility is the four months of missing data from February through May of 1964 after the breakdown of the CO_2 analyzer. Perhaps Keeling was reluctant to display a time series chart with more than four months of missing data in the final year.

Pales and Keeling (1965) include half a page of acknowledgments at the end of their paper that reveals the complexity of the CO_2 project by listing its many participants. In view of the international significance the work was to assume in coming years, the acknowledgments are worthy of inclusion here.

> The inauguration of the program was aided by the Departments of Oceanography of the University of Washington and Texas Agricultural and Mechanical College, the Lamont Geological Observatory, and Woods Hole Oceanographic Institution, who generously donated equipment purchased from their own funds for the International Geophysical Year.
>
> The Scripps program to monitor CO_2 in the atmosphere and oceans was conceived and initiated by Dr. Roger Revelle who was director of the Scripps Institution of Oceanography while the present work was in progress. Revelle foresaw the geochemical implications of the rise in atmospheric CO_2 resulting from fossil fuel combustion, and he sought means to ensure that this "large scale geophysical experiment," as he termed it, would be adequately documented as it occurred. During all stages of the present work Revelle was mentor, consultant, antagonist. He shared with us his broad knowledge of earth science and his appreciation for the oceans and atmosphere as they really exist, and he inspired us to keep in sight the objectives which he had originally persuaded us to accept.
>
> We are indebted to Dr. Les Machta and the late Dr. Harry Wexler whose faith in the Mauna Loa Observatory as a high-altitude meteorological station and whose resolution in the face of financial and logistic obstacles led to the establishment and continued operation of the observatory by the Office of Meteorological Research of the U.S. Weather Bureau; to Dr. Norris Rakestraw who gave generously of his time and energy at Scripps in the initial planning and execution of the program; and to Mr. Saul Price, of the U.S. Weather Bureau, Honolulu, whose broad interest in Mauna Loa Observatory helped us in many ways. We thank Mr. Lee Waterman for critically reading the several manuscripts which led to the present version. Personnel of the 55th Weather Reconnaissance Squadron of the United States Air Force collected the aircraft samples under the guidance of Major Bernard M. Rose and Captain E. B. Huber. Messrs. Harry Arashiro, Howard Ellis, Colby Foss, Cliff Kutaka, and Robert Williams of the staff of Mauna Loa Observatory operated the gas analyzer, processed data, and collected flask samples at the observatory near Hilo. Messrs. Eugene Collias, Frank Maurer, Lee Waterman, Mrs. Denise Phleger, and Mrs. Joan Soutar of the Scripps staff and Dr. Eugene Wilkins and Mr. Thomas Harris of the U.S. Weather Bureau further processed the data. Mr. Stig Fonselius assisted on many phases of the program at Scripps while on leave from the International Institute of Meteorology, Stockholm. Mr. Harris and Mrs. Blanche Christman analyzed the flask samples.
>
> We gratefully acknowledge financial support from the U.S. National Science Foundation.

The Lunar Astronauts Visit MLO (1965)

On August 26, 1964, Dr. Earl G. Droessler of the National Science Foundation visited MLO. After signing in, he wrote "A station with rich possibilities for atmospheric research." On October 28, 1964, Rosalie Sisson of Albuquerque, New Mexico, signed in at MLO and wrote, "Who needs to go to the moon!"

Going to the moon was much on the mind of a group of astronauts who were even then planning a stop by the station on their way to the moon. During January 1965, a group of NASA astronauts and geologists visited Hawai'i as part of a training exercise to experience firsthand the inhospitable environment of Mauna Loa. Bernard Mendonca recalls their visit:

> They thought we were nuts for working at the station, and we thought they were even crazier for thinking that MLO resembled the moon's surface. We got along fine. (Mendonca 2006)
>
> We were particularly amused to see the astronauts chipping away at the different kinds of lava with little hammers. We silently wondered about how this contrasted with their high tech preparations for the journey to the moon. (Mendonca, telephone interview with author, June 9, 2008)

For those who had grown accustomed to living and working at MLO, the astronaut visit was a reminder of the extraordinarily rugged and remote setting of their workplace, as was made clear in NASA's contemporary summary of the exercise:

> Jan. 19–24, 1965. Hawaii Aldrin, Anders, Armstrong, Bassett, Bean, Chaffee, Collins, Conrad, Cunningham, Eisele, Gordon, Schweikart, Scott, See, Williams
>
> Incomparable display of recent basaltic volcanic features. Fresh, recent, and ancient lava-flow surfaces could be compared and related to possible lunar surface features. . . . The first day consisted of a bus trip to several volcanic features including lava tree molds, gas and lava vents, spatter ramparts and cones, flank eruptions of a'a and pahoehoe flows, fissure eruptions and beach sands of eroded lava. The second day included foot and bus traverses to observe lava tubes, lava lakes, lava blisters, Pele's tears and hair, Halemaumau fire pit, and pit craters. In the evening a briefing utilized photos of the lunar surface that were analogous to features seen during the past 2 days. The third day started with fly-overs in a 12-passenger aircraft to observe the volcanic features from the air. During the afternoon a lecture was given on the various geophysical measurements that are used to keep track of volcanic activity and determine the structure of the subsurface volcanos. On the 4th day the crews were taken to the saddle area between Mauna Loa and Mauna Kea where problems were posed by the geologists for the crews to solve. They then went to the summit of Mauna Loa for a view of the summit crater and its features. (NASA 1964–1972)

Many of the astronauts and geologists signed the *Visitors* book during their visit on January 14, 1965, including Charles Conrad Jr., Roger B. Chaffee, Edwin E. Aldrin Jr., Alan L. Bean, and R. T. Gordon. Most of the astronauts who visited MLO were destined to eventually orbit or land on and explore the moon in the most costly and daring achievement in the history of exploration. The cost of sending a chunk of Mauna Loa lava to the moon and back would greatly exceed MLO's annual budget, but in terms of the observatory's significance to our home planet, a strong case could be made that the practical value of the science flowing from MLO would far exceed that which would emerge from all the lunar missions. Many years later Charles Keeling (1998) addressed what he viewed as the low priority given environmental monitoring when he reflected back on his own pioneering research:

> I learned a lesson that environmental time-series programs have no particular priority in the funding world, even if their main value lies in maintaining long-term continuity of measurements.

The Atomic Energy Commission (1965)

The Atomic Energy Commission (AEC) was the first government agency outside the Weather Bureau to establish a major, albeit temporary, research program at the Mauna Loa Observatory after the full-time Weather Bureau program was begun in 1957. The AEC used MLO as a site from which to monitor Pacific nuclear tests, and the *Visitors* book was signed by dozens of personnel from the AEC, the Los Alamos Laboratory, Sandia Laboratories, and various contractors, especially during the early to middle 1960s. Most signers listed addresses in New Mexico, the state where in 1955 a serendipitous encounter of Robert Simpson with Ralph Stair led directly to the construction of the main building at MLO.

The AEC left behind a large building east of the main building that is still used by visiting science programs. They also left behind a metal building west of the main building that had been equipped with sophisticated cameras. Today that building houses the MLO sulfur dioxide project and a USGS seismometer. The AEC also left an even more important contribution in the form of financing for the new route to Saddle Road negotiated by Howard Ellis in 1963.

The Mauna Loa High Altitude Observatory (1965)

The High Altitude Observatory (HAO) was established by the National Center for Atmospheric Research in 1965 and is still in operation at MLO today. The original site for the solar coronameter telescope of the HAO was to have been on the summit of Haleakalā on the nearby island of Maui, a project coordinated with the help of Walter Steiger of the University of Hawai'i. As explained by former director Charles J. Garcia, it became necessary to find an alternate site for the telescope, whose purpose was to image the solar corona, the ionized streams of gas continuously released into space by the sun. As recalled by Garcia (video interview with author, Hilo, Hawai'i, December 2006), his colleague Richard Hansen made several trips to MLO to discuss arrangements for the move with Howard Ellis:

> When it was determined that we were not able to operate our telescopes at Haleakala because it was a single mount telescope, we started looking for another site in Hawaii. Dick Hansen made several trips to Mauna Loa [and] met Howard. . . . About that time, Howard said, if you want to move the telescope over here, you just put it in the property over there on top of the hill. And then negotiations proceeded from that point, and we built a small dome. . . . I do not know if he had to go further up the ladder to get this okayed or not. But it was still [through] Howard that the whole thing was started. He actually pointed out [where] we should put the dome.

Hansen signed the *Visitors* book on September 26, 1964, and listed his affiliation as Haleakalā Observatory. Garcia, the HAO founding director, first signed in at MLO on July 18, 1965. Rebecca M. Garcia also signed in and wrote "Gracias! for the nice hospitality. Enjoyed it immensely."

In his role as physicist in charge of MLO, Howard Ellis worked with the National Center for Atmospheric Research to select an appropriate site for the HAO, which was approved by Lester Machta. The HAO was constructed at the upper edge of the four-acre site nearly due south of the main building and its air sampling ports. Ellis later said that this location might have been a bad choice, for it placed the large HAO structure and dome directly in the path of the nocturnal downslope winds representative of the free atmosphere above the mountain.

Yet at the time there was little choice about where to place the HAO. The lower (north) slope of the site was already occupied by the main building, the Butler Building, the AEC buildings, various other structures, and the access road. This left the upper (south) slope of the site. A 50-foot (15-meter) tower for collecting air for the carbon dioxide project was located at the southeast corner of the site, so this area was unsuitable. Perhaps it might have been best to place the HAO closer to the middle of the south boundary, but in the end there was not an ideal location, and this did not please Charles Keeling:

> Placing the HAO facility in the clean air (up-slope) section on the mountain was another point of contention between Keeling and Ellis. Keeling thought strongly that the HAO facility should have been situated downslope of the MLO and that he should have been consulted in the siting decision. In the long run the present sighting of the HAO had no negative impact on the

> CO_2 measurements nor any of the other observatory measurements. (Mendonca, telephone interview with author, June 9, 2008)

Garcia remained the director of the HAO until 1991 and developed a close working relationship with MLO veterans Howard Ellis, John Chin, Alan Yoshinaga, Bob Uchida, and others. The HAO was expanded in 1997 and may yet be expanded again. It is now known as the Mauna Loa Solar Observatory.

Garcia remembers that Howard Ellis would sometimes spend nights sleeping on the lava instead of in the main building. Ellis (video interview with author, Hilo, Hawai'i, December 6, 2006) explains why:

> I did not like to be at the end of a road which people were not supposed to travel. Although nobody ever showed up, I had bad dreams about them showing up, and I was just plain scared. So I went out to the . . . east [side] of my concrete solar pier, and I just stretched out with my feet pointed to the North Star. [I] slept with two sleeping bags, and I could see the Big Dipper. I would check it when I was dozing off to sleep, and if I woke up, I could see it again and know what time it was. I always got up as soon as it was light.

The First Decade Ends "In the National Interest"

On July 13, 1965, the Weather Bureau, which had supervised MLO since its inception by Robert Simpson, was merged with the Coast and Geodetic Survey in a major reorganization of scientific activities within the Department of Commerce. The two agencies became known as the Environmental Science Services Administration (ESSA 1965), which was the predecessor of the National Oceanic and Atmospheric Administration (NOAA). MLO came under new management that suggested impending changes in the status quo on the mountain.

In spite of MLO's precarious status, somehow the ambitious predictions made by Saul Price, who revisited MLO on November 19, 1964, and Jack Pales in the conclusion of their paper about the first five years at MLO seemed as valid at the end of 1965 as when they were published at the end of 1963:

> It is certain that the Observatory will continue to attract projects as numerous and diverse as those listed earlier in this report. . . . But perhaps more valuable than any of these endeavors, actual or potential, is the Observatory's role as a bench mark station.
>
> But beyond that time Mauna Loa Observatory will continue to regard as its primary responsibility the careful and uninterrupted surveillance of carbon dioxide, solar radiation, total and tropospheric ozone and, at intervals, the atmospheric electric variables—the significant indices of global atmospheric change.

The extended International Geophysical Year ended at the close of 1959, and the key measurements begun during the IGY were still underway. Most of the observatory's IGY legacy instruments were performing well, and the cranky Scripps APC Model 70 infrared analyzer continued to measure a relentless climb in carbon dioxide.

As 1965 came to an end, John Chin remained on duty at MLO working for Scripps, but Mike Keyes had been transferred to a new Weather Bureau position earlier in 1965. The MLO staff was reduced to only Howard Ellis and Bernard Mendonca, one less than the crew Jack Pales had begun with in July 1957. In spite of MLO's growing reputation and scientific significance, 1965 ended much as 1957 had, with no Hilo office and a meager budget for fuel, supplies, and salaries.

Yet Howard Ellis and Bernard Mendonca were determined that the observatory would survive so that its measurements could be continued. Ellis' mind-set is best illustrated by his reaction to generator troubles that occurred when he was manning the observatory on Thanksgiving Day of 1964 (Ellis, telephone interview with author, June 12, 2008). One generator was running rough, so Ellis switched on the second unit and turned off the first one. This meant the observatory no longer had a backup generator, so Ellis telephoned Wendell Leite, the supervisor of

Hilo Transportation Terminal and Company, the firm that maintained the generators. Leite asked why Ellis expected him to interrupt his Thanksgiving dinner with his wife Shirley and their family, and Ellis calmly replied, "Well, it's in the national interest." Leite then interrupted his meal to telephone Santiago "Spike" Cordero, one of his technicians, who drove up the mountain to fix the generator.

Wendell and Shirley Leite (telephone interview with author, June 12, 2008) still remember the Thanksgiving that Howard Ellis, whom Leite describes as a "really good man," called "in the national interest." But the future of both the lonely mountain station and the physicist in charge who had kept it alive were very much in doubt as the world's most unique atmospheric observatory entered its tenth year.

References

APC. 1954. *Model 70 Infrared Analyzer (Modified Shell Design)*. Applied Physics Corporation. (The MLO edition with calibration and other data is dated November 23, 1957.)

Brown, Craig W., and Charles D. Keeling. 1965. The concentration of atmospheric carbon dioxide in Antarctica. *Journal of Geophysical Research* 70: 6077–6085.

Chin, John. 2001. Unpublished retirement scrapbook, prepared by Leslie Pajo. Copy in author's files.

Ellis, Howard. 1978. The observatory. In John Miller, ed., *Mauna Loa Observatory: A 20th Anniversary Report*. National Oceanic and Atmospheric Administration Special Report, September: 24. Boulder, CO: NOAA Environmental Research Laboratories.

———. 1988. The observatory: On their own. (Accessed at http://www.mlo.noaa.gov/webmuseum/publications.html.)

———. 2005. My *Mauna Kea Connection*. Unpublished manuscript.

ESSA. 1965. Reorganization Plan No. 2 of 1965. Environmental Science Services Administration, Department of Commerce, 3 C.F.R., reprinted in 5 U.S.C. app. at 1517 (1994), in 79 Stat. 1318–1320 (1965), and in 30 Fed. Reg. 8819 (1965). (Accessed at http://www.history.noaa.gov/legacy/act5.html.)

Fujisawa, Adeline. 2006a. Former MLO staff information sheet. In author's files.

———. 2006b. Unpublished memorandum. In author's files.

Grinter, Kay. 2000. NASA Gemini missions. Kennedy Space Center. (Accessed at http://www-pao.ksc.nasa.gov/history/gemini/gemini-manned.htm.)

GSA. 1981. Transcript/Statement of Federal Service for Howard T. Ellis. General Services Administration. Cited with permission of Ellis.

Jefferies, John T. 2001. Dawn of a brilliant opportunity. (Accessed at http://www.ifa.hawaii.edu/users/jefferies/Dawn_of_a_Brilliant_Opportunity.htm.)

Keeling, C. D. 1960. The concentration and isotopic abundances of carbon dioxide in the atmosphere. *Tellus* 12: 200–203.

———. 1965. Carbon dioxide in surface waters of the Pacific Ocean: 2. Calculations of the exchange with the atmosphere. *Journal of Geophysical Research* 70: 6099–6102.

———. 1978. The influence of Mauna Loa Observatory on the development of atmospheric CO_2 research. In John Miller, ed., *Mauna Loa Observatory: A 20th Anniversary Report*. National Oceanic and Atmospheric Administration Special Report, September: 36–54. Boulder, CO: NOAA Environmental Research Laboratories.

———. 1998. Rewards and penalties of monitoring the Earth. *Annual Review of Energy and the Environment* 23: 25–82. (Accessed at scrippsco2.ucsd.edu/publications/keeling_autobiography.pdf.)

Keeling, Charles D., Robert B. Bacastow, Arnold E. Bainbridge, Carl A. Ekdahl Jr., Peter R. Guenther, Lee S. Waterman, and John F. S. Chin. 1976. Atmospheric carbon dioxide variations at Mauna Loa, Hawaii. *Tellus* 28: 538–551.

Keeling, Charles D., and Jack C. Pales. 1965. *Mauna Loa Carbon Dioxide Project*. Report No. 3. American Documentation Institute, Document 8530, June 15.

Keeling, Charles D., Norris W. Rakestraw, and Lee S. Waterman. 1965. Carbon dioxide in surface waters of the Pacific Ocean: 1. Measurements of the distribution. *Journal of Geophysical Research* 70: 6087–6097.

Keeling, C. D., T. P. Whorf, and the Carbon Dioxide Research Group. 2005. Atmospheric CO_2 concentrations (ppmv) derived from in situ air samples collected at Mauna Loa Observatory, Hawaii. La Jolla: Scripps Institution of Oceanography, University of California. (Accessed at http://cdiac.esd.ornl.gov/ftp/trends/co2/maunaloa.co2.)

Mendonca, Bernard G. 1964. Memorandum for the record to physicist-in-charge, MLO, thru research meteorologist and meteorologist-in-charge, PSO (January 16).

———. 1978. The first twenty years: An unscientific remembrance. In John Miller, ed., *Mauna Loa Observatory: A 20th Anniversary Report*. National Oceanic and Atmospheric Administration Special Report, September: 17–23. Boulder, CO: NOAA Environmental Research Laboratories.

———. 2006. Former MLO staff information sheet. MLO archives.

NASA. 1964–1972. Appendix E: Geology field exercises: Early training, field training schedule for the first 3 groups of astronauts (29). (Accessed at http://www.hq.nasa.gov/alsj/ap-geotrips.pdf.)

Pales, Jack C. 1964. Letter of recommendation on behalf of Adeline Fujisawa, January 17. Copy in author's files.

Pales, Jack C., and Charles D. Keeling. 1965. The concentration of atmospheric carbon dioxide in Hawaii. *Journal of Geophysical Research* 70: 6053–6076.

Parker, Barry R. 2001. *Stairway to the Stars: The Story of the World's Largest Observatory*. New York: Basic Books.

Price, Saul, and Jack C. Pales. 1959. The Mauna Loa High Altitude Observatory. *Monthly Weather Review* 87: 1–14.

———. 1963. Mauna Loa Observatory: The first five years. *Monthly Weather Review* 91 (October): 665–680.

———. 1964. Ice nucleus counts and variations at 3.4 km and near sea level in Hawaii. *Monthly Weather Review* 92 (May): 207–221.

Scripps. 1964a. MLO CO_2 data (time series). Keeling Papers, Box 31, Mauna Loa Project plotted hourly CO_2 indices, January 1, 1964–December 31, 1964. Scripps Oceanographic Institute Library.

———. 1964b. John Fee S. Chin, Job Description Card, May 1964. Keeling Papers 2003–18, Box 8, Personnel, John Fee S. Chin [December 1963–June 6, 1966], Scripps Oceanographic Institute Library.

———. 2008. Data, in situ CO_2, Mauna Loa record. Scripps CO_2 Program, Scripps Institution of Oceanography. (Accessed at http://scrippsco2.ucsd.edu/data/data.html.)

Steiger, Walter. 2001. Origins of astronomy in Hawai'i. Epilog, Institute for Astronomy, University of Hawai'i. (Accessed at http://www.ifa.hawaii.edu/users/steiger/epilog.htm.)

Wilson, Robert W. 1978. The cosmic microwave background radiation. Nobel Lecture, December 8. (Accessed at http://nobelprize.org/nobel_prizes/physics/laureates/1978/wilson-lecture.pdf.)

CHAPTER TEN

The Second Decade (1966–1975)

The Mauna Loa Observatory quietly entered its second decade staffed only by Howard Ellis, who was still officially listed as supervisory physicist, and Bernard Mendonca. Johnny Chin was taking care of the carbon dioxide measurements for Scripps with Ellis' help. The budget was still minimal, but major changes were in the wind now that the Weather Bureau had become part of the Environmental Science Services Administration (ESSA) of the Department of Commerce.

Overload at MLO (1966)

The old MLO correspondence files reveal that scientists and supervisors on the mainland sometimes didn't fully understand the difficulties Ellis faced in running a remote, high-altitude observatory with only two men. Their only office was on the mountain, and both vehicle and generator maintenance still posed major problems. There was little storage space for new equipment at MLO, and there were problems distributing electrical power to all the experiments.

Late in March 1966, Ellis wrote Helmut Weickmann, director of the Atmospheric and Chemistry Laboratory at ESSA, to explain why some new equipment received in November had not yet been placed in operation. In a one-page letter typed on a plain sheet of stationery without a Weather Bureau letterhead, Ellis explained his predicament:

> My problems have been compounded since then and the equipment has not yet been unpacked. My thinking continues to be that it is better to conduct programs well than to attempt many programs, any of which may be questionable. . . . However, as I wrote to you last week, we have just about finished a major effort to solve our problems of space and electrical load distribution.

Ellis was also overwhelmed by the carbon dioxide program that he managed prior to the breakdown of the analyzer in 1964 and the subsequent hiring of John Chin by Charles Keeling, and in the same letter he explained his concerns to Weickmann:

> Dr. Weickmann, I am very concerned about the Carbon Dioxide program for the next couple of months, until Lothar Ruhnke [Ellis' successor as MLO director] can grasp with the problem. I think that I can handle it and that there will be no break in the data, but, perhaps since I have become so personally involved with Dr. Keeling—

> I feel that he has damaged my career, completely without basis—I am afraid that we might have an equipment breakdown that will not be anyone's fault and that I will be blamed. It has been a real hard fight with the CO_2 program. I feel very strongly about the importance of the work and I feel that I understand the equipment and program as well and in many of its operational aspects better than anyone at the present time. Because of this I have given several thousand hours of my own free time to keep this program going over the years. . . . There is nothing I can ask of you at the present time that will help except to mentally cross your fingers that the equipment will continue to work.

Ellis would soon learn that he was right to be concerned about his relationship with Keeling. Meanwhile, John Chin was growing weary of the difficult schedule on the mountain. He still had to depend on Ellis or Mendonca for transportation to and from MLO. And he didn't enjoy having to spend some nights alone on the mountain.

> The CO_2 project [air sample lines] needed to be switched manually [each evening] so I still needed to stay overnight. I drove up to the Yamanouchi gas station on Kaumana Drive where I parked my car and waited for a ride from the MLO observers. It wasn't until Scripps was able to automate their switching that I no longer had to spend lonely night[s] on MLO. (Chin 2001)

In his letter to Weickmann, Ellis (1966) wrote that not having someone on the mountain every night "continues to be an unsolved problem." He believed it was necessary for MLO to be manned overnight to take care of generator breakdowns and unexpected problems.

Shortly before writing Weickmann, Ellis had learned that his days in charge of MLO were numbered. The new administrative lineup at ESSA was bringing change to the long neglected observatory, and lightning expert Lothar Ruhnke was being assigned to take over as director. Ellis closed his letter to Weickmann by recommending that the unpopular all-night stays on the mountain be continued:

> I hope this can be discussed with Lothar [Ruhnke] before he comes out—should we plan to stay overnight. I think we should, but only if we have people and a schedule that allows this to work without creating more problems than it is worth.

MLO Director Lothar Ruhnke (1966–1968)

Before moving to ESSA, Helmut Weickmann had worked as a scientist specializing in precipitation at the US Army Signal Research and Development Laboratory, Fort Monmouth, New Jersey. There he came to know Lothar H. Ruhnke, a scientist who specialized in lightning and related phenomena. When Weickmann became the new director of ESSA's Atmospheric and Chemistry Laboratory in 1965, MLO came under his supervision. Weickmann was well aware of the significant research opportunities afforded by MLO and the fact that the staff was down to only two men. He contacted his friend Ruhnke, who had become a senior atmospheric scientist at Litton Industries in Minnesota, and offered him the position of director of MLO. Ruhnke (telephone interview with author, August 8, 2007) later recalled that he was excited about the prospect of moving from snow country to Hawai'i. He accepted the position without even visiting Weickmann in Washington.

On March 1, 1966, Ralph Stair called Ellis (telephone interview with author, June 8, 2008) to inform him he was planning a trip to MLO to continue his solar constant measurements. The same day, Weickmann called Ellis to inform him that Ruhnke was being sent to Hilo to take over the observatory. In early April, Ellis flew to Honolulu to meet the Ruhnkes and escort them to Hilo, where he had arranged temporary accommodations for them. On April 10, 1966, Ellis' title was changed back to physicist from supervisory physicist. The *Visitors* book records that Lothar and Gerda Ruhnke signed in at MLO on April 12, 1966. MLO still had no office in Hilo, and Ruhnke recalls that his first night on the mountain was unpleasant and that "I couldn't sleep very well." He recalled that Ellis explained that the altitude made it difficult to work when making daily trips, but staying overnight on the mountain allowed one to eventually acclimate to the altitude.

The nature of signatures in the *Visitors* book provides clues about visitors who might have been experiencing a reaction to the altitude when they signed the book, and Ruhnke's squiggly signature fits the pattern. Three decades later, Russell Schnell (1996), a future MLO director, would note in his monthly report that some signatures could not even be read:

> The other names were so badly written we are unable to determine who they are. The high altitude does strange things to some people like making them forget names and slurring their writing, not to mention their speech.

Ruhnke's solution to working on the mountain was to quickly find office facilities in Hilo. His stay on the mountain also provided good reason to give high priority to improving the road and to eliminate the noisy, dirty generators by providing commercial electrical power to the observatory (Ruhnke, MLO 50th anniversary, e-mail to Darryl Kuniyuki, September 14, 2007).

MLO and the Cloud Physics Observatory (1966)

Noted cloud physicist Dr. E. J. Workman retired in 1964 as president of New Mexico Tech. The university, officially known as the New Mexico Institute of Mining and Technology, is located just 106 miles (171 km) northwest of Sunspot, where plans for the slope unit of the Mauna Loa Observatory were begun when Robert Simpson met Ralph Stair in 1955. During his eighteen years as president, Workman transformed what had been a mining school into a science and engineering university, and on retiring he moved to Hilo to help establish the Cloud Physics Observatory at the University of Hawai'i at Hilo. Workman added to the list of New Mexico connections to MLO when in 1966 Lothar Ruhnke contacted him about the need for an MLO office. Workman offered the small space intended for the office of the newly constructed Cloud Physics Observatory building on the campus. Construction of the new building had been directed by Bob Uchida, an electronics technician who in June 1965 had worked on a cloud physics project at MLO and had befriended Howard Ellis there. (Uchida became an MLO employee in 1991.) In May 1966, shortly after the new MLO office opened, Charles M. Fullerton first visited the Cloud Physics Observatory (CPO), where he was newly assigned. He describes the facility in a memoir:

> We drove into the campus along the Puna-side of the gym (now known as the old gym), past the caretaker's cottage, and then followed a circuitous gravel road up a small hill. At the top of the hill, I was surprised to find a brand new concrete block building, 130 x 25 feet—the Cloud Physics Observatory. It contained a number of shops, labs, and offices. . . . As I faced the building, on the left end there was a very well-equipped machine shop with a large metal lathe and a smaller precision lathe. Next, there was a large and equally well-equipped woodworking shop with double doors to the parking lot. Observatory vehicles could be driven into the shop and worked on there. There was a small office next to the shop that was designed as the Observatory office, but instead it was occupied by personnel from the Mauna Loa Observatory (MLO). Dr. Workman had invited the MLO folks to set up their Hilo office in the CPO building because of the similarity of scientific interests of the two groups. . . . (It is interesting to note that there was no designated women's restroom because, I suppose in the chauvinistic thinking of the time, all scientists were men!)
>
> At the right end of the building, there was a large room that was marked on the building plans as "bedroom." The idea was that this room would serve as living quarters for short-term visiting scientists. There was a bathroom, complete with shower, off this room. Well, in fact, this room became the CPO office and the adjacent bathroom the women's room since there *were* females on the staff of both MLO and CPO. (Fullerton 2001)

For the first time since January 1964, when Howard Ellis closed the original MLO office in the Air Cargo Terminal near the Weather Bureau at the General Lyman Field, the observatory once again had a real office. And this time it was at a university, not a cargo terminal. Bernard Mendonca (video interview with author, Boulder, Colorado, April 27, 2007) was especially glad to have office space after more than two years without any: "We had our own desks, we could put things away, we had a telephone, and things were good."

His sleepless night at MLO and Ellis' and Mendonca's work schedule helped Ruhnke understand the difficulties the two men had endured driving back and worth to MLO and living on the mountain. Ruhnke told Ellis that he deserved to be promoted, and his promotion to GS-12 took effect on June 5, 1966. His annual salary was increased from $7,560 to $9,810 when Ruhnke took over and to $10,797 with his promotion to GS-12 (GSA 1981). Mendonca also received a promotion. Both men continued spending nights on the mountain, but a major new development would soon make the long tradition optional.

Meanwhile, Ellis remained concerned about the future of the Scripps carbon dioxide program, and he persuaded Ruhnke to visit Charles Keeling at Scripps to learn more about the project that demanded so much of the observatory's time. Ruhnke (telephone interview with author, August 8, 2007) recalls that while Keeling's determination was responsible for the success of the CO_2 program, he had nothing good to say about the man who had done more than anyone to keep MLO and the carbon dioxide project alive during the budget crisis: "Keeling warned me about Howard Ellis. He said he is no good. . . . Keeling was not the easiest man to work for."

When Ruhnke returned to Hilo, he informed Ellis that he was being taken off the carbon dioxide project that had so dominated his time on the mountain since his arrival in 1961. Yet when John Chin needed help with certain chores, such as doing the time-consuming checks of the reference gas cylinders, Ellis continued to quietly help.

Judy Bright (1966–2000)

Mauna Loa Observatory had not had secretarial or administrative help since Adeline Fujisawa left during the budget crisis in January 1964. Shortly after moving into the new office in the Cloud Physics Lab, Ruhnke asked Ellis if he knew of any potential secretarial candidates he could interview. Ellis suggested Judith "Judy" Bright, a nineteen-year-old sophomore majoring in speech therapy at Hilo College, which in 1970 became part of the University of Hawai'i at Hilo. The youngest of ten children, Bright's father was Hawaiian and English, and her mother was of Portuguese ancestry. When she was a senior at Hilo High School, she won the Miss Aloha Hawai'i competition and became the official greeter for the Big Island. Her most memorable assignment was to accompany Lynda Bird Johnson, the daughter of President Lyndon Johnson, during her visit to Hawai'i. She also greeted visiting US Navy vessels and presented leis to the captains and first officers. And she gave talks at meetings of Hilo civic organizations. But her biggest opportunity occurred in June 1966 when Howard Ellis asked his wife Lena to call Bright and ask if she might be interested in working at MLO. She recalls what happened next:

> Howard Ellis was like an uncle to me because his kids were my age. We grew up together, and my brother married their cousin. So, family ties there. When he asked me, I was thrilled. At the time I never [had] worked in my life, because I was the last child, and my dad took care of everything I needed. So, I excitedly said, "Oh, yes! When do you want us to show up?" He said, well, how about tomorrow? I said, "Great!" I hung up, called my friends, and the next day after school we were up there. (Judy Pereira, video interview with author, Boulder, Colorado, April 27, 2007)

The next day Ellis picked up Bright and two of her friends at the college and drove them up the rough gravel road to the new Cloud Physics Observatory. Ellis had told Ruhnke the night before that Judy Bright had been Miss Aloha Hawai'i, and he recalls that he had decided even before the interview he would hire her as his secretary. Bright (who later became Judy Pereira) describes the job interview at the Cloud Physics Lab in almost the same words that Ruhnke related forty years later:

> I walked in with my friends, and . . . Mr. Ellis said, " . . . I want you to meet Mr. Ruhnke." He turned and he looked at all of us, and he said, "Which one of you is Miss Aloha Hawaii?" I said, "I am." He said, "Well, you're my secretary." I said, "Okay, that's fine with me." (Pereira, video interview with author, Boulder, Colorado, April 27, 2007)
>
> We laughed (unaware at that time of the political incorrectness of that statement!), and he promptly put us all to work doing data reduction from chart rolls and data sheets. I had to type letters using an antiquated manual typewriter and carbon paper, but I was thrilled to be working for such an important facility, one preparing to celebrate its 10th anniversary in a few months. (Pereira 1996)

Bright may have been inexperienced in the workplace, but Bernard Mendonca (video interview with author, Boulder, Colorado, April 27, 2007) recalls that she was a fast learner:

> It didn't take long before we all realized Judy was a diamond in the rough. I mean, she was just incredibly capable. So we kind of worked to get a position, and we offered it to her, and she took it. The rest is history. She just got better and better and better and never stopped. She learned everything on the job. There was no training or anything; she was capable enough. She'd call Boulder or Honolulu or whoever [and] got to know the government intricacies and bureaucracy. She mastered it with one hand tied behind her back. She was capable, that was the point; she was very capable.

Howard Ellis, who had to adapt to the music from Bright's transistor radio, advised her to not become indispensible, but that's what she became within two weeks. In the same interview, Mendonca reflected on this a moment before offering his opinion:

> Well, no one is indispensable, but if anyone could be, I suppose it was her. Yes, she was good. She was good. And . . . she was not only capable as an administrator, office worker, stenographer, secretary, she could [also] handle personalities very well, and she kept the office.

After six months as a temporary employee, Judy Bright took and passed the Civil Service clerk-typist exam and became a full-time federal employee of the Mauna Loa Observatory. It was her very first job, and though she originally planned to return to college, it was a position she would keep for thirty-three years. When asked during the 2007 interview if her role as Miss Aloha Hawai'i affected her future work at MLO beyond helping her get the job, she responded,

> It did indeed. . . . I made friends that I later was able to communicate with about getting certain things that the Observatory needed and [it helped with] making contacts for all kinds of things, from hotel rooms to road work. . . . I made a lot of friends that I was able to call upon when I worked at Mauna Loa.

Ralph Stair Returns to MLO (1966)

In the summer of 1966, the indefatigable Ralph Stair returned to the slope unit of MLO that he and Robert Simpson founded a decade before. This time he was sponsored by NASA. Wearing the same hat, coat, and shoes he had worn a decade earlier, Stair arrived with the same solar spectroradiometer he had used in New Mexico in 1955 and at MLO in 1957. He was accompanied by his second wife, Angie, and William R. Waters, his assistant from the National Bureau of Standards (NBS). The three signed the *Visitors* book on July 28, 1966, and moved into the concrete block building Stair had designed in 1955.

Stair and Waters set up their bulky apparatus on the concrete pad outside the building and began the same measurements of the solar constant that he was making in June 1955 at Sunspot when he first met Robert Simpson and learned that the sky was much cleaner and drier over Hawai'i. Angie Stair, who had visited MLO with a group from Hilo during Stair's 1957 stay on the mountain, was not pleased by the condition of the kitchen and bedrooms in the building:

> She stayed, and she fixed the rooms and fussed about why the place was so untidy, and she tidied it up quite a bit. I guess she could not reach these pictures that I had all jumbled up there [shown hanging on the wall in a photograph]. She cleaned the place very good. (Ellis, video interview with author, Hilo, Hawai'i, December 2006)

Angie Stair served grapefruit for breakfast each morning and prepared full meals for lunch and supper. Howard Ellis provided major assistance to Stair, who, Ellis recalled, liked being addressed as Dr. Stair. A photograph made at lunch the first Sunday after their arrival shows the Stairs, Waters, and Ellis seated at the dining table in the MLO kitchen. Ellis later recalled that the onions in the frying pan smelled just like home. Lothar Ruhnke arrived with his young son shortly after the photo was taken, a visit that was punctuated by an admonition from Stair when the curious boy got too close to his apparatus. There was good reason for Stair's concern, for, as shown in Figure 10.1, his spectrometry equipment was quite complicated. Figure 10.2 shows Stair pointing to sunlight intensity measurements recorded by his spectrometer's chart recorder.

The best observing days during the Stairs' visit were August 15–19, 20–21, and 23. In late August the team completed its mission and packed for home. As in all his prior research involving the solar constant, Stair made new findings and prepared a detailed paper describing them. Ellis had provided essential assistance to the project, and Stair named him the second author of the paper (Stair and Ellis 1968). The paper was published after Stair had moved from NBS to Neotec Corporation.

Ellis drove the Stairs down to Hilo several times during their stay, where they met Judy Bright at the Cloud Physics office. The Stairs became friends of the Ellises and spent time in their home. Years later, the Ellises visited the Stairs in Washington. Other well-known visitors signed the *Visitors* book during the summer and fall of 1966, including Jack Pales and his daughter Deborah, who made a return visit to Hilo in July. Pales had been unable to work at MLO after his heart attack, but he and his daughter visited the mountain without incident and signed in on July 18, 1966. There was talk that he planned to return as director someday.

MLO director Lothar Ruhnke signed in for the second time on June 28, 1966, the tenth anniversary of the dedication of MLO. So did Ray Busniewski of the Hilo Weather Bureau office, Howard Pierce of the *Hilo Tribune-Herald*, and Jack Bryan of the *Honolulu Star-Bulletin*. Judy Bright, who was assigned a desk at the Cloud Physics office in Hilo, signed the book for the first time on October 25, 1966.

The Winds of Mauna Loa

The trade wind inversion layer along the slopes of Mauna Loa and the wind circulation around the big mountain are keys to many of the measurements made at MLO. The trade wind inversion separates cool, moist air of the lower elevations from the dry air of the free atmosphere. The wind circulation at MLO is determined by the trade winds and the localized up- and downslope winds generated by the presence of the mountain itself. Measurements of sunlight are best accomplished during the morning when the inversion layer is well below MLO. Sampling of carbon dioxide and other trace gases is best done during the night and morning when the downslope wind channels clean air from the free atmosphere over MLO.

Charles Wilkes of the US Exploring Expedition studied the wind circulation and clouds of Mauna Loa in 1841. In 1897, Henry Guppy provided a detailed overview of the mountain's circulation. Saul Price and Jack Pales (1963) conducted various studies of the seasonal and diurnal (daily) winds with the help of Bernard Mendonca, who processed and analyzed much of their data when he worked for Price in Honolulu and assisted with papers that Price and Pales wrote. After he was assigned to MLO, Mendonca expanded on the work of Price and Pales and published two significant papers of his own about the trade wind inversion layer and the wind on Mauna Loa (Mendonca 1969; Mendonca and Iwaoka 1969). Judy Pereira typed the papers, and Mendonca wrote thank-you notes on copies of each paper he gave her after they were published.

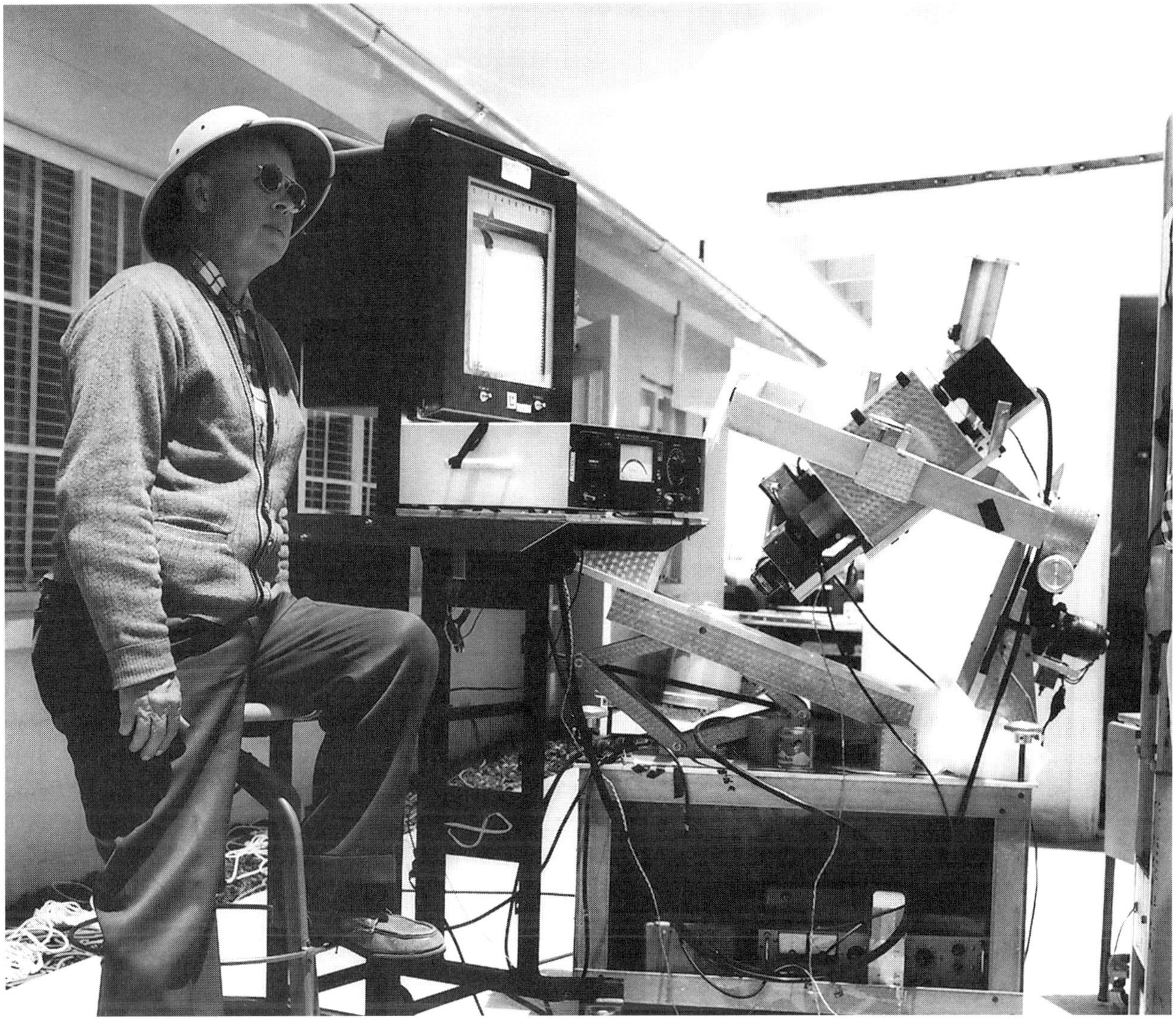

FIGURE 10.1. Ralph Stair in August 1966 with the sun-tracking spectroradiometer with which he measured the solar constant at Sunspot, New Mexico, in 1955 and at MLO in 1957. *(MLO archive.)*

FIGURE 10.2. Ralph Stair points to scans of sunlight measured by his spectrometer at MLO during August 1966. *(MLO archive.)*

Some of the field research for Mendonca's trade wind inversion paper, which was developed with the assistance of new MLO employee Wayne Iwaoka, involved holding a psychrometer outside the open window of a car while driving the lonely road to MLO. A psychrometer measures relative humidity, and the measurement transects along the road provided a detailed cross section of the trade wind inversion along the slope of the mountain that could be compared with upper air soundings made by the twice-daily weather balloon flights from the Hilo Weather Bureau. Mendonca also measured the condensation nuclei in the air along the road using a portable instrument inside the car and an air sampling tube outside the car. He found a close similarity in the elevation of the inversion layer along the slope of the mountain with that over the ocean measured by balloon-borne radiosondes (Mendonca and Iwaoka 1969).

For his second study, Mendonca used a tethered balloon to measure the nocturnal and morning downslope breeze from the free atmosphere flowing across MLO and the upslope air arriving from below. He discovered that the downslope wind is a layer only around 180 feet (55 meters) thick, while the upslope wind is a layer about 2,000 feet (610 meters) thick (Mendonca 1969).

Mendonca's findings were made possible by instruments and methods that were unavailable to Wilkes and Guppy. They provided important information for the timing and the location of sampling ports for the measurements of carbon dioxide and other trace gases. They also provided new insight into the clarity of the atmosphere at MLO at night and during the early morning hours. Thirty-eight years later, Mendonca (personal communication with author, March 18, 2008) reflected back on his early studies at MLO:

These two studies helped to confirm with observational evidence that Mauna Loa Mountain was an excellent vantage point from which it was possible to continuously monitor the free troposphere far removed from local sources of manmade pollutants that could have an impact on climate. And more importantly, it was a place where it would be possible to test the hypotheses that there was an accumulation of manmade pollutants with time that could have far reaching consequences on the earth's climate. Later on it would become clear that sporadic inputs from natural phenomena, such as dust veils from explosive volcanic eruptions that spread globally, could be detected from the mountain.

More Carbon Dioxide Breakdowns

By the end of 1965, John Chin was weary of coping with the temperamental carbon dioxide monitor and working on the mountain. As noted earlier, his final entry in the CO_2 *LOG* was a list of problems he titled "Major Breakdown." In April of 1966, an assistant meteorologist position opened at the Cloud Physics Observatory of the University of Hawai'i:

> This position offered me an increase in pay along with the benefit of being able to stay in town. I was really getting tired of spending my nights on the mountain. So I left Scripps and joined the gang at Cloud Physics, where I met and worked side by side with Bob Uchida for the first time. It never occurred to me that he would become my partner again many years later. [Bob Uchida joined MLO in 1991.] (Chin 2001)

Chin's last official day of work for Scripps at MLO was April 30, 1966, the same day that his replacement, Ronald C. Priebe, arrived at the airport at 5:57 p.m. Priebe had worked for the University of Washington as a micrometeorologist who studied Arctic sea ice. He signed in at MLO on May 2, 1966, and made his first entry in the CO_2 *LOG* the same day: "Changed freeze out coil." Priebe made many entries in the CO_2 *LOG* describing numerous tests and problems with the analyzer, some of which involved routine tests at late hours: "November 25, 1966. Ran tank check for 1200 psi. Working tank from 1805–2320."

Two days later trouble developed:

> November 27, 1966. Replaced one OD3 tube in invertron. It had been causing occasional power fluctuations that resulted in noise and shift of the CO_2 trace [on the chart recording]. Changed freeze out coil at 2025.

Sometimes human error or shipping delays intervened, and Priebe entered in the CO_2 *LOG* an especially serious incident when the reference gas supply was exhausted before a replacement tank arrived from Scripps:

> November 2, 1966. Ran tank check for 400 psi. Working tank [of reference gas] from 0540–1240. No more working tanks at station. Shut recorder off.

The carbon dioxide program was completely shut down, for the analyzer could not perform properly unless it periodically measured a sample of gas having a known concentration of CO_2. The replacement gas did not arrive until nine days later, and Priebe was up past midnight installing the new cylinder and reactivating the analyzer and chart recorder: "November 11, 1966. Installed new working tank at 0400. Program back in operation."

The Power Line (1967)

Since MLO opened in 1956, maintaining and providing fuel for the diesel generators was a major headache. Like his predecessors, Lothar Ruhnke was concerned about how the exhaust from the generators was sometimes contaminating the carbon dioxide measurements. He was also concerned about the dust from the road, and he soon addressed both problems. The road was closed to routine visitors to reduce the dust, and an annual open house was begun to allow residents and others to visit MLO in a preplanned fashioned. In an arrangement with the Kūlani Prison, the final mile of the road to MLO was paved. Obtaining electrical power was a much more complicated procedure, and Judy Pereira (video interview, Boulder, Colorado, April 27, 2007) recalls some of the details:

> Because of the problems that the generators were producing, the smoke and all that kind of stuff, we knew we'd have to get commercial power. Without that, the observatory would eventually die, because there would be too much contamination [from] the generators. Lothar Ruhnke . . . really hustled the state government, the county government, the power company, to please consider putting commercial power out there. He even set up a program where we would prepay a certain amount of it and then, over the years, $10,000 for ten years would satisfy our commitment. That was the best deal anybody could have made, because to do that would be millions of dollars today. But we got it for, like, $100,000.

At 3:00 p.m. on July 21, 1967, electrical power was finally brought to MLO by the Hawaiian Electric Company (Keeling et al. 1976). The arrival of commercial electricity was a revolutionary development, for it meant that for the first time the staff did not have to work nights at MLO to keep the generators working. But this was a mixed blessing from Howard Ellis' perspective. According to Ellis (telephone interview with author, June 11, 2008), Cliff Kutaka, one of the original MLO staff members under Jack Pales, told him, "This place will never be the same with a power line." Kutaka was right.

Helmut Weickmann Visits MLO (1967)

Dr. Helmut Weickmann, director of ESSA's Atmospheric Physics and Chemistry Laboratory in Boulder, had hired Lothar Ruhnke to take over as MLO's director, and he visited MLO on June 21, 1967. While in Hilo he was photographed with the staff (Fig. 10.3), which included Ruhnke, Howard Ellis, Bernard Mendonca, Judy Bright, and instrument technician Josef Mueller. Also in the photograph were summer employee Wayne Iwakoa, who worked with Mendonca on the inversion layer study, and part-time student employee Sharon Nagata (Plate 9), who first signed in at MLO on June 15, 1967. The Mauna Loa Observatory staff had returned to the level it was prior to the 1964 budget crisis.

John Chin and Bill Cobb Return to MLO (1968)

Because the CO_2 project was so intimately tied to MLO, a major decision was made for the observatory to return to the days when its staff was in charge of the maintenance and operation of the CO_2 analyzer. John Chin (2001) explains what happened next:

> While I was working for Cloud Physics, Mauna Loa Observatory rented space for an office in our building. One day Mr. Lothar Ruhnke approached me and asked if I would like to join the Mauna Loa Observatory staff to maintain the Scripps CO_2 program. He said that I would be the perfect person for the job since I had worked with the Scripps program before. I jumped at the chance because it would mean that I would become a Federal Government employee and not employed by Scripps Institution of Oceanography.

John Chin signed in for his third assignment at MLO on May 10, 1968; it would be by far his longest. Three days later, Bill Cobb signed in, adding "home again." Cobb was among the MLO staff who had been reassigned when the 1963 budget crisis arrived. His new assignment at MLO would lead to a paper on the effect of the sun on the atmospheric electricity measured on Mauna Loa (Cobb 1967.) The paper was published after he was assigned to ESSA's Atmospheric Physics and Chemistry Laboratory in Boulder.

The MLO Family

During the time when only Howard Ellis and Bernard Mendonca were assigned to MLO, there was no free time for recreation and socializing. This changed after Lothar Ruhnke arrived with access to funds to hire more staff. The

FIGURE 10.3. The Mauna Loa Observatory staff on June 21, 1967, included *(left to right)* Wayne Iwaoka, Sharon Nagata, Bernard Mendonca, Helmut Weickmann (director, APCL), Howard Ellis, Judy Bright, Lothar Ruhnke (director, MLO) and Josef Mueller. *(MLO archive.)*

return of a feminine influence also made a difference, as recorded in the memoir by Judy Bright (Pereira 1996), who married Jerry Pereira, her childhood sweetheart, in 1967:

> People like Howard Ellis, Bernard Mendonca, Bill Cobb, and John Chin (who worked for the Cloud Physics group when I first started) became my work family. Mauna Loa was a place where co-workers worked and played together. I guess working in a small office was conducive to everyone being close knit. We all ate lunch at the office together, and the first time I saw Johnny Chin eat his chicken salad sandwiches, banana, and carrots, I had no idea that I'd still be watching him eat that same type of lunch thirty years later, and every work day in between! If we could talk him into coming out to lunch with us, he'd bring his brown bag with him and eat that. Bernard's wife and children; Johnny; Barry Bodhaine, a graduate student working at the Cloud Physics Observatory; Charlie Garcia of the High Altitude Observatory co-located at the MLO site, his family; and my fiancé and

> later husband Jerry, and I, became avid ping pong players, poker players, potluck groupies, and Hapuna Beach campers. Lothar, Howard, Bill and their families occasionally joined us for these activities.

Interviewed in Boulder, Colorado, on April 27, 2007, Judy Pereira added more details about what people still call the "MLO family":

> We became a family, sort of—I mean, everything we did was, well, what's going on this weekend, what's anybody doing? . . . [On] some long weekends we'd camp. We'd pack all our food and go out to Hapuna Beach. At that time it wasn't like it is now; we could camp, pitch tents and camp, and just have a good old time. It was, like I say, it was a family-type thing, and we enjoyed working together and playing together.

The family connection sometimes extended to the observatory itself, as when kindergartner Karl Mendonca signed the *Visitors* book with neat block letters on June 12, 1967. Young Karl had arrived in Hilo in February 1964 with his parents, Bernard and Clara Mendonca, when Bernard was assigned to MLO.

Howard Ellis Again Takes Charge (1968–1970)

After his first year at MLO, Lothar Ruhnke spent most of his time working on his doctorate at the University of Hawai'i on O'ahu. Although his tenure at MLO was brief, Ruhnke accomplished much. He hired Judy Pereira, brought in Josef Mueller as electronics technician and instrument maker, and returned John Chin and Bill Cobb to MLO. He worked with Charlie Garcia to arrange for paving the final mile to MLO to cut down on dust. He organized an annual open house. And he found a way to finally eliminate the noisy, polluting, and maintenance-prone diesel generators. In her memoir, Pereira (1996) describes the latter as Ruhnke's most significant contribution to MLO:

> His one stellar accomplishment, which set the observatory on its way to being an upgraded world class facility, was his role in acquiring commercial power to span the 18 miles of isolated road between [Hilo] and the observatory. Gone were the days of generator power whose pollution concerned everyone who knew of our work there, not to mention the enormous work load for staff members to keep it running. Lothar Ruhnke spent his second year commuting to the University of Hawaii in Honolulu, and left Mauna Loa in 1968 as Dr. Ruhnke.

Howard Ellis was again placed in charge of MLO, and on July 28, 1968, his position was changed to supervisory physicist. Without a strong advocate back in Boulder, MLO again found itself with a depressed budget.

> The funding was cut; we were left with really no budget, except operating budgets. . . . And, so, money was always a problem. (Pereira, video interview with author, Boulder, Colorado, April 27, 2007)

At least this time the staff was not reduced, and the work of the observatory continued. A photograph of the staff (Fig. 10.4) shows Howard Ellis, Bill Cobb, Bernard Mendonca, Johnny Chin, Judy Pereira, and Janice Sias. Though they were not trained scientists, both women followed the tradition begun by Adeline Fujisawa before them and assisted with some of the measurements, as Judy related in the same interview:

> Howard still cranked out his solar radiation numbers on an old Friden clunker type calculator, and Bernard wielded his slide rule like a sword. I assisted in field work by making Gardner counter measurements in Hilo, near the erupting Kilauea Mauna Ulu vent, and at Mauna Loa whenever I went up. I changed the rain gauge chart at the Cloud Physics building and the hi vol [high volume] sampler bubblers behind the building.

In May 1969, Judy Pereira took six months of maternity leave to have her son Monty, who was born on June 28,

FIGURE 10.4. The Mauna Loa Observatory staff in May 1969 included *(left to right)* Bill Cobb, Bernard Mendonca, Janice Sias, Judy Pereira, Howard Ellis, and John Chin. *(MLO archive.)*

MLO's birthday. Her childhood friend Janice Sias filled in during her absence, and Sias (2006) recalls, "My boss was Howard Ellis, and the 'team' included Bernard Mendonca and Johnny Chin." Shortly after she arrived, Sias learned that her job required much more than secretarial duties when she was enlisted by Mendonca to help make particle measurements very near the major eruption of Mauna Ulu that began on May 24 (Plates 10 and 11). Sias visited MLO and signed the *Visitors* book on July 24, 1969 (Plate 12).

"Excellent Observatory—Well Run"

In 1969, a steady trickle of visitors from around the world drove the bouncy road to MLO and 112 signed the *Visitors* book. The visitors that year included an eclectic blend of scientists, reporters, tourists, hikers, local residents, and a smattering of children. Some complained about dizziness and altitude sickness. Others arrived on very cold days. Most commented on the beauty of the mountain, as on April 5, when a group of Peace Corps volunteers signed in on their way to the summit. The next day one wrote, "Beautiful trip to the top of Mauna Loa." On June 28, Joan Moyers wrote, "The view of Mauna Kea is marvelous!"

The condition of the road improved dramatically after it was first paved in July 1969 (Keeling et al. 1976), as noted by Eleanor Brown of Hilo, who signed in on July 7 and wrote, "Great view and a smooth trip!" The thin coating of asphalt applied to the lava road greatly simplified the drive for the staff members who made daily trips up the mountain. It also reduced the stress on their vehicles.

The number of visitors who signed the *Visitors* book increased after the road was paved. Based on their names and relationships, about thirty-one visitor cars or vehicles drove the road from January to June and about forty-six from July to December.

Charlie Garcia of the MLO Solar Observatory sometimes brought family members to the observatory, and on July 27, his son Thomas, who was no stranger to the mountain, signed in and wrote "It's my birthday." Years later, Tom Garcia would join the staff of MLO.

On August 15, 1969, Jody Bennet of Honolulu wrote, "I'll take the beach!" On August 21, 1969, the Allen M. Bassett family of La Mesa, California, wrote, "Wonderful variety of rock textures en route to top of Mauna Loa but petrologically monotonous." On August 25, 1969, Al Perez of Rialto, California, who was Howard Ellis' brother-in-law, wrote, "One can feel so small up here if one does not realize his worth."

Judy Pereira and Janice Sias visited the mountain together on November 12 after Pereira returned from maternity leave. After their signatures in the log, Pereira added "Sec. MLO," and Sias wrote "Ex-sec MLO." There were no more signatures until Robert Grass from ESSA in Boulder stopped by on his way back from Antarctica on November 22. Grass was a specialist in calibrating and servicing Dobson spectrometers that measure the ozone layer, including the one at MLO.

Not everyone was happy about the visitor traffic:

> [The power line] has nearly eliminated local production of CO_2 by the station itself, but, with the completion of a paved public road to the station in July 1969, daytime automobile pollution has become a problem. (Keeling et al. 1976)

Of all the visitors' comments in 1969, Howard Ellis was probably most pleased by a notation on May 5 by Pete Kuhn of ESSA in Boulder, who wrote, "Should be exploited much more. Excellent observatory—well run."

As for Judy Pereira (video interview with author, Boulder, Colorado, April 27, 2007), her administrative skills blossomed during Ellis' tenure as director:

> Howard, I think, was a real southern gentleman, and didn't like the idea of telling people what to do. He would ask people, you know, do you think you have enough time to do this, or, do you think you need something else that maybe we could get for you; really, really, a very, very nice man. But, he was not dynamic in the sense that he would call up the powers that be and say, we need this, or, we want that. No, that was not his style at all. He would write letters asking maybe for something, or he'd ask me to write for him. . . . Howard depended on me a lot, and I appreciated that, because I did have a chance to grow, and I grew administratively, and I began to learn a little bit more about the science that was going on at the observatory. He was such a nice, personable person to work with, and he never questioned anything that I did that was within the realm of the observatory mission.

Fountains of Fire (1969)

With but one brief exception, Mauna Loa had remained quiet since 1949, when MLO founder Robert Simpson (telephone interview with author, March 17, 2007) was mesmerized by the glow from its summit crater:

> It was a very spectacular eruption. You could see the glow on the horizon from Honolulu quite a good distance away, about 200 miles I believe. I flew over Mauna Loa in a navy aircraft, flying at different heights above the erupting volcano to measure how the heat plume from the volcano was spreading out as we went up above it.

> It went up to a considerable height without much heating occurring laterally and was the subject of some interest when I attended the Pacific Science Congress during 1949 in New Zealand.

Even though the 1949 eruption lasted 145 days, the second longest since eruptions of the big mountain were first recorded in 1843, Simpson knew that the enormous size of the mountain left plenty of room for a carefully located observatory.

Mauna Loa had erupted thirty-one times between 1843 and 1950, the last being a one-day event (Lockwood and Lipman 1987). Kīlauea, the hyperactive volcano southeast of Mauna Loa and well away from MLO, had erupted fourteen times since the MLO Slope Observatory was dedicated in 1956 (Macdonald et al. 1986). When Kīlauea roared back to life in May 1969, some of the MLO staff made atmospheric measurements in the vicinity of the eruption. An especially memorable occasion occurred on December 30, 1969, when Bernard Mendonca took a close look at the Mauna Ulu lava fountain accompanied by Judy Pereira and Lasaro R. Waipi, who was visiting from the weather bureau of Truk and the Caroline Islands (Plates 13 and 14).

Six days earlier, Waipi had signed in at MLO and added "Very cold!" But on December 30, he and his hosts found themselves much too close for comfort to a fearsome plume of molten lava roaring more than 1,500 feet (457 meters) into the Hawaiian sky. Mendonca and Pereira recalled their volcano adventure during a video interview with the author at Boulder, Colorado, on April 27, 2007:

> Judy Pereira: Bernard thought it would be a great idea to take [Lasarao Waipi] out to where the active volcano was going on. I mean, this thing was shooting up over 1,000 feet in the air. He apparently had never seen an active volcano, so Bernard said, "Why don't we go? We can take some measurements while we're out there." So, we go out there, we talk somebody from the volcano observatory into taking us in his truck to the site where we thought we'd be close to the volcano. Well, we went out there and when I started hearing the lava falling on the roof of the cars, [I said] "Bernard, let's get out of here! Hurry up, we're going to get killed!" It was scary, definitely scary; it was an experience I will never forget.
>
> Forrest Mims: Did Bernard want to leave, or what?
>
> Pereira: No.
>
> Bernard Mendonca: I kept saying, "Wait, wait, just a minute!" . . .
>
> [When we arrived, the volcano was] burbling and gurgling, and then we drive down there and then, wow! This monster eruption just happens while we're there. I remember . . . two big fountains, and we stood there and our eyes got bigger and bigger and bigger, and then it hit the car. Judy, the one with the common sense, says, "I think we better go." I said, "Well, wait a minute," and [Lasaro], he was just—he could not believe this, he just could not believe it.
>
> Pereira: Yes, and he was on my side anyway.
>
> Mendonca: We got in the car, reversed, screamed out, and as we were going out, the park guys [arrived], all kind of excited, and asking us what we were doing [there].

Other MLO volcano adventures were to follow, but none were to provide the close call experienced at Mauna Ulu by Bernard Mendonca, Judy Pereira, and their hapless visitor from Truk. As Mendonca later recalled, "Science is wonderful!"

Farewell to Jack Pales (1969)

Jack Pales was responsible for setting up all of the original International Geophysical Year measurement programs at MLO when the Weather Bureau began full-time work on the mountain in July 1957. The measurements of sunlight, surface ozone, the ozone layer, and carbon dioxide were all begun on his watch, and he was deeply involved in all of those legacy programs. He coauthored the first papers with Saul Price on the status of the MLO observing programs (Price and Pales 1959, 1963), and he coauthored with Charles Keeling the first major paper on the carbon dioxide findings (Pales and Keeling 1965).

Pales was known for the many hours he spent on the mountain, often going without adequate sleep. Howard Ellis took over much of the work at MLO after Pales had a heart attack in February 1963, and when Pales requested to be reassigned to the Weather Bureau at the Nevada Test Site after the 1963 budget crisis, Ellis took over his position. In 1966 Pales and his daughter Deborah visited MLO. In 1968, Pales received an NOAA Meritorious Service medal, and, according to Ellis and others, he was planning to resume work at MLO. He never got the chance. As Lester Machta (1972) relates, "[Jack Pales] worked just as hard for the Weather Bureau at the Nevada Test Site and unfortunately had a fatal heart attack in 1969."

An Invitation to Do Science at MLO (1970)

The International Geophysical Year (IGY), which sparked construction of what was then called the slope unit of the Mauna Loa Observatory, attracted considerable public attention from 1957 to 1959. The IGY also resulted in the launching of the first artificial Earth satellites, and public interest in science eventually shifted from earth to space, especially after July 20, 1969, when Neil Armstrong and Edwin Eugene "Buzz" Aldrin Jr. became the first men to walk on the moon. In some ways, the terrain around the little observatory on a volcano in the center of the Pacific Ocean resembles the surface of the moon, and as previously noted, NASA astronauts had even briefly trained there during Howard Ellis' watch. But when it comes to science applicable to life on earth, there is no comparison, for MLO offers a research platform that far exceeds even that which was available to the Apollo astronauts.

In May 1970, former MLO director Lothar H. Ruhnke and Joann T. Dennett published a detailed report about MLO in an ESSA technical memorandum. The report updates earlier papers by Saul Price and Jack Pales and closes with "An Invitation from Hawaii's Science Peak" and hints of things to come:

> In its charter, ESSA was charged with increasing man's ability to adapt to and live with his physical environment. Thus, the ESSA Research Laboratories encourage observatories and research units such as Mauna Loa to participate fully within the science community. Qualified researchers will find willing cooperation, research freedom, and support at Mauna Loa Observatory.

The record clearly shows that this and other invitations to perform research at MLO bore fruit over the years, for many scientists responded. While the road was still unpaved, their research was made much simpler by the dependable, quiet, and pollution-free electricity provided by the power line installed under Lothar H. Ruhnke's tenure at MLO (Plate 15).

MLO Director Rudolf Pueschel (1970–1972)

Howard Ellis kept MLO alive for more than twenty-six months during the budget crunch of 1964 to 1966 before being replaced by Lothar Ruhnke. After Ruhnke left in 1968, Ellis was placed in charge of MLO for the next twenty-five months. Then on September 6, 1970, he was again shunted aside. This time he was replaced by Dr. Rudolf Pueschel, who, like his predecessors, took his wife Brigitta on a get-acquainted visit to MLO. Brigitta Pueschel signed in on October 3, 1970.

The same day that Mrs. Pueschel signed in at MLO, the observatory came under the administrative chain of command of the newly formed National Oceanic and Atmospheric Administration (NOAA). In July 1970, President Richard Nixon had proposed to the Congress the creation of NOAA within the Department of Commerce "for better protection of life and property from natural hazards . . . for a better understanding of the total environment . . . [and] for exploration and development leading to the intelligent use of our marine resources" (NOAA 2006). ESSA was absorbed into NOAA, and MLO continued operation under the Atmospheric Physics and Chemistry Laboratory in Boulder, Colorado.

Rudolf Pueschel brought a scholarly perspective to MLO reminiscent of Saul Price's influence in the earliest days of the observatory. During his time, the scientific mood of the observatory blossomed, and the number of scientific papers by staff and visiting scientists greatly increased. The origin of Pueschel's interest in science was complex, for he was not drawn to the sciences during his youth. Instead, his scholarly approach to science evolved from a sequence of events that began during his childhood during World War II.

> I am born German. I [was] exposed to political turmoil in World War II and getting expelled from [Czechoslovakia] because we were German. We were expelled to East Germany, which became communist dominated, which gave me cause to leave for [freedom in] West Germany and, finally, the United States of America as a scientist. In fact, my first job was at Caltech in Pasadena in the United States of America.
>
> I wish I could say [my interest in science] was passion, but it was politically, ideologically influenced . . . it [was] a decision based on ideology [and my exposure to] fascism or national-socialism and, thereafter, communism. I learned quite early that humanity's history . . . is [more] easily distorted than science, so I decided to become a natural physical scientist. (Pueschel, video interview with author, November 30, 2007)

During college in Frankfurt, Pueschel studied physics, mathematics, and chemistry, and he left for the United States in 1962 after receiving a diploma in physics (equivalent to an MS degree). There he found work studying the health effects of air pollutants, which led to a stipend to pursue graduate studies at the University of Washington, where he earned a doctorate in 1969. He then received a one-year postdoctoral appointment to the Environmental Services Sciences Laboratories in Boulder, Colorado, as noted in the same interview:

> It was a one year, extendable for two years, assignment, and my boss told me if I want to have a permanent job, I have to go out to Mauna Loa and prove myself.

There were important parallels in the arrival of Ruhnke in 1966 and Pueschel in 1970. Ruhnke completed his doctorate while assigned to MLO, and Pueschel had just completed his when he arrived. Both men were assigned during a reorganization of the observatory's administrative chain of command. And both arrived with a significantly enhanced budget that allowed improvements and changes only dreamed of by MLO veterans Howard Ellis and Bernard Mendonca. Judy Pereira's memoir (1996) succinctly summarizes the significance of Pueschel's arrival: "With him came a resurgence of budget monies and increased staff."

Howard Ellis was left to wonder why this latest change in the bureaucracy, which finally brought more funding and staff, again cost him his position as director, the assignment that Harry Wexler had given him when he received orders to go to Hilo in 1960. But he bore no resentment toward Pueschel, who clearly recalls that when he first visited MLO, Ellis shook his hand and announced in his polite, southern style, "Welcome to the observatory. It's all yours now." Thirty-seven years after his arrival as the new director, Pueschel expressed a high regard for Howard Ellis and the MLO staff.

> Personally, first, I was very much touched by the reception I received from the staff at Mauna Loa Observatory. I felt acknowledged. I felt appreciated and the feeling was mutual. Scientifically, I realized the uniqueness of the observatory [because of] its remoteness, remote from continents by being located on Hawaii and remote from self-spaced emissions along the shore by the trade inversion, which prevents exchanges vertically. So I realized a tremendous potential, as a scientist, working at Mauna Loa Observatory. . . .
>
> What I noticed, first of all, was records of scientific data of carbon dioxide measured by Johnny Chin [Plate 16], administered by Dave Keeling at Scripps; of ozone measured by Bernard Mendonca, administered by Walt Komhyr at Boulder; and of solar radiation measured by Howard Ellis and also administered by Howard. And those three records were quite impressive when I got here. . . .

FIGURE 10.5. The Mauna Loa Observatory staff on September 24, 1970, included *(left to right)* John Chin, Rudolf Pueschel (director), Bernard Mendonca, Judy Pereira, Robert Knecht (deputy director, ERL), and Howard Ellis. *(MLO archive.)*

> Howard Ellis was in charge of the observatory when I got here. Howard welcomed me, like other members of the staff, quite heartily. I remember Howard Ellis took me up to the mountain for the first time. He explained various bends and lava strata along the way. We arrived at the observatory after a strenuous—what I considered a strenuous—ride [being my first] and [he] shook my hand and welcomed me to my observatory, which I found very, very warm and was received very favorably. (Pueschel, video interview with author, November 30, 2007)

Shortly after Rudolf Pueschel arrived in Hilo, Robert Knecht, deputy director of ESSA's Environmental Research Laboratories, visited MLO. This provided the occasion for a new staff photograph at the Cloud Physics Observatory office (Fig. 10.5).

Where's the MLO Deed? (1971)

During June 1971, questions were raised about the existence of the old Executive Order Ralph Stair had secured from Governor King in 1956. The order granted to the Department of Commerce the four acres where the Mauna Loa Observatory is located, and Judy Pereira (video interview with author, Boulder, Colorado, April 27, 2007) recalls what happened next:

> Well, actually, all the time people were asking, "Where is that deed? Oh, it does exist, I've seen it before," you know. Everybody knew that it existed, [but] nobody could put their hands on it.

On June 10, Pereira wrote a letter to the Hawai'i Land Office in Honolulu that concluded, "We seem to have misplaced our copy and would like to have one on hand. Your earliest reply to our request will be very much appreciated."

When the Land Office sent an authorized copy of the deed, Pereira stashed it in a folder she kept in a special place on her desk, because, otherwise "everybody would touch it or lose it or want to see it." For those who wanted to see the deed, she made copies that she filed away for anyone to view, and they still remain in a file cabinet in the Hilo office.

The MLO Staff Expands (1970–1972)

An increase in the MLO staff began shortly before Pueschel arrived when Dr. H. James "Jim" Simpson, who had just earned his doctorate from Columbia University, reported for one year of postdoctoral work. He signed in at the mountain station on August 23, 1970. More key MLO people were soon to follow.

Barry Bodhaine (1971–1976)

Jim Simpson was followed in June 1971 by Barry Bodhaine, a meteorologist with a cross-disciplinary background that was to prove very useful in his work and research at MLO. Bodhaine's father had been a hydraulic engineer, and by high school he knew he also wanted to go into engineering.

> Originally, I always wanted to be an electrical engineer. So I majored in electrical engineering [at the University of Washington], got my bachelor's degree, and then accepted a scholarship at the University of New Mexico to work at Sandia Laboratories and get a master's degree in electrical engineering. While I was working there, Jack Reed . . . convinced me to go into the atmospheric sciences and get a Ph.D. in meteorology. So when I finished my tenure there in Albuquerque [in 1965], I moved out to Hawaii [and] enrolled in a Ph.D. program at the University of Hawaii, Manoa Campus. . . .
>
> I had to take some course work in meteorology, so I actually lived on Oahu for the first two years to take courses in meteorology. But I also needed to develop a dissertation project, and it was my good fortune that Dr. E. J. Workman was on the Big Island at the Cloud Physics Observatory in Hilo. He was doing work in thunderstorm electricity, believe it or not, and with my background in electrical engineering it seemed a natural thing, so I went down to Hilo [during Christmas vacation of 1965] and met Dr. Workman at the Cloud Physics Observatory [and] he became my thesis adviser. So that is how I got to Hilo. (Bodhaine, video interview with author, Kona, Hawai'i, November 30, 2007)

Bodhaine moved to Hilo during the summer of 1966 to work on his dissertation at the Cloud Physics Observatory. Only a few months before, Lothar Ruhnke had arrived to take over as MLO director and arrange for office space at the Cloud Physics Observatory. The Bodhaine interview continues:

> I met the people who were working there . . . Lothar Ruhnke was the director, and Howard Ellis was there [along with] Bernard Mendonca and Judy Pereira. . . . Johnny Chin and Bob Uchida were working with Dr. Workman at the Cloud Physics Observatory. So I was actually working with Dr. Workman, Bob Uchida and Johnny Chin on my thesis project.

Ruhnke's tenure at MLO ended while Bodhaine was completing work on his doctorate.

> It was my good fortune that . . . during that time Dr. Rudolf Pueschel became the new director at Mauna Loa Observatory, and I got to be friends with him. And about that time, I finished my degree, and I needed to find a job, so he sponsored me as a National Research Council resident research associate. And so I had the good fortune of working with Rudi for two years under that program, and I have to add that it was a great working relationship, and I consider Rudi my mentor. [He] really introduced me to the field of science, and I learned how to do research and write papers [while] working with Rudi.

On August 4, 1971, Dr. Bodhaine signed in at MLO for the first time during a picnic visit with Judy Pereira and four others from Hilo.

During Bodhaine's year at MLO, Rudolf Pueschel arranged for an instrumented house trailer to be set up at Cape Kumukahi on the southeast coast of the Big Island to make aerosol and meteorological measurements close to the Pacific Ocean. The trailer provided important data before being gradually and significantly vandalized.

On July 22, 1972, Bodhaine and his wife Terri signed in at MLO, adding, "New address: Boulder. Last trip to MLO for an unknown time." The move to Boulder (with Rudolf Pueschel) was so Bodhaine could finish his second year as a National Research Council resident research associate. In June 1973 he was hired as a staff member at MLO, and he and his wife Terri moved back to Hilo. Bodhaine worked as a regular observer at MLO, taking his turn on the mountain, until September 1976 when he was transferred to Boulder and assumed duties as project leader for the aerosol program until 1995. He then worked with the solar radiation program until retiring on December 31, 2001 (Bodhaine, personal communication with author, May 3, 2008).

Al Shibata (1972–1990)

Honolulu native Al Shibata was an electronic technician who had worked for the Weather Bureau in the South Pacific, Panama Canal Zone, and the Air Research Laboratory in Las Vegas before being assigned to MLO in 1972 as senior electronics technician. He began work on March 13, 1972, and signed in at MLO the next day. Bob Grass, then NOAA's expert at maintaining Dobson spectrometers that measure the ozone layer, was visiting at the time, and he signed in on March 14. Shibata would go on to serve eighteen years at the Mauna Loa Observatory.

Mark Goldman (1972–1973)

Dr. Mark Goldman, who was an assistant professor of chemistry and physics at the University of Hawai'i, replaced Barry Bodhaine as a research associate at MLO. He arrived as a presidential intern and signed in on May 23, 1972. Like Bodhaine, Goldman already knew some of the MLO staff, for his office was also in the Cloud Physics Observatory. After he began work at MLO, he became the first person to begin placing the observatory's data into a digital format that could be processed by a computer. MLO hired a student to help write a FORTRAN program to process the data that was entered into punched cards that could be read into a computer at the University of Hawai'i.

Like most everyone who worked at MLO, Goldman's year was not without adventure. Once while returning to Hilo he was sideswiped by a US Army vehicle on the narrow Saddle Road. He also experienced a major earthquake while working at the observatory:

> You could hear that metal [roof] like people banging the tops of old trash cans together. And I ran out into the parking lot, and you could see the cars just going up and down on their springs, and that was the first time I was ever really afraid, I think, because my thought was that Mauna Loa was erupting, and there I was, stuck on the mountain with the eruption. . . . There were landslides on the road so . . . I had to get out of the car and physically remove the rocks from the road. (Goldman, video interview with author, Hilo, Hawaii, December 4, 2006)

Alan M. Yoshinaga (1972–2007)

Hilo native Alan Yoshinaga had a unique connection with the Mauna Loa Observatory, for his uncle had worked for contractor Jas. W. Glover, Ltd., when the firm constructed the original building back in 1956. After serving in the US Army for three years, Yoshinaga joined MLO as staff chemist on June 12, 1972, following a job interview with Rudi Pueschel at the University of Hawai'i Library in Honolulu. After his signature in the *Visitors* book on the mountain, he added, "1st day of work." During a video interview with the author (Hilo, Hawai'i, December 4, 2006), Yoshinaga recalled the interview with Pueschel:

> He promised me that the job would be interesting; it will be challenging, and that I would learn a lot. And as it turned out, that is exactly what it is; I have learned a lot. I have learned a lot about the environment even though my background is chemistry. I did not know much about global effects on climate and so forth.

Yoshinaga also noted that he was impressed by Pueschel's ability to motivate the MLO staff to do science:

> He was very animated. He was very interested in the work. He sort of—how can I say it—he built up the enthusiasm for what they were studying at Mauna Loa.

When Yoshinaga first arrived at MLO, Marge Kealanahele was the temporary secretary while Judy Pereira was on maternity leave. He was surprised to see photographs of Judy in the office, and for good reason:

> Judy was my classmate. We went to school together. . . . Intermediate school [and] I went through high school with her. And then also through college, I think she went to college for maybe a year or so. . . . And after I started working for several months . . . Judy came back from her maternity leave and that was when we met. . . . I said, "Hey, you used to sit behind me."

Yoshinaga was to eventually become an MLO old-timer and spend a thirty-five-year career working on the mountain before retiring in 2007.

Geophysical Monitoring for Climatic Change (1972)

In 1972 Mauna Loa Observatory was placed under Geophysical Monitoring for Climatic Change (GMCC), a new program established by NOAA to expand the ongoing tracking of carbon dioxide, aerosols, and solar radiation. GMCC was founded in part by Dr. Lester Machta, who many years earlier had founded NOAA's Air Resources Laboratory (ARL) in Silver Spring, Maryland. GMCC was an amalgam of some ARL offices in Silver Spring and NOAA's Atmospheric Physics and Chemistry Laboratory in Boulder, Colorado. GMCC's first director was ARL's Donald Pack, and the first MLO monthly *Activities Reports* were addressed to Pack.

The *Activities Reports* became a monthly tradition that has continued to the present. From the beginning they were loaded with information about the latest measurements, equipment breakdowns, new projects, weather events, visitors, and so forth. They provided plenty of resources for the production of a new annual NOAA publication: *Geophysical Monitoring for Climatic Change*. The new publication was edited by Dr. John M. Miller, who would become the MLO director in 1976. Contributors to the first issue included a cross section of MLO staff, including Barry Bodhaine, Ronald Fegley, and Bernard Mendonca. GMCC ozone scientists Walter Komhyr and Robert Grass and solar physicist Douglas Hoyt also contributed, as did Sam Oltmans, Thomas Harris, Gary Herbert, and GMCC director Donald Pack. In a foreword to the new publication, Environmental Research Laboratories director Dr. Wilmot N. Hess wrote, "The documentation and preservation of observational data has a special place in science" (Miller 1972). Hess also announced that the publication would be the first in an annual series.

Editor John Miller closed his introduction to the new publication with a relevant quotation from Lt. P. H. Ray (1885), leader of the first US scientific expedition to Point Barrow, Alaska, in 1882–1883:

> . . . for the work of scientific observation in these high latitudes is one of patient endurance on the part of the observer confined as he is, within narrow limits.

The annual *Summary Reports* soon became a coveted resource for detailed information about the findings at GMCC's baseline observatories at Mauna Loa, South Pole, Point Barrow, and Samoa. Twenty-five years after the first volume was published, Dr. David Hofmann (1997) wrote,

> Some things that have not changed include the annual summary report with No. 1 appearing for the year 1972.

There has been considerable demand for these publications as data references. However, with increased use of electronic data dissemination and increased costs of producing a hard copy annual report, the publication became biennial with the 1994–1995 issue, which was also made available on CMDL's home page on the internet [http://www.esrl.noaa.gov/gmd/], and will continue in biennial hardcopy and electronic versions as long as demand is present and resources allow.

Protecting the Mauna Loa Environment

From the time he arrived in Hawai'i, Rudolf Pueschel was highly concerned about protecting the environment and the clean air at the Mauna Loa Observatory. A thick folder in the MLO files labeled "MLO Environment Protection I" includes considerable correspondence related to this issue.

Pueschel was especially concerned about any developments on Mauna Loa that would increase automobile and truck traffic on the mountain, and he asked prominent scientists around the world to join him in protesting plans to install commercial radio and television transmitters near the summit.

In the fall of 1971, Pueschel sent a two-page letter to Sunao Kido, chairman of the Hawai'i Board of Land and Natural Resources, in response to Kido's letter entitled "Conservation District Use Application for FM Radio Transmission Use at Mauna Loa, Hawaii." Pueschel first explained his frustration that the issue remained on the table after his previous protests. He then addressed the specific threats to MLO:

> The attached copies of letters from internationally renowned scientists prove that our concern is shared worldwide. . . . Automotive exhaust contains an amount that is 10 million times the concentration detectable by our sensors [of] carbon dioxide and particulate matter, two of the atmospheric constituents which conceivably affect the earth's climate through their effects on the planet's temperature. . . .
>
> . . . We know enough about Mauna Loa's environment to firmly state that any activity south, that is uphill, of the Observatory will severely jeopardize our activities for the most part of any day. Therefore, *no operation of a transmitter above the Observatory location ever can be tolerated*.
>
> . . . Scientific evidence exists that the operation of a single automobile in the vicinity of Mauna Loa Observatory can impose limitations on its functioning. Such an interference will be multiplied in proportion to the number of cars operated. The problem facing us here is that once permission has been granted to one company, other applicants will follow with certainty. I have been approached in recent months by the representatives of two other companies who intend to broadcast from Mauna Loa.
>
> I wish not to leave the impression that personally I oppose the cultural progress the island of Hawaii will experience from FM reception. I greatly enjoy good music and am the owner of a stereo console and a FM car radio, both of which are useless unless FM reception is provided. The question before the Board and every citizen of Hawaii is, however, whether FM reception depends entirely on the operation of a powerful FM transmitter near the summit of Mauna Loa. Chances are that alternative locations can be found.

On August 5, 1971, Donald H. Pack, deputy director of NOAA's Air Resources Laboratories, wrote Pueschel about his calculations of automobile pollution on Mauna Loa. He proposed a delay in circulating his findings until the calculations were refined by the latest information on automobile emissions. On August 31, 1971, Judy Pereira sent letters on behalf of Pueschel to Hawai'i Senators Daniel K. Inouye and Hiram L. Fong and Representative Patsy Mink requesting that they consent to meeting with Pueschel to discuss the threat to MLO posed by the proposed radio transmitter. Pueschel also planned meetings for September 22–23 with his supervisors at the Environmental Research Laboratories in Boulder.

Pueschel began proposing that a portion of Mauna Loa be officially declared a scientific reserve zone that would be off-limits to commercial traffic and development, an idea still advocated by some today. He demonstrated his concern about the impact of vehicular pollution on the measurements at MLO by acquiring a propane-powered car that emitted less pollution than a conventional car fueled by gasoline. The High Altitude Observatory (HAO) also obtained one. Judy Pereira (video interview with author, Boulder, Colorado, April 27, 2007) recalls the switch from gasoline to propane:

> Rudi thought that it'd be really odd that we work so hard to get commercial power in there in order to be pollution free, and drive vehicles right up to the station with gasoline emissions. So, he worked with the GSA fleet management car people and got two vehicles, two full-sized sedans that were run on propane. We had a huge propane tank in the back of our Cloud Physics Office. Guys would come down the mountain, drive around the building, fill up with propane. That stayed with us until shortly after Rudi left.

The cars were regularly used by the MLO and HAO staffs during their daily trips to the mountain. Mark Goldman (video interview with author, Hilo, Hawai'i, December 4, 2006) recalls that "Unless somebody told you it used propane, I do not think you would know any difference."

Based on their replies to Pueschel, officials with the Hilo Electric Light Company and the Hawaiian Telephone Company were quite surprised when he requested that these companies allow only propane-powered cars and trucks on the mountain. In October 1971, the director of the fledgling Mauna Kea Observatory across the Saddle from Mauna Loa also received a memo from Pueschel, in which he proposed that the new observatory replace its diesel generator with commercial electric power as Mauna Loa Observatory had done in 1966. Pueschel included in his three-page letter detailed calculations of the pollution emitted by the Mauna Kea generator and concluded, "The calculations presented above show that we have a 'borderline' case of interference." While the scientists at Mauna Kea did not agree, they did say they hoped to soon install an electric power line.

Checking the Air (1972)

Rudi Pueschel's concern about maintaining the purity of the air at MLO was vindicated during a pioneering study that he conducted in 1972 with Barry Bodhaine (Plate 17) and Bernard Mendonca (Plate 18). Microscopic aerosols in air were traditionally sampled using filters with very fine pores. Pueschel, Bodhaine, and Mendonca passed sampled air through a tube and heated it while illuminating the air and measuring any changes in the light scattered by whatever was suspended within it. Using this method, they could eliminate organic aerosols by simply heating them until they vaporized and no longer scattered light. Heating also vaporized the water that coats marine aerosols found over the ocean. Thus, these aerosols could be analyzed by measuring changes in the light scattered by the aerosols after the water is removed. In the end, the three investigators made fundamental discoveries about the aerosol content of the marine air at Cape Kumukahi and in the pristine air that flows down the slopes of Mauna Loa each night. They proved that "The pure nighttime air MLO atmosphere indeed provides a natural clean laboratory." And they accomplished this using methods that were much simpler than the former filter method. Their findings were published in the *Journal of Applied Meteorology* (Pueschel et al. 1973).

The Cooling

Rudolf Pueschel's tenure at MLO occurred during a flurry of stories in the press about the cooling of earth and even the prospect of an impending ice age. Following a period of unusually high temperatures during the 1930s, the temperature measured around the Northern Hemisphere began to decline during the 1940s. During the 1960s, the decline had lasted long enough to attract the attention of meteorologists and climatologists, who had no explanation for its cause and worried about its potential consequences. The cooling could not be seen in the temperature records from Honolulu, but it was evident in the plots of temperature at Hilo and several Pacific islands (GISS 2008). A subtle cooling was also evident in the data from MLO.

Was the cooling merely a natural cycle lasting only a few decades, or was the world in the process of leaving behind the temperate interval between ice ages known as the interglacial? Was a new ice age imminent? And why would Earth be cooling when carbon dioxide, the major greenhouse gas after water vapor, was steadily increasing year after year?

J. Murray Mitchell Jr. of the Office of Climatology of the US Weather Bureau in Washington addressed the cooling in his 1961 paper, "Recent Secular Changes of Global Temperature," in which he notes the well-known transient temperature reduction that occurs when aerosols from major volcanic eruptions block sunlight for a few years. Only two years after publication of Mitchell's paper, a major eruption of the Indonesian volcano Agung injected a large volume of sulfur dioxide into the stratosphere, where it eventually formed an aerosol veil over much of the earth. This volcano-induced cooling lasted only a few years, which was much too brief to account for the observed changes in temperature.

Robert A. McCormick was known to MLO, for he visited in May 1965 to give Howard Ellis some freshly calibrated Volz sun photometers for measuring haze. In a provocative paper published in 1967 in *Science,* one of the world's leading scientific journals, McCormick and John H. Ludwig of the National Center for Air Pollution Control offered an intriguing hypothesis for the ongoing cooling. In "Climate Modification by Atmospheric Aerosols," they proposed that haze caused by air pollution provides a plausible explanation for the observed cooling:

> Theoretical considerations and empirical evidence indicate that atmospheric turbidity, a function of aerosol loading, is an important factor in the heat balance of the earth-atmosphere system. Turbidity increase over the past few decades may be primarily responsible for the decrease in worldwide air temperatures since the 1940's.

McCormick and Ludwig further proposed that the cooling resultant from the reduction in sunlight caused by haze might have masked any warming caused by CO_2. They closed their paper by citing Roger Revelle's grand experiment:

> The emission of long-lived aerosol, keeping pace with the accelerated worldwide production of CO_2, (3) may well be leading to the decrease in worldwide air temperature in spite of the apparent buildup of CO_2. In any case, it is clear that in this "large-scale geophysical experiment" in which human beings are engaged (1), the course of atmospheric turbidity must be documented with concern.

These and other papers about aerosols were based on limited observations of the solar dimming effect of aerosols, measurements that were often made in or near cities. Urban areas are often capped by domes of polluted air that include a complex mix of aerosols and gases that reduce sunlight and can cause cooling that somewhat mitigates the urban heat island effect. But were these effects local or global?

James T. Peterson and Reid A. Bryson (1968) of the University of Wisconsin analyzed changes in the intensity of sunlight measured at Mauna Loa from 1957 to 1967 and concluded that the aerosol effect was probably global in nature. "However," they observed, "these observations (on which our report is based) are tentative and subject to future correction." A better answer was needed, and the required data had already been collected at MLO. The time had arrived to implement the "future correction" mentioned by Peterson and Bryson.

Howard Ellis' Atmospheric Transmission Curve (1971)

Very soon after he arrived at MLO, Rudolf Pueschel understood the implications of the sunlight measurements so carefully acquired by Ellis and Pales on the growing concerns about the potential impact of aerosols on climate:

> As I mentioned, one of the records which impressed me when I got here was the solar radiation record, which Howard Ellis diligently acquired over some [13 years].

> I told Howard right away [that the data] are of importance, they have to come out. And Howard told me that he tried but [was unable] to get those data approved by his superiors in Boulder, Colorado. So we got together, and we thought about it and re-wrote the paper and thought a little more about it. And, finally, we had [a] paper . . . which I thought was publishable, and [Judy Pereira typed the draft and] we presented it first to our superiors in Boulder. They accepted it. (Pueschel, video interview with author, November 30, 2007)

In this era of concern about climate change, what happened next is a story worth telling. The sunlight measurement program was begun at MLO by Jack Pales in November 1957, and data from those measurements were used by Peterson and Bryson to blame the cooling on aerosols. However, the results of their analysis, hence their conclusions, were problematic, for they were based solely on the raw voltage output from the Eppley Laboratory pyrheliometer (Plate 19) at four specific times during the morning and again in the afternoon. Any drift in the calibration of the instrument and the consequences of occasionally replacing the instrument were completely unknown. Pueschel knew that much better MLO data were available.

In his memoir, Howard Ellis often refers to people working "on their own" to accomplish things at MLO. Working on his own, Ellis overcame the limitations of the pyrheliometer data by devising an elegantly simple, self-calibrating method for processing the sunlight data. His method extracted from the raw data on the chart recordings the *atmospheric transmission*—the fraction of direct sunlight that penetrates the atmosphere on its way to MLO. The method was self-correcting and preferentially sensitive to aerosols that block sunlight. Solar radiation expert Doug Hoyt (2000) concisely summarized Ellis' method:

> The transmission is monitored by the air mass ratioing technique invented by Howard Ellis. This technique is insensitive to any calibration changes and insensitive to changes in amounts of ozone and water vapor. It is sensitive to changes in aerosol loading, so the results are probably quite robust.

Not everyone was as quick as Hoyt to recognize the elegance of Ellis' method:

> I had a lot of trouble convincing people that I knew Beer's Law better than they thought I did, because I was using the Beer's Law technique for total spectrum measurement [instead of the usual narrow bands of sunlight]. But I found out that by using the least squares method . . . you could extract the transmission, and it sure made a lot of sense. I'm very proud to have Doug Hoyt say that of me. (Ellis, video interview with author, Hilo, Hawai'i, December 2006)

Ellis' method of calculating atmospheric transmission yielded a continuous plot going back to November 1957. It is today the longest time series of the International Geophysical Year legacy measurements from MLO and is proposed here as the Ellis curve, a designation agreed to by MLO founder Robert Simpson, Ellis coauthor Rudolf Pueschel, Ellsworth Dutton, the current keeper of the atmospheric transmission record, and solar expert Doug Hoyt. While Hoyt (personal communication with author, December 16, 2006) is concerned about scientists who seek fame and recognition, he excludes Ellis:

> Howard is not in that crowd and is a modest guy. It seems to me he should be recognized for what he did and people should be aware of it. Having an Ellis curve would do it, but I wish there were another way.

Even today, Judy Pereira and other MLO veterans remember the seemingly endless hours Ellis invested in processing the solar irradiance data used for the atmospheric transmission project. Pereira (video interview, Boulder, Colorado, April 27, 2007) echoes the view of the scientists that the fifty-year MLO atmospheric transmission chart be known as the Ellis curve:

> Yes, because Howard Ellis worked so hard on the solar radiation program and developed this curve that started back [in 1957], I think it would be very appropriate that it be called the Howard Ellis curve, because he's the one that really initiated that program.

How the Ellis curve is generated is rather technical but worth an explanation for readers who are interested. Calibrating optical instruments that measure sunlight is sometimes as much an art as a science, and even instruments calibrated side-by-side may exhibit differences of several percent or more in their response to light when pointed at the sun. Worse is the fact that calibrations often drift over time, and instruments sometimes cease to operate and must be repaired or replaced. Ellis' method did not require calibration. Instead, he looked only at measurements of the intensity of sunlight acquired when the light arriving at the instrument had passed through each of four specific thicknesses of the atmosphere. When the sun is at the zenith position straight overhead, its light passes through one thickness of atmosphere. As explained earlier, this is known as one air mass, which is the reciprocal of the sine of the sun's angle over the horizon (when corrected for atmospheric refraction and the Earth's curvature). Ellis analyzed only the morning data for clear days when the sunlight had traversed 5, 4, 3, and 2 air masses, which occur, respectively, when the sun rises through about 11.5, 14.5, 19.5, and 30 degrees over the horizon. He then computed the average of the ratios of the sunlight intensity at adjacent pairs of air masses: 2:3, 3:4, and 4:5. The result was the transmission in percent of a beam of sunlight through the atmosphere. Since the instrument was compared against itself, the method was self-calibrating. It was also very sensitive to the reduction of direct sunlight caused by dust and other aerosols in the sky.

Publishing the Ellis Curve (1971)

As noted earlier, soon after becoming director of MLO, Rudolf Pueschel met with Howard Ellis (Plate 20) to discuss a response to the global cooling papers, especially the one by Peterson and Bryson that was based on uncalibrated sunlight data from the MLO Eppley pyrheliometer. The instrument had been mounted on a sun tracker to measure direct solar radiation by Jack Pales in November 1957. Ellis and Pueschel classified 410 of 4,473 days (9 percent) from 1957 through 1970 as pristine control days. When the atmospheric transmission on those days was deduced using Ellis' self-calibrating method and plotted on a chart, no statistically significant change was evident aside from the reduction and subsequent recovery caused by the eruption of the Agung volcano on March 17, 1963. Pueschel then used Ellis' findings to draft the final paper. He listed Ellis as senior author due to his long-term work in servicing the instrument, breaking the data from the chart recordings, deriving the atmospheric transmission, and plotting the chart proposed here as the Ellis curve. The paper's abstract draws a line in the sand on the aerosol question:

> Measurements of solar radiation made at Mauna Loa, Hawaii, over a period of 13 years give no evidence that human activities affect atmospheric turbidity on a global scale. Short-term fluctuations in insolation appear to be associated with naturally produced tropospheric aerosols. The intrusion of volcanic dust into the stratosphere results in prolonged increases in atmospheric opacity due to the extended residence times of aerosols in the stratosphere. (Ellis and Pueschel 1971)

After their paper was acceptable to the managers in Boulder, it was time to decide where it should be submitted. Papers about aerosols and global cooling had been published in the prestigious journal *Science*, so Pueschel decided to go straight to the top:

> We chose *Science* as one possibility to get those data out, and it was accepted. The importance of it was not like carbon dioxide and [the] increase in that global pollutant but the absence of an increase in aerosols as a global pollutant. (Pueschel, video interview with author, November 30, 2007)

The editorial staff at *Science* received the paper on January 5, 1971. After Ellis and Pueschel made changes recommended by the reviewers, the revised paper was submitted on March 1 and published in the May 21, 1971, edition of *Science*. The paper specifically acknowledges that, while significant aerosols are found over industrialized regions,

> The occurrence of air pollution, however, to which these turbidity increases can be attributed, is a property of the lower troposphere and thus subject to temporal and spatial variations within short intervals.

In short, Ellis and Pueschel asserted that the air pollution suspected of causing global cooling was regional and not global in nature, a point that would later be clarified in studies of sunlight measurements from other locations using the Ellis ratio method and by satellite observations. Thirty-seven years later, Pueschel said of the atmospheric transmission record that "It's as important as the carbon dioxide trend" (from notes by author of conversation of Pueschel and Howard Ellis at Ellis' home in Hilo, Hawai'i, December 2, 2007). Donald Pack, director of NOAA's Division of Geophysical Monitoring for Climatic Change (now the ESRL), visited Hilo and MLO for the fifteenth anniversary of the observatory on January 9, 1972, and Ellis recalls that he told him the atmospheric transmission paper was "worth a million dollars" to the GMCC research budget.

Bernard Mendonca (personal communication, March 19, 2008) describes another significant spinoff that resulted from Pueschel's collaboration with Ellis:

> When Dr. Rudolf Pueschel arrived at the Observatory as the new director, he immediately recognized the uniqueness and importance of Howard's transmission record and what it could say about the changes in atmospheric transmission with time and he strongly encouraged Howard to get his record published. . . . This new kind of leadership was noticed by everyone on the staff, and we all fell in with Dr. Pueschel to begin our own publications from the MLO data that we were all collecting, with his help and encouragement. Dr. Pueschel encouraged all of us to publish and get the Mauna Loa data out to the scientific community. It was a great time for all of us and a very prolific time for publications of Mauna Loa data. It was a critical turning point for the scientific development and maturity of the observatory and for the staff.

Triggering an Ice Age?

On July 9, 1971, only two months after *Science* published the paper by Ellis and Pueschel refuting a global aerosol blanket, the same journal published a report that predicted severe cooling of the planet and even an ice age should aerosols continue to accumulate in the atmosphere. The authors were S. Ichtiaque Rasool and Stephen H. Schneider of NASA's Institute for Space Studies, now the Goddard Institute for Space Studies (GISS), in New York City. Rasool and Schneider analyzed the effects on climate of both CO_2 and aerosols from a theoretical perspective. They were apparently unaware of the findings by Ellis and Pueschel at MLO when they submitted their paper, for they concluded that sunlight-reducing aerosols presented a far more significant problem than any increase in warming that might be caused by CO_2:

> We will report here on the first results of a calculation in which separate estimates were made of the effects on global temperature of large increases in the amount of CO_2 and dust in the atmosphere. It is found that even an increase by a factor of 8 in the amount of CO_2, which is highly unlikely in the next several thousand years, will produce an increase in the surface temperature of less than 2°K. However, the effect on surface temperature of an increase in the aerosol content of the atmosphere is found to be quite significant. An increase by a factor of 4 in the equilibrium dust concentration in the global atmosphere, which cannot be ruled out as a possibility within the next century, could decrease the mean surface temperature by as much as 3.5°K. If sustained over a period of several years, such a temperature decrease could be sufficient to trigger an ice age! (Rasool and Schneider 1971)

While the paper by Ellis and Pueschel was based on 13 years of carefully processed data, the report by Rasool and Schneider was based on an early global climate model that dismissed a significant role for CO_2 in climate change: "On the short time scale, if CO_2 is augmented by another 10 percent in the next 30 years, the increase in the global temperature may be as small as 0.1°K." Their concern that increasing aerosols might trigger an ice age turned the attention of some climate change researchers to monitoring aerosols, experimentally determining their radiative properties, and modeling their impact on climate.

The paper by Ellis and Pueschel, which had refuted the notion of a global aerosol blanket, went unnoticed as the papers by Bryson, Rasool and Schneider, and others triggered worried articles in the popular press, including the *New York Times* and the *Christian Science Monitor*. *Time* magazine (1974) published a 1,050-word article titled "Another Ice Age?"

> As they review the bizarre and unpredictable weather pattern of the past several years, a growing number of scientists are beginning to suspect that many seemingly contradictory meteorological fluctuations are actually part of a global climatic upheaval. . . . the atmosphere has been growing gradually cooler for the past three decades. The trend shows no indication of reversing. Climatological Cassandras are becoming increasingly apprehensive, for the weather aberrations they are studying may be the harbinger of another ice age.

Lowell Ponte's *The Cooling: Has the Next Ice Age Already Begun?* (1975) includes a preface by Bryson that claims the majority of scientists who believed climate was changing "believe that the longer trend will be downward." The back cover of the book includes an endorsement by Schneider that concludes, "Massive world-wide actions to hedge against that threat deserve immediate consideration."

A recent paper concludes that the global cooling scare was more myth than reality and that from 1965 to 1979, more scientific papers predicted warming (forty-four) than cooling (seven):

> An enduring popular myth suggests that in the 1970s the climate science community was predicting "global cooling" and an "imminent" ice age, an observation frequently used by those who would undermine what climate scientists say today about the prospect of global warming. (Peterson et al. 2008)

Yet the era was indeed characterized by a prolonged temperature downturn that a number of scientists attributed to particulate matter in air pollution. Finding the origin of the cooling was a valid scientific exercise, and Ellis and Pueschel, who made no predictions about future cooling or warming (their paper and its sequels are not cited by Peterson et al.), proved a global aerosol blanket was likely not the reason.

A Voice of Reason

There is great irony in the fact that the discovery of both the increasing concentration of atmospheric carbon dioxide and the stability of atmospheric transmission—and thus the absence of a global aerosol trend—were made at the Mauna Loa Observatory. While the cooling was being measured around the Northern Hemisphere, the MLO data, at least to that point in time, showed that the role played by aerosols was not global. On the other hand, the rise in carbon dioxide was indeed a global phenomenon. There was simply insufficient data to attribute the global cooling phenomenon to aerosols or to assert it might be even more pronounced were it not for increasing carbon dioxide. Climate scientists would have to look elsewhere for an explanation of the cooling.

Lester Machta, director of NOAA's Air Resource Laboratories, was well aware of the complexity of these factors and how they could be superimposed on natural warming and cooling cycles. In a paper published in the *Bulletin of the American Meteorological Society*, Machta (1972) discusses the difficulty of making climate predictions "due to increased man-made dustiness":

> It may not now be possible to state unequivocally whether added dust in the lower atmosphere will warm or cool the ground-level air on a global scale. It is true that estimates have been made, mostly leading to cooling, but I urge that such predictions be treated with caution.

Machta's sage advice was proved correct, for the ice age scare would eventually be left in the dust as the era of cooling came to an end. The *New York Times* eventually placed its stamp of approval on the aerosol observations at Mauna Loa in the form of an article by science reporter Walter Sullivan, who first visited MLO on September 4, 1961. Sullivan returned on June 29, 1978, and John Miller (personal communication with author, April 23, 2008) recalls driving him to MLO. Sullivan wrote in the *Visitors* book, "Much changed since 1961!" Sullivan's article, "Studies Hint Dust in Air Won't Curb Rising Temperature; Little Pollution Recorded," which was published on August 9, 1978, announced to a much wider audience than the Ellis-Pueschel paper in *Science* that no evidence of a global aerosol blanket had been found over Hawai'i:

> MAUNA LOA OBSERVATORY, Hawaii—Although observations here document a long-term rise in carbon dioxide content of the atmosphere presumably caused by worldwide fuel burning, there is no evidence of a parallel rise in dust pollution that many climatologists have thought would temper the carbon dioxide's warming effect.

The reason or reasons for the cooling of the 1960s–1970s remain uncertain. The Intergovernmental Panel on Climate Change concluded that "Reported decreases in solar radiation at the earth's surface from 1970 to 1990 have an urban bias," a conclusion seemingly in agreement with the findings at MLO (IPCC 2007). Yet in a 2007 interview conducted by NASA writer David Herring, climate modeler James Hansen, who was acknowledged by Rasool and Schneider (1971) for his contributions to their paper in *Science*, suggested that both natural cycles and anthropogenic aerosols may have been responsible:

> If greenhouse gases are to blame then why did Earth's average temperature cool from 1940–1970? And why has the rate of global warming accelerated since 1978? [James] Hansen's answers to these questions brought him full circle to where he began his investigation more than 40 years ago.
>
> "I think the cooling that Earth experienced through the middle of the twentieth century was due in part to natural variability," he said. "But there's another factor made by humans which probably contributed, and could even be the dominant cause: aerosols."
>
> . . . .
>
> After World War II, the industrial economies of Europe and the United States were revving up to a level of productivity the world had never seen before. To power this large-scale expansion of industry, Europeans and Americans burned an enormous quantity of fossil fuels (coal, oil, and natural gas). In addition to carbon dioxide, burning fossil fuel produces particulate matter—including soot and light-colored sulfate aerosols. Hansen suspects the relatively sudden, massive output of aerosols from industries and power plants contributed to the global cooling trend from 1940–1970.
>
> "That's my suggestion, though it's still not proven," [Hansen] said. (Herring 2007)

While urban aerosols may have played a role in regional cooling, Pueschel defends the conclusion he and Ellis reached that the MLO data showed there was no global aerosol blanket:

> It was postulated by quite popular, famous scientists and politicians who said you could compensate the warming potential due to carbon dioxide and the greenhouse gases by an increased layer of aerosols to block solar radiation. This concept is unacceptable, because it is very hard to control the effect of aerosols. It is a short-lived substance; it is very sporadic; it is very hard to control. And I found out through the grapevine not too long ago, some scientists admitted NASA aims to discuss just this possibility again. I was not involved so I cannot tell you anything else, but I would be [strongly] against it. (Pueschel, video interview with author, November 30, 2007)

Beginning soon after publication of the Ellis-Pueschel paper, Glenn Shaw of the University of Alaska made many visits to MLO to measure aerosols over the observatory, research that led to his discovery of Asian dust drifting from Asia to North America. He agrees with Pueschel about the possible role of aerosols during the cooling:

> I don't believe there is indication that aerosols were thicker in the 1940s through '60s. Maybe in the northeast USA and in LA [Los Angeles], because these places were cleaned up a little after passage of the Clean Air Act. But these are a paltry contribution to the planet's aerosol load. . . . But just in general, I think there is considerably more aerosol pollution now than in the 1940s. . . . If [scientists] said that aerosols were responsible for this cooling, I think this would just be utterly wrong. I would bet a great sum of money on it. (Shaw, personal communication with author, January 26, 2008)

Shortly after Sullivan's 1978 report in the *New York Times*, Bernard Mendonca, Kirby Hanson, and John DeLuisi of NOAA's Air Resources Laboratories in Boulder, MLO's parent organization, updated the 1971 Ellis-Pueschel paper in *Science* with a new paper describing twenty years of atmospheric transmission results at MLO (1978). Like Ellis and Pueschel, they found significant volcano effects in the data but no obvious clues about manmade air pollution. Mendonca (personal communication with author, March 19, 2008) reflects on the significance of the paper:

> It revived the interest in the transmission record and documented what explosive volcanic eruptions could do to the atmospheric transmission even years after such explosive eruptions occurred. After I moved on into a more supervisory role in the laboratory, Ellsworth Dutton followed up researching this unique and valuable record with more publications using new data that became available from our four observatories after I had established a new hand-held transmission measurement program in conjunction with the Dobson measurements at all of the four NOAA stations at Barrow, Mauna Loa, Samoa, and the South Pole. These measurements continue to this day to document and characterize the sporadic injections of volcanic dust veils high into the stratosphere as they spread globally. This data record, along with other measurements from satellites and other ground networks, was used to compute the temporary cooling of the planet that occurred after major volcanic eruptions ejected fine particles into the stratosphere as they spread poleward. [The data also showed] how such coolings could temporarily counter greenhouse warming until the [particles] fell to earth or were chemically scrubbed from the stratosphere.

Meanwhile, back in 1971 it was science as usual at the Mauna Loa Observatory, which continued its 13-year series of measurements of both aerosols and CO_2 along with dozens of other observations. Rudolf Pueschel continued his drive to protect MLO from encroachment. Howard Ellis continued the laborious and tedious process of extracting atmospheric transmission data from lines traced on charts by the sun-tracking pyrheliometer, unaware that thirty-seven years later he would visit the observatory and see a wall chart depicting half a century of atmospheric transmission measurements extracted from the raw solar data using the method he had devised (Plate 21).

Fifteenth Anniversary Open House (1972)

On January 9, 1972, MLO celebrated its fifteenth anniversary with an open house. Around two hundred visitors were present, and some sixty-two of them signed the *Visitors* book. The guests were welcomed by Donald Pack, director of NOAA's Division of Geophysical Monitoring for Climatic Change. MLO director Rudolf Pueschel and staff members Howard Ellis, Bernard Mendonca, John Chin, and Barry Bodhaine were on hand to answer questions and explain the operation of the instruments that monitored sunlight, carbon dioxide, ozone, and other atmospheric parameters. Charles Fullerton of the Cloud Physics Observatory was also present.

The Visitors Book (1972)

The *Visitors* book provides a unique historical record of MLO's many visitors. Judy Pereira took a second six-month maternity leave on April 17, 1972, to give birth to her daughter Wendy, and Marge Kealanahele filled in during her absence. During a visit to MLO on May 19, 1972, Marge signed the *Visitors* book and entered a phrase in Hawaiian that she translated as "Instruments of life!" Her husband, the Reverend E. G. Kealanahele, also signed in and added "*Malia ka la*," which he translated as "Clear day." When Marge signed in again on September 16, she wrote, "It's another beautiful day—*he ui o ka la!*"

On March 10, 1972, NOAA ozone specialist Walter Komhyr signed in, and he was followed by Bob Grass on March 14. Their purpose was to install the Dobson spectrophotometer that measured the ozone layer from inside the newly built dome facility that replaced the wood shed previously used to house the Dobson.

Sam and Kay Oltmans arrived from Boulder on May 30, 1972, to calibrate the newly installed Dobson spectrophotometer on clean, clear mornings. Their names joined the short list of married couples who had previously stayed at MLO. At the time, Walter Komhyr was refining NOAA's Dobson network that measured the ozone layer over the United States, but his Langley calibrations of a Dobson in Virginia were unsatisfactory due to the poor sky conditions. Oltmans, who had been an Air Force meteorologist for the Strategic Air Command, was then a graduate student at the University of Colorado, and Komhyr asked if he would be willing to go to MLO and calibrate Instrument 83 (I83), a Dobson that Komhyr hoped could become a standard against which other Dobsons could be compared. Julie London, Oltmans' faculty advisor who had referred him to Komhyr, liked the idea, and the Oltmans headed for Hawai'i on an adventure they would never forget:

> We stayed up at the observatory. We'd go up Monday morning with the crew from Mauna Loa Observatory and come back Friday afternoon, and we lived up at the Observatory during the week so we could get up . . . early in the morning to do the Langley plot measurements, which essentially are measuring ozone as the sun rises through the day. . . . It wasn't a particularly good year; we had quite a few cirrus clouds, and that sort of thing, but we . . . got a good sample of measurements [and] were able to establish a strong calibration for Instrument 83. Subsequent to that, Instrument 83 was established as the world standard instrument for Dobsons. My wife and I lived in what's now called the Keeling Building. It was the main observatory building at that time, still had facilities for people to stay overnight. There was a bunkroom and a kitchen in the building so you could stay overnight there. . . . It was relatively primitive living circumstances up there. Some days in the afternoon when it would cloud up and we couldn't make measurements, we would hike around the trail up to the summit. (Oltmans, video interview with author, Boulder, Colorado, April 27, 2007)

While the Oltmanses were staying at MLO, sun photometer pioneer Frederick Volz visited from June 3 to 6, 1972, after attending a conference in Japan. MLO is famous for its clear skies, but only around 10 percent of the days are pristine. Volz visited during one of the cloudy periods that interfered with Oltmans' Dobson calibrations and wrote in the *Visitors* book, "Wishing unsuccessfully for a clear sky." Solar and climate expert Douglas Hoyt visited on June 23. Sam and Kay Oltmans stayed until June 27 and flew home wearing Mauna Loa sunburns.

MLO Director Ronald Fegley (1972–1976)

Dr. Rudolf Pueschel left MLO for the Atmospheric Physics and Chemistry Laboratory in Boulder on August 16, 1972. In his last MLO bimonthly Activity Report, he welcomed Dr. Mark Goldman, Alan Yoshinaga, and his successor, Dr. Ronald Fegley. He also summarized the major developments during his tenure, which shows he did not let his campaign against possible encroachment on the observatory by commercial broadcasters interfere with his scientific duties:

> The MLO budget increased by 85%; the number of positions increased from 4 full time and one post doc appointee to 7 full time, one part time and two post doctoral fellows. Floor space [at the Cloud Physics Observatory] was more than doubled. The scientific output in the past two years is reflected in three publications,

> two papers accepted for publication, three papers in the review process, and three more papers in preparation. (Pueschel 1972)

Physicist Ronald Fegley earned his doctorate from the University of California in 1970 and spent two years at the Wave Propagation Laboratory in Boulder prior to his MLO assignment. Fegley went through an orientation period at MLO beginning in June 1972 and succeeded Pueschel as director of MLO in August. Soon thereafter, he stood with his new staff outside the Cloud Physics Observatory for a group photograph (Fig. 10.6).

Dr. Fegley (1972a) expressed his appreciation to Dr. Pueschel and the "excellent condition of the Observatory" in his first MLO Activity Report. In September, Fegley and Charles Garcia of the Mauna Loa High Altitude Observatory visited the Kūlani Prison Honor Camp to have lunch with the warden and discuss maintaining the MLO road (Fegley 1972b). Though the MLO staff stopped driving through Kūlani after the 1964 road to Saddle Road was constructed, the prison was still actively involved in maintaining the road. Judy Pereira (video interview with author, Boulder, Colorado, April 27, 2007) recalls the importance of the prisoners and how the arrangement eventually came to an end:

> They enjoyed getting out of where they were to work on the road. The road gang . . . became friendly with our people, and we even had a big party for them when the roadwork was done. And, actually, the roadwork never was done, but their commitment to us was over, because the state was concerned, and the US government was concerned, that there might be liability problems. The state owned the mountain road. They didn't want anything to happen; they would be responsible. We had to let go of the prisoner labor. After that, it was, like, doing pieces of road at a time just to keep it so that we could get up to the observatory.

Bernard Mendonca (video interview, Boulder, Colorado, April 27, 2007) remembers a colorful side of the collaboration with the prisoners:

> As only Judy could do, [she organized] a kind of a luau up at the station at the end, and we'd all be over there on the tables . . . [eating and] having a good time. No alcohol, of course. And, so, we'd start the conversation, "Thanks a lot" . . . and "What are you in for?" "Oh, I beat my wife to death." . . . But these were real people, and . . . we wanted to reward these guys, to tell the community about the good work they'd done. So I said, . . . I'll take their pictures and everything, and we'll have it in newspapers and tell them [about] all the good work. This big, old prisoner guy said, "I don't want my picture in a paper! What do you think I am?" So that was the end of that idea. They didn't want people to know that they were up there in the prison.

Wilmot Hess, director of NOAA's Environmental Research Laboratories, visited MLO on October 12, 1972, and posed for a photograph with Ronald Fegley, interim secretary Marge Kealanahele, and secretary Judy Pereira, who had just returned from a five-month maternity leave (Fig. 10.7).

Fegley's wife Susan signed in at MLO on October 14, 1972, during an open house celebration of NOAA's second birthday on a beautiful sunny day. John Chin organized the event, and he was assisted by Marge Kealanahele, who had completed her employment as interim secretary the previous day. Marge wrote in the *Visitors* book, "Aloha MLO! Mahalo for a beautiful view." Susan wrote, "Nice hail storm—almost like Colorado!" While their comments seem to be describing different days, the weather can change very quickly at MLO when the inversion layer rises over the observatory. In this case, a thunderstorm suddenly arrived and sprayed millions of ice pellets across the lava. Fegley (1972b) described the lightning, hail, and rain that arrived that day in the bimonthly Activity Report.

The First MLO Lidar (1972)

Ronald Fegley was sent to MLO with a specific assignment. No major new experiments had been introduced since the International Geophysical Year, when MLO became fully operational in July 1957, and Fegley's assignment was to

FIGURE 10.6. The Mauna Loa Observatory staff in the summer of 1972 included *(standing, left to right)* Howard Ellis, John Chin, Marge Kealanahele, Alan Yoshinaga, Ronald Fegley (director), and *(kneeling)* Mark Goldman and Al Shibata. Not pictured is Judy Pereira, who was on maternity leave. *(MLO archive.)*

begin operation of a new apparatus that probed the stratosphere with brilliant pulses of coherent red light emitted by a laser made from a rod of synthetic ruby. Any aerosols high in the sky scattered some of the laser light back toward the surface, and a telescope mirror adjacent to the laser intercepted some of the light and focused it onto a highly sensitive photomultiplier tube. The resulting signal was painted on the screen of an oscilloscope as an electronic diagram that showed a cross section of the layers of dust and other aerosols high in the sky. All this took place

FIGURE 10.7. The Mauna Loa Observatory office staff on October 12, 1972, included *(left to right)* MLO director Ronald Fegley, interim secretary Marge Kealanahele, Wilmot Hess (director, NOAA Environmental Research Laboratories), and secretary Judy Pereira. *(MLO archive.)*

only when the system worked as designed, which was easier to accomplish back in the labs in Boulder than on a remote mountain in Hawai'i.

Laser atmospheric probes of this nature are known as LIDARs (or lidars, an acronym for light detection and ranging), and the one at MLO was designed and built by Earl A. Barrett and Dennis Wellman of the Atmospheric Physics and Chemistry Laboratory while Rudolf Pueschel was director of MLO. On February 23, 1972, Dr. Wellman and Richard Proulx signed it at MLO, with Wellman adding "To install lidar for MLO." Wellman returned on April 27, and wrote after his signature in the *Visitors* book, "To finish lidar installation (I hope!!)." On July 13, Wellman returned still again, this time accompanied by Thomas J. Kiyama of Apollo Lasers, the company that built the lidar system's powerful ruby laser. The frustrated Barrett wrote in the book, "3rd try at lidar."

Meanwhile, Ron Fegley began a two-week orientation on June 26 and was appointed as the new MLO director in July "because of his lidar expertise so that he could take over the project and develop the program" (Bodhaine, personal communication with author, May 3, 2008). Donald Pack visited the Hilo office during Fegley's orientation on June 28, and by August, Fegley was in charge of MLO.

Fegley's work led to a sophisticated atmospheric sounding program that remains a major program at MLO today. But the work was not easy, for lidar was a comparatively new technology that was less than a decade old. Fegley selected Howard Ellis to be his primary lidar assistant, and the two had their hands full keeping the lidar working properly. The ruby laser's mirrors required very careful alignment to be made perfectly parallel with one another so that optimum beam quality and power output could be achieved. The ruby rod and the mirrors at its two ends that provided the optical feedback for laser action to occur were damaged or even destroyed if the laser was allowed to emit excessive power. Experienced ruby laser operators (including the author of this book) were well aware that excessive power from the laser could blow small chips off the polished face at the end of the ruby rod or even propel the end of the expensive rod across a room. The powerful flash lamp could also be destroyed, as in January 1974 when "the flashlamp in the ruby laser disintegrated" (Fegley 1974a). Obtaining spare parts was often time consuming, and in this instance two weeks passed before the lidar was back in service.

The laser was invented only twelve years before the early lidar work at MLO, which was not nearly as advanced as the lidar research performed there today by MLO station chief John Barnes. The first MLO lidar is described in detail in *Geophysical Monitoring for Climatic Change* (Miller 1972), the new annual publication of the Environmental Research Laboratories. According to the report, "The lidar system was not really operational in 1972. The data presented here are from 1973 and are included in this annual report to illustrate the method." The "method" was quite primitive by comparison with the current lidar, for the data were in the form of Polaroid photographs of an oscilloscope trace:

> The photos are now being analyzed manually, although this will soon be done by a computer. Each trace is [manually] digitized at 4-usec intervals using a ruler. The upper and lower edges of the trace are read and the result averaged.

Howard Ellis (video interview, Hilo, Hawai'i, December 2, 2006) reflects on how the data were extracted from the photographs in the same way data were broken from chart recordings and without the help of a modern computer and sophisticated data analysis software:

> We would take a Polaroid picture automatically of an oscilloscope [screen] so it would show the trace of the lidar beam as it went up. So if I had had a traveling microscope, I could have done better, but I just had to measure [the trace on the oscilloscope] with an ordinary ruler and a handheld glass.

NOAA Begins to Measure Carbon Dioxide (1974)

In April 1974, a URAS-2 nondispersive infrared CO_2 analyzer manufactured by Hartmann and Braun was taken to MLO by John Chin. Also taken to the mountain was a water vapor freeze out trap, a chart recorder, air pumps, and 2,000 feet (610 meters) of aluminum tubing. The decision to use the URAS-2 was apparently not arbitrary, for the same model analyzer had been used to monitor atmospheric CO_2 in Germany since 1972 (Schmidt et al. 2003).

During May, Tom Harris and John Chin installed the URAS-2 system in the main building at MLO. They also installed three aluminum tubing air lines for sampling air at three different points away from the observatory buildings. In an initial comparison of CO_2 measured simultaneously by the aging Scripps APC and the new NOAA URAS analyzers, the URAS results were approximately 0.13 to 0.36 parts per million (ppm) lower than the APC results (Chin 1974). Because the total CO_2 concentration at MLO that month was 333.16 ppm (Keeling et al. 2001), the agreement between the two instruments was within a tenth of a percent. This began a remarkable comparison of measurements by two independent systems of the concentration of carbon dioxide in the air at the observatory that had earned world-class status for the high quality of its findings. The simultaneous Scripps–NOAA measurements were to become a long-term comparison that would continue to the present and provide high scientific credibility for both programs.

Bumps in the Road

Judy Pereira (1996) observes in her memoir that Ron Fegley's tenure at MLO wasn't accompanied by the budget increases that arrived with Lothar Ruhnke and Rudi Pueschel:

> Ron's appointment at MLO came at a time when budget problems once again stifled the growth of the observatory. A few new pieces of equipment and our first HP calculator (which then cost us $2000!) were purchased.

The Hewlett-Packard electronic calculator was a boon to Howard Ellis' atmospheric transmission work. For a decade he had relied on a clumsy mechanical calculator to do the complex calculations.

Another bump in the road occurred in early 1974 when MLO's propane car was wrecked (Fegley 1974a). Then there was a severe snowstorm on March 16–17, 1974, that caused a variety of problems and left behind unwanted condensation inside the dome of an ultraviolet meter. And there was some friction between Fegley and some of the more experienced staff members. Judy Pereira (1996), who recalls Fegley as "very much a gentleman," describes his four years as "bumpy, to say the least." Bernard Mendonca (video interview, Boulder, Colorado, April 27, 2007) agrees with Pereira's assessment: "It got to the point that some of the staff were so exasperated we concocted a scheme where we all applied for jobs elsewhere."

Howard Ellis got along well with Ron Fegley, and the two often went to MLO together for lidar sessions. In a telephone interview with the author on June 7, 2008, Fegley recalled how they would leave Hilo in the late afternoon and return late at night after the lidar session. Fegley and Ellis wrote several reports and papers about their lidar findings, which became the first of many lidar papers that have emerged from MLO over the years. Of special interest was the detection by the lidar of layers of volcanic material in the stratosphere from the eruption of Fuego in Guatemala in 1974. Fegley recalls this as a highlight of his time at MLO:

> We were one of the first facilities to make regular measurements of the atmosphere using lidar. We began to understand [aerosols in] the stratosphere.

Fegley was a pilot who enjoyed flying in his spare time, and Judy Pereira (1996) recalls the time her husband flew with him:

> Ron was an avid flyer and owned a percentage of a Cessna airplane that he flew every chance he got. Although my husband Jerry enjoyed flying with him, I never went, and told him I didn't care how good he was I wasn't getting into any toy airplane to go flying!

Airborne Bacteria (1974)

Fegley's monthly Activity Reports provided a close-up view of the wide variety of science activities at MLO, including an early project to sample airborne bacteria. In March 1974, Bruce Lighthart of the National Ecological Research Lab of the US Environmental Protection Agency sent a bacteria sampler apparatus for a three-month test to determine if trends could be found in the presence of bacteria in the air at MLO (Fegley 1974a). The first sample was taken on April 16 and yielded only two colonies in two liters of clean, downslope air. Only three samples were successfully collected:

> These have typically produced concentrations in the neighborhood of 1 to 10 bacteria per cubic meter of air sample. We do not have enough samples yet to systematize the results, but we are learning about the problems of taking bacteria samples in a clean air location like Mauna Loa. (Fegley 1974b)

Skirmishing over Carbon Dioxide (1974)

Charles Keeling's carbon dioxide program at the Scripps Institution of Oceanography had been threatened with a string of potentially fatal budget cuts almost since the beginning. The quality of Keeling's science was never in question. The major issues involved the grant-making policy of the National Science Foundation (NSF) and whether or not NOAA should assume the major responsibility for routine CO_2 monitoring, particularly since NOAA had begun measuring CO_2 in 1974.

A key justification required for work sponsored by the NSF was that a project must be based on pioneering research and not routine data collection, and in the eyes of the NSF and NOAA, Keeling's CO_2 program had long since passed from new to routine. Keeling passionately disagreed, and he related his perspective on the battle to keep his CO_2 program alive in *Rewards and Penalties of Monitoring the Earth* (1998), a candid memoir that provides fascinating details about how and why the study and precise measurement of carbon dioxide became his lifelong passion. Keeling was especially concerned about long-term data reliability, and he could not abide scientists he considered novices who had little background in measuring trace gases. He eventually developed a reputation for firmly and successfully challenging government officials whom he felt wanted to close him out of the CO_2 monitoring loop.

As noted earlier, even before the CO_2 program was begun at MLO in 1958, the Weather Bureau had considered beginning its own measurements, and later ESSA had considered taking over the CO_2 program at MLO and expanding the program elsewhere. Keeling was well aware of the potential threat this placed on his work, and he writes about it in his memoir (1998):

> Lester Machta, following a meeting in 1966 which I attended, laid plans for the US Weather Bureau to begin a Bureau-run program to measure atmospheric CO_2 at fixed locations.

When, under Machta's leadership, NOAA finally began making its own measurements of carbon dioxide at MLO on May 17, 1974, NOAA continued to conduct Keeling's program on the mountain at no cost to Scripps. But Keeling was unhappy about NOAA's new CO_2 analyzer, for he had yet to fully address a major new question involving the proper mixture of reference gas for his own program. The issue was triggered by differences in CO_2 concentrations that were measured when the reference gas standards used CO_2 diluted with nitrogen and those in which air was used.

There had been instances of tension between the Scripps Institution of Oceanography and the Weather Bureau scientific leadership during the beginning of the CO_2 program in 1957–1958, and there was friction between Charles Keeling and Howard Ellis in 1964–1966. By 1974 the trouble had returned to a much higher level, and Keeling began a fight for his program that would last for years to come. Physicist and science historian Spencer R. Weart, director of the Center for History of Physics of the American Institute of Physics in College Park, Maryland, is author of *The Discovery of Global Warming* (Harvard University Press, 2003). Weart (2007) has produced a 250,000-word online supplement to his book that includes a paragraph that sums up the tension that enveloped Keeling and the agencies that funded his pioneering carbon dioxide research:

> By 1970, severe cuts again threatened [Keeling's] program. This weak position was further weakened, as so often in human affairs, by personality issues. Some officials would have preferred to pursue CO_2 monitoring without Keeling himself. "Keeling's a peculiar guy," Revelle later remarked. "He wants to measure CO_2 in his belly. . . . And he wants to measure it with the greatest precision and the greatest accuracy he possibly can." [Roger Revelle, interview by Earl Droessler, February 1989, American Institute of Physics.] As he stubbornly sought more money to improve his results and complained angrily about shortfalls, Keeling struck some people as a demanding "know-it-all." But when it came to monitoring CO_2, Keeling really did know it all better than anyone else. There was nobody so reliably skilled nor so dedicated. Although a few other scientists in the United States and around the world also undertook studies of CO_2 levels, in the end Keeling and his staff at Scripps kept control of the core of the monitoring work.

A few weeks after NOAA began making its own measurements of CO_2, Ralph Stair returned to MLO and spent a week in the building he designed in 1955. The *Visitors* book showed that he stayed from April 30 to May 7, 1974. After his signature he wrote, "Much change for the better since my first trip in 1955 (only lava)." Bernard Mendonca (1974) includes in his part of the monthly Activity Report, "Talked to Dr. Ralph Stair regarding solar radiation work he did at MLO during his recent visit here" (Mendonca 1974).

Mauna Loa Awakens (1975)

The first weekend of 1975 was marked by the heaviest snowfall on record for the observatory. On January 9, the snow at MLO was 15 inches (38 cm) deep, and vehicular access was not returned until January 11 (Mendonca 1975a).

A much more significant natural event occurred on July 5, 1975, when Mauna Loa awoke and staged its first eruption in a quarter of a century. A newspaper headline assured Hilo citizens there was no immediate danger, at least to the population:

> A paved access road to several scientific installations at Mauna Loa Observatory was threatened with an overflow of lava from Pohaku Hanalei, an old crater which appeared to be the source of molten rock after an initial outburst at the summit.
>
> The observatory road, reached via the Saddle Road, which weaves between Mauna Loa and Mauna Kea, was roadblocked by police through last night. . . . As a precautionary measure, Hawaii Electric Light Co. shortly after dawn cut its power to the Mauna Loa Observatory, which essentially suspended operations because the fumes and smoke from the eruption had completely ruined the high altitude pollution experiments carried on there. (Clark 1975)

Fortunately, the eruption lasted only one day. But it was a harbinger of things to come. Bernard Mendonca (1975b), who was acting director while Ron Fegley was on vacation, added a "Special Note" about the eruption in the July Activity Report that noted that MLO was undamaged.

The MLO Office Space Battle Begins (1975)

While Charles Keeling was fighting to keep his carbon dioxide program going at Scripps, MLO was about to become unwillingly engaged in what came to be known as the office space war with the University of Hawai'i at Hilo. On December 10, 1975, W. E. Alexander of the Procurement and Management Office of the university sent a letter to the National Weather Service Pacific Region in Honolulu, which began as follows:

> In accordance with the terms and conditions of U.S. Government Contract No. N-28-0002-72(N) with the University of Hawaii, we hereby advise you of the University's intent to terminate its lease with you effective July 1, 1976.

The letter was apparently sent to the National Weather Service and not forwarded to MLO for more than a month, and Ronald Fegley was caught by surprise when it finally arrived. MLO was about to be evicted from its Hilo offices at the Cloud Physics Observatory, and there was no place to move.

References

Bodhaine, B. A., and R. F. Pueschel. 1974. Source of seasonal variations in solar radiation at Mauna Loa. *Journal of the Atmospheric Sciences* 31: 840–845.

Chin, John. 1974. May activity report, June 6. In *Geophysical Monitoring for Climatic Change,* Mauna Loa Observatory, Monthly Reports, Hilo, Hawai'i, 1974–1975.

———. 2001. Unpublished retirement scrapbook, prepared by Leslie Pajo. Copy in author's files.

Clark, Hugh. 1975. No immediate danger, officials say. *Honolulu Advertiser,* July 7.

Cobb, William E. 1967. Evidence of a solar influence on the atmospheric electric elements at Mauna Loa Observatory. *Monthly Weather Review* 95: 905–911.

CO_2 LOG. 1960–1967. Unpublished.

Ellis, Howard T. 1966. Letter to Helmut K. Weickmann (March). In MLO archive.

Ellis, Howard T., and Rudolf F. Pueschel. 1971. Solar radiation: Absence of air pollution trends at Mauna Loa. *Science* 172: 845–846.

Fegley, Ronald W. 1972a. July–August 1972 activity report, September 15. In *Geophysical Monitoring for Climatic Change,* Mauna Loa Observatory, Monthly Reports, Hilo, Hawai'i, 1971–1973.

———. 1972b. September–October 1972 activity report, November 15. In *Geophysical Monitoring for Climatic Change,* Mauna Loa Observatory, Monthly Reports, Hilo, Hawaii, 1971–1973.

———. 1974a. March 1974 activity report, April 11. In *Geophysical Monitoring for Climatic Change,* Mauna Loa Observatory, Monthly Reports, Hilo, Hawai'i, 1974–1975.

———. 1974b. June 1974 activity report, July 11. In *Geophysical Monitoring for Climatic Change,* Mauna Loa Observatory, Monthly Reports, Hilo, Hawai'i, 1974–1975.

Fullerton, Charles M. 2001. "Some thoughts on my association with the University of Hawai'i of Hilo." In Frank T. Inouye and Edward J. Kormondy, eds., *The University of Hawai'i–Hilo: A College in the Making,* Honolulu: University of Hawai'i Press.

GISS (Goddard Institute for Space Studies). 2008. GISS surface temperature analysis, station list search: 19.8° N,157.0° W. (Accessed at http://data.giss.nasa.gov/cgi-bin/gistemp/findstation.py.)

GSA. 1981. Transcript/statement of federal service for Howard T. Ellis. General Services Administration. Cited with permission of Ellis.

Herring, David. 2007. Earth's temperature tracker: NASA Earth Observatory, November. (Accessed at http://www.giss.nasa.gov/research/features/temptracker/.)

Hofmann, David. 1997. Climate monitoring & diagnostics laboratory: Science review, Mauna Loa Observatory. 40th Anniversary Commemoration and Dedication of the NDSC Building, November 20–21, Hilo, Hawai'i.

Hoyt, Douglas V. 2000. E-mail to John L. Daly, March 4. (Accessed at http://www.john-daly.com/nuke-em.htm#hoyt2.)

IPCC. 2007. *Climate Change 2007: The Physical Science Basis.* Technical Summary 41. Contribution of Working Group I to the Fourth Assessment Report of the Intergovernmental Panel on Climate Change [S. Solomon, D. Qin, M. Manning, Z. Chen, M. Marquis, K. B. Averyt, M. Tignor, and H. L. Miller, eds.]. Cambridge, UK, and New York: Cambridge University Press.

Keeling, Charles D. 1998. Rewards and penalties of monitoring the Earth. *Annual Review of Energy and the Environment* 23: 25–82. (Accessed at scrippsco2.ucsd.edu/publications/keeling_autobiography.pdf.)

Keeling, Charles D., Robert B. Bacastow, Arnold E. Bainbridge, Carl A. Ekdahl Jr., Peter R. Guenther, Lee S. Waterman, and John F. S. Chin. 1976. Atmospheric carbon dioxide variations at Mauna Loa, Hawaii. *Tellus* 28: 538–551.

Keeling, C. D., S. C. Piper, R. B. Bacastow, M. Wahlen, T. P. Whorf, M. Heimann, and H. A. Meijer. 2001. Exchanges of atmospheric CO_2 and $13CO_2$ with the terrestrial biosphere and oceans from 1978 to 2000. I. Global aspects, SIO Reference Series, No. 01-06. San Diego: Scripps Institution of Oceanography. (Data accessed at http://scrippsco2.ucsd.edu/data/data.html.)

Lockwood, J. P., and P. W. Lipman. 1987. Holocene eruptive history of Mauna Loa Volcano. In R. W. Decker, T. L. Wright, and P. H. Stauffer, eds., *Volcanism in Hawaii.* U.S. Geological Survey Professional Paper 1350, v. 1, 518. (Summary accessed at http://hvo.wr.usgs.gov/maunaloa/history/historytable.html.)

Macdonald, G. A., A. T. Abbott, and F. L. Peterson. 1986. *Volcanoes in the Sea: The Geology of Hawaii,* 2nd ed. Honolulu: University of Hawai'i Press, 80–81.

(Accessed in revised form at http://hvo.wr.usgs.gov/kilauea/history/historytable.html.)

Machta, Lester. 1972. Mauna Loa and global trends in air quality. *Bulletin of the American Meteorological Society* 53: 402–420.

McCormick, R. A., and J. H. Ludwig. 1967. Climate modification by atmospheric aerosols. *Science* 156: 1358.

Mendonca, Bernard G. 1969. Local wind circulation on the slopes of Mauna Loa. *Journal of Applied Meteorology* 8: 533–541.

———. 1974. May activity report, June 6. In *Geophysical Monitoring for Climatic Change,* Mauna Loa Observatory, Monthly Reports, Hilo, Hawai'i, 1974–1975, Air Resources Laboratories, NOAA.

———. 1975a. January activity report, February 7. In *Geophysical Monitoring for Climatic Change,* Mauna Loa Observatory, Monthly Reports, Hilo, Hawai'i, 1974–1975, Air Resources Laboratories, NOAA.

———. 1975b. July activity report, August 8. In *Geophysical Monitoring for Climatic Change,* Mauna Loa Observatory, Monthly Reports, Hilo, Hawai'i, 1974–1975, Air Resources Laboratories, NOAA.

Mendonca, Bernard G., Kirby J. Hanson, and John J. DeLuisi. 1978. Volcanically related secular trends in atmospheric transmission at Mauna Loa Observatory, Hawaii. *Science* 202: 513–515 (November 3). (Accessed at http://www.sciencemag.org/cgi/reprint/202/4367/513.pdf?ck=nck.)

Mendonca, Bernard G., and Wayne T. Iwaoka. 1969. The trade wind inversion at the slopes of Mauna Loa, Hawaii. *Journal of Applied Meteorology* 8: 213–219.

Miller, John M., ed. 1972. *Geophysical Monitoring for Climatic Change No. 1* (Summary Report, 1972), 40–41. Boulder, CO: NOAA.

Mitchell, J. Murray, Jr. 1961. Recent secular changes of global temperature. *Annals of the New York Academy of Science* 95: 235.

NOAA. 2006. NOAA celebrates 30 years of service. National Oceanic and Atmospheric Administration. (Accessed at http://www.history.noaa.gov/legacy/30year.html.)

Pales, Jack C., and Charles D. Keeling. 1965. The concentration of atmospheric carbon dioxide in Hawaii. *Journal of Geophysical Research* 70: 6053.

Pereira, Judy. 1996. Mauna Loa Observatory: A 30-year personal overview. (Accessed at www.mlo.noaa.gov/pdf/memorabilia/JUDY%20BKLT.doc.)

Peterson, James T., and Reid A. Bryson. 1968. Atmospheric aerosols: Increased concentrations during the last decade. *Science* 162: 120–121.

Peterson, Thomas C., William M. Connolley, and John Fleck. 2008. The myth of the 1970s global cooling scientific consensus. 20th Conference on Climate Variability and Change, American Meteorological Society, January. (Accessed at http://ams.confex.com/ams/pdfpapers/131047.pdf.)

Ponte, Lowell. 1976. *The Cooling: Has the Next Ice Age Already Begun?* Englewood Cliffs, NJ: Prentice-Hall.

Price, Saul, and Jack C. Pales. 1959. The Mauna Loa high altitude observatory. *Monthly Weather Review* 87: 1–14.

———. 1963. Mauna Loa Observatory: The first five years. *Monthly Weather Review* 91: (October): 665–680.

Pueschel, R. F., B. A. Bodhaine, and B. G. Mendonca. 1973. The proportion of volatile aerosols on the Island of Hawaii. *Journal of Applied Meteorology* 12: 308–315. (Accessed at http://ams.allenpress.com/archive/1520-0450/12/2/pdf/i1520-0450-12-2-308.pdf.)

Pueschel, Rudolf. 1972. May–June activity report. July 12. In *Geophysical Monitoring for Climatic Change,* Mauna Loa Observatory Monthly Reports, Hilo, Hawai'i, 1971–1973, Air Resources Laboratories, NOAA.

Rasool, S. I., and S. H. Schneider. 1971. Atmospheric carbon dioxide and aerosols: Effects of large increases on global climate. *Science* 173: 138–141.

Ray, P. H., Lt. 1885. *Report of the International Polar Expedition to Point Barrow, Alaska.* Washington, DC: Government Printing Office.

Ruhnke, Lothar H., and Joann T. Dennett. 1970. *Mauna Loa Observatory: High Altitude Science in a Tropical Maritime Environment.* ESSA Technical Memorandum ERLTM-APCL 10, Atmospheric Physics and

Chemistry Laboratory, Department of Commerce, May: 32–33.

Schmidt, M., R. Graul, H. Sartorius, and I. Levin. 2003. The Schauinsland CO_2 record: 30 years of continental observations and their implications for the variability of the European CO_2 budget. *Journal of Geophysical Research* 108 (D19): 4619.

Schnell, Russell. 1996. April activity report. Mauna Loa Observatory, Monthly Reports (CD-ROM), Hilo, Hawai'i.

Sias, Janice. 2006. MLO alumni mailing list response. Copy in author's files.

Stair, Ralph, and Howard T. Ellis. 1968. The solar constant based on new spectral irradiance data from 310 to 530 nanometers. *Journal of Applied Meteorology* 7: 635–644.

Sullivan, Walter. 1978. Studies hint dust in air won't curb rising temperature; little pollution recorded. *New York Times*, D18 (August 9). (Accessed at http://select.nytimes.com/gst/abstract.html?res=F30B13FE3D5511728DDDA00894D0405B888BF1D3.)

Time. 1974. Another ice age? June 24. (Accessed at http://www.time.com/time/magazine/article/0,9171,944914,00.html.)

Weart, Spencer. 2007. Money for Keeling: Monitoring CO_2 levels. In *The Discovery of Global Warming*, Center for History of Physics of the American Institute of Physics. (Accessed at http://www.aip.org/history/climate/Kfunds.htm.)

CHAPTER ELEVEN

The Third Decade (1976–1985)

MLO did not receive notice about the termination of their office space lease at the Cloud Physics Observatory for more than a month after the notification letter was sent. Then on January 15, 1976, Judy Pereira received a copy of the letter forwarded from the National Weather Service. Both MLO and the Cloud Physics Observatory were happy with the office arrangements and the scientific collaborations they had with one another, but the university insisted that the lease would be terminated. Pereira describes what followed as "the office space war."

The Office Space War (1976)

On January 27, MLO director Ronald Fegley sent a "Request for Space" form to the General Services Administration. The next day, W. E. Alexander of the Procurement and Management Office of the University of Hawai'i at Hilo sent a second letter to the National Weather Service Pacific Region in Honolulu. The letter repeats the lease termination language of the first letter and concludes with a shot fired over Ron Fegley's desk:

> We would appreciate your official acceptance of this termination, however, if we do not hear from you by February 10, 1976, we will unilaterally end our agreement as provided for in the above contract and requested in our two letters.

Fegley replied on February 3 and asked Alexander to reconsider:

> We feel that our work will be seriously disrupted if we have to move out because much of our work is done in close cooperation with the Cloud Physics staff and their facilities. They in turn have the use of our equipment and supplies. We also have a working relationship with other campus divisions (i.e. chemistry, physics, Beaumont Ag. Station) that has been mutually beneficial and will be difficult to maintain if we are not located on campus.

Meanwhile, Dr. John Miller, who was slated to take over as the new MLO director, learned of the office space war and joined the fray in a letter dated February 9 to Fegley:

> I feel that we must make every effort to remain in the present offices. Though there are disadvantages involved, I think the contact between the MLO staff and the University personnel is scientifically useful and should be maintained at all costs. Since I will have to live with whatever decision is made, please let me know how things go.

Charles Fullerton, director of the Cloud Physics Observatory, joined with MLO in a detailed, three-page appeal (February 26, 1976) to the president and chancellor of the University of Hawai'i at Hilo. Fullerton vigorously protested the seemingly abrupt manner in which the lease termination was being pursued. He noted that MLO office space included lab and storage space that would not be suitable for office use. He then launched into a lengthy summary of the beneficial nature of the ten-year relationship with MLO, including providing jobs to university students, joint research projects, sharing of equipment, facilities, and science periodicals, and MLO–sponsored seminars in which internationally renowned scientists made presentations. Fullerton described all these collaborations as a "critical mass" of scientific personnel, and he copied his appeal to a dozen officials at the University of Hawai'i and NOAA.

Alexander sent the university's final decision on April Fools' Day (April 1, 1976):

> After careful consideration . . . the University of Hawaii regrets to inform you that its request for your lease termination effective July 1, 1976 . . . must prevail. The University appreciates the cooperation and benefits derived from our mutual use of facilities, equipment, and expertise, however, it has to consider all priorities within the system and unfortunately the Mauna Loa Observatory must suffer the consequences of our limited resources.

Judy Pereira (1996) took on the responsibility of finding new office space, and the candidate sites were the downtown Federal Building and the former customs building at the Hilo docks. With help from the General Services Administration, she and Fegley eventually found new office space in the Federal Building at 154 Waiānuenue Avenue in downtown Hilo (Plate 22). The main part of the two-story building, which is one of the most distinguished structures in Hilo, was built in 1910 to serve as the Hilo Post Office, and two facing wings were added to the ends of the original building in 1936. The observatory that had originally been headquartered in an office in the Air Cargo building at the airport was about to move into the second floor of the classiest government office space on the island of Hawai'i. At least that was the outward appearance of the structure.

Meanwhile, the observatory on the mountain was firmly anchored in place. An aerial photograph (Fig. 11.1) made twenty years after the main building was completed in 1956 shows many new structures, most of which were tied together with a network of cables and air sampling lines laid across the lava. Wood boardwalks simplified walking between buildings, and a paved parking area and access road greatly reduced dust.

A Microprocessor Arrives at MLO (1976)

In January 1975, the hobby computer revolution began in earnest when *Popular Electronics* magazine published a cover story with detailed instructions on how to build the Altair 8800 microcomputer from a kit that used an Intel 8080 microprocessor. The computer was designed by Ed Roberts of MITS, Inc. in Albuquerque, New Mexico. (The author, who cofounded MITS, wrote the operator's manual.) Soon Paul Allen and Bill Gates moved to Albuquerque to work with Roberts developing software for the Altair. They named their tiny company Micro-Soft (Mims 1984).

In June 1976, Ron Fegley began experimenting with a microprocessor kit to determine if it might be possible to automate the lidar system without the help of a NOVA minicomputer. After much study, he was able to operate the system successfully in July, and he laid plans for "incorporating a microprocessor into the lidar program in the near future" (Fegley 1976).

An Old-Timer Visits MLO (1976)

The digital revolution would eventually transform the various ways in which MLO collects and processes scientific data on the mountain. James W. Steiner of the Weather Bureau's Hilo office had worked only with basic instruments that recorded data on paper charts when he serviced the original summit observatory. Steiner left Hawai'i

FIGURE 11.1. Aerial view of MLO in May 1976. The High Altitude Observatory is at left. The main building, Dobson ozone dome, and Butler Building are at top center. The Atomic Energy Commission buildings are at right. *(J. P. Lockwood, USGS Hawai'i Volcanoes Observatory.)*

in January 1957 and returned with his wife Dolores in 1976 two weeks before the twentieth anniversary of the opening of the slope unit of the Mauna Loa Observatory. The couple signed the *Visitors* book on June 14, 1976, with Steiner giving his address as the National Weather Service in Fresno, California. Later he recorded his impressions about the visit:

My wife and I visited the observatory . . . just 2 weeks to the day short of the anniversary of dedication. I found some aspects of the observatory unchanged, but others quite different. The instrumentation and the vast array of projects are a far cry from our early beginnings. But the awesome beauty of the mountains remains the same. I want you to know how I look forward to reading

> articles on GMCC and Mauna Loa's part in it. I swell with pride knowing I was there at the beginning. My congratulations to the observatory on this anniversary. I know there will be many more. After a 20-year marriage, Madam Pele has accepted you and will do you no harm. (Steiner 1978)

A week after the Steiner visit, MLO old-timer Bernard Mendonca left his good-bye in the *Visitors* book:

> Last duty day at MLO. Strange—sad—feeling. Part of me will be here for a long time along with all the other[s in the] long line of sturdy men who did work here. For the good times, hard work and accomplishments, I feel satisfied and for the people who were there with me my thanks. Aloha.

MLO Director John Miller (1976–1978)

On July 1, 1976, MLO completed moving from the Cloud Physics Observatory to the Federal Building. The staff found itself working in more formal surroundings than before, as can be seen in a photograph of Judy Pereira typing in her new office space (Plate 23). But the move was not simple and involved remodeling part of the basement of the old building into a chemistry lab and electronics shop, a project that was not completed when the move was made (Pereira 1996).

Dr. John Miller arrived in Hilo several weeks later to transition into the director's position. He signed in at MLO on July 21 and added "First day on job!" Two days later Susan Fegley signed out at MLO and wrote, "A farewell visit." On July 29, 1976, the Fegleys returned to Boulder, and Miller took over as MLO director.

John Miller grew up on a farm in Virginia, where weather and rain were crucial aspects of his family's livelihood. It was an interest that stayed with him through his time at Virginia Tech, where he received a physics degree before becoming a meteorologist in the US Air Force, an assignment that was preceded by three semesters of study at St. Louis University. After six years in the Air Force, Miller enrolled in a graduate program at Pennsylvania State University and earned a masters and a doctorate in meteorology with a focus on atmospheric chemistry. In 1972, he became a presidential intern at NOAA's Air Resources Lab in Silver Spring, Maryland. This was when Lester Machta and Don Pack were organizing the Global Monitoring for Climatic Change (GMCC) program that took over responsibility for MLO, and Miller (video interview with author, Keauhou Bay, Hawai'i, November 30, 2007) accepted the assignment to become MLO's director.

Shortly after his arrival, John Miller joined the MLO staff for a traditional group photograph (Fig. 11.2). In his first director's report, Miller established several priorities. Mauna Loa had briefly erupted a year before his arrival, and Miller's first concern was to develop a contingency plan in the event of a future eruption: "One of our main priorities will be to investigate alternate locations of MLO in case of volcanic activity." He also planned to reduce the clutter of old, unused equipment that had been accumulating on the mountain:

> Now that we have excellent facilities at Hilo, I would like to clean up the landscape at the observatory. If anyone in GMCC has any equipment, materials, etc. at the observatory unknown to the MLO staff that he wants preserved, please let us know. (Miller 1976a)
>
> On [August] 24 and 31 the whole MLO staff went to the mountain and loaded a sixteen foot van with furniture and junk. The furniture can be used in the Hilo office. The junk, which included boxes, old equipment, beer cans, etc., was taken to the Hilo dump. Since sufficient junk is left to fill the van again, a third clean up day is scheduled on September 21. (Miller 1976b)

Two days after the third cleanup day, Barry Bodhaine was on the way to a new assignment in Boulder. He signed in at MLO and prophetically wrote, "My last day at the mountain but I promise I'll be back (some day)!" Meanwhile, the GSA was moving much too slowly in completing the basement laboratory and shop facilities badly

FIGURE 11.2. The Mauna Loa Observatory staff during the summer of 1976 included *(left to right)* Director John Miller, Richard Cram, James Jordan, Judy Pereira, Leona Ferreira, Charmaigne Makanui, and John Chin. Prof. Kinsell Coulson *(at right)* became director of MLO in 1979. *(MLO archive.)*

needed by MLO staff, and Miller (1976c) wrote, "Judy [Pereira] has been doing an excellent job in keeping GSA on their toes."

Meanwhile, major changes were in store for the observatory. This was unknown to the various visitors and tourists who visited, some of whom described their difficult hikes to the summit in the *Visitors* book. On February 20, 1977, a hiker wrote, "Made it to the top but never again." On February 27, Donald and Carrie Woods wrote, "Rain, sleet, wind fog. Nice hike." The next day, a hiker wrote, "Never again!!!!!" and on March 2, a pair of hikers wrote, "Never again" and "I loved it!?"

On January 15, 1977, Dr. Kinsell L. Coulson (Plate 24) of the University of California arrived at MLO. His purpose was to install his polarimeter, a sophisticated optical device that measures the polarization of the sky, thereby enabling the detection of aerosols. It may have been during this visit that vandals struck the mountain station:

> He had to set up his instruments in the south-facing bedroom. And unfortunately, he was one of the first few guys who got ripped off. We never used to lock the buildings. And I think someone stole his computers and his equipment. We do not know who it was. We just used to leave it open for hikers and for stranded people, and, after that, we started locking it up and installed the emergency phone. (Alan Yoshinaga, video interview with author, Hilo, Hawai'i, December 4, 2006)

At the time, Coulson didn't know that he would eventually return as MLO's director. But first, John Miller had much more work to do.

A Very Close Call with the Lidar (1976)

Howard Ellis continued as lidar operator after Ronald Fegley left MLO, and the evening of January 31, 1976, he was excited when the lidar's laser beam detected a stratospheric layer of material from what was likely a distant volcanic eruption. That night the lidar system failed, and the next morning Ellis asked John Miller if he could drive up to MLO on his own to attempt a repair of the laser. Miller had wanted Ellis to accompany him and Mendonca to a meeting with the contractor who was going to remodel the observatory, but he agreed to allow Ellis to return to MLO.

After Ellis arrived at the mountain station, he entered the telescope dome that housed the Apollo ruby laser system and opened the drawer that held the laser's electronics. In the center of the drawer was an aluminum hammer-like mechanism known as a dead man's switch designed to safely discharge the system's huge capacitors that were charged to hundreds of volts prior to being dumped through the laser's xenon flash lamps.

> I wanted to see what would happen when I fired the laser with the button. I thought that the electronics were set up to hit the laser rod with just sufficient power to make it fire and then dump the remaining large amount of voltage to ground. . . . When I pressed the button, nothing happened. I believe that Fegley and I had done this same thing when it was working well and observed the hammer falling. I guess I thought that I would give it a little help and slapped it with the ends of my right hand fingers.
>
> The first sensation was my hair going straight up. Then my legs just disappeared, and I believe that I went straight down. My only thought at that time was that this was one incident that I would not be able to tell anyone about. Then there was this long sense of moving fast to my right as on a train, and I was looking out the left side only. There was a steady electronic-like buzz. It was in color and the vegetation was like very thick sorghum cane that had tasseled out evenly, and there was pale blue sky above. (Ellis, telephone interview with author, June 12, 2008)

As Ellis recovered from the jolt, which left a 0.6-inch (16 mm) circular burn in his right forearm, he saw his feet and legs seemingly stretched across the floor of the dome as they appeared to gradually move toward him.

> The last thing that came to life was my [right] hand. I ran outside the dome and screamed for happiness, because I didn't think I was going to live. When I went back inside [the dome], it really stunk [from the burnt flesh].

Ellis found some salve to spread over his burned right hand and lower right forearm. He wrote in the green lidar log, "Shocked," and then called the Hilo MLO office. The time was about 9:30 a.m. Judy Pereira (telephone interview with author, June 12, 2008) took the call and recalls, "He tried to minimize what had happened to him." Pereira immediately called for medical help, but the responders waited for Ellis to return to Hilo and did not drive up the mountain. Meanwhile, Ellis spent the rest of the day working on the recalcitrant lidar system, which now bore a fresh quarter-inch crater in its aluminum frame where Ellis' forearm was resting when the capacitor bank discharged through his hand and arm. He didn't return to Hilo until about 5:00 p.m., where a doctor asked lots of questions and placed more salve on his burns, which required several weeks to fully heal.

The next day Curt Shoemaker, a line supervisor at the telephone company where Ellis' wife Lena worked, visited Ellis at MLO and spotted the holes left in his clothing by the shock:

> He pleaded with me to let him talk to a doctor friend [Dr. Samuel Gingrich] who was his neighbor in Waimea. He set up an appointment to see him early next morning at the new hospital in Waimea. I drove out from home in our own car. . . . At the end of the corridor was Dr. Gingrich in a white smock. . . . He told me that he was going to make me a necklace with all his telephone numbers

> on it, and I was to call him whenever I felt like doing something like that. (Ellis, telephone interview with author, June 12, 2008)

Later a technician from the laser company told Ellis that a shock such as the one he experienced should have been lethal. It would have, had Ellis' left hand also been touching the high-voltage power supply or the metal frame. He was not the last MLO scientist to be on the receiving end of the wrath of the lidar.

Modernizing the Observatory (1977)

Removing large amounts of unused equipment and junk was only the first phase of John Miller's ambitious plans for upgrading the observatory. The arrival of electricity in 1966 had long since eliminated the need for MLO staff to stay on the mountain overnight, and when Miller arrived the bedrooms and kitchen in the main building were not often used. Space was badly needed for instruments, and Miller began making plans to remodel the building and modernize the entire observatory.

On March 14, 1977, Yilman Lee of Lee Construction Inc. of Hilo visited MLO, signed the *Visitors* book, and began preparing an estimated cost for remodeling the observatory. In April, the Kūlani Honor Camp removed the old diesel storage tanks used to store fuel for the generators prior to the arrival of electricity in 1966. They also took away an unused trailer and storage container and demolished two unused electronics and camera buildings (Miller 1977a).

The remodeling work began in earnest in May after instruments and furnishings inside the main building were removed. The Scripps Applied Physics CO_2 analyzer and the associated sampling pumps and cylinders of calibration gas were moved into the Butler Building on the south side of the main building. The staff worked fast and managed to reconnect the complete system and the air sampling lines with the loss of only one day of data (Miller 1977b).

The bedrooms and kitchen walls inside the main building were removed (Plate 25), and the kitchen fixtures and appliances were moved into the Atomic Energy Commission (AEC) Building. A false ceiling was suspended 16 inches (41 cm) below the original ceiling, which was left in place complete with the original lightbulbs. This arrangement provided new space for routing air intake lines and cables. The windows in the building were sealed to block air leaks, and an air conditioner was installed to keep the Scripps CO_2 analyzer happy. The red linoleum floor was replaced, new wiring was installed, and the inside of the building was painted (Miller 1977c). After the project was complete, only those who knew where to look could see traces in the floor that outlined the former kitchen and bedroom walls.

The upgrading of the building included a new air sampling pipe extending well above the roof from almost exactly where Jack Pales had installed a thin glass tube to sample ozone from the outdoor air two decades before. Installation of the new sampling assembly proved difficult (Plate 26), and it was finally in place on June 24 (Plate 27). Construction of a new and much higher solar radiation tower was begun on July 5.

While work was going on at the observatory proper, Willocks Construction completed a small visitors parking lot below the observatory (Miller 1977d). The purpose of the visitor's lot was to reduce vehicle exhaust at the observatory and to provide a place for Mauna Loa hikers to leave their cars.

By September, the contracted portion of the remodeling effort was entirely completed, and the staff continued the project by making alterations and needed repairs. The instruments were returned to the main building, and during October the staff removed unused equipment racks and wiring from the AEC Building east of the main building. They then installed the cabinets, stove, beds, and other furnishings removed from the main building. This provided the AEC Building with facilities for overnight stays. By fall, the major modernization effort was largely completed (Plate 28).

MLO began to return to a more traditional schedule until the Scripps CO_2 analyzer began acting up again. Problems with the aging vacuum tube instrument were by now traditional, but it was hoped that the new air-conditioning

FIGURE 11.3. Interior of the remodeled main building at MLO in 1978. The NOAA *(left)* and Scripps *(center)* carbon dioxide analyzers were separated by a freezer that removed water vapor from the sampled air. The plastic over the Scripps analyzer reduced the instrument's temperature sensitivity. *(MLO archive.)*

system would eliminate problems with the instrument's temperature sensitivity. But even small shifts in the building's temperature caused problems:

> Scripps has agreed to purchase the equipment necessary to modify our AC for better temperature control. At present the main building temperature is cycling at about 3°–5° C around 20°. Scripps believes that this is affecting the Applied Physics CO_2 measurements. We have not experienced a problem with the other equipment. (Miller 1977e)

During December, the staff did what they could to resolve the temperature problem with the Scripps CO_2 analyzer:

> As a result of wrapping 6 layers of heavy plastic sheets on the Applied Physics analyzer [see Fig. 11.3], the excessive thermal drifts which previously appeared on the data

> disappeared. SIO [Scripps Institution of Oceanography] wanted to have more sensitivity on the [chart recorder] data traces by replacing some amplifier tubes in the analyzer. Of the 8 tubes received from SIO, 2 were bad tubes, 3 were weak tubes, and only 3 tubes were good. (Miller 1978)

Acquiring replacement vacuum tubes became more difficult over the years as the solid state electronics revolution supplanted tubes with transistors and integrated circuits. Old-timers at MLO speculate that Scripps would sometimes send used tubes from their own instruments simply because they had trouble finding new ones. The MLO staff began hoarding a stash of vacuum tubes in an upper shelf of the storage cabinet where replacement parts for the CO_2 project were kept.

The Brown chart recorder had recorded data from the Scripps CO_2 analyzer on paper rolls for twenty years. In December Scripps send a refurbished Brown recorder to replace the first. Unfortunately, it produced erratic traces, so the MLO staff set to work correcting the problem.

In July 1977, Cynthia Aki was hired to work with John Chin on carbon dioxide data reduction, which then meant extracting data by hand from chart recordings. She worked at MLO until 1981 and recalls the friendships she made there. The downside of her job was eye problems caused by the tedious task of breaking the data from the lengthy chart recordings, and during her time at MLO she went through five pairs of eyeglasses (Cynthia Chow, video interview with author, Hilo, Hawai'i, December 4, 2006).

International Activities (1977)

As former MLO director Dr. John Miller (personal communication with author, April 30, 2008) observes, "Every country interested in background monitoring knew about MLO." MLO began developing its reputation within the international scientific community even before the slope building was dedicated in June 1956. In September 1955, Dr. Ukichiro Nakaya of Hokkaido University at Sapporo, Japan, visited the Weather Bureau office in Honolulu to make plans for a study of snow from the original MLO summit building. The snow study was postponed until after the MLO slope unit was opened, and Ukichiro Nakaya and Mikio Shoda, who became informally known as the "snowflake people," stayed at MLO from December 7, 1956 to January 21, 1957.

International collaborations continued after the Weather Bureau moved to MLO in 1957, and the annual *Geophysical Monitoring for Climatic Change* Summary Report No. 6 for 1977 was the first to include a section entitled "International Activities." Editor James T. Peterson (1978) listed eight international activities, including one that was to lead to a long-term collaboration with the World Meteorological Organization (WMO):

> J. Miller participated in a consultants' meeting for the coordination of WMO Background Air Pollution Monitoring [Network] and International Atomic Energy Agency (IAEA/WMO) Isotopes in Precipitation networks at the IAEA Headquarters, Vienna, Austria, October 4 to 6, 1977.

The Twentieth Anniversary Celebration (1978)

Modernizing MLO required six months of hard work. The project was completed in January, and on January 28, 1978, the remodeled and modernized observatory was dedicated during the GMCC annual meeting, which was held in Hilo that year. The annual meetings, which have continued to the present, brought together the key scientists and staff involved in the entire GMCC observatory program. Those who visited MLO saw how the removal of the bedrooms and kitchen had transformed the main building into an efficiently organized scientific research station.

The dedication ceremony on the mountain was presided over by the Rev. Eddie Kealanahele (Plate 29), whose wife Marge had filled in for Judy Pereira during two of her maternity leaves. The ceremony was attended by Ralph Stair, who had designed the main building back in 1955 and supervised its construction. Representative Daniel Akaka, Mayor H. Matayoshi, Lester Machta of NOAA's

GMCC, and numerous others also attended. The *Visitors* book lists at least sixty-two signatures, including Ralph and Angie Stair ("Back for the dedication"), Bernard Mendonca ("20 year dedication. Good to be back!!"), Eddie and Marge Kealanahele ("It's just wonderful to be back and see the growth!"), Barry Bodhaine ("Good to be back!"), ozone scientist Walter Komhyr, and aerosol scientist Glenn Shaw ("Radiation measurements"). Shaw was from the University of Alaska and had made prior trips to MLO to measure the clarity of the atmosphere.

Although Judy Pereira experienced a mild case of altitude sickness, she wasn't too ill to enjoy the proceedings, which she describes in her memoir (1996):

> At the precise moment of cutting the maile lei [made from the leaves of a Hawaiian vine] that was strung across the front door and officially open the building, and with news cameras rolling as the ceremony proceeded, Glenn Shaw of the University of Alaska Geophysical Institute opened the door from the inside and walked right through the lei and the dignitaries gathered at the door. A very solemn occasion turned suddenly into a hilarious situation with everyone bursting into laughter at the awkwardness of the moment. Glenn will never forget that, I'm sure, and none of us at MLO will ever forget it either!

Shortly before the dedication ceremony, John Miller, Kirby Hanson, and Fritz Hertzel visited the summit of Mauna Loa to see the site of the original summit observatory founded by Robert Simpson. The day was sunny, and the looming summit of Mauna Kea punctuated the northern horizon. As Miller gazed at the bare concrete slab atop the giant mountain, he reached for his camera and captured a photograph of the site where it all began in 1951 (see Fig. 4.8).

Federal Building Headaches (1978)

Before he arrived in Hilo, Miller had campaigned to keep MLO headquartered in the Cloud Physics Observatory to continue the ten-year collaboration with the University of Hawai'i. While the Federal Building was a far statelier edifice, it introduced a completely new and unexpected set of problems, some of which Miller describes in a letter of July 24, 1978, to Richard J. Nee, the building manager of the General Services Administration (GSA) in Honolulu:

> After several months of working in our new facilities we have become very concerned about a hazardous situation that exists in our basement area. The air conditioner from the chemistry laboratory is vented directly into a working area.

Miller's letter describes the high humidity and unbearable heat caused by the air conditioner and expresses concern about the possibility of fire that might be caused by venting hot air through the MLO library's accumulation of books, journals, and chart rolls. Nee replied the next day (July 25, 1978):

> I don't believe that you should have your people working in that area. The "hot" air from the air conditioner does not represent a fire hazard and if it did that problem will be cured by the new sprinkler system.

More letters about this and other problems were sent by Miller to Nee, and problems inside the Federal Building and vandalism of the MLO vehicles parked outside were to continue for years to come.

Mauna Loa Observatory: A 20th Anniversary Report (1978)

John Miller's vision to modernize MLO included a plan to document the slope observatory's twenty years of monitoring the atmosphere. Miller solicited contributions from a cross section of MLO pioneers, including founders Robert Simpson and Ralph Stair. Former MLO directors Howard Ellis, Lothar Ruhnke, Rudolf Pueschel, and Ronald Fegley made contributions. Charles Keeling provided a detailed report on his CO_2 measurements since 1958. Bernard Mendonca, Charlie Garcia, J. P. Lockwood, Barry Bodhaine, and others submitted papers and memoirs. Judy Pereira describes the development of the book in her memoir (1996):

> John's crowning glory, however, was the *Mauna Loa 20th Anniversary Report*. His wife was a JGR [*Journal of Geophysical Research*] editor and assisted him in editing all the papers, scientific and reminiscent, that were submitted for the volume. Using an electric typewriter (word processors and computers had not yet come on the scene at MLO), I typed, re-typed, and re-typed again all of the submissions. John worked with the publication specialist in Boulder via mail and telephone (yup, no fax here either), sending galley proofs back and forth, and setting up the layout of the book. Finally it went to press and has been a widely requested and distributed publication. All copies are now gone [but the text is available at the MLO Web site].

Lester Machta's foreword ends by summarizing the significance of MLO:

> To mark the observatory's 20-year existence, this volume describes the milestones of its history, highlights work done in the past, and outlines its present purposes and goals. (Machta 1978)

Mauna Loa Observatory: A 20th Anniversary Report remains the definitive history of the founding of the observatory and its early years, and the contributions by the various MLO pioneers provided valuable sources for the present book.

While work on the book was underway, Miller and the MLO staff gathered on the steps of the Federal Building for a group photograph (Fig. 11.4). Miller enjoyed his work at MLO and fully intended to stay past 1978, but after two years he received a new assignment as deputy director of NOAA's Atmospheric Research Laboratory. On August 20, 1978, John and Sylvia Miller signed out at MLO. Sylvia drew a smiley face and wrote, "Looks great now!" and John wrote "Last time up (maybe)!" As they drove down the newly paved road, they left behind a modernized observatory, a better organized staff, and a comprehensive book about the observatory's twenty-year history.

MLO Acting Director Howard Ellis (1978–1979)

John Miller's final monthly Activity Report for July 1978 reported that Kinsell Coulson of the University of California at Davis had been appointed the new director of MLO. Coulson was no stranger to MLO, for he had considerable experience making skylight measurements at the mountain station, including ninety-six days of observations during the first half of 1977. On August 16, 1978, Coulson signed in at MLO during a house-hunting visit to Hilo. But he was unable to assume the directorship of MLO for nearly a year due to his university responsibilities. Howard Ellis was designated acting director until Coulson took over in July 1979.

Ellis took over a completely refurbished and cleaned-up observatory that allowed the staff to concentrate on their measurements. One of his first tasks was to remove a new Park Service sign on the Mauna Loa Observatory road that included the observatory's unpublished telephone number. Ellis treated unexpected visitors with respect, sometimes even sharing some of his food. But he and other MLO staff members didn't like taking time from their duties to entertain unannounced visitors, and he didn't relish the prospect of people telephoning the observatory merely to ask about the weather or the presence of snow. So he hauled the sign down the mountain and left it at a Hilo landfill. Soon afterwards, the Park Service called to ask about their missing sign, and Ellis admitted to disposing of it. A Park Service official told Ellis he would have to charge him with destroying the sign if it wasn't replaced, and Ellis replied that it was too late to retrieve the sign. The incident was soon forgotten, and the sign was never replaced (Ellis, personal communication with author, April 9, 2008). John Miller related a very similar version of this incident.

FIGURE 11.4. The Mauna Loa Observatory staff in March 1978 included *(left to right)* Howard Ellis, Mamoru "Al" Shibata, Judy Pereira, Alan Yoshinaga, Cynthia Aki, John Chin, Dick Cram, Duane Harding, and MLO director John Miller. *(MLO archive.)*

The High Altitude Observatory staff had added a linoleum floor and some rugs to the old Atomic Energy Commission Building in May, which was where overnight visitors stayed after the main building was remodeled. They even included a "Home-Sweet-Home" welcome mat. Robert Evans stayed in the AEC Building from July 9 to August 28, 1978, while calibrating the Dobson spectrophotometer that remains the world-standard ozone measuring instrument to this day. On August 28, Evans signed out with "Dobson 83 cal. No longer Pele's next door neighbor."

John Miller visited MLO on October 18 and wrote in the *Visitors* book, "Everything looks great. Good to be back for a visit."

In spite of his previous unpleasant encounters with Charles Keeling, Ellis found himself actively involved in the CO_2 program at MLO. During his first month as acting director, the Scripps APC CO_2 analyzer began responding to changes in the humidity of the main building. This was surprising, for the APC analyzer had been overhauled the previous month by Scripps. NOAA's URAS CO_2 analyzer was unaffected by the humidity changes, so Ellis (1978) installed a double freeze trap in the air line to the Scripps analyzer to make sure any water vapor was completely removed. This had no effect, so the original freeze trap from 1958 was connected to the air line. For unknown reasons, the Scripps analyzer required frequent adjustments during October. By December, Ellis finally reported good news: "Both the URAS-2 and Applied Physics analyzers were in normal operation" (Ellis 1979a).

Both gas analyzers were still working normally in February when a lightning bolt disabled a brand new data acquisition system for the CO_2 program that had just been installed the same day by Dick Cram. The system was back in operation two days later. Meanwhile, the trusty old chart recorders saved the data that couldn't be saved electronically (Ellis 1979c). The lightning strike also damaged a number of other systems. A friendlier form of anomalous weather occurred in the form of snow, and Ellis reported that many snow people drove the observatory road in January and February. On January 16 a visitor from Lawrence Berkeley Laboratory wrote in the *Visitors* book, "Just wait—the CO_2 conc[entration] will drop when we nonsmokers prevail!"

Problems with the lidar computer also occurred in 1979. In those days, programs were entered into a computer via a paper tape punched with patterns of holes that formed specific instructions. Before the paper tape could be loaded, the computer had to be manually programmed to know how to read the tape. This required a tedious process of manually toggling on-off patterns into a row of switches, flipping a switch to load the instruction, and then repeating the process to enter the next instruction. Glitches in the computer's power supply and other problems could cause a computer to lose its program and require a restart procedure, as sometimes occurred with the lidar computer:

> The computer lost its program on four of the evenings. Each time it was brought back up by manual switch settings and feeding in paper tape. This process consumes over an hour of time for each failure. (Ellis 1979b)

On May 16, 1979, during Ellis' final month as acting director, MLO experienced a major electrical power surge that caused the power company to shut down electricity to the observatory. This outage affected all the station's observations, and the following day instruments had to be reset. Fortunately, no equipment was damaged by the surge and the outage. But five days later another kind of surge did cause problems. This surge took the form of a lightning strike that tripped circuit breakers and shut down much of the observatory. Several days were required to bring everything back to normal status.

The Discovery of Asian Dust Over Hawai'i

The earliest explorers of the summits of Mauna Loa and Mauna Kea reported extraordinarily clear skies, and those reports continued with the arrival of the slope observatory on Mauna Loa in 1956. The *Weather Log* kept during 1957–1958 consistently indicated exceptionally high visibility, with no mention of atmospheric haze on days when clouds were below the observatory (Pales et al. 1957–1958).

The instrumental record, however, provided clues that something was indeed in the air, for seasonal fluctuations in direct solar irradiance measured at MLO since 1958 indicated the presence of enhanced aerosols during the spring months. The first paper about the relative stability of direct sunlight and thus atmospheric transmission at MLO discussed these fluctuations and suggested that they were

> most likely the result of increased worldwide photochemical aerosol formation caused by the oxidation of volatile

> organic materials of plant origin in the atmosphere (8), or the result of the seasonal variations in general atmospheric circulation, or both. (Ellis and Pueschel 1971)

During his postdoctoral time at MLO, Jim Simpson studied samples of particulate matter collected in Hawaiian air and published his findings (Simpson 1972). According to Barry Bodhaine (personal communication with author, July 2, 2008), "There is evidence of Asian dust in his data, but it may not have been recognized as such at the time." The possibility that dust from the Gobi Desert could drift a quarter of the way around the globe to Hawai'i and cause the fluctuations in atmospheric transmission measured at MLO had intrigued Bodhaine, and in the same communication he reflects on the fate of his scientific paper that attempted to propose what was then considered a rather radical hypothesis:

> Rudi Pueschel and I tried to attribute the solar transmission annual cycle on springtime Asian dust in the early 1970's, but the reviewers shot us down, and we had to take it out of our paper (Bodhaine and Pueschel 1974). Turned out we were right, but it didn't get into print. . . . It took us about two years to get it published. It was rejected the first time because the reviewer didn't believe there was transport in the upper troposphere from Asia. We rewrote the paper and submitted it to *Journal of Atmospheric Science,* and we were unlucky enough to get the same referee. Ralph Cicerone was editor, and he finally agreed to publish it. But we had to take out Asian dust as the aerosol source. There was a prevailing opinion that water vapor variations were the cause of the annual cycle in solar transmission, but our paper proved that it couldn't be water vapor. So instead we speculated that it was biogenic aerosol, even though we thought it could be Asian dust.

This was not the first time that reviewers of a scientific paper would be skeptical about findings at MLO, and it would not be the last, as Glenn Shaw of the University of Alaska would soon find out. Even before Bodhaine and Pueschel were precluded from publishing their Asian dust hypothesis, Shaw initially met resistance when he proposed that Arctic haze was coming from power plants and industrial activity in Asia in 1972. But he persevered and managed to publish his findings (Shaw 1995). Shaw would be the next scientist to ponder the Asian dust hypothesis.

Glenn Shaw Verifies Asian Dust over Hawai'i (1979)

Dr. Glenn Shaw of the Geophysics Institute of the University of Alaska made many visits to MLO to calibrate his precision solar radiometer and to determine the presence of aerosols in the air over Mauna Loa by measuring changes in the intensity of various wavelengths of direct sunlight. As recalled by Judy Pereira (video interview with author, Boulder, Colorado, April 27, 2007) and others, Shaw had a reputation for doing good science and getting along well with the MLO staff:

> Glenn Shaw had been friendly with every director that was there. He was welcomed to Mauna Loa by everyone.

The record of Shaw's cordial interaction with the observatory staff over the years can be found in his letters that remain in the MLO files. Shaw spent considerable time on the mountain making his measurements, and he was well acquainted with the difficulty of maintaining the station and the road and working at MLO's high elevation. But some cooperative observers who rarely or never visited MLO did not fully appreciate the difficulties of working at an alpine station using instruments that worked fine in a lab but not at high altitude. Charles Keeling was the classic case, for he was not fond of air travel and, as John Chin explains, he had a back problem that made long flights difficult. Thus he rarely visited MLO. During the early years of his carbon dioxide program, he seemed not to understand how difficult it was for the staff to stay on the mountain twenty-four hours a day to be available to service the CO_2 analyzer in the event of a problem. Then there was the aerosol scientist whose only known visit to MLO was marred by clouds and who later left his tact behind when he wrote MLO about problems that the mountain crew was experiencing with one of his instruments:

> These things induce me to conclude:
>
> . . .
>
> 2. A little study of the instructions (or common sense for that matter) should have revealed that the timer motor is not essential to the twilight operation.
> 3. Putting the 6V motor into the Timer box was a mindless exercise. Fortunately, I have several proper spares.
> 4. A complete lack of concern for the whole twilight operation. . . .
>
> . . . I kindly acknowledge that the operation of the photometer is voluntary. (Frederick Volz, letter to director, Mauna Loa Observatory, two pages, December 7, 1976)

Howard Ellis (video interview with author, Hilo, Hawai'i, December 2006) recalls the high expectations of this scientist, who was quite friendly in person and who had published a number of significant scientific papers about volcanic particles in the upper atmosphere:

> [The sun photometer] was always there, and that was one of the things that I had to do early in the morning. He always was fussy; he wanted a perfectly clear sunrise and sunset but [that was] just never going to happen. Right at sunrise, that was what he wanted.

Glenn Shaw was cordial in his dealings with the MLO office and mountain staff. If he or his coworkers used the MLO telephone, he made sure the University of Alaska paid the bill. When he needed help, he asked politely. And he never failed to send thank-you letters. The result was full cooperation from the MLO directors with whom he collaborated, as when Kinsell Coulson sent a long "Dear Glenn" letter (September 5, 1979) that began,

> Thanks for your letter. We shall be glad to cooperate on your project to the fullest extent possible commensurate with the regular operations of the observatory.

From March to August 1976, Shaw used his radiometer to measure sunlight at MLO and detected traces of the aerosol cloud from the eruption of Mt. Augustine, an Alaskan volcano that had begun erupting the previous January. He returned during January and February 1977 and continued his program of making measurements during every clear day.

Shaw wrote a paper about his findings at MLO that covers both the Augustine eruption and technical details about the occasional presence of dust from North America and Asia in the ordinarily pristine air over MLO. Howard Ellis (video interview with author, Hilo, Hawai'i, December 2006) had observed dust at MLO before Shaw's time, but he was unaware of its source:

> There were times when there were periods of four or five days when the wind was very high, and when you have a high wind, it is always out of the east or the west, one or the other . . . there were lots of dust rays, and you could see veils of dust across [the saddle], not so much stratified, but it fades out everything.

Shaw's discovery of Asian dust over Hawai'i was by far his most important finding, yet he mentions it only in the abstract and the last sentence of the paper:

> It is deduced that aerosol from North America and Asia occasionally reaches the Hawaiian Islands. . . . [This work] supports the hypothesis that the excess aerosol consists of particles that had been carried in the middle troposphere from sources in Asia and North America. (Shaw 1979a)

There was good reason for Shaw to barely mention that he had detected Asian dust over Mauna Loa. Others had previously discovered that dust from the Sahara Desert blows across the Atlantic to North America. But no one had yet proposed in a scholarly paper the transport of dust across the Pacific all the way to Hawai'i. Thus, like Bodhaine and Pueschel before him, Shaw's claim was met with considerable skepticism by some within the atmospheric sciences community.

Unless otherwise credited, all photographs taken by the author.

PLATE 52. MLO physicist Steve Ryan walks toward the Pacific while opening the valve of an evacuated glass flask to collect an air sample.

PLATE 53. Steve Ryan using an automated air sampler to collect air in two flasks.

PLATE 54. Ryan prepares to launch an ozonesonde from the National Weather Service facilities at the Hilo Airport on December 12, 2006.

PLATE 55. This ozonesonde reached 113,934 feet (34,727 meters) before the balloon burst.

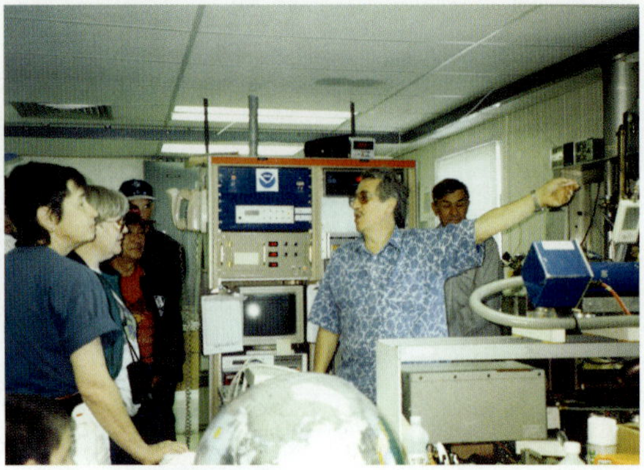

PLATE 56. MLO director Russell Schnell explains instruments to visitors. *(Darryl Kuniyuki, MLO archive.)*

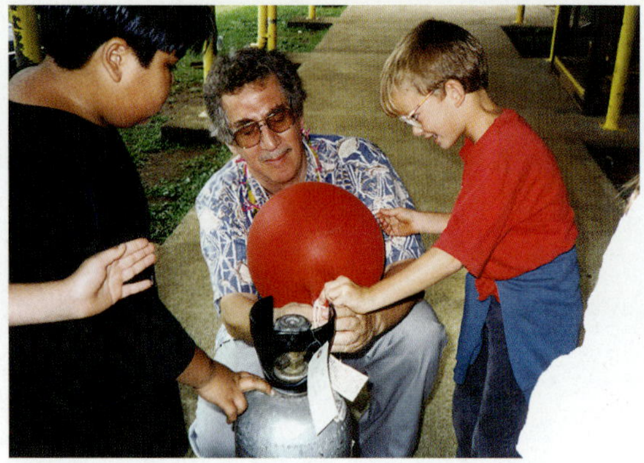

PLATE 57. MLO director Russell Schnell helps students inflate balloons with helium. *(Tracie Kuniyuki, MLO archive.)*

PLATE 58. Ozone scientist Brooke Walsh explains ozone observations to a University of the Nations class on June 27, 2003.

PLATE 59. MLO station chief John Barnes speaks to a University of the Nations class on September 30, 2004.

PLATE 60. Darryl Vierra *(left)*, Don Maclean *(center)*, and Mike Gillette cycled to MLO on June 3, 2006.

PLATE 61. Dr. Derek Chignell (by Dobson dome) leads University of the Nations students in a blessing for MLO on May 11, 2007.

PLATE 62. Philip Russell *(left)* and Yohei Shinozuka of Ames Research Center calibrate an airborne sun photometer on May 18, 2008.

PLATE 63. Prof. Joseph Shaw of Montana State University calibrated sun photometers on May 22, 2008.

PLATE 64. Forrest Mims calibrates a Microtops sun photometer on May 8, 2009, next to a NASA robotic sun photometer.

PLATE 65. Sun Young Kim *(left)* and Cynthia Anani of University of the Nations with GLOBE sun photometers on June 15, 2001.

PLATE 66. Geologists Erika Ronchin (with sample) and Julie Herrick mapped lava during May 2005 while staying at MLO.

PLATE 67. Yuji, a cameraman for Japan Productions, films a Brewer spectrophotometer on the solar deck on April 20, 2003.

PLATE 68. The International Space Station passes over the Mauna Loa Solar Observatory after sunset on May 8, 2009.

PLATE 69. The Mauna Loa Solar Observatory telescopes stare at the sun 363 days of the year. Founded in 1965, the observatory is manned by Eric Yasukawa, Darryl Koon, and Allen Stueben; Allen is shown here on December 1, 2007.

PLATE 70. Stuart McDermid *(left)* and Thierry Leblanc of NASA's JPL ultraviolet lidar facility were at MLO for the 50th Anniversary of the Global Carbon Dioxide Record Symposium MLO field trip on December 1, 2007.

PLATE 71. The brilliant laser beam from the lidar built by John Barnes probes the stratosphere for aerosols over MLO on December 5, 2006.

PLATE 72. During a lidar session on May 23, 2005, Dr. Barnes' computer displays the presence of high-altitude water vapor and any aerosols over MLO.

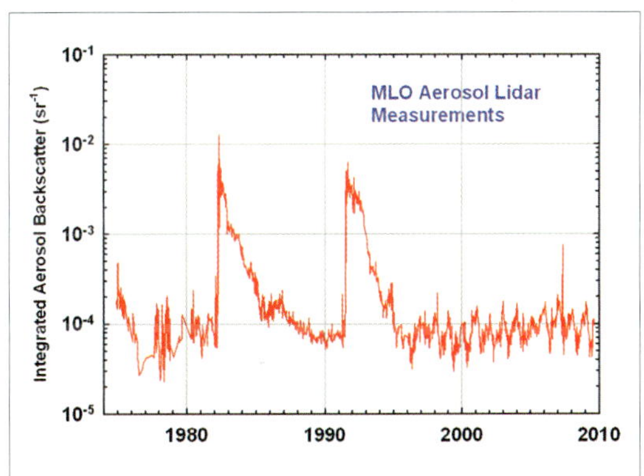

PLATE 73. This lidar plot shows laser light scattered from stratospheric aerosols. The initial spike was caused by the volcanic eruption of Fuego (1974). The two large spikes were caused by eruptions of El Chichón (1982) and Pinatubo (1991). *(John E. Barnes.)*

PLATE 74. The sun as seen by the Dobson spectrophotometer when it measures the ozone layer from inside its protective dome.

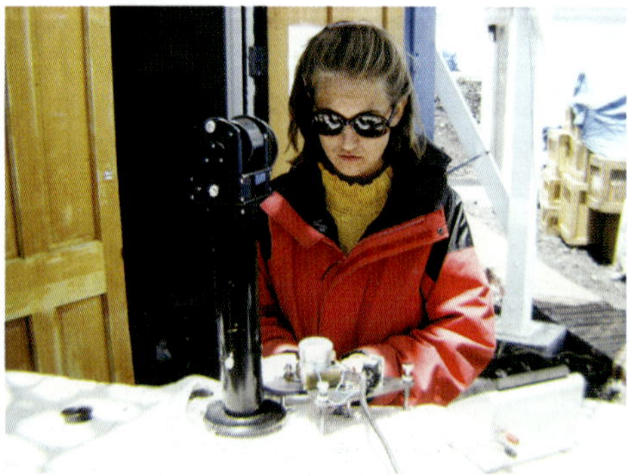

PLATE 75. Grae Roth measures the ozone layer using the world-standard Dobson spectrophotometer on August 9, 1992.

PLATE 76. Leslie Pajo measures the ozone layer with the MLO Dobson spectrophotometer on October 5, 2004.

PLATE 77. Charles Keeling (1928–2005) and his son Ralph, who took over the Scripps carbon dioxide program after his father died. *(Scripps Institution of Oceanography. Used with permission.)*

PLATE 78. Charles Keeling signed in at MLO for the first time on November 22, 1958.

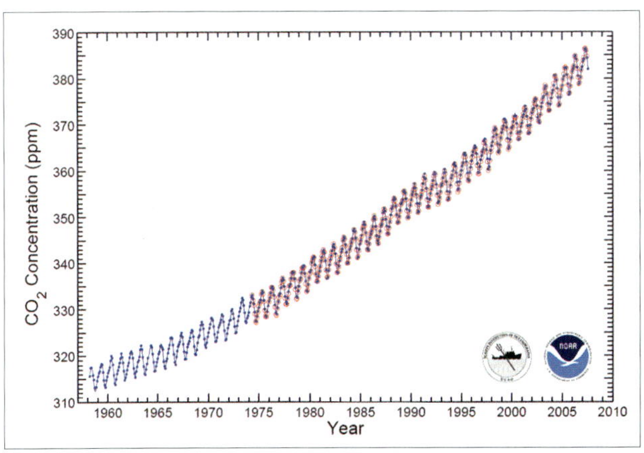

PLATE 79. Carbon dioxide measured at MLO since 1958 by Scripps (blue) and by Scripps and NOAA (red) since 1978. *(Ralph Keeling, Scripps Institution of Oceanography.)*

PLATE 80. John Chin with the APC carbon dioxide analyzer that he maintained for Charles Keeling for most of his MLO career.

PLATE 81. MLO as viewed from the Mauna Loa Solar Observatory on December 10, 2005. The boardwalk points to the Keeling Building.

PLATE 82. The Keeling Building and the Dobson spectrophotometer dome as viewed from the solar deck over the NDSC Building on December 10, 2005. The inactive Hualālai volcano is in the background.

FIGURE 11.5. Twenty-eight years after Glenn Shaw discovered layers of Asian dust over MLO, the dust layers shown here appeared over the cloud-draped Pacific and the observatory on May 21, 2007. This dust event covered most of the North Pacific and North America.

Shaw was at MLO to make still more measurements during the twentieth anniversary dedication of the remodeled main building on January 28, 1978, and he returned on January 22, 1979. Ginnie Ferrell, also of the Geophysics Institute, arrived at MLO to continue Shaw's work on March 16. After her signature in the *Visitors* book, she wrote, "Come on sun!"

Shaw brought his family to Hawai'i for a spring vacation in late April 1979. The Soufrière volcano on the island of St. Vincent, which has a long history of deadly eruptions, had experienced a major series of eruptions from April 13 to 26, 1979 (Shepherd et al. 1979). Would the ash and aerosols from the eruption travel a quarter of the way around the world to Hawai'i? To find out, Shaw left his family at the coast and made the long drive to MLO on April 26. He wrote in the *Visitors* book, "Here to look for Soufrière's cloud." Instead, he found something far more important:

> In 1979 [I] happened to be visiting the Mauna Loa Observatory and noticed some dark bands in the sky [Fig. 11.5]. I basically ruined our family vacation driving up and down the mountain making optical measurements. It turns out these were dust layers that had crossed the Pacific Ocean from China. Now it's recognized that such ocean-sized clouds of dust also affect climate, sometimes cooling and sometimes heating. We are beginning to understand that the modulation of climate by dust layers complicates climate models. The Earth is experiencing global warming from rising carbon dioxide and possibly cooling, but sometimes also warming from sunlight interacting with dust and pollution blown across oceans. (Shaw 2007b)

Shaw published his findings in a remarkable paper that includes both measurements and photographs of a thick layer of dust across the western horizon and a bright aureole around the solar disk. Back trajectories of the movement of air provided by Japan helped establish that the dust originated from the Gobi Desert (Shaw 1979b).

Shaw's assertion that dust had arrived at Hawai'i from Asia created a stir among the scientists who reviewed his paper:

> In fact there was a great deal of hostility, when I wrote a paper in 1979, proposing that desert dust blown into the air over the Gobi Desert crosses the Pacific Ocean and makes its way to the Hawaiian Islands. This was an unpopular idea and it made a lot of people angry. One reviewer even wrote, "This idea is obviously erroneous and I reject the paper unequivocally."
>
> But in spite of that, we got the paper published and now we realize that the dust not only commonly reaches Hawaii, but California and Alaska, also. Additionally, this dust contains pollutants in increasing amounts as China expands its industrial activities. The dust impacts climate. (Shaw 2007b)

Subsequent studies at MLO transformed the Asian dust hypothesis into reality. There was finally an explanation for the seasonal cycles in atmospheric transmission observed by Howard Ellis during the 1960s. A California professor was not surprised by any of this, for he had observed and measured thick layers of dust over Hawai'i during May 1977, two years before Glenn Shaw's discovery. Unfortunately, he apparently did not publish his findings promptly, for he was conducting a major experiment at MLO using a new kind of automated instrument that measures the intensity of direct sunlight at a range of wavelengths. The professor was Kinsell Coulson, and he was next in line to direct the Mauna Loa Observatory

MLO Director Kinsell Coulson (1979–1983)

Howard Ellis' third assignment as acting director of MLO ended on June 11, 1979, when Dr. Kinsell L. Coulson officially took over. Coulson and his wife Vivien had lived in Hilo during the first six months of 1977 while Coulson was working on the mountain making measurements of skylight, and Judy Pereira (1996) and most of the MLO staff had gotten to know and like them:

> It was so nice to have people here who were probably not going to stay for only a couple of years. And Kin was a known quantity; previously there was always the concern of "What is this new director going to be like?" Kin was a very likeable person whom the staff immediately embraced. He pulled duty on the mountain just like every observer at MLO. He had an open door policy and encouraged everyone to see him about anything that bothered them. The budget was still a problem, but we were getting used to being told to watch our spending, and sometimes we'd even listen.

Coulson was an internationally known solar radiation expert, and in 1975 he had published a book that describes both historical and modern methods for measuring sunlight (Coulson 1975). Among his specialties was the effect of aerosols on the polarization of the sky, a phenomenon that can be observed by gazing at the sky through polarized sunglasses or the viewfinder of a camera equipped with a polarizing filter. If the sky is clear and blue, a broad, dark band will be seen as an arc in the sky 90 degrees opposite the sun. The degree of polarization of this arc is reduced by dust and haze, and Frederick Volz and others before them had devised simple methods for measuring the reduction. Coulson went much farther and built a computer-controlled instrument for measuring sky polarization. It is this instrument that he had used at MLO during his six-month visit in 1977 (Coulson 1988: 272–273).

It was also this instrument that allowed Coulson to measure the optical properties of thick bands of dust he observed over MLO during May 1977, which he described a decade later:

> A second and more persistent type of turbidity at Mauna Loa is that due to dust transported across the Western Pacific during the spring months from the large desert areas of northern China and Mongolia. At first glance, it seems unlikely that significant quantities of dust would be carried a quarter of the way around the world, but computations of atmospheric trajectories show exactly that to be the case. Further evidence of eastern Asia as the source of the dust comes from examinations of the dust particles themselves, and from the correlation between dust episodes in Hawaii with observations of dust storms in northern China a few days previously.
>
> These episodes of high-level dust in the atmosphere are almost universally confined to the spring months of March, April and May. Although they do not necessarily occur annually, they do occur most years, and some years see more than one dust intrusion. [Hawai'i is often south of the principal Asian dust corridor to North America.] A dusty period during the polarization measurements of 1977 at Mauna Loa occurred during the first few days of May [two years before Glenn Shaw's observations]. Distinct dust layers were visible in all directions from the Observatory, and the fact that the peak of Mount Haleakala on the island of Maui at a distance of about 140 km was obscured by the dust indicated that the dust was in the middle troposphere. (Coulson 1988: 289)

Coulson's appointment as the new MLO director would allow him to become increasingly familiar with the annual invasion of Asian dust. Meanwhile, other new personnel arrived soon after Coulson:

> [Lily Kam completed two years as a part-time student employee, and 14-year old Albert Bungula arrived to spend seven weeks working at MLO.] We had our final post-doc to date, Dr. Allen (Al) Dittenhofer, join us for a year and a half. He collected rain water samples and analyzed them with an electron microscope owned by and used at the University of Hawaii-Hilo campus. (Pereira 1996)

More Federal Building Headaches (1979)

The Federal Building offices continued to present problems to the MLO staff. On May 9, 1979, Dr. Coulson wrote MLO veteran Bernard Mendonca, who was then the deputy director of GMCC in Boulder, about the situation:

> All of the books, journals, and other reference materials available for use by the observatory are still in the basement. As a result of the excessive humidity which exists there, the books are getting moldy and rapidly deteriorating so for the mere preservation of this valuable resource some other location must be found.
>
> A second consideration for a new location of the library is that the basement is an unpleasant, and even oppressive place in which to use the library. The air is dank and musty, the lighting is bad, and there is no place for a library table at which the reader may sit. The unpleasantness of the location has recently been further exacerbated by hot air from an air conditioner being discharged directly into the library space. The net result of all these features is that the library is virtually unused and unusable in its present location.

Coulson went on to request that MLO be allowed to rent a recently vacated room adjacent to their office space on the second floor of the Federal Building, a request that was eventually approved.

Carbon Dioxide Issues (1979–1983)

In July 1979, the temperature sensitivity of the Scripps carbon dioxide analyzer was once again a high priority. The temperature regulation of the air conditioning system installed during the 1977 remodeling project allowed fluctuations of several degrees, and this was enough to cause problems. Coulson described the latest attempt to resolve the problem, and an unexpected consequence, in the July 1979 Activity Report:

> The air conditioning system in the main Observatory building was modified during the month to achieve a more stable temperature for the Scripps carbon dioxide analyzer. Dave Moss [of Scripps] spent several days at the Observatory overseeing the work of the air conditioning company. . . . After the modification the room temperature is found to vary by less than one degree, which is adequate for the operation of the Scripps instrument. Unfortunately, the method requires continuous operation of the compressor on the air conditioner, thereby involving a continuous power drain of nearly 8 kilowatts. At the present price of electricity in Hilo, the cost of running the air conditioner alone is about $630 per month. Such large expenditures of both energy and dollars [are] obviously excessive for the task, so efforts are under way to develop an alternative method of achieving the required temperature stabilization. (Coulson 1979a)

A year later, the temperature stability of the main building was again an issue, and Dave Moss visited MLO twice in July 1980 to add new air intakes to the Scripps system and again work on stabilizing the building's temperature. An agreement was reached with Moss that the temperature control system would be revised by October (Coulson 1980c).

NOAA's URAS-2 CO_2 analyzer did not have the frustrating temperature sensitivity of Scripps' much older APC analyzer, which required what an MLO staffer described as "babysitting." The MLO Activity Reports described very few problems with the URAS-2 until June 15, 1983, when the instrument's internal infrared source ceased working. In an ironic coincidence, Charles Keeling visited MLO on June 13, where after his signature in the *Visitors* book he wrote, "1st visit since 30 June 1965!" The next day Keeling was in Hilo for a rare three-day visit to discuss with Coulson possible modifications of the Scripps CO_2 system, which was working well during his visit. The URAS-2 analyzer was promptly fixed, but it failed again the day after Keeling left. The Activity Report notes that the URAS-2 problem was traced to a defective transistor, but it does not disclose whether Dr. Keeling was informed that his competition had finally developed a bug (Coulson 1983d).

Tom DeFoor Arrives at MLO (1979–1993)

Electronics engineer Richard Cram was transferred in June 1979, and Tom DeFoor, who was also an electronics engineer, arrived as his replacement in December. DeFoor was perfect for the job, for he had prior experience with the NOAA Corps and with the Climate Monitoring and Diagnostics Laboratory in Boulder. He arrived in Hilo immediately after a two-year stint in Barrow, Alaska, and moments after becoming acquainted with his new desk the phone rang:

> I was down at the office getting introduced to everybody and finding my desk and putting my pencils in the drawer, that kind of stuff. And they called down from the observatory and said the computer was down. They could not get it up. Can they ask me if I could go up there and help them? The [Hilo staff] kind of pointed me in that direction.
>
> I had not been up there yet. But you [Mims] have been up there, it is pretty simple. You go one way for 27 miles; you make a left, and stay on that for 17 miles or whatever. But I got up there.
>
> Now, I had worked at the Barrow Observatory for two years, and I had loaded that . . . bootstrap [program into the computer by toggling the panel switches] hundreds of times, and I could do it with my eyes closed. And I got up there after driving up there as fast as I could,

> and I tried toggling it in, and my fingers were not doing what they were supposed to be doing. And I would have to just really slow down and concentrate on what every finger was doing. I had to concentrate on it even though it was an automatic thing. And that, to me, dramatized the facts of the altitude, because that was a trivial task. (DeFoor, video interview with author, Boulder, Colorado, May 3, 2007)

The altitude symptoms DeFoor experienced during his unexpectedly quick introduction to MLO are well known to the staff, and even those who never had symptoms notice the effect of altitude on those who do. The altitude effects are exacerbated by the stress of driving the MLO road, as DeFoor asserts:

> Some days you hardly noticed it; other days you just did not feel good. And I think, again, the reduced oxygen impact was different, depending on what condition your body was in at the time. The more often I went up . . . the less impact it seemed to have. But the drive was long; it was an hour each way, and that is a long time to spend, two hours is a long time to spend in your car to go to work every day. . . . There was hardly any point [on that road where] you left the steering wheel in the center position. You were constantly turning one way or the other, braking and accelerating. So it was wearing on you.

It was about this time that Judy Pereira began experiencing altitude problems on the mountain. She had been able to attend the twentieth anniversary dedication, but she left a hint of what was to come after her signature in the *Visitors* book: "Yuk!"

> Well, at that time it was fine, I had no problem with altitude or anything. I was able to make sure that all the food and things that were there for the party were ready. Subsequently, it was more and more difficult for me to go up there, I began getting dizzy, headaches, and I developed sinusitis, which causes pressure in the head as you go up, but worse coming down. So, I'm glad I had the opportunity to go up as often as I did in the early years, because after that, the last 15 years, I think, 10 or 15 years, I never went up. (Pereira, video interview with author, Boulder, Colorado, April 27, 2007)

The Mauna Loa Museum (1979–1983)

MLO founder Robert Simpson, who had become a professor at the University of Virginia, visited the Hilo MLO office on November 2, 1979, and signed in at the observatory for the third time on November 5. Coulson, who was fascinated by history—especially the history of meteorological instruments and those that measure all aspects of the light from the sun and sky—was then planning a museum for MLO. On December 17, room 205 of the Federal Building was assigned to MLO, and several days later some cabinets assembled by the Hilo Community College were delivered:

> The cabinets will be used for displaying the instruments, books, and other materials being assembled for a meteorological instruments museum at MLO. (Coulson 1980a)

A cross section of atmospheric scientists made donations to the museum. NOAA's Kirby Hanson donated the slide rule used by the late Harry Wexler, the Weather Bureau scientist who had played a key role in the establishment of MLO. Professor J. Sauer of UCLA donated a 1930 barograph. Many donations were made by Robert Roosen, who specialized in evaluating old solar measurements made since the turn of the century by the Smithsonian Astrophysical Observatory (APO). Roosen donated early books, photographs, and letters, and he and Ronald Angione of San Diego State University donated a large glass plate on which was recorded sunlight intensity measured by the Smithsonian APO at its solar observatory at Mt. Harqua, Arizona, on February 2, 1922. The collection grew so large that in April 1983 a catalog of the many items was prepared. "The catalog includes the history of the various items, as well as the names of their donors" (Coulson 1983b).

The Major Storm of 1980

While the weather at MLO is generally moderate, thunderstorms with dangerous lightning can arrive, snow can occasionally block the road, and major windstorms sometimes occur. An especially severe storm occurred from January 8 to 12, 1980. The storm brought wind gusts of up to 85 miles per hour (137 km per hour), more than 5 inches (127 mm) of rain, and a power outage. The roof gutter along one side of the main building was damaged, and considerable water entered the main building, the AEC Building, and the lidar building. Fortunately there was very little water damage (Coulson 1980b), but reaching the observatory required a major detour, as Tom DeFoor (video interview with author, Boulder, Colorado, May 3, 2007) recalls:

> We had horrendous rainstorms. . . . I was going to the mountain with Dr. Coulson, and . . . we got above Upper Kaumana, and all of a sudden there was a big water puddle. And we drove into it very slowly, and when the water got up to the door level we went back to the office and we waited an hour and tried it again. I think we tried it three times that day. On the third time we met somebody . . . who had swum across it, and he said, "This is over six feet deep!"
>
> And that lasted for [about] a week, and I think I went up every day during that time. . . . We drove all the way around the mountain [and] came up from the Kona side. . . . Basically it was damage control—trying to keep everything running. There were power outages, of course. The rain—it was heavy rain at the observatory level, consistent heavy rain for days. . . . And so we kept things running the best we could. Obviously, we were not making any Dobson or lidar observations or anything like that. But we were keeping the CO_2 running and everything else that was practical.

A few weeks after the big storm, Ginnie Ferrell of the University of Alaska returned for a one-year stay to operate a 12-wavelength solar radiometer developed by Glenn Shaw. Ferrell signed in at MLO on January 29 and wrote, "Here to operate the radiometer." The radiometer worked well during its year at MLO, and on January 30, 1980, the project was completed, and the radiometer was packed for its return to Fairbanks. Shaw (personal communication with author, April 2, 2008) reports that he hopes to modify the instrument for unattended operation and return it to MLO for a new set of measurements to be compared with those made three decades before.

Tracking Volcanic Eruptions (1980–1983)

Since Ronald Fegley installed the MLO lidar in 1972, several volcanic aerosol clouds had been detected as they drifted over Hawai'i in the lower stratosphere.

On May 18, 1980, the north face of Mount St. Helens in Washington collapsed during a nine-hour eruption of the volcano (Forest Service 2007). Some of the volcanic debris ended up in the stratosphere, and on July 8 the MLO lidar detected a layer of Mount St. Helens debris. NOAA issued a press release about the detection, after which Coulson fielded inquiries from newspapers and radio stations in Honolulu and Hilo (Coulson 1980c). The frequency of lidar observations was increased, and the volcanic cloud was detected throughout August. It had largely faded away by the following March.

During June 1981, Coulson spent considerable time helping design a new building for the lidar system, which was slated to be moved from the dome it occupied. When August arrived, no government-approved contractors were willing to build the structure for the available funds. In September a builder in Oakland, California, was selected, but the company's blueprints for the building didn't fully meet the specifications of the original plans. The project was turned over to the purchasing office in Boulder, and the building was finally manufactured and shipped in pieces to Hilo in January 1982. But it remained in storage, since the bids for its erection on the mountain were higher than anticipated. Meanwhile, the lidar found an aerosol cloud of unknown origin on January 27, 1982:

> MLO personnel granted a great number of telephone, radio, and television interviews regarding the cloud

(which eventually was termed the "mystery cloud"). (Coulson 1982)

By April the "Mystery Cloud" was supplanted by the "Monster Cloud," both of which were named by Brian Toon (Ellsworth Dutton, personal communications with author, May 20, 2008). Coulson provides details about this multilayered event in the April 1982 Activities Report:

> 4/30 Cloud density had increased again, with multiple layers between 21.4 and 28.3 km. The lower portion of the "Monster Cloud" was superimposed on the remainder of the "Mystery Cloud" in the 18 km region. A layer with a BR [backscatter ratio] = 9 was centered at 28.3 km, the highest altitude of any layer yet observed at MLO.

The "Monster Cloud" was indeed monstrous. It was formed by the eruption of El Chichón volcano in Mexico, and its eruption plume passed directly over MLO. Tom DeFoor (video interview with author, Boulder, Colorado, May 3, 2007), who had inherited the lidar responsibilities from Howard Ellis, was on the mountain when the El Chichón cloud arrived:

> I was up there working on the lidar, and it was getting dark and looked like it might be a little cloudy. I was not paying much attention, because I was working on the system and then I fired it up after it got dark. Well, it got dark and I saw stars, and it did not look like it was cloudy. I fired it up, and I think I had saturated signals at 28 kilometers. . . . Initially, I thought I had misconfigured the system, and so I checked after I saw the return on the oscilloscope. I rechecked everything . . . several times . . . before I trusted what I was seeing. . . .
>
> I came down at probably one, two o'clock in the morning, came in the next morning and then showed them what I had. I think [Dr. Coulson] came out with me the next night. We did it again, and then we . . . checked with other people and nobody was seeing anything and nobody expected to see what we were seeing here. So . . . we stepped up the observing schedule.

The El Chichón eruption proved the scientific value of the MLO lidar and provided Coulson with a perfect opportunity to compare the lidar measurements with his method of determining the height of aerosol layers by means of changes in skylight polarization shortly before sunrise or after sunset. Coulson also made a series of spectacular photographs of brilliantly colorful volcano twilights that occurred following the eruption (Plates 30–32). In the same interview, Tom DeFoor recalled the El Chichón twilights:

> They were spectacular. . . . You had just brilliant, brilliant orange and red sunsets, which you normally never get up there. But you could also see lots of structure in it. The crepuscular rays were much brighter than normal. Yeah, it was beautiful. Kin Coulson took some pictures up there one evening when we were up there together, and they were spectacular pictures.

By April 1982 the new lidar was complete except for minor modifications to correct a design flaw in the movable roof opening. The roof was modified in May, and the new building was ready to receive the lidar.

While the MLO lidar played a significant role in tracking the presence and elevation of the El Chichón aerosol cloud, Dr. Ellsworth Dutton (personal communications with author, May 20, 2008) notes that a far simpler, much older instrument at MLO also provided almost daily observations of the cloud:

> It is unfortunate that the role of the Ellis curve in both the Mystery and Monster volcanic clouds is not better documented. Since the SAGE satellites were not operational in 1982, I believe that both the Mystery and Monster clouds were simultaneously discovered in the Ellis curve and the lidar record at MLO. There was knowledge of the El Chichón eruption before the April transmission data were processed, but I think Bernard Mendonca saw an anomaly in the curve in the 1982 January transmission data that later became the Mystery Cloud. For months afterwards it was the Ellis curve that put the eruptions into perspective relative to other historic erup-

tions, primarily Agung, which was captured in the curve given in Mendonca et al.'s paper in *Science* (1978).

International Activities (1980)

MLO's role in tracking giant aerosol plumes from major volcanic eruptions would soon enhance its reputation as one of the world's premier atmospheric monitoring observatories. Meanwhile, the World Meteorological Organization added to MLO's international responsibilities by designating GMCC in Boulder as the World Dobson Spectrophotometer Calibration Center. GMCC's Dr. John DeLuisi (1981) describes the details:

> A program that began in 1977 under the auspices of the WMO Global Ozone Research and Monitoring Project to upgrade Dobson spectrophotometers in use throughout the world was continued. At the request of WMO, [Walter] Komhyr prepared *Operations Handbook—Ozone Observations with a Dobson Spectrophotometer* (Komhyr, 1980) to standardize Dobson instrument operating and calibrating procedures throughout the world. World Primary Standard Dobson Instrument no. 83 was recalibrated at Mauna Loa Observatory, Hawaii.

Various other international activities were also described by DeLuisi, for GMCC and MLO scientists were traveling around the world to assist other countries with a wide variety of atmospheric monitoring projects.

The Professor Tries to Buy a Computer (1980–1981)

Being a retired university professor, Kinsell Coulson was an advocate of formal training, as he demonstrated when he took a course on understanding human behavior in 1979. In October 1981, he arranged for the MLO staff to attend a six-session writing course taught by two professors at the University of Hawai'i at Hilo. He also arranged for Tom DeFoor to take five days of lidar training in Boulder. Later he arranged for the staff to take a video course on the BASIC computer language, for Coulson very much wanted MLO to enter the computer age.

During the fall of 1980, student employees came and went while a frustrated Coulson worked hard to order a Digital Equipment Corporation LSI-11 microcomputer for the observatory. The experience demonstrated as well as any other why, as Judy Pereira (1996) notes, "It took time for him to adjust to the mega-bureaucracy that we dealt with as a matter of course."

In April 1981, the computer was finally shipped to Hilo, but not in the manner specified by Coulson, whose frustration escalated to exasperation:

> The computer system, the delivery of which is long overdue, was shipped from New York by surface freight instead of by air freight as had been specified. Thus, further delay in delivery was created, apparently by incompetent personnel in the company. (Coulson 1981c)

The LSI-11 finally arrived on May 5, but MLO's attempt to fully enter the computer age received still another major setback:

> Inspection of the LSI-11 microcomputer system from Cyberchron after the system was finally delivered to MLO on May 5 determined the system to be unsatisfactory. In brief: 1) The system's disk unit arrived with damaged heads; 2) The computer circuit boards were shipped in the computer chassis (without benefit of adequate packaging support) and as a result were all torn loose during shipment and spent at least part of the surface journey crashing around the bottom of the chassis. Because the disk unit is inoperative it is not known if any modules were damaged to the extent that the electrical performance might be affected; 3) The system was not in compliance with the requirement that the system: a) be housed in a rack or roll-around cabinet; and b) the microcomputer enclosure include front panel Reset, Halt/Enable and Boot switches. (Coulson 1981d)

Coulson sent a list of the many deficiencies back to the purchasing office in Boulder, and it was decided that the company would be given thirty days to repair the damage and to add the missing items.

> However, as of this writing Cyberchron has not been notified officially in writing. This is mostly the result of delays in Purchasing and a "lost in the mail" letter to Cyberchron. (Coulson 1981d)

The entire computer system was returned to the supplier in June with appropriate paperwork from NOAA requiring complete repairs and full compliance with the specifications agreed to when the computer was purchased. Cyberchron was given thirty days to comply, and Coulson made sure to note that the deadline would expire in mid-July. The computer was "at long last" received in time, in proper working condition and in compliance with the purchase order's specifications.

Coulson was so optimistic about the potential role the LSI-11 and its ten-megabyte drive would play in supporting the observatory that he continued working on the purchase of a second computer, this one specifically for the lidar system. This time he was better prepared when he ran into the paperwork bureaucracy in Boulder.

The Lava Barrier (1980)

As the significance of MLO's observations increased over the years, concerns about a volcano eruption increased. Under John Miller, the observatory prepared contingency plans for fleeing a future eruption. Coulson added a new strategy: diverting a future lava flow around the observatory by means of a huge V-shaped barrier pointed upward toward Moku'āweoweo, the enormous crater at the summit of the mountain. In November 1980, Coulson reported that a topographical map of the observatory area had been completed in preparation for the possible construction of a lava diversion barrier:

> The map was made from aerial photos using the latest mapping capability of the Lunar Laboratory of the USGS [US Geological Survey] in Flagstaff, Arizona. According to Dr. J. Lockwood, who is responsible for arrangements for the map, the accuracy is an order of magnitude better than that for any other area of the state, and the MLO buildings are located more accurately on the map than would be possible by any conventional type survey. Dr. Lockwood has already indicated the most appropriate configuration for the barriers, so actual construction can be started as soon as official approval is obtained. (Coulson 1980d)

In December 1980, ARL provided $5,000 to continue the lava barrier study, and by the following July Coulson reported significant progress:

> Dr. Jack Lockwood is progressing well on plans for construction of lava diversion barriers above the observatory. An expert on modeling of lava flows (Dr. Moore, USGS, Menlo Park) will spend two weeks in Hawaii in early August for inspection of the site and refinement of the model. It is anticipated that a special allocation of about $20,000 will be requested in FY 1982 for actual construction of the barriers. (Coulson 1981e)

The project moved along at a snail's pace, and by January 1983 the lava above MLO was staked out with the outline of the lava barrier. Jack Lockwood was ready to begin directing the contractor's bulldozers, but Coulson was about to encounter an unexpected delay in the project that might one day save the observatory from destruction. His previous bureaucratic adventures in buying computers and finding employees were child's play compared with what he was about to experience:

> The documentation necessary for obtaining authorization from the State of Hawaii for construction of the lava diversion barriers above the observatory was submitted to the Hilo office of the State Department of Land and Conservation. Granting of the authorization will be a long and involved process, requiring perhaps two months delay. In the meantime, preliminary documents have been sent to the contracting office in Seattle to permit them to look into the legal ramifications of the U.S. Government

> making changes (the term they use is "improvements") to State property. It is expected that the paperwork can proceed by May 1. There is some urgency regarding the barriers because of the significant and continuing increase of seismic activity on the mountain. (Coulson 1983a)

Reality sank in during May, when Coulson increased from two to six months the predicted delay in receiving approval from the State. There was good reason to be concerned about the delay:

> The time requirement, however, is indeed a problem in view of the fact that the mountain just now shows pronounced seismic activity with the inflation of the summit region, and the volcanologists are predicting an eruption within a year. (Coulson 1983c)

In September the Board of Land and Natural Resources finally approved construction of the lava barriers at MLO, but it was contingent on still one more detail:

> Final approval is contingent on a ruling by the office of the Attorney General on liability responsibility of the State for diversion of the [lava] flow. The fact that any diversion of the flow could not possibly imperil any structures or private property seems to be immaterial. Lava diversion is a very sensitive issue in Hawaii, and great care must be taken in the setting of precedents. There is presently no indication as to when the Attorney General will rule on the application, but probably not for some months. (Coulson 1983e)

The lava barrier was Kinsell Coulson's most important MLO project, but he would retire before a final decision was reached about its future. Another retirement would precede his.

Coping with Bureaucracy and Managing People (1980–1983)

More than six months had elapsed between Richard Cram's departure from Hilo and Tom DeFoor's arrival, and Kinsell Coulson was concerned about how Cram's absence required extra observing duties by the remaining staff. He included the need for a replacement as the first item in each installment of the monthly Activities Report until December when DeFoor arrived. There were several other instances in which there were lengthy delays in receiving replacement staff, and Coulson dutifully reported the details about unfulfilled promises and unmet deadlines. Then there were events that even Coulson and the bureaucracy could not anticipate, as when Al Shibata fell ill and was unable to work. Shibata eventually returned to work after heart bypass surgery.

John Bortniak, who had been assigned to the NOAA ship *Albatross*, had left MLO on February 26, 1982, leaving the mountain staff shorthanded. Arne Austring was hired on an emergency, temporary basis to replace Bortniak on May 10. Nearly a year and a half later, Austring was finally elevated to a permanent, full-time position—or so he thought:

> The employment forms received from the Western Area Service Center on the appointment of Arne Austring indicated that his appointment was on a permanent full-time basis, in spite of previous information that the appointment was for part-time. As a result of the personnel document, Mr. Austring started working full time. A few weeks later, however, personnel discovered that they had made a mistake and that the part-time proviso had been correct after all. As a result, Mr. Austring was again put on part-time, but hopefully the appointment can be upgraded to full-time early in the new fiscal year. (Coulson 1983e)

Austring finally received a full-time appointment on November 28, 1983, two days before Coulson's retirement party. He would go on to make major contributions to MLO's use of computers.

Judy Pereira had a close-up view of Coulson's willingness to confront government bureaucracy, which she discusses in her memoir (1996):

> Kin was a kindly country boy at heart, and everyone loved him and his anecdotal stories which were always

> reminiscences of his youth and were always appropriate for the occasion. But he had trouble shifting from university mentality to government mentality. It took time for him to adjust to the mega-bureaucracy that we dealt with as a matter of course. Kindly as he was, he always stood up for what he believed was right, and several times I witnessed the "wrath of Kin" in his dealings with cooperators and GMCC personnel as well. Placing bias aside, I can honestly say if he was angry enough to make an issue of something, he was nearly always right.

Because of his academic background, Coulson was troubled by the MLO budget process, and Pereira (video interview with author, Boulder, Colorado, April 27, 2007) helped him understand and cope with the process:

> At the university they were given so much freedom, budget was not a problem. So when he came here he was like, well, why can't we just buy this? I said, "Well, at the university it's done a little differently. Here we have to go through several offices before we can get final approval. . . . Ridiculous it may be, but that's what we have to do." And, it took him a while to get used to that mentality, but he did, and pretty soon we were doing things the way they were supposed to be done. . . . I was able to help him through that. I prepared all paperwork that needed to be done for his signature [and] explained why we were doing it, because it did seem like nonsense.

Often Pereira, who had taught herself how to perform most of her duties, took care of business entirely on her own, as when she sent an official memo about a space overcharge:

> 6/5/80
> To: Betty Perry
> From: Judy Pereira
> Mauna Loa Obs.
>
> This GSA 1476 shows the exact space we presently occupy. The GSA space assignment sheet (attached) is incorrectly charging us for 772 [square feet] more than we occupy. I have talked to our local GSA Bldgs. Mgr and he is working to straighten this out. Aloha!

John Chin (video interview with author, Hilo, Hawai'i, December 4, 2006) remembers management changes that Coulson brought to the observatory, some of which reminded him of Jack Pales:

> He had a different type of operation system. He also was another hard worker. He insisted [that everyone] take observations up the mountain on his turn. So everybody working [at the] observatory had to take his turn to do observation work at least once a week. The rest of time you can spend . . . on your own project. So that was how the daily observations for observers started.

Alan Yoshinaga (video interview with author, Hilo, Hawai'i, December 4, 2006) remembers that Coulson did not expect more from his staff than he himself was willing to perform:

> Kin Coulson, at first, when I met him, I was kind of in awe of him. I have read some of his accomplishments, I said, "Boy, what has this guy come to Mauna Loa for?" He is "the guy" in solar energy, and he came here and he talked to us; he got us going, and I thought to myself, "Oh, what is this guy going to do? Sit in the office all day?" . . . [But] he took a day on the mountain; he took an observing day.

Coulson's penchant for working on the mountain concerned his wife but earned him the respect of his staff, as Yoshinaga recalled in the December 4 interview:

> I used to have rain collectors [for an acid rain project], about ten rain collectors along the whole Saddle Road from Hilo to Mauna Loa. And we had to walk through rocky lava terrain. And at the 19-mile marker . . . we had to walk about 150 yards over the lava up and down ridges and that was one of our daily collections. [One day Coulson fell in the lava] and he came back, and he was just ripped through his hands, elbow, and his knees. He did not complain; he suffered for many weeks. And yet

> the next mountain day he . . . collected my sample again. [He was] a real hard worker; he stayed up at the mountain at nights. . . . [He was] just a great guy . . . a true scientist and was just awesome. He was there whenever things had to be done or interesting items to be done. He was there. And he was not the kind of guy who would say, "I'm the boss. You go up and you do it." He would lead by example. . . . And so whenever we had to do stuff that was out of the ordinary, we would not think twice about it.

Coulson knew he needed help in coping with the human side of his new management responsibilities and understanding how best to adapt to the differences between academic and government bureaucracies. He reported in an Activity Report (Coulson 1979b) that he had attended a three-day course entitled "Understanding and Managing Human Behavior" during November 1979.

While Coulson had a low tolerance for bureaucratic ineptitude, he had a soft heart for people in need, including a scientist who had published important papers in atmospheric and climate science and shared his interest in the history of atmospheric measurements:

> Kin had a habit of befriending vagabonds. He became benefactor of a former scientist who had left the real world to buy property and live in Kalapana (an area since destroyed by lava) to commune with the psyche of the earth. This man became quite dysfunctional in the traditional sense of normal life, and quickly found himself penniless. But Kin, over the objections of his wife, would take him food and loan him money to get him through the tough times. As it turned out, he sold his property, moved back to the mainland, and became a working scientist again. Kin had a way of knowing how to deal with people that others considered lost causes. (Pereira 1996)

The former scientist recently verified Judy Pereira's account and said of Coulson, "He was so good to me." He responded to Coulson's generosity by contributing some rare artifacts and papers to the Mauna Loa Museum. After spending time on the mainland, he returned to Hawai'i to reside in a tent and resume his itinerant lifestyle.

Howard Ellis Retires (1981)

Howard T. Ellis directed MLO from 1964 to 1966, the most critical time in the observatory's history. Major budget cuts nearly closed the observatory and reduced the staff to just Ellis, Bernard Mendonca, and Malcolm "Mike" Keyes. After a year Keyes was reassigned, leaving only Ellis and Mendonca on the staff. It was widely agreed by MLO veterans that Ellis was responsible for the survival of the observatory during this crucial period, and Bernard Mendonca and John Chin summed up their views as follows:

> He [did] more than anyone else to establish and keep MLO going in the early days. Without his efforts . . . MLO may just as easily have suffered another fate and not survived. (Mendonca, personal communication with author, March 19, 2008)

> We have to give credit to Howard Ellis. He was the one that was hard headed. He kept [MLO] going. Otherwise there [would] be no more observatory at all, because nobody [was] actually interested. . . . He was the one that kept it going. (Chin, video interview with author, Hilo, Hawai'i, December 4, 2006)

Ellis had a love-hate relationship with Mauna Loa, for while he loved the challenges of his job, in the old days he hated spending nights alone at MLO. He did so anyway and spent countless nights on the mountain to tend Charles Keeling's temperamental carbon dioxide analyzer and to make sure the MLO generator was running. Ellis was responsible for the shortcut road that greatly simplified travel to MLO, and the observatory would very likely have been closed without his leadership in 1964. He played a key role in Gerard Kuiper's initial explorations of Mauna Loa and Mauna Kea as potential astronomical observatory sites. He coordinated the placement of the High Altitude Observatory at MLO. He developed the elegant methodology that led to the Ellis curve, a measure of the transmission of sunlight through the atmosphere that dates back to November 1957 and is the longest series of data collected by the International Geophysical Year instruments at MLO. The paper that Ellis and Pueschel (1971) published in *Science,* the world's leading scientific journal, on the long-term stability of the atmospheric transmission at MLO was among the most important reports to come from the observatory. The Ellis curve led to the discovery of changes in atmospheric transmission caused by various volcanic eruptions.

Most of his former coworkers described Ellis as a polite southern gentleman who led by suggesting rather than by ordering, and most enjoyed his company during the drives up and down the mountain. John Chin remembered him as "a very pleasing, easy-going fellow, easy to talk to." Mike Keyes (video interview with author, Keauhou Bay, Hawai'i, November 29, 2007), who helped keep MLO alive during the 1964–1965 budget crisis, remembers that he "was dedicated to the point where it hurt." Alan Yoshinaga (video interview with author, Hilo, Hawai'i, December 4, 2006) says, "I really got along with him. He was a giant of a man, real mellow, real nice guy." Former MLO director Rudolf Pueschel had great respect for Ellis and his science:

> His record shows how dedicated a scientist he is, how precise his measurements were and what potentials they had, and we kept a fairly good relationship along the way, and I hope the feeling has been mutual. (Pueschel, video interview with author, Keauhou Bay, Hawai'i, November 29, 2007)

But some found Ellis inscrutably enigmatic, for he often had his own ideas about how best to resolve a problem, conduct a program, or end his participation in a project. When his supervisors far away on the mainland sent new projects and special requests during the time only he and Mendonca were left to staff the observatory, Ellis sometimes ignored or delayed their requests while he struggled to take care of the responsibilities he already had. Then there were the unpredictable situations and events for which no one had budgeted or planned, like fixing broken vehicles, repairing the road, and caring for visitors and scientists who arrived at MLO with no food or supplies. "On Their Own," the title of his MLO memoir, sums up Ellis' perception of getting things done on the mountain, and he reflected on this in an interview with the author (Hilo, Hawai'i, December 2006):

> That was all I ever heard or thought about. . . . I wanted to do all kind of things "on their own." It was dealing with local people to get fuel and automobile repair. And this is kind of a public relations thing. And I guess you would call that "on your own." . . . And, of course, the road; I fooled around with the road. So I think it helped acclimate me to change in elevation. It is only about 50 minutes when I go up, and when I came down I tried to throw rocks [away from] the worse places. Some people would help me. Some people said I was crazy.

Ellis' coworkers at MLO understood what he meant by "On Their Own":

> I gather that was his way of letting people know that a lot of the work was done with a lot of personal sacrifice. . . . Funding was very poor; it was very irregular. And so when people came over here to work, they were promised many things, and after they got here, it was not as they were promised. So a lot of things were done on their own, on their own time, on weekends. (Yoshinaga, video interview with author, Hilo, Hawai'i, December 4, 2006)

Ellis invested thousands of hours of his own time maintaining MLO in the early days, and in the 2006 interview he reflected back on his long-term solar radiation study and his role in operating Charles Keeling's carbon dioxide project:

> Yes, it turned out that it was just really a very important thing. And this was a time when people called that . . . oddball stuff. It is not oddball anymore.

Ellis was willing to fill in when needed, and Judy Pereira (video interview with author, Boulder, Colorado, April 27, 2007) points out that he spent more time on the mountain than most of the staff:

> He would go up more often than most people. Everybody had at least one day that they had to man the observatory

> for the regular observations. Howard Ellis had one, but he would go up, he would take the place of other people if they were on vacation, if they were sick. He was happy to do that.

Besides directing MLO through its major crisis from 1964 to 1966, Ellis also directed MLO between directors Lothar Ruhnke and Rudolf Pueschel and again between directors John Miller and Kinsell Coulson, the latter three of whom held the doctorate that Ellis lacked. As he had been replaced as director twice before, it was natural for the MLO staff to assume that a change in Ellis' work habits (primarily that he stopped going in to the MLO Hilo office), which occurred shortly after Coulson's arrival, was due to his disappointment in being passed over still again. Judy Pereira reflected in the same interview on what was the common view of Ellis' coworkers at the time:

> Howard was going through some internal issues about wondering why he wasn't being named director. Why was this period, this long period necessary when he was the logical choice? He had been waiting to be named director since he came in the early '60s. When they named Kin Coulson, Howard withdrew. . . . He really believed that he was sent out here to eventually become director, and this was the logical time. It had been hard for him with the previous directors, but now that there was a break—this was the time that he felt was his. And it turned out not to be. . . . And, so, every day for [about a year] he went to work, I mean, he came to the building, got into the car that goes up the mountain with whoever was going up, and worked up there. Every day.

Ellis had always maintained that Harry Wexler had sent him to MLO to take over the directorship from Jack Pales in 1961, and Pereira and others on the MLO staff assumed that Ellis' withdrawal was due, at least in part, to his being passed over as director again. But Bernard Mendonca (telephone interview with author, June 7, 2008) explains that Ellis had accepted his status at MLO by the time Coulson arrived. As Ellis himself explains (telephone interview with author, June 8, 2008), he stopped going into the Hilo office for reasons of his own, including the fact that Coulson had assigned to other staff some of his main responsibilities on the mountain. Most of the staff had mountain duty only one or two days a week, but, as Pereira notes, Ellis made the trip up and down almost every workday. While some of his coworkers wondered why Ellis stayed away from the office, nearly all of them enjoyed his company and reported that he was a hard worker who made a good work partner on the mountain.

When Coulson spent the last six weeks of 1980 in California during an illness, John Chin served as acting director of MLO. After Coulson returned to duty, he announced that Howard Ellis was planning to retire:

> Efforts were initiated to abolish the position held by Howard Ellis, due to a reorientation of the observational program at MLO. If the effort is successful, Mr. Ellis expects to take early retirement from government service. (Coulson 1981a)

The effort was successful, and Ellis retired in January 1981. MLO director Kinsell Coulson praised his contributions to MLO in the January 1981 Activities Report:

> At his own request, Howard Ellis retired after approximately nineteen years at the Mauna Loa Observatory. As everyone knows, he made major contributions to the observatory programs throughout his tenure in Hilo. A fact that may not be so well known, however, is that Howard Ellis single handedly kept the programs going during some periods of budget cuts during the early 1960s. If he had not been sufficiently dedicated to actually live on the mountain during long periods, it is entirely possible that the observatory would not now be in existence. Thus both MLO and GMCC are deeply indebted to Howard Ellis for his foresight and dedication to the observatory programs. Howard's date of retirement is 1/17/81. (Coulson 1981b)

While Coulson's acknowledgment of Ellis' role at MLO was well deserved, there's more to the retirement story.

Ellis did retire at his own request, but he felt he had no other choice. While he was popular with his coworkers, his independent nature annoyed Coulson and some of the NOAA hierarchy in Boulder. His position had been officially abolished by those who felt it was time for him to go. Unfortunately, the timing would preclude him from completing the government service that would entitle him to full retirement benefits.

Howard Ellis was replaced by John Bortniak, who began work at MLO on March 2, 1981. On August 31, 1981, NOAA's Ellsworth Dutton signed in at MLO and wrote "Installed H_2O meter in NCAR dome. First trip to station." Dutton was to later play a prominent role in analyzing the MLO atmospheric transmission record using the method developed by Howard Ellis and in writing numerous scholarly papers about his findings.

Lisa McPherson (1981–1984)

MLO had a long record of hiring student workers, some of whom stayed on at the observatory after completing college. Lisa McPherson (now Wallace) was a chemistry major at the University of Hawai'i when she began working for the observatory during the summer of 1981. Among her duties was processing lidar data acquired by Tom DeFoor. She continued working at MLO and stayed through graduation in 1984. The highlight of Wallace's experience at MLO was meeting the many scientists who passed through. She also credited MLO for providing a stepping-stone for her career. While she originally planned a career in geology, her atmospheric work at MLO impressed the Air Surveillance and Analysis Section of Hawai'i's Department of Health, where she has worked to the present time (Wallace, telephone interview with author, April 9, 2008).

Cynthia McFee (1982–1985)

Several full-time women employees stand out in the history of the Mauna Loa Observatory, notably Adeline Fujisawa, the secretary who worked for Jack Pales, MLO's first director, and Judy Pereira, who was hired by Lothar Ruhnke in 1966 and who worked for eight MLO directors before retiring in 1999. Though they were not trained scientists, both Fujisawa and Pereira extracted data from chart recordings and performed various other scientific tasks. Judy Pereira also helped make particle measurements during eruptions at Kīlauea and helped prepare and launch ozonesonde balloons. In later years, Leslie Pajo managed the MLO office while also taking part in mountain observations.

On May 17, 1982, NOAA Corps officer Cynthia McFee became the first woman scientist to be assigned to MLO. McFee was a trained geologist who had spent a year at NOAA's South Pole Observatory prior to her assignment to Hawai'i and had visited MLO the first time on December 7, 1981. Prior to her assignment to Antarctica, she visited the Scripps Institution of Oceanography to meet Charles Keeling and his staff and to learn the proper procedure for collecting samples of air in evacuated glass flasks. While assigned to MLO, McFee performed observations on the mountain one day a week. She also prepared and launched ozonesondes once a week in Hilo, a task that then required three people due to the size of the balloons needed to lift the ozonesondes. Judy Pereira sometimes assisted McFee.

> Judy and I came up with checklists to make sure we got everything right. . . . She came out and helped me with the ozonesondes . . . in the beginning, because it was a new thing. . . . We just kind of figured it out. (McFee, video interview with author, Boulder, Colorado, April 27, 2007)

Tom Garcia (1982–1987)

Tom Garcia was no newcomer to Mauna Loa, for his father Charles was the director of the High Altitude Observatory. The Garcia family had moved to Maui when Tom was four so that Charles could open the solar observatory at the summit of Haleakalā. After Charles opened the High Altitude Observatory on Mauna Loa in 1965, he moved his family to Hilo, where Tom eventually graduated from Hilo High School

During his youth, Tom Garcia (video interview with author, Hilo, Hawai'i, December 2006) visited MLO many times to assist his father with basic maintenance tasks, such as painting and constructing boardwalks across the lava. He first signed the *Visitors* book on October 3, 1971.

> My dad worked at the Mauna Loa Solar Observatory, so we had gone up quite a bit as kids and helped him. Oh, it was pretty interesting. I remember getting sick sometimes. The road was a little windy, and it was quite a drive up there at that time. [It was] cold and [there was a] lack of oxygen, but it was a really neat place to go.

After graduating from the University of Colorado with a degree in environmental population and organism biology, Garcia returned to Hawai'i and was offered a summer job measuring ozone at MLO. This interested him, for from his studies he was well aware of the connection between the ozone layer and solar ultraviolet at the surface. Garcia took the job, and he and Charles Garcia became the first father-son pair to work at the observatory. Besides measuring the ozone layer and calibrating the Dobson spectrophotometer at MLO, Garcia (video interview with author, Hilo, Hawai'i, December 2006) became involved in preparing and launching ozonesondes from Hilo, sometimes with help from Judy Pereira:

> I think she enjoyed it a lot. And there is just something different maybe, a chance for her to get out and get in the field and see what we do actually. She had a great time; a lot of different people helped.

Darryl Kuniyuki (1982–Present)

Darryl Kuniyuki was born on O'ahu and raised in Hilo. MLO had long hired student workers, and Kuniyuki was brought on board under the new US government Junior Fellowship Program. Kuniyuki's high school science grades and his science fair projects played a major role in his selection:

> I had just attended the state science fair; I had a project on *Phytophtora palmivora*, which is a little fungus. Then I also had . . . an electronic project. So I had just finished the science fair and I won some awards for the *Phytophtora* project. Actually, I probably took whatever science classes I could in high school. So I took everything from biology, chemistry, physics; I had a lot of science classes. (Kuniyuki, video interview with author, Hilo, Hawai'i, December 7, 2006)

Judy Pereira (video interview with author, Boulder, Colorado, April 27, 2007) describes Kuniyuki as "a very bright high school senior":

> Darryl was the state champion science fair winner. So he was definitely the first choice as far as we were concerned. But, we interviewed several people, there were a lot of good people who, you know, could have become Junior Fellows, but we chose Darryl, and that was the best move we made. He worked at MLO during summers and vacations, and eventually moved into a permanent position after he graduated from the University of Hawaii at Manoa as an electrical engineer [in 1987]. (Pereira 1996)

Kuniyuki's high school science teacher had nominated him for the fellowship program, and prior to going to work for MLO he knew nothing about the science programs on the mountain. But that soon changed when his first work assignment was to assist MLO director Kinsell Coulson:

> My first job was plotting the polarimeter data for Dr. Coulson. When I came, there were four students. Back then, the data had to be reduced by the students, and they had one student doing the CO_2 data, one student doing the ozonesonde data, another one did the lidar data, and I think the fourth one did the chem lab data with Alan [Yoshinaga]. So I guess Dr. Coulson did not want to give more work to these other students because they were really busy. So when I came along I became his

student aide. So I sat in his office, and my first job was to just plot all of the polarimeter data, since, I think, when he had started [acquiring data at] Mauna Loa. I think he was able to just do the first year by himself, because there were so much data to be plotted. So that summer he gave me this stack of the data, just tables of data, and then I had to plot, just plotting dots. That was what I did all day. (Kuniyuki, video interview with author, Hilo, Hawai'i, December 7, 2006)

Donald Pack's Challenge (1983)

Academic institutions usually require that their most senior professors possess an earned doctorate. This isn't always the case in government, and a classic case was Donald Pack, who was the first director of the Geophysical Monitoring for Climatic Change program from 1971 to 1975. Pack arrived at a time when the need for long-term, global atmospheric monitoring was more obvious than ever before, and he responded to the challenge in a manner that was to have a lasting impact on MLO and its companion observatories. Dr. Kirby J. Hanson, Pack's successor, summarizes Pack's role in establishing global atmospheric monitoring:

> Mr. Pack's unique contribution was to develop within the National Oceanic and Atmospheric Administration (NOAA) a timely response to this need. Thanks to his foresight and hard work, the global background monitoring effort was established early, on firm ground. (Hanson 1977)

Pack became a consulting meteorologist after leaving GMCC, and in 1983 Joyce Harris and Barry Bodhaine, who were editing the *Geophysical Monitoring for Climatic Change* Summary Report for 1982, invited Pack to write the foreword. He begins with a brief retrospective of what had been achieved at GMCC's four observatories, reserving special attention for the carbon dioxide record at MLO and the increasing concentration of a variety of other trace gases:

> The unique CO_2 time series at Mauna Loa Observatory, now covering 25 years, stimulated the inclusion of changing gas concentrations in global climate models. Now we find apparently, that tropospheric methane is on the increase. Also, the chlorofluorocarbons continue to increase as does trichloromethane. One could add to the list many other compounds whose concentrations and trends appear to be changing. Many of these compounds absorb irradiation and add to the potential CO_2 "greenhouse effect," but their concentration histories are much less known. . . . The role of aqueous chemistry in cloud water is also being recognized. When one thinks that about one-half of the globe is always cloud covered, the necessity of understanding these processes is obvious. (Pack 1983)

Pack's forward goes on to discuss the role of measurements made from ships and aircraft to provide more global coverage. He then addresses the role of the major surface observatories and challenges those who would staff them in years to come:

> To date, our documentation of clean-air tropospheric chemistry is from only a handful of permanent observatories, less than 10 globally. . . . Yet it is primarily from the fixed sites that we find diurnal, seasonal, and interannual variability of concentration. . . .
>
> Answers to these challenges and these questions (and many others not stated here or not yet known) will alter the GMCC program by 1992. . . .
>
> What will not change is the continuing need for precise and accurate gas and aerosol measurements from which radiation and cloud behavior can be derived. . . .
>
> A new generation of observers and scientists will enter the program in the next decade. They, too, must carry on this tradition and meet these challenges.

MLO and John Chin Receive Awards (1983)

Mauna Loa Observatory's pioneering role in measuring carbon dioxide since 1958 and its discoveries of various

FIGURE 11. 6. The Mauna Loa Observatory staff in February 1983. Seated *(from left)*, Cynthia McFee, Judy Pereira, Stephanie Matthews, and Lisa McPherson. Standing *(from left)*, MLO director Kinsell Coulson, Alan Yoshinaga, Tom DeFoor, Franz Kraintz, Arne Austring, and John Chin. *(MLO archive.)*

volcanic aerosol clouds by means of its ruby lidar had earned the observatory considerable scientific and media attention. In the January 1983 Activities Report, Coulson reported that Lester Machta had accepted on behalf of the observatory a special award for "Twenty five years of outstanding service to the scientific community" at the January 1983 annual meeting of the American Meteorological Society. A reception was scheduled for the MLO staff in Hilo on February 11 so they could be presented with their award, which reads as follows:

> **American Meteorological Society**
> Special Award is conferred upon NOAA Mauna Loa Observatory for service to the scientific community by providing measurements of the chemical and physical background state of the atmosphere. The continuous records of CO_2 and atmospheric transmission in particular, are internationally unique and especially important as basic information applicable to understanding climate variability.
>
> January 1983
> R. E. Hallgren
> President

Fig. 11.6 shows the MLO staff at this time. The photograph is from February 1983 and might have been made on the occasion of the reception.

In October 1983, John Chin received a major NOAA award, and Coulson describes the particulars in his final Activities Report:

> John Chin of the observatory staff received a signal honor during the month for his outstanding work over the years in the carbon dioxide monitoring program. He was awarded the second highest award given by the Department of Commerce—the Silver Medal. The ceremony, which Mr. Chin attended, was held in the Department of Commerce building in Washington, DC on 10/26. While in Washington, he was hosted also by Hawaii legislators Senator Matsunaga and Representative Akaka, and given a guided tour of the Capital building. (Coulson 1983f)

To date, Chin is the only MLO veteran who has received a NOAA Silver Medal for his work at the observatory, but he doesn't advertise the honor. During an interview with the author (Hilo, Hawai'i, December 4, 2006), he skipped past the page about the award in his retirement scrapbook prepared for him by Leslie Pajo and later explained that everybody at the observatory played a role in his carbon dioxide work. Chin believes he would never have earned the award had not "Howard Ellis insisted on continued mountain observations [that] kept this program going." Thereafter, all his coworkers played a role:

> Everybody got involved with this program, but it so happened that I was the one in charge of the program. But everybody did their work, and [made] some kind of contribution to that.

MLO Acting Director John Chin (1983–1984)

Though he was the MLO director, Dr. Kinsell Coulson spent many days working on the mountain alongside the day crews. He was especially fascinated by atmospheric aerosols and devoted considerable time to his personal project on the impact of volcanic aerosols on sky polarization. As related earlier by Alan Yoshinaga, Coulson fell on the lava while helping remove an acid rain collector. Though both his arms were bloodied by the fall, he kept on working. Incidents like this and frequent stays on the mountain impressed the MLO staff but troubled his wife:

> In 1983 Vivien finally decided that Kin was getting just the slightest bit too old to be going up the mountain as often as he was, and she convinced him that it was time for him to retire and move back to California. They had been here over 5 years and it was difficult to see them leave. (Pereira 1996)

As Judy Pereira observes in her memoir, Coulson was popular with the MLO staff. Tom DeFoor (video interview with author, Boulder, Colorado, May 3, 2007) reflects on his time working for Coulson:

> I consider him one of my best bosses. . . . I liked him very much as an individual. He was obviously very smart and he was inspiring. He was great to work for.

John Chin was appointed acting director after Coulson left in December 1983. Chin inherited one of Coulson's major projects, the MLO lava barrier, and he reported in his first Activity Report that the attorney general had finally approved the project. There was still more paperwork to be prepared, reviewed, and approved before work could be commenced, but the world's largest volcano had other plans in mind.

Mauna Loa Erupts (1984)

For months, Mauna Loa had signaled its intentions through earthquakes and an expansion of its summit. Then on March 25, 1984, lava began erupting from the giant Moku'āweoweo Crater atop the mountain. Within hours lava began emerging from the northeast flank of the mountain exactly 8 miles (13 km) east of and about 1,500 feet (457 meters) below the elevation of MLO. The lava headed directly for Hilo, and on its way the flow, which was 1 mile (1.6-km) wide, took out thirty-one of the power line poles along the old MLO road between the Kūlani Prison and the Kūlani Mauka junction. The power line had provided electricity to MLO since 1966, and the destruction of the poles cut electrical power to the observatory and ended the nonstop collection of data. For a time there was confusion over whether or not the main road to MLO had been blocked. Worse, no one knew if the lava would reach Hilo or if a new lava outbreak might occur closer to the observatory.

The staff who worked at MLO were taken by surprise when Mauna Loa erupted. Charles Garcia (video interview with author, Hilo, Hawai'i, December 2006) of the High Altitude Observatory recalls receiving a predawn phone call from Judy Pereira:

> And I had been out partying the night [before] because a friend of mine's daughter graduated. So I got home by three o'clock, and it was boring, kind of too much partying. . . . She called me at 4:30: "Charlie, the volcano is erupting! You've got to be up there!" So I get in the car, and I go up that same day, and we were still able to drive up there.

In an interview in Boulder, Colorado, on April 27, 2007, Pereira referred to the eruption as "very, very probably the worst time of my career at Mauna Loa, because suddenly we had no observatory." She captures the concerns of the MLO staff in her memoir (1996):

> Unfortunately for us the eruption of Mauna Loa occurred when we were between directors. It was a difficult time for all of us because of the uncertainty of the duration of the eruption, the possibility of new outbreaks closer to the observatory, and the concern for getting power back to MLO since the power lines were crossed by the lava and cut. Not having a plan in place when the eruption first broke out, we had to depend on Boulder to make decisions for us and to quickly organize a measurement schedule to maintain. The stress of the situation bore heavily on us, tempers were short and instructions from Boulder were resentfully received. Comments from staff members were heard such as: "Taking Gardner counter measurements using a gas mask is ridiculous" and, "Those guys in Boulder are just coming up with busy work for us to do to make sure we're not goofing off here."

The eruption continued, though greatly diminished, until April 14, and one of the first concerns thereafter was the status of the road and power line. Fortunately, the main MLO road was unaffected by the lava flow, for the eruption had occurred east of Kūlani Mauka where the old road between the Kūlani Prison and MLO was linked to Saddle Road by the spur to the Road to Nowhere and the shortcut road arranged by Howard Ellis and Tom Vance in 1963. Jack Lockwood, who had designed the as yet unapproved lava barrier, visited the site for a firsthand inspection:

> The HELCO power line poles (nos. 2 thru 32, approx. 1 mile) to the observatory were destroyed. According to Jack Lockwood, a volcanologist with Hawaiian Volcano Observatory, the lava across the damaged area varied from 2 to 20 meters in thickness and for the most part [was] less than 6 meters thick. Most of the flows were a'a. The surface temperatures of the lava, as of May 3, were below 50°C in most areas and few areas exceed 100°C (Jack Lockwood). (Chin 1984)

Tom DeFoor (video interview with author, Boulder, Colorado, May 3, 2007) tried to drive to MLO, but for a number of days the road was blocked by the police:

> My concern was the instrumentation. I was afraid that the sulfur fumes and sulfuric acid droplets . . . could impact some of the instruments. So it was interesting getting up there and getting some of the more fragile stuff down.

Judy Pereira (video interview with author, Boulder, Colorado, April 27, 2007) recalls, "Boulder was after us to get things back online. We had no generator there to run things." Some work resumed at MLO as early as April 6, including the collection of acid rain samples on Mauna Loa and air samples at both MLO and at a similar elevation on Mauna Kea. The Dobson spectrophotometer was powered by a battery beginning April 10 so ozone measurements could be resumed. Virtually everything else at MLO was shut down, including both the URAS and the Scripps carbon dioxide analyzer systems.

When Charles Keeling learned about the problem, he called Boulder and, as a scientist who overheard the conversation recalled, "screamed and yelled about why he wasn't getting data." Keeling was especially unhappy to learn there was no backup power generator on the mountain; he was losing data and wanted a quick solution. As a former MLO scientist recalled, "Dave Keeling was being his typical, unreasonable self." Because of the uncertainty about when power would be restored, Keeling decided to act unilaterally and instructed Dave Moss at Scripps to arrange for a temporary generator. Finding and arranging the wiring for a 10-kW generator became a major project for Moss.

Meanwhile, Charlie Garcia of the High Altitude Observatory managed to borrow a large 90-kW generator that had once been used on Mauna Kea and now belonged to the county of Hawai'i, and he and MLO electronics engineer Tom DeFoor began work on its installation:

> The generator was serviced and checked out, a weather shelter was built, and two 550 gal. fuel tanks were borrowed from Mauna Kea. It was decided to install the generator system at the Kulani Mauka site [at] 8300 [feet] to keep generator contamination away from the observatories and make fuel hauling logistics easier. (Chin 1984)

After the power company installed a special transformer for the generator and the county signed off on the installation, the generator was installed and ready for testing on April 23. The next day the generator was successfully tested, but it failed after rain entered the shelter. The generator was repaired, and on Saturday, April 28, power was restored and gradually applied to the various systems at the two observatories, "thanks to Tom DeFoor's great efforts and engineering knowledge" (Chin 1984). For the first time since March 26, Scripps' APC and NOAA's URAS carbon dioxide analyzers began providing data. The Scripps analyzer soon began acting up when its thermal controller failed. The system was repaired and back in operation two days later, the last day of April. The NOAA carbon dioxide record for April 1984 is –99.99, the code for missing data, for there were insufficient days to provide a monthly average (Tans, telephone interview with author, August 13, 2008; Tans 2009). Scripps applied a different protocol for data acceptance, and its record gives 347.07 ppm for April (Keeling et al. 2001).

Four days later, Dave Moss (telephone interview with author, August 12, 2008) and his generator arrived on the same plane. Moss had arranged in advance for a flatbed truck and a forklift. After loading the generator onto the truck, he drove to MLO and somehow managed to push and slide the heavy generator along the boardwalk west of the main building to just upslope from the shed that once held the original MLO generators. Moss had just completed what he later described as one of his most difficult assignments. On May 4 the Scripps generator was installed, but the MLO staff was concerned that its exhaust would contaminate the carbon dioxide sampling program. As DeFoor (telephone interview with author, August 13, 2008) recalls, "I don't think that generator was ever used because of the possible contamination problem."

Only the lidar was still not working. It had been shut down for repairs and realignment in March shortly before Mauna Loa erupted. It was returned to service on May 30, 1984.

MLO Director Elmer Robinson (1984–1991)

John Chin's brief but excitement-filled tenure as acting director of MLO ended when Dr. Elmer Robinson became the eighth director of MLO. He came from the Washington State University at Pullman and reported for duty on July 2, 1984.

The Mauna Loa Observatory had operated at a frenetic pace during Kinsell Coulson's tenure. The staff attended courses on technical writing and computer programming using the BASIC language. The new lidar building was finally built. The LSI-11 computer was finally acquired. And a few years of groundwork were invested in the lava barrier project. Then, after John Chin took over as acting director, Mauna Loa erupted and cut power to the observatory for more than a month.

The pace lightened up considerably when Elmer Robinson arrived. Like Coulson, he was a retired professor who didn't completely understand the ways of government. But he was more flexible than Coulson had been when the bureaucracy failed to deliver. Just three weeks after Robinson arrived, Hawai'i Electric Light Company (HELCO) completed their repairs to the power line, the temporary generator system at Kūlani Mauka was disconnected from the power line, and the observatory was back in normal operation.

> The 90 kW generator proved to be a very reliable power source. Over the three months of operation only two outages occurred, both in mid-July just before it was taken out of service. (Robinson 1984)

The observatory was in good repair. The lidar system was being upgraded with help from the EPA. And even the Scripps vacuum tube carbon dioxide analyzer was working properly. All was well at MLO, and Elmer Robinson was on a first-name basis with the staff from day one. The only problem was his academic background. Judy Pereira (video interview with author, Boulder, Colorado, April 27, 2007), having worked closely with Prof. Kinsell Coulson, knew exactly what to do:

> Basically, I had to teach him how to do government. He had been in the university system all his life, and the government bureaucracy just about made him sick, really. I used to tell him what needed to be done for anything we bought, any person we wanted to hire, a student helper or anything like that, and [he would say], oh, at the university we would do this so much easier. I said, well, welcome to the government, because everything is done so-so, you know. It has to be done in a certain way.

In a telephone interview with the author on April 9, 2008, Robinson agreed with Pereira's characterization of the response to bureaucracy exhibited by him and Coulson. "We had not had the indoctrination of the Federal bureaucracy," he said. Pereira simplified the process by preparing paperwork for Robinson's signature—"and sometimes she would sign it." Robinson developed his own style of administration in a sometimes circuitous effort to get the funds he needed for various projects.

> Eventually he would get what he wanted, but he never came right out, like Kin Coulson would, . . . and ask for something. And he was oftentimes put down. . . . Kin would come back with a vengeance, [but] Elmer got along better with the administration. He would subtly ask for things, and eventually he got them. . . . He loved being at the Observatory. We loved him. It was a good time, it was a good period, because although he never fought for the budget, you know, at that point money was being given to Mauna Loa by cooperators, people who wanted to do work at Mauna Loa. They were building things up there, giving money for the personnel to run their programs and things like that. So, it was easier financially for Elmer than any of the previous directors. (Pereira, video interview with author, Boulder, Colorado, April 27, 2007)

Closing the Third Decade at MLO

On January 16, 1985, Lt. Cynthia McFee completed her MLO assignment. She returned to Colorado, where she left the NOAA Corps and entered medical school.

> The MLO staff who were present during the 1984 Mauna Loa eruption were presented with special service awards for above-the-call-of-duty service during this period. The awards included a certificate and a check. (Robinson 1985a)

Al Shibata completed thirteen years of service with MLO on May 15, 1985. He was awarded a twenty-five-year NOAA service pin and certificate and treated to an office party.

Recognition extended beyond the MLO staff to include Janet Chin (Mrs. John Chin), who was named the outstanding public school science teacher for the island of Hawai‘i. She was recognized at the annual Sigma Xi dinner at the University of Hawai‘i at Hilo on May 9.

Darryl Kuniyuki, who was first hired by Kinsell Coulson in 1982, returned to the MLO summer work program in June 1985. Besides observing duties on the mountain, he and Hank Walther of MLO and Steve Garcia of HAO made repairs to the deteriorating MLO road.

During August, MLO veterans Kinsell Coulson, Barry Bodhaine, and Bernard Mendonca visited the observatory. Coulson and Bodhaine checked on their instruments, and Mendonca reviewed observatory operations in his capacity as deputy director of NOAA's observatories.

One of the few negative events during Robinson's first two years at MLO occurred during the summer of 1985:

> On July 1 hikers on the Mauna Loa trail made a forced entry into MLO to use the telephone to request emergency help from the police. Fortunately the lost member of their party walked in shortly thereafter, but considerable damage was done to the MLO door. We are requesting payment for the damage. We are also planning to place a phone outside at MLO "under glass" for emergency use. (Robinson 1985b)

In September the hikers reimbursed the cost of repairing the door.

After he became director in July 1984, Elmer Robinson continued the complex series of bureaucratic steps necessary for the approval and construction of the lava barrier. The process was begun by Kinsell Coulson in 1980, and Robinson had good news to announce in December 1985:

> Construction of the MLO lava diversion embankment was begun by J. Glover, Ltd. on 12/23. We are concerned with possible adverse effects from dust, but so far these have appeared to be minimal. . . . When the work is finished a special notice will be sent to program managers and cooperative investigators so that they can flag any unusual results that might have been due to construction activity. (Robinson 1986)

The third decade at the Mauna Loa Observatory had come to a mellow end.

References

Bodhaine, B. A., and R. F. Pueschel. 1974. Source of seasonal variations in solar radiation at Mauna Loa. *Journal of Atmospheric Science* 31: 840–845.

Chin, John F. S. 1984. April 1984 activities report, May 23. In *Geophysical Monitoring for Climatic Change*, Mauna Loa Observatory, Monthly Reports, Hilo, Hawai‘i.

Coulson, Kinsell L. 1975. *Solar and Terrestrial Radiation*. New York: Academic Press.

———. 1979a. July 1979 activity report, August 13. In *Geophysical Monitoring for Climatic Change*, Mauna Loa Observatory, Monthly Reports, Hilo, Hawai‘i.

———. 1979b. November 1979 activity report, December 17. In *Geophysical Monitoring for Climatic Change*, Mauna Loa Observatory, Monthly Reports, Hilo, Hawai‘i, 1980.

———. 1980a. December 1979 activity report, January 10. In *Geophysical Monitoring for Climatic Change*, Mauna Loa Observatory, Monthly Reports, Hilo, Hawai‘i, 1980.

———. 1980b. January 1980 activity report, February 11. In *Geophysical Monitoring for Climatic Change*, Mauna Loa Observatory, Monthly Reports, Hilo, Hawai'i, 1980.

———. 1980c. July 1980 activity report, August 8. In *Geophysical Monitoring for Climatic Change*, Mauna Loa Observatory, Monthly Reports, Hilo, Hawai'i, 1980.

———. 1980d. November 1980 activity report, December 5. In *Geophysical Monitoring for Climatic Change*, Mauna Loa Observatory, Monthly Reports, Hilo, Hawai'i, 1980.

———. 1981a. December 1980 activity report, January 9. In *Geophysical Monitoring for Climatic Change*, Mauna Loa Observatory, Monthly Reports, Hilo, Hawai'i, 1980.

———. 1981b. January 1981 activity report, February 5. In *Geophysical Monitoring for Climatic Change*, Mauna Loa Observatory, Monthly Reports, Hilo, Hawai'i, 1981.

———. 1981c. April 1981 activity report, May 12. In *Geophysical Monitoring for Climatic Change*, Mauna Loa Observatory, Monthly Reports, Hilo, Hawai'i, 1981.

———. 1981d. May 1981 activity report, June 8. In *Geophysical Monitoring for Climatic Change*, Mauna Loa Observatory, Monthly Reports, Hilo, Hawai'i, 1981.

———. 1981e. July 1981 activity report, August 6. In *Geophysical Monitoring for Climatic Change*, Mauna Loa Observatory, Monthly Reports, Hilo, Hawa'ii, 1981.

———. 1982. February 1982 activity report, March 18. In *Geophysical Monitoring for Climatic Change*, Mauna Loa Observatory, Monthly Reports, Hilo, Hawai'i, 1982.

———. 1983a. March 1983 activity report, April 11. In *Geophysical Monitoring for Climatic Change*, Mauna Loa Observatory, Monthly Reports, Hilo, Hawai'i, 1983.

———. 1983b. April 1983 activity report, May 12. In *Geophysical Monitoring for Climatic Change*, Mauna Loa Observatory, Monthly Reports, Hilo, Hawai'i, 1983.

———. 1983c. May 1983 activity report, June 13. In *Geophysical Monitoring for Climatic Change*, Mauna Loa Observatory, Monthly Reports, Hilo, Hawai'i, 1983.

———. 1983d. June 1983 activity report, July 12. In *Geophysical Monitoring for Climatic Change*, Mauna Loa Observatory, Monthly Reports, Hilo, Hawai'i, 1983.

———. 1983e. September 1983 activity report, October 12. In *Geophysical Monitoring for Climatic Change*, Mauna Loa Observatory, Monthly Reports, Hilo, Hawai'i, 1983.

———. 1983f. October 1983 activity report, November 14. In *Geophysical Monitoring for Climatic Change*, Mauna Loa Observatory, Monthly Reports, Hilo, Hawai'i, 1983.

———. 1988. *Polarization and Intensity of Light in the Atmosphere*. Hampton, VA: A. Deepak Publishing.

DeLuisi, John J. 1981. *Geophysical Monitoring for Climatic Change No. 9* (Summary Report, 1980), 149. Boulder, CO: NOAA.

Ellis, Howard T. 1978. August 1978 activity report, September 7. In *Geophysical Monitoring for Climatic Change*, Mauna Loa Observatory, Monthly Reports, Hilo, Hawai'i, 1978.

———. 1979a. December 1978 activity report, January 5. In *Geophysical Monitoring for Climatic Change*, Mauna Loa Observatory, Monthly Reports, Hilo, Hawai'i, 1979.

———. 1979b. January 1979 activity report, February 7. In *Geophysical Monitoring for Climatic Change*, Mauna Loa Observatory, Monthly Reports, Hilo, Hawai'i, 1979.

———. 1979c. February 1979 activity report, March 7. In *Geophysical Monitoring for Climatic Change*, Mauna Loa Observatory, Monthly Reports, Hilo, Hawai'i, 1979.

Ellis, Howard T., and Rudolf F. Pueschel. 1971. Solar radiation: Absence of air pollution trends at Mauna Loa. *Science* 172: 845–846.

Fegley, Ronald W. 1976. June 1976 activity report, July 7. In *Geophysical Monitoring for Climatic Change*, Mauna Loa Observatory, Monthly Reports, Hilo, Hawai'i, 1976.

Forest Service. 2007. Mount St. Helens National Volcanic Monument. USDA Forest Service. (Accessed at http://www.fs.fed.us/gpnf/mshnvm/.)

Hanson, Kirby J. 1977. *Geophysical Monitoring for Climatic Change No. 9* (Summary Report, 1976), vii. Boulder, CO: NOAA.

Keeling, C. D., S. C. Piper, R. B. Bacastow, M. Wahlen, T. P. Whorf, M. Heimann, and H. A. Meijer. 2001. Exchanges of atmospheric CO_2 and $13CO_2$ with the terrestrial biosphere and oceans from 1978 to 2000. I. Global aspects. SIO Reference Series No. 01-06. San Diego: Scripps Institution of Oceanography.

Machta, Lester. 1978. Foreword. In John Miller, ed., *Mauna Loa Observatory: A 20th Anniversary Report,* September. Environmental Research Laboratories, US Department of Commerce. Boulder, CO: NOAA.

Mendonca, Bernard G., Kirby J. Hanson, and John J. DeLuisi. 1978. Volcanically related secular trends in atmospheric transmission at Mauna Loa Observatory, Hawaii. *Science* 202: 513–515.

Miller, John. 1976a. July 1976 activity report, August 11. In *Geophysical Monitoring for Climatic Change,* Mauna Loa Observatory, Monthly Reports, Hilo, Hawai'i, 1976.

———. 1976b. August 1976 activity report, September 3. In *Geophysical Monitoring for Climatic Change,* Mauna Loa Observatory, Monthly Reports, Hilo, Hawai'i, 1976.

———. 1976c. September 1976 activity report, October 7. In *Geophysical Monitoring for Climatic Change,* Mauna Loa Observatory, Monthly Reports, Hilo, Hawai'i.

———. 1977a. April 1977 activity report, May 9. In *Geophysical Monitoring for Climatic Change,* Mauna Loa Observatory, Monthly Reports, Hilo, Hawai'i, 1977.

———. 1977b. May 1977 activity report, June 9, In *Geophysical Monitoring for Climatic Change,* Mauna Loa Observatory, Monthly Reports, Hilo, Hawai'i, 1977.

———. 1977c. June 1977 activity report, July 11. In *Geophysical Monitoring for Climatic Change,* Mauna Loa Observatory, Monthly Reports, Hilo, Hawai'i, 1977.

———. 1977d. July 1977 activity report, August 5. In *Geophysical Monitoring for Climatic Change,* Mauna Loa Observatory, Monthly Reports, Hilo, Hawai'i, 1977.

———. 1977e. October 1977 activity report, November 7. In *Geophysical Monitoring for Climatic Change,* Mauna Loa Observatory, Monthly Reports, Hilo, Hawai'i, 1977.

———. 1978. December 1977 activity report, January 4. In *Geophysical Monitoring for Climatic Change,* Mauna Loa Observatory, Monthly Reports, Hilo, Hawai'i, 1978.

Mims, Forrest M. 1984. The Altair story: Early days at MITS. *Creative Computing* 10 (November): 17–27. (Accessed at http://www.forrestmims.org/biography.html.)

Pack, Donald H. 1983. *Geophysical Monitoring for Climatic Change No. 11* (Summary Report, 1976), v–vi. Boulder, CO: NOAA.

Pales, Jack, Clifford Kutaka, and Robert Williams. 1957–1958. *Weather Log.* Unpublished log in MLO files.

Pereira, Judy. 1996. Mauna Loa Observatory: A 30-year personal overview. (Accessed at www.mlo.noaa.gov/pdf/memorabilia/JUDY%20BKLT.doc.)

Peterson, James T. 1978. *Geophysical Monitoring for Climatic Change No.* 6 (Summary Report, 1977), 134. Boulder, CO: NOAA.

Robinson, Elmer. 1984. July 1984 activity report, August. In *Geophysical Monitoring for Climatic Change,* Mauna Loa Observatory, Monthly Reports, Hilo, Hawai'i, 1984.

———. 1985a. February 1985 activity report, March 11. In *Geophysical Monitoring for Climatic Change,* Mauna Loa Observatory, Monthly Reports, Hilo, Hawai'i, 1985.

———. 1985b. July 1985 activity report, August 13. In *Geophysical Monitoring for Climatic Change,* Mauna Loa Observatory, Monthly Reports, Hilo, Hawai'i, 1985.

———. 1986. December 1985 activities report, January 10. In *Geophysical Monitoring for Climatic Change,* Mauna Loa Observatory, Monthly Reports, Hilo, Hawai'i, 1986.

Shaw, Glenn. 1979a. Aerosols at Mauna Loa: Optical properties. *Journal of the Atmospheric Sciences* 36: 862–869.

———. 1979b. Transport of Asian desert aerosol to the Hawaiian Islands. *Journal of Applied Meteorology* 19: 1254–1259.

———. 1995. The Arctic haze phenomenon. *Bulletin of the American Meteorological Society* 76: 2403–2413.

———. 2007a. Impact of global warming looms on the horizon. *Fairbanks Daily News-Miner,* April 22. (Accessed at http://newsminer.com/2007/04/22/6603/.)

———. 2007b. Mauna Loa and the greenhouse planet. Unpublished manuscript. Copy in author's files.

Shepherd, J. B. , W. P. Aspinall, K. C. Rowley, J. Pereira, H. Sigurdsson, R. S. Fiske, and J. F. Tomblin. 1979. The eruption of Soufrière volcano, St. Vincent, April–June 1979. *Nature* 282 (November): 24–28.

Simpson, H. J. 1972. Aerosol calculations at Mauna Loa Observatory. *Journal of Geophysical Research* 77: 5266–5277.

Steiner, James W. 1978. (Letter following Foreword.) In John Miller, ed., *Mauna Loa Observatory: A 20th Anniversary Report.* National Oceanic and Atmospheric Administration Special Report, September: vii. Boulder, CO: NOAA Environmental Research Laboratories.

Tans, Pieter. 2009. Trends in atmospheric carbon dioxide—Mauna Loa. (Accessed at http://www.esrl.noaa.gov/gmd/ccgg/trends/.)

CHAPTER TWELVE

The Fourth Decade (1986–1995)

The Mauna Loa Observatory entered its fourth decade without a crisis and in good condition. MLO director Elmer Robinson wrote:

> The observatory program continued in a generally satisfactory manner with only the usual type of maintenance problems and without any prolonged downtime on any instrument. The weather was generally clear and dry with only a trace of precipitation recorded on 1/9. (Robinson 1986)

The new decade also arrived with good news. On February 1, 1985, Hawai'i's Deputy Attorney General William Lau signed the easement for the lava barrier that had been approved by the Board of Land and Natural Resources on December 16, 1983. Bernard Mendonca signed his copy on February 20 and wrote, "Whew." Construction of the long-delayed lava diversion embankment planned by Kinsell Coulson and designed by Jack Lockwood was formally completed on January 15, 1986. The structure took the form of a huge, inverted V-shaped lava barrier that pointed directly toward the summit crater. Lockwood and Robinson inspected the project from the air on January 12, 1986 (Fig. 12.1). Later Robinson quipped that "We have not yet arranged for a test of the lava diversion system" (Robinson 1986).

More good news came in the form of a computerized bibliography of MLO–related papers prepared by data clerk Barbara Kerecman. The list included 234 scientific papers published from 1956 to 1984, an average of more than eight papers per year.

In May a special exhibit celebrating MLO's thirtieth anniversary was displayed at Hilo's Lyman House Museum. Other exhibits about science in Hawai'i were also displayed. The MLO staff celebrated the thirtieth anniversary with a picnic on June 8. Judy Pereira was presented with an orchid lei in honor of her twentieth anniversary at MLO.

While MLO was humming along with only minor problems, Hilo took it on the chin in February when a major storm with near hurricane-force wind caused major damage. Parts of Hilo had power outages lasting more than four days, but MLO was without power for only six hours. A second natural event with potential for disaster occurred on May 7 when Hilo came under a tsunami alert. The small wave that arrived caused no damage.

FIGURE 12.1. The completed lava diversion barrier forms an inverted "V" whose apex points toward the summit crater of Mauna Loa. This photograph was probably made on January 12, 1986, when Jack Lockwood and Elmer Robinson inspected the newly constructed barrier from the air. *(J. P. Lockwood, US Geological Survey.)*

Installation of the 38-Meter Walk-Up Tower (1986–1987)

On October 16, 1986, construction was begun on a new 126-foot (38.4-meter) meteorological and atmospheric sampling tower at the Mauna Loa Observatory (Plate 33). The tower would be much higher than any previous sampling tower at MLO and provide for the sampling of air high above any possible pollution emitted by vehicles and other sources at the observatory. It would feature steps inside its frame that would allow MLO staff to more easily and safely ascend to any point in the structure to install and remove equipment.

The tower's foundation was poured on October 27–28. In early December the tower itself was erected in only four days by Ken Thaut, Tom Garcia, and three Hilo construction workers. Jim Wendell then arrived from GMCC in Barrow, Alaska, to help wire the tower and install some of the instruments (Robinson 1987a). On January 14, the railing at the top of the tower was installed, thus completing its construction and allowing the installation of instruments and the air inlet lines (Plate 34) that were connected to pumps inside the main building (Plate 35) . The new tower transformed the appearance of the observatory. It loomed high over one of the two original buildings from 1956, the "Clean Restrooms" outhouse (Plate 36).

The University of Arizona Solar Variability Building (1986)

The University of Arizona had long been interested in sunlight and atmospheric transmission. In 1986 the university began a study of solar variability in a new building at MLO. Darryl Kuniyuki (personal communication with author, June 9, 2008) helped maintain the instrument and recalls details about the project:

> The project was run by Meg Whittle, an electronic technician who was hired by University of Arizona. She made observations and did maintenance on the equipment. After many instrument problems and lack of funding, the project was turned over to NOAA in April 1, 1988. The project was run by Tom Defoor and me, [and we were] trained by Meg before she left. The project under NOAA lasted until November 1989, when it was terminated. The building became a place for visiting scientists to use until the arrival of the University of Denver FTIR program in 1991.

Darryl Kuniyuki Takes Over from Tom Garcia (1987 to Present)

During the entire month of June 1987, Tom Garcia and Ron Thorne conducted nearly daily calibration measurements of the world standard ozone spectrophotometer, Instrument 83 (I83). This and helping with the assembly of the walk-up tower were Garcia's last major projects at MLO, for he had new plans. He moved to the mainland and eventually became an earth science and biology teacher at Cheyenne High School in North Las Vegas, Nevada. He took with him many memories of his adventures on the big mountain, one of which especially stands out:

> I think the neatest thing I remember was one night I went up ahead of time, and I decided to stay overnight there at the observatory when we were first calibrating the Dobson. And just walking out there in the middle of the night, a star-filled night, you would swear something was going to come down from heaven and get you. It was just so dark in the sky. It was like you were right up there in the universe. . . . I know it is maybe not really related to ozone or meteorology, but just being up there, I think, is a neat experience.

Darryl Kuniyuki, the science fair winning high school student who had begun working at MLO during the summer of 1982, received a degree in electrical engineering from the University of Hawai'i in 1987 and became a full-time employee of the observatory on September 15. The transition was smooth, for Kuniyuki was already working full time during summers and winter breaks under the student fellowship he had held since 1982. During the summer of 1987, he was trained to become Tom DeFoor's engineering and lidar assistant and to take over Tom Garcia's ozone work. Both men were working at MLO when a staff photograph was made in front of the door of the MLO office at the Federal Building (Fig. 12.2). For the next eight years, Kuniyuki's principal assignments were to measure the ozone layer and service the surface ozone instruments at MLO and launch ozonesondes from Hilo, one of which crashed in the bay. Kuniyuki also wrote many computer programs to assist the MLO staff in analyzing their data.

Instrument Problems

Surface ozone measurements were among the original International Geophysical Year (IGY) measurements made

FIGURE 12.2. The MLO staff during the summer of 1987 included *(standing, from left to right)* Tom DeFoor, Judy Pereira, MLO director Elmer Robinson, Barbara Kerecman, John Chin, Alan Yoshinaga, and Lynne Ashman; and *(kneeling, from left to right),* Mamoru "Al" Shibata, Darryl Kuniyuki, and Thomas Garcia. *(MLO archive.)*

at MLO since 1957. During March 1987, the Dasibi ozone analyzer began producing noisy data, and by April 7 the data were so noisy that they were useless. The instrument stopped working on April 10 and was shipped to Boulder to be repaired. Jack Pales, MLO's first director, would have understood, for he had many problems with the first ambient ozone instruments back in the late 1950s. The ozone analyzer was out of service until July 25. Thereafter the full suite of original IGY measurements was back in operation.

On January 3, 1988, the repaired analyzer measured an ozone concentration of about 90 ppb, which was much higher than usual. Robinson offered no explanation for the spike in ozone in the January Activities Report, which was not to be the last such spike in the MLO record. Surface ozone at MLO can reach much higher concentrations than usual when ozone from the stratosphere is occasionally circulated down to the lower atmosphere. High surface ozone at MLO can also accompany polluted air arriving from Asia.

The monthly Activities Report included many notices about unusual circumstances that might alter the many observations and measurements conducted at MLO. For example, Robinson included many notes about the impact on observations caused by volcanic smog events, including a string of such episodes in October 1989:

> Probably the most notable features of the October data are the numerous periods of heavy aerosol contamination. . . . These are probably the result of widespread haze events driven by Kilauea volcanic emissions and a general weakening of the trade winds across the state. This was a frequent situation affecting the Hawaiian Islands during this month. (Robinson 1989f)

Dust and exhaust from construction equipment occurred during April and May of 1989, and powerful wind gusts on August 11 caused an automated tipping bucket–style rain gauge to indicate 2.49 inches (63 mm) of rain that never fell.

A very different kind of instrument problem occurred with the discovery of unsafe conditions in the observatory's electrical wiring:

> Herman Chan, EE., WASC, is working on improvements in the MLO electrical power distribution. His preliminary findings are that a large number of very unsafe and non code conditions are present at MLO and that a complete rebuilding of the system will probably be necessary. Most problems are outside the buildings. (Robinson 1990b)

Several years were to pass before the wiring problems were corrected.

New Programs

Mauna Loa Observatory had long hosted experiments and research projects organized by universities and other government agencies, including Scripps Institution of Oceanography, the Mauna Loa Solar Observatory, NASA, and the EPA. Early in 1988 two new cooperative programs were added. In January, Tony Clarke of the University of Hawai'i began a brief program to measure aerosols and their size by means of an automated apparatus. The experiment ended on March 30. Dr. Tom Cahill of the University of California at Davis began a program to collect particulate matter in the clean, downslope air on filters. The project was based on the Clean Air Act's requirement to determine the air quality in national parks (Robinson 1988a).

Visualizing the MLO Data (1987)

Since 1952, most of the data collected at the Mauna Loa summit and slope observatories had been recorded on paper rolls by strip chart recorders. In the 1970s, computer technology was introduced by Ronald Fegley to control the lidar system. Kinsell Coulson ordered a powerful LSI-11 microcomputer for MLO in 1981. And in April 1987 Elmer Robinson's monthly Activities Report began to include two pages with some twenty rows of computer-generated plots of key MLO data collected during the month. The plots, which were developed by computer whiz Arne Austring, represented a major milestone for MLO, since for the first time much of the key meteorological and air quality data for an entire month could be quickly compared:

> These plots are based on the hourly CAMS [Control and Monitoring System] data and are produced from copies of the CAMS tapes kept in the MLO office. The program to generate this plot was developed by Arne Austring. The daily mountain wind cycle is clearly evident on most days in the wind direction, temperature, and dew point (TD) plots. The CN [condensation nuclei] data show that the upslope flow is usually marked by a significant CN increase. The CO_2 record shows that upslope air usually has lower CO_2 concentrations. There are a few occasions where the CO_2 downslope data were apparently influenced by volcanic outgassing, e.g., 3rd, 7th, 13th, etc. Breaks in the CO_2 record occur when the regular weekly calibration is run. . . . The [URAS] CO_2 record seems to have changed both in the general concentration level and in the sensitivity or response at the time of the calibration on the 15th–16th. . . . We looked into this apparent problem, and through a comparison of URAS, SIO, and GC CO_2 data determined that this

> apparent change was real and not a result of a URAS analyzer problem. (Robinson 1987b)

The value of the monthly Activities Reports was enhanced by the new data plots. Weather events were clearly evident as sharp changes in temperature and gentle changes in pressure. When the inversion layer passed over the observatory, the condensation nuclei count skyrocketed. The back-and-forth changes that accompany the night and morning downslope wind and the afternoon upslope wind were clearly visible in the plots. Clear days were indicated by smooth bell curves of solar radiation, and clouds were indicated by downward spikes in the daily curve. And so it went. The new data plots were important for individuals at MLO who collected the data and those elsewhere who analyzed it. The parallel rows of data traces also provided a cross-disciplinary summary of the data, and Elmer Robinson (1987b) was pleased by the new addition to his monthly reports:

> We will generate these monthly plots on a regular basis and will use them to review instrument performance and correlations between meteorology and air quality, as has already been the case with the CO_2 record as outlined above. Arne Austring is to be congratulated for the work put into the generation of this record.

Beginning with his May report, Robinson included an explanation and discussion of the data plots produced by Austring in each report. Occasionally gaps can be seen in the data plots, such as the mysterious break in the meteorology data that appears in the plots for September 3–4, 1987. The gap was caused by lightning during a thunderstorm that dumped 1.48 inches (37.6 mm) of rain on the observatory. Breaks in data due to lightning strikes are not uncommon at MLO, and the new data plots in the monthly activity reports made them much more obvious.

Volcanic Activity

Volcanic activity and earthquakes were a regular topic of the monthly Activities Report, and the former sometimes caused sharp spikes, dips, or undulations in the data plots. For example, "sharp, clearly defined spikes" in the carbon dioxide record occurred on several days in October 1988 and again in November:

> These are undoubtedly due to small volcanic vent emissions imbedded in the nocturnal downslope flow, not the southerly winds at these times [from the direction of the summit crater]. (Robinson 1988e)

Aerosols from the ongoing eruption of Kīlauea formed a natural layer of air pollution that Big Island residents call *vog* (volcanic fog or smog). While the Kīlauea volcano is southeast of MLO and on the other side of Mauna Loa, vog sometimes forms a visually obvious aerosol layer across part or all of the saddle between Mauna Loa and Mauna Kea. Then as now, it sometimes reached the observatory. When that occurred, vog episodes were very obvious in the monthly data plots, for they caused huge increases in the particle count. Robinson included comments about vog in his Activities Reports so that scientists back on the mainland would understand the reason for unusual data from some of their instruments:

> As a probable result of the ongoing Kilauea eruption the Big Island has been frequently bothered by "volcanic haze" or as it is called locally, "vog." From time to time this haze reaches the Observatory as part of the upslope wind. (Robinson 1989d)

NOAA Replaces its CO_2 Analyzer (1987)

In August 1987, NOAA's URAS CO_2 analyzer was replaced by a newer Siemens ULTRAMAT 3 CO_2 analyzer. This was a radical step when compared to the Scripps program run by Dave Keeling, who was intent on using as long as possible the aging APC analyzer installed at MLO in March 1958. NOAA was careful to compare the URAS with the new Siemens instrument before allowing the latter to take over:

> As described in a subsequent section, considerable comparison data were taken with the two instruments. Our initial checks of these data show that the Siemens results seem to be about 0.3 to 0.5 ppm lower than the URAS data. (Robinson 1987d)

More comparisons during September showed that the Siemens indicated 0.3 ppm less CO_2 than the URAS. The new Siemens CO_2 instrument was designated the official MLO CO_2 analyzer on August 6, 1987, and, like the URAS instrument it replaced, for years to come its measurements of CO_2 would almost perfectly track the data from the much older Scripps analyzer.

The Soviets Visit Hilo (1987)

International cooperation at MLO began with the Japanese snowflake scientists in 1956 and had continued ever since. From April 28 to May 1, the USSR research ship *Akademik Korolev* (Plate 37) made an official visit to Hilo as part of the Soviet/American Gas and Aerosol Expedition II (SAGA II). Judy Pereira (1996) describes the visit in her memoir:

> One of the main events that occurred during Elmer's tour was MLO's participation in the Russian research vessel *Korolev's* cooperative research effort with NOAA in the Pacific. We greeted the vessel, provided supplies and other requested items, and played host to not only the Russian scientists and crew, but the many NOAA and NSF officials, and the American science crew as well. It was a week of hustling and bustling and preparation of on-board facilities to accommodate the instruments from CMDL for the voyage. We had a huge celebration and feast for everyone [Plate 38] and enjoyed the camaraderie with our foreign visitors. The *Korolev* returned for a second cooperative voyage two years later. By then we were old hands at this and everything went like clockwork.

Thirteen scientists and many instruments were loaded onto the Soviet ship during its visit, and the *Akademik Korolev* left port for a three-month science cruise on May 1, 1987 (Plate 39). Notable among the participating scientists during the Soviet visit was former MLO director Dr. John Miller.

On May 30, 1987, eight scientists from NOAA's research ship *Researcher* toured MLO. Their visit wasn't as exciting as the Soviet visit and rated only one sentence in the monthly Activities Report. More about this atmospheric intercalibration mission would be included here, but more could not be found. American scientists were just not as exciting as Russian scientists from the other side of the world.

The *Korolev* returned to Hilo on July 28, 1987, and received due notice in the Activities Report:

> The Soviet research ship *Akademik Korolev* returned to Hilo on 7/28 to conclude the joint US/USSR SAGA II operation. Scientists and equipment were off loaded at this time. MLO personnel provided considerable assistance to this operation. Soviet and U.S. scientists and other visiting U.S. people were taken on a tour of MLO (7/30) and to the Kilauea volcano sites (7/29). A party/reception was held during the evening on 7/30. (Robinson 1987c)

Judy Pereira (video interview with author, Boulder, Colorado, May 2, 2007) reflects on the Soviet visitors, and her account suggests that they did not soon forget their time in Hawai'i:

> That was one of the most exciting things that happened during my career at Mauna Loa. . . . They arrived in Hilo, and maybe two or three of them could speak English. The leader, of course, the captain of the vessel, and the first assistant scientist, they spoke very well in English. But it was such a treat. We got treated to food on the vessel. We got tours and got to know them really well. We put on a potluck for them, and a potluck in Hawaii means tons of food . . . they tried everything, and they were happy to try everything. My husband and I took several of them on sightseeing trips, and they were very grateful. . . . They arrived in port [with] lists of things,

> things that they couldn't get in Russia. . . . Office Depot [had] everything that they asked for on that list. . . . They got pencils, papers, printer papers, everything that you could find in the stationery store. They just didn't have it. They didn't even have toilet paper. We brought dozens and dozens of toilet paper [rolls]. So, it was an experience, and it was really, really great. We made friends there.

The collaboration continued in July 1990 when MLO's Tom DeFoor visited a lidar laboratory in Obminsk near Moscow and attended the International Lidar Conference in Tomsk, Siberia.

Breakdowns, Accidents, and Vandalism

The rough road to MLO had taken a toll on MLO vehicles for many years. After the road was paved in 1969, cars could travel faster, and there were a number of accidents and even some collisions in the lower, shortcut section of the single-lane road where several blind curves occur. The propane-fueled car acquired under Rudolf Pueschel's time as director was lost in an accident in the early 1970s. John Chin (video interview with author, Hilo, Hawai'i, December 4, 2006) describes an accident on Saddle Road that occurred while he was sleeping in the back seat of an MLO vehicle late one afternoon:

> Three of us were in the car coming down from the mountain, and we were kind of late. And, of course, during the afternoon [Saddle Road] was always very foggy, and also, the road was wet. . . . We had a visitor from Boulder. I let the fellow sit on the passenger side, and I stayed in the back, and they all got the seatbelt on. And I fell asleep on the way down, and all of a sudden, the next thing I knew, the car was rolling over—a couple of times. By the time I woke up, the car was sliding [and] hit a tree. . . . The driver injured his leg, and the fellow from Boulder hit his head on the windshield, and I was sitting in the back [and] I got a cut. . . . So that was a very horrifying experience. . . . We had to wait on the side of the road until somebody came down from the Pohakuloa Military Camp. . . . We did not have a cell phone. We did not have any communication to contact anybody at all.

Mark Goldman (video interview with author, Hilo, Hawai'i, December 2006), who worked at MLO in 1972–1973, experienced a different kind of accident on Saddle Road:

> Coming back one time, an army truck was coming up and we sideswiped each other. I proceeded back to Hilo, and it was foggy at that time also, and informed the police. I remember the police officer heading back up with me about 5 p.m. as it was starting to get a little dark saying, "Will you remember the site where it happened?" Then I said, "Uh, I think so." And as it got darker and foggier, he kept repeating the question to me, and finally I took the hint and told him, "No, I do not think I can find it in the dark." He seemed quite relieved; he did not have to drive up the Saddle. Then weeks went by, and Judy was pushing me to get the police report for her forms, somebody is going to get a ticket, what was going on. Finally, the police came back, and they said, "We measured the cars. We measured the road." I do not remember the exact numbers but they said, "Since the car is 5 feet 8 inches wide and the lane is 5 feet 2 inches wide, we cannot give anyone a ticket."

Then there were the accidents that should have happened but never did, as when Mike Keyes once raced down the unpaved mountain road to Hilo in only sixty-five minutes, and the time Bernard Mendonca received a surprise when he pressed the brake pedal and nothing happened:

> Countless, countless times our cars broke down. Thanks to the supreme efforts of Mr. Yamanouchi, we were able to negotiate the road with our cars, government and rental cars. My best breakdown was when I had to drive a car to the Yamanouchi Garage from the Kulani Mauka turnoff [8,300 feet elevation] with no brakes. Never forgot that. (Mendonca 2006)

The first accident during Robinson's tenure occurred in September 1987:

> September 25th spoiled our automobile safety records when the MLO station wagon and the HAO [High Altitude Observatory] sedan caught each other in the middle of a blind S-curve on the MLO road. Our damage was left-front fender and door. HAO lost a left front fender and headlight. There were no injuries except to the drivers' pride. (Robinson 1987e)

When MLO was headquartered at the Cloud Physics Observatory on the University of Hawai'i campus, the observatory cars were generally safe from vandals, but the vehicles experienced occasional vandalism when the office was moved downtown to the Federal Building. In July 1988, the MLO pickup was broken into and the steering column was badly damaged in an unsuccessful attempt to hot-wire the vehicle. Earlier in July the trunk of the MLO sedan was broken into while it was parked at a Hilo hotel where a temporary worker was staying.

Vandals struck again in November 1989, and every detail was noted in the Activities Report:

> Over the weekend of the 10th–12th someone got into the [Federal Building] main office area in Hilo and made off with several small, hand-carryable items. Missing were 6 hand calculators (personal and MLO) and the MLO 35 mm camera. Also lost were 2 cans of V-8 juice, single serving size, which were taken from the office refrigerator, consumed in the office, and cans put in the waste basket.
>
> A more serious, and costly, event occurred over the Nov. 25–26 weekend when the MLO Ford Ranger pickup truck was stolen. It was recovered [on the opposite side of the Island] in Kona badly trashed and missing the camper shell and parts of the 2-way radio. Also, apparently in the same incident the right side door lock of the station wagon was punched out but entry was not obtained. (Robinson 1989g)

Repairing the MLO pickup truck required several weeks and cost more than $3,000, all of which came from the MLO budget.

Rocks in the Mail (1988)

Superstitious tourists were known for mailing MLO and other organizations on the Big Island lava souvenirs that they took home:

> In early December [1988] we received a piece of lava rock from an unknown tourist, presumably to return to Mauna Loa and thus placate [the volcano goddess] Pele. (Robinson 1989a)

MLO director Robinson had some fun when another shipment of lava arrived in January 1989:

> On about the 3rd or 4th either the Hawaiian menehunes [mythical little people] or Mdm. Pele were apparently at work at MLO and both water storage tanks drained mysteriously through a loosened drain plug on the lower water tank line. Human vandalism can probably be ruled out. . . . More lava rocks were received to return to Pele; this time from a young lady in Baton Rouge, LA. (Maybe if we had returned them sooner we would not have lost our water supply.) The note accompanying these rocks is attached.
>
> I did not know about the Goddess of Pele curse until I came home. Please give the volcano rock back to Pele. I don't need or want it. Thank you. Rose ______. (Robinson 1989b)

The loose drain on the water system was tightened, and subsequent rain filled the tanks with more than a thousand gallons of water by March.

MLO's Thirtieth Anniversary: Lightning Creates "Problems Galore!" (1988)

While lightning is uncommon at MLO, it can occur quite unexpectedly and can strike both power line poles and structures, as Rudolf Pueschel learned while installing some equipment in the AEC Building during his tenure as director (1970–1972). Large clouds began building up west of the observatory, blocking the sun and appearing like what he described as a woman with flowing blond hair. Suddenly Pueschel was startled by a loud boom of thunder. He rushed to the door, saw that a nearby power pole was in flames, and decided it was time to leave (from notes by Forrest M. Mims III of conversation of Pueschel and Howard Ellis at Ellis' home in Hilo, Hawai'i, December 2, 2007).

Atmospheric electricity specialist Bill Cobb entered a lightning observation in the *Visitors* book on January 16, 1976, during one of his many visits: "Looks to me like the lightning stroke hit the outhouse." Cobb (personal communication with author, April 10, 2008) recalls that the lightning that day also destroyed several thousand dollars of equipment he used to study atmospheric electricity. Bob Uchida (video interview with author, Hilo, Hawai'i, November 30, 2006) once observed lightning striking the parking area:

> Lightning storms can be vicious up there. Like I have been up there where I saw lightning hit the parking lot, and you can see the lightning bolt running down the road where the water is flowing.

One of the worst lightning events at MLO occurred in 1988, and Robinson describes it in the Activities Report:

> Problems galore! MLO had several short power outages (totaling about 7 minutes) from 3/27 thru 3/29 caused by lightning strikes on the power lines during a week of some of the worst thunderstorms ever experienced at the observatory. Lightning induced surges damaged a not-insignificant portion of the equipment. . . . A total of 7 circuit boards was lost including 3 A/D boards [that converted analog signals from sensors into digital form]. CO_2 data was interrupted only briefly with no significant loss of data. . . . The surface ozone data was lost until 4/06. . . . Air temperatures were lost from 3/27 to 4/06. (Robinson 1988b)

The freak storms occurred at a most unusual time, since the GMCC annual meeting was taking place in Hilo, and a field trip to the observatory to celebrate the thirtieth anniversary of the Keeling CO_2 record was accompanied by an unexpected weather event:

> The observation program proceeded in a generally uneventful fashion until about the last week in the month when a prolonged period characterized by deep, unstable air masses moved in over the area. This unstable air mass brought snow, hail, lightning, and rain showers, frequent cloud cover, below normal temperatures, and strong winds to MLO. It also coincided with the GMCC Annual Meeting being held in Hilo from [March] 28th to the 31st. On the 28th a short program was held at MLO to commemorate the 30th anniversary of the SIO [Scripps Institution of Oceanography] CO_2 observations under the direction of Dave Keeling. The program was attended by about 100 visiting scientists and friends. A brisk hail storm occurred during a portion of the planned program. (Robinson 1988b)

The thirtieth anniversary hailstorm was reminiscent of one that occurred on October 14, 1972, during a celebration of NOAA's second birthday (Fegley 1972). In both cases the visitors made it safely down the mountain, but a week was required to get most of the lightning-disabled systems up and running again after the 1972 event.

The thirtieth anniversary event was notable for the presence of Dr. Charles David Keeling, who wrote after his signature in the *Visitors* book, "30 years since CO_2 measurements began." Keeling's wife Louise wrote, "First trip." Timothy Whorf, one of Keeling's colleagues at the Scripps Institution of Oceanography, signed in next and wrote, "Brr! Hail the station's 30th! (Hail last 3 days. No hike today.)" The old Scripps gas analyzer was sampling the outside air from only a few steps away from where they signed the book.

Dr. Keeling's visit had special significance for Judy Pereira (video interview with author, Boulder, Colorado, May 2, 2007), who had never met him in all her years at MLO:

> Everybody talked about Dave Keeling, because he was the Mr. CO_2, and every CO_2 person at Mauna Loa then

> and until now remembers Dave Keeling as being the founder of the program. It was years before I actually met the man. In my mind I pictured [him resembling] the Wizard of Oz with all this smoke in front of him. Nobody really knew him or knew what he looked like, because he was just someone who wrote to us, talked to the directors on the phone. But we never saw him; we never knew, really, who he was. The excuse was that he had a horrible back [and] couldn't ride an airplane for five or six hours to get out to Mauna Loa. But, for the 30th anniversary he came out. He made it, and he wasn't the wizard that I thought he was going to be. He was a regular guy, and he was very personable.

Computer Progress (1989)

In March 1989, shortly before the invention of the World Wide Web, MLO expanded its participation in the personal computer revolution with its acquisition of an IBM PS/2 Model 80 desktop computer. As with the LSI-11 acquired by Kinsell Coulson in 1981, the PS/2 arrived damaged. A replacement was received on March 31, but it arrived without its DOS operating system. The laid-back Robinson wrote, "This will not cause any major problems" (Robinson 1989d).

The acquisition of the first personal computers for the Hilo office streamlined chores that once took considerable time. Pereira, who used a typewriter to type the monthly Activities Reports, describes in her memoir (1996) the significance of the arrival of desktop computers for the MLO office environment:

> Computers were introduced to Mauna Loa for the purpose of word processing (data acquisition computers were already in place) and a whole new world opened up for me. What an incredible invention! What I would have given to have this when we did the *20th Anniversary Report!* Arne Austring became mentor and guru of the PC system and passed on to me, and anyone else who wanted to learn, some of his knowledge of word processing, spreadsheets, and graphics. Because the computer took on additional applications, more PC's were purchased, and the rest is history.

Arne Austring (1982–1989) Replaced by Steve Ryan (1989–Present)

Arne Austring was well known for his versatility at MLO. Among his most significant contributions was developing the computer-generated plots of key MLO data that were included in each of the monthly Activities Report beginning in April 1987. Austring completed his work at MLO in 1989:

> April 14 was Arne Austring's last day on the MLO staff after being a permanent staff member for almost seven years. He was active in all areas of the MLO program, but especially in the area of solar radiation. We wish much success in his mainland endeavors. (Robinson 1989e)

Steve Ryan arrived to replace Austring on March 27. Ryan's interest in science during his teenage years well prepared him for his atmospheric science assignments later in life:

> Well, I was always kind of a science nerd I guess you would say. When I was in high school, I built a domed astronomical observatory with a 10-inch telescope. I took daily sunspot observations for four years. So I guess I was just sort of cut out to do this kind of work. (Ryan, video interview with author, Hilo, Hawai'i, December, 2006)

A graduate of the University of Colorado, Ryan's assignment to MLO made him the first person to serve at all of NOAA's monitoring observatories: South Pole (1982); Barrow, Alaska (1984); Samoa (1985–1989), and MLO (1989 to 2011). A desk job with NOAA's Space Environment Center, the group that studies the sun and provides forecasts about its activity, launched him on his career at the observatories, as he noted in the same interview.

> I spent a couple of years working with magnetospheric data, and the story I like to tell is, one day, here I was in

> my mid-twenties, it was a nice summer day in Boulder, and I was sitting in front of my computer looking out the window thinking, "What am I doing in an office? I should be doing something adventurous." So that eventually led me to my year at the South Pole, which was when I started working for [what is now the Global Monitoring Division].

Like most other permanent staff, Ryan was a cross-disciplinarian. He could service and troubleshoot recalcitrant instruments inside MLO buildings in the morning, make two climbs to the top of the forty-meter tower to install a new air collecting hose in the afternoon, and then go home and practice playing his banjo. He did this in spite of suffering with altitude sickness during about 40 percent of his workdays on the mountain. On about 5 percent of his workdays, the altitude sickness was severe enough to last overnight.

Volcanoes emit carbon dioxide, and one of Ryan's first assignments when he arrived at MLO was to determine how many hours per year the CO_2 data were influenced by emissions from Mauna Loa. The 2006 interview continues:

> That number was put in each annual report. So I did that for 1989, and I happened to look at '88, '87, '86 and noticed there was a geometric progression on those numbers. Each year was . . . not quite half as much as the year before, but there was a clear progression. So then I looked at the whole record and found that the CO_2 coming out of Mauna Loa had a very predictable nature. Whenever there was an eruption, there would be a great increase in CO_2 after that, and then in subsequent years there would be a very predictable exponential decrease. . . . I published those findings. There have been two different publications now. And the volcanologists were thrilled by that. Jack Lockwood said that he had always known about these CO_2 events, but he never realized that they would show any sort of a systematic behavior.

Ryan's studies, which go back to early CO_2 measurements at MLO in 1959, provide what is by far the longest known time series of carbon dioxide emitted by a volcano. By carefully tracking wind direction during CO_2 events measured at MLO, Ryan was able to estimate the location of the vents from which most of the CO_2 was emitted. Of special interest is that Ryan discovered that volcanic CO_2 can be entrained within the nighttime surface wind flowing down the slope of the mountain. As described by Ryan (1997),

> The nighttime surface wind is composed of two components. These are a radiation wind, which is the downslope gravitational flow of radiatively cooled air near the ground, and a barrier wind that results when the winds in the free troposphere flow around the mountain obstacle.

The radiation wind forms a layer over the lava with an average thickness of around 165 feet (about 50 meters). Ryan measured the thickness of the nighttime radiation wind by means of temperature monitors on the MLO atmospheric sampling tower. For his sulfur dioxide (SO_2) study, Ryan installed air sampling ports at four heights on the tower and found that SO_2 at MLO is well correlated with volcanic CO_2. This allowed him to estimate the thickness of CO_2–bearing radiational air flowing down from the summit.

Fortunately for the carbon dioxide monitoring program at MLO, volcanic CO_2 events are easily detectable and relatively infrequent. For example, during the first three hundred days of a study by Ryan in 1998–1999, CO_2 emissions from Mauna Loa were present during twenty-one nights (7 percent) between midnight and 7:59 a.m. However, during and after eruptions of the volcano, CO_2 at Mauna Loa can be much higher. For example, volcanic CO_2 was present at MLO for about half the time shortly after the major 1984 eruption, and its concentration sometimes exceeded the upper detection range of the MLO carbon dioxide analyzers (Ryan 2001).

While performing his detailed studies of MLO CO_2 data, Ryan (video interview with author, Hilo, Hawai'i, December 2006) learned about factors other than volcanic emissions that could contaminate the data:

> So in . . . working with the CO_2 events from the volcano, I became intimately involved in looking at all

> those hours where the CO_2 trace was variable. . . . Most of those were due to volcanic CO_2, but a lot of those were due to other kinds of instrumental problems. There were all kinds of things—radio frequency interference, a leaky line, maybe a leaky line where someone would come into the room and their breath would get into the air stream and make a little bit of increase.

This and other research performed by Ryan led to a string of scholarly papers published in leading scientific journals (Ryan 1990, 1995, 1997, 2001; Ryan et al. 2006). Of all his many studies, Ryan considers his carbon dioxide findings the most meaningful achievement of his scientific career.

Among Ryan's recent research projects is the measurement of atmospheric pressure over a network of nine miniature pressure recording data loggers that he installed across the island of Hawai'i. The pressure data he collected from the sensors played a key role in validating a new atmospheric model that simulated daily atmospheric tides caused by heating due to sunlight. According to a paper about the model,

> In addition, we present new determinations of the semidiurnal pressure oscillation based on barometric data from a special array of 9 sensors established on the island of Hawaii, representing an unprecedented sampling of surface tides over a large range of elevations in a relatively small geographical region. The results of the AFES simulations agree quite well with these new detailed observations in Hawaii. (Hamilton et al. 2008)

While this book was being prepared, Ryan's regular duties included preparing and launching ozonesondes and collecting air samples at MLO and Cape Kumukahi.

Other Arrivals, Departures, and Events

Kinsell and Vivien Coulson visited Hilo during the first two weeks of October 1986, and Dr. Coulson visited MLO on October 3.

On March 5, 1987, two official visitors found the air at MLO to be too clean, at least for their purpose:

> Dr. William Malm, National Park Service, Fort Collins, Colorado [and] Doug Latimer SAI visited MLO on 3/5 to assess it as a possible background transmissometer site [to measure the clarity of the atmosphere]. Tentative conclusion: it may be too clean, i.e. a small fraction of Rayleigh scattering. (Robinson 1987b)

Glenn Shaw made another in his long series of visits to MLO on April 6, 1987, to calibrate sun photometers, and former MLO director Rudi Pueschel returned for the same purpose in April 1989 and in March and June 1991.

On May 6, 1987, hikers stopped by MLO to report a missing companion. The lost companion arrived soon after they used the emergency telephone mounted on the main building.

A more serious situation involving hikers occurred in February 1989 during a time when construction dust affected some measurements:

> A problem of more serious note was that a hiker with frostbitten feet who was stranded at [the Mauna Loa cabin] near the volcano's caldera in quite heavy snow (for Hawaii, anyway). After five days, she was rescued by helicopter when her boyfriend was able to hike to the observatory for help. We all hope she didn't lose any toes. Apparently, they expected a search and rescue party to be sent out automatically when they did not check back in with the National Park Service after 2 days; however, that is not the policy of the NPS [National Park Service], because sometimes hikers return from their treks without bothering to notify the Park that they have returned. (Robinson 1989c)

Nature Takes Charge (Again)

Elmer Robinson arrived at MLO shortly after the Mauna Loa eruption of 1984. While he missed out on that misadventure, his tenure as director was accompanied by a string of unusual and even rare natural events, including the volcanic eruption of Mount Pinatubo in 1991, the solar eclipse of 1991, and record rain and snow events.

> January 1990 was the wettest MLO January on record with 13.10 in of rain recorded! (Robinson 1990a)

The heavy rains of January were followed by frozen precipitation during March:

> March was the month of the big snow at MLO. The storm hit on the 2/27 and continued through 3/1. It left 2–3 ft. drifts across the upper 3 miles or so of road. This is above the 10,000 ft level. Several attempts to reach the observatory by us and the HAO [High Altitude Observatory] people (even by 4-wheel drive vehicles) were unsuccessful until Tuesday, 3/6, when Darryl [Kuniyuki] and Steve [Ryan] hiked in about 2 miles and were able to check station conditions [Plate 47]. The station facilities and instrumentation were in generally good condition and many observations had continued under CAMS control throughout the storm isolation period. . . . Chart [recorder paper] rolls had run out, operating filters had run extra time, moisture traps on the CO_2 analyzers had frozen up, etc. These were not unexpected. On Friday, 3/9, John [Chin], Steve and Darryl used a 4-wheel drive truck and got within 100 yds of the lower (visitor) parking area and then walked the rest of the way in. At this time most of the remaining instruments were reset and restarted at a normal operational routine. On Monday, 3/12, it was now possible to drive to the observatory in our station wagon although it was necessary to use tire chains (obtained by emergency shipment from Boulder) to get up the last hill above the lower parking area. Full operation of the MLO program was resumed on the 12th. This storm was, by far, probably the most severe in the 34-year history of MLO. The "old-timers" cannot remember a situation where snow blocked access to MLO for more than two consecutive days. It is encouraging to note that the MLO operations survived this event in good shape and with a high percentage of data recovery. (Robinson 1990b)

Soon after the snowstorm, Robinson was happy to report that "the legendary MLO clear rainless sky conditions returned."

The Climate Monitoring and Diagnostics Laboratory (1990)

Since the founding of the slope unit in 1956, Mauna Loa Observatory had fallen under a series of agencies within the Department of Commerce. It had been under the Geophysical Monitoring for Climatic Change (GMCC) program since 1972, and in January 1990, GMCC was merged with the newly formed Climate Monitoring and Diagnostics Laboratory (CMDL), whose first director was Dr. Eldon Ferguson, the former director of the NOAA Aeronomy Lab, which, like CMDL, was headquartered in Boulder, Colorado. Ferguson came out of retirement to head CMDL. Dr. David Hofmann (1997), who would eventually become CMDL director, describes the new organization:

> CMDL as it exists today is mainly an expanded, modernized version of GMCC with several important differences, including an augmentation of its global sampling program of greenhouse gases, extension of baseline, clean air monitoring to regional locations important for climate change (e.g., sulfate aerosol effects), a highly enhanced capability in detecting ozone-depleting trace gases, and an increased emphasis on data analysis and publication as well as archiving. An airborne and balloon borne capability was developed for ozone and ozone-reactive gas measurements. Major building programs at the Baseline Observatories also improved the infrastructure conducive to successful monitoring.

Another significant milestone for both the observatory and MLO veteran Bernard Mendonca occurred in 1990. Previously, each monthly Activities Report was addressed to Jim Peterson, director, GMCC. Beginning in May 1990, the reports were addressed to Bernard Mendonca, chief, Observatory Operations. John Chin also passed a significant milestone in 1990:

> An important MLO milestone slipped past us unrecognized in the latter part of 1990 when on October 22, 1990, John Chin completed 25 years of NOAA service. All of John's NOAA time has been spent as a member of the Mauna Loa Observatory staff and as a principal scientist in the CO_2 monitoring program. John first joined the MLO staff as a "temporary" employee in 1960 to 1962. Then after a short time with NASA in Alabama, he returned to MLO as an employee of Scripps Institut[ion] of Oceanography in Dave Keeling's CO_2 program. In May 1968, John rejoined the MLO NOAA staff and has been the MLO principal scientist for CO_2 and other related sampling programs since that time. In 1982, John was awarded the Department of Commerce Silver Medal in recognition of his service at MLO and in the CO_2 monitoring program. Congratulations, John, on your 25 years at MLO! (Robinson 1991a)

Beginning with Jack Pales and Howard Ellis, MLO directors had often become involved in Hilo activities. A milestone for Director Elmer Robinson occurred in June 1990 when he was named president-elect of the 110-member Rotary Club of Hilo, the second oldest in Hawai'i. He assumed the office in June 1991.

Still another 1991 milestone occurred on Christmas Day when Mikhail Sergeyevich Gorbachev resigned the presidency of the USSR. The nation that had launched the first artificial Earth satellites and the first man into earth orbit during the International Geophysical Year was no more.

David Hofmann Arrives at CMDL (1990–Present)

In 1990, Dr. David Hofmann became chief scientist of NOAA's new Climate Monitoring and Diagnostics Laboratory in Boulder. A US Navy veteran, Hofmann studied physics at the University of Minnesota, where he earned a doctorate in 1966. Hofmann's graduate research involved cosmic ray studies in a pioneering balloon program supported by the Office of Naval Research, and this background led to an appointment as an atmospheric scientist at the University of Wyoming. There he continued his high-altitude research by flying ozone and particle detecting instruments to the edge of space suspended under large balloons. The Mauna Loa Observatory and NOAA's other remote meteorological stations provided the scientific resources that attracted Hofmann (video interview with author, Boulder, Colorado, May 1, 2007) to the position at the newly launched CMDL:

> At the University of Wyoming, we were reasonably limited. We could do some things related to climate, like measuring particles in the stratosphere, going to Antarctica, measuring the ozone hole, and things like that. But I really wanted to get more involved in the broader issues of the climate program. And that meant monitoring greenhouse gases all over the world.

Hofmann was soon to find himself immersed in the broader issues that so intrigued him, for a volcano halfway around the world from Boulder was about to awaken and spew millions of tons of particulate matter and sulfur dioxide high into the stratosphere. At the same time, a new war in the Middle East would lead to the torching of hundreds of oil wells that would create tons of smoke ladened with soot and a soup of undesirable gaseous compounds. The smoky plumes were so large they would be easily visible from satellites. Hofmann's expertise in studying aerosols made him the right man for the job.

A huge jump in technology accompanied Hofmann's first year at CMDL, for in October 1990 Tim Berners-Lee developed the World Wide Web. The Web would change how the world communicates and learns about the news. It would also provide unprecedented access by scientists and the general public to the data flowing from MLO.

Mamoru "Al" Shibata Retires and Lynne Hanzawa Arrives (1990)

Some former MLO staff are universally admired by the surviving old-timers, and among them is electrical engineer Mamoru "Al" Shibata:

FIGURE 12.3. The MLO staff during the retirement party for Mamoru "Al" Shibata on November 30, 1990, included *(from left to right)* Shibata, Darryl Kuniyuki, Lynne Hanzawa, Steve Ryan, Judy Pereira, John Chin, Tom DeFoor, Alan Yoshinaga, and MLO director Elmer Robinson. *(MLO archive.)*

> November 30, 1990 was a memorable day in MLO history as it marked the retirement of Al Shibata after almost 30 years of government service, the last 18 of which were spent at the Mauna Loa Observatory. Al has decided to "take it easy" and enjoy the good life. . . . We at the observatory will miss his many technical skills, as well as his warm disposition and subtle humor. We at MLO wish him a long and happy retirement. (Robinson 1990c)

A retirement party was held for Shibata, and a group photograph was made of him and the rest of the MLO staff (Fig. 12.3). Darryl Kuniyuki took over Shibata's projects on the mountain, which included aerosols (nephelometer, cloud condensation nuclei, Pollak counter, and aethalometer), meteorology, and the solar ultraviolet projects (Smithsonian Institution, University of Hawai'i, and erythemal irradiance, the solar UV that causes sunburn).

Meanwhile, Lynne Hanzawa was hired by Judy Pereira as a data clerk in 1990. Her primary responsibility was working with scientists on the preparation of data and charts. While she worked at MLO for only a year and a half, she left with fond memories of the staff. "It was like a home to me, and Judy Pereira was like a mother to me," Hanzawa (video interview with author, Hilo, Hawai'i, December 4, 2006) remembers. She also left with memories of the July snowstorm on the mountain.

Bob Uchida Arrives (1991–2006)

Al Shibata was replaced by Bob Uchida on July 29, 1991:

> It was good to be back to a full strength staff and to have the ET [electronics technician] in the hands of someone

> as highly qualified as Bob. Bob most recently [was] on Oahu with the NOAA Tsunami Warning Center. (Robinson 1991d)

Uchida was from Hilo. After service in the US Army, he studied electronics at Oregon State University. Uchida was more than highly qualified in electronics, for he had considerable construction experience and had built the Cloud Physics Observatory building where MLO had its Hilo offices before the move to the Federal Building. How did he learn about construction? "Well, there are a lot of books out there and . . . you learn by doing things" (Uchida, video interview with author, Hilo, Hawai'i, November 30, 2006).

Uchida also had prior experience at MLO, where he had spent nights and where he had met Howard Ellis while working on projects for the Cloud Physics Laboratory in the 1960s. (He signed the station *Visitors* book on June 8, 1965.) Besides taking on the maintenance of the observatory's aerosol and meteorology instruments and several of the cooperative programs, Uchida played the major role in maintaining and upgrading the buildings at MLO and planning the location of the Network for the Detection of Atmospheric Composition Change (NDACC) Building. He also repaired instruments and was a regular part of the observing program on the mountain. While most of the MLO staff experienced occasional or even frequent altitude symptoms, Uchida had no problems working on the mountain. His skills were so useful that he continued working part time for MLO after his retirement in 2006.

The Kuwait Oil Field Fires and the Eruption of Mount Pinatubo (1991)

Two major atmospheric aerosol events in 1991 were of special interest to CMDL's new chief scientist and aerosol specialist, David Hofmann. The burning of the Kuwait oil fields by the Iraq army in April 1991 caused enormous columns of smoke that were clearly visible on satellite images. Did some of the soot reach the skies over Hawai'i? The frequency of lidar measurements at MLO was increased in an effort to find traces of smoke clouds drifting overhead:

> Results from the enhanced lidar observations are inconclusive. The big problem is that we have no historic or background data below [an altitude of] 11 km. Comparison of the Apr '91 average measured profile with recent monthly average profiles suggest[s] some possible enhanced backscatter below 12 km. . . . However, the comparison of 1985 through 1991 average measured April profiles indicates no enhancement of tropospheric aerosol backscatter above 10 km. . . . Also there is no way of determining whether a particular layer of a profile is the result of Kuwait oil fires, Asian dust or sub-visible cirrus. (Robinson 1991b)

While the ruby laser lidar was trying to detect tiny particles of soot high in the sky, an instrument called an aethalometer detected high carbon levels in the air at MLO during several days in April. Back in Boulder, David Hofmann, MLO veteran Barry Bodhaine, and others studied the back trajectories to determine from where the air containing the carbon had come. They wrote a scientific paper suggesting that some Kuwait smoke did indeed reach Hawai'i and that the overall impact of the smoke was a net cooling (Bodhaine et al. 1992).

Two months after the Kuwait oil field fires were ignited, Mount Pinatubo, a volcano near Manila in the Philippines, showed signs of activity. On June 15, 1991, Pinatubo erupted in a massive explosion that propelled some twenty million tons of sulfur dioxide high into the stratosphere. The gas eventually became sulfate particles that blanketed much of the Earth. The aerosol cloud caused brilliant twilights much like those that Kinsell Coulson had photographed from MLO in 1983. The cloud was so vast that it also caused a global cooling of around a degree Celsius for a few years. Smoke plumes from the Kuwait oil field fires were still visible in satellite images, but they were dwarfed by the vast cloud of Pinatubo aerosols, which formed a veil over satellite images of the oil field smoke plumes. MLO's lidar was put to work probing the skies for volcanic aerosols:

> The plume is expected to be detectable with the MLO lidar in early July after traveling around the globe in the equatorial regions. In anticipation of the arrival

> of the Pinatubo cloud weekly [launches of balloon-carried] ozonesondes were resumed on June 26 and the frequency of lidar observations has been increased. (Robinson 1991c)

When a very rare solar eclipse arrived at MLO on July 11, instruments were not needed to know that something was very different about the sky, which seemed to be obscured by a thick veil of whitish volcanic aerosols. On some days, a straw-tinted ring around the sun could be seen. The phenomenon was first described in the journal *Nature* by the Rev. Sereno Bishop of Hawai'i after the eruption of Krakatau in 1883 and is known as a Bishop's ring.

As when the El Chichón aerosol cloud arrived over Hawai'i in 1982, the Pinatubo cloud caused a huge increase in backscatter measured by the MLO lidar and a significant reduction in atmospheric transmission, which was clearly indicated as a sharp downturn in the long-running Ellis curve. Observations of the Pinatubo cloud continued into 1992 and beyond.

Of special interest was Ellsworth Dutton's observation (personal communication with author, May 20, 2008) that the very simple MLO atmospheric transmission measurements were as important to tracking the Pinatubo cloud as they were when Agung and El Chichón erupted:

> The Ellis curve provided the only continuous, consistent, and inherently adequately stable record during the second half of the 20th century. Unfortunately, climate models were not capable of ingesting the mean apparent transmission measured by this method, and other atmospheric optical depth records were pieced together to provide what is now considered the authoritative aerosol loading record for the period (GISS aerosols).

The Total Solar Eclipse of 1991

In March 1991, MLO and Charlie Garcia of the High Altitude Observatory (now the Mauna Loa Solar Observatory) began making plans for the July 11 total solar eclipse. The moon's shadow would be almost centered over Mauna Loa at 7:30 a.m., and the period of totality would be approximately four minutes and one second (Espenak 1991).

Elmer Robinson (1991a) reported that "HAO has several very important 'once-in-a-lifetime' experiments and calibrations planned for the eclipse period." During June, research teams and their equipment began arriving in Hilo, and the days prior to the eclipse were quite busy.

Many of the tourists who flocked to Hawai'i for eclipse day were disappointed by clouds that spoiled the view of the eclipse itself—but not the darkness it brought. Before and during the eclipse, the MLO road was blocked at Saddle Road, and a security guard was posted to prevent the entry of any vehicles, which might cause vibration, pollution, and other problems during the sensitive eclipse observations. An exception was made for MLO staff and families, who were allowed to gather at a turnout in the lava about 4 miles (6.4 km) from the observatory.

> In general, the special scientific eclipse investigators at MLO seemed satisfied with the eclipse; although there was some disappointment that the Pinatubo haze and some patches of thin cirrus reduced the sky contrast and limited resolution of some details of the event. (Robinson 1991d)

The July Activities Report includes a list of nine scientists and organizations that conducted a wide variety of special experiments. The report concludes with a two-page paper by MLO scientist Steve Ryan, excerpts of which follow:

> The Effect of the July 11, 1991, Total Eclipse on Meteorological Conditions at Mauna Loa Observatory. By: Steve Ryan
>
> The solar eclipse of July 11, 1991, began with first contact at 0631 LST and ended at 0836 LST. Totality lasted 4 min and 1 sec, from 0728 LST to 0732 LST. Sky conditions during the eclipse were mostly thin cirrus covering 5/10 or more of the sky and a thick haze from the Pinatubo eruption. . . .
>
> At maximum eclipse, the 2 m[eter] temperature was 2°C–3°C lower than it would have been on a normal day. Observers reported feeling much colder beginning about 15 min prior to totality. This was apparently

> due more to the loss of heat from solar radiation than to the slight drop in temperature. . . . Observers at 3000 m on the observatory road reported the formation of a thin ground-fog higher on the slope at about 0800 LST which seemed to be moving downslope and persisted for perhaps 5 min. This phenomenon is highly unusual and probably eclipse related. (Ryan 1991)

Ryan closes his report with a full page of superimposed plots showing solar radiation, temperature, dew point, and wind speed and direction during the eclipse. The solar radiation bottomed out at totality and provided a marker for comparing the other effects.

Notwithstanding the clouds that spoiled the eclipse for many residents and tourists below the mountain and the thin cirrus and volcanic haze that interfered with the spectacle at MLO and atop Mauna Kea, the solar eclipse was truly a once-in-a-lifetime experience for the residents of Hawai'i. Steve Ryan (video interview with author, Hilo, Hawai'i, December 2006) considers it the most unique experience of his time at MLO:

> [It] was a very spectacular thing to see the moon's shadow racing from the west and then covering this side; and it was dark. When you looked to the west, you could see the horizon was dark, and that darkness kind of got wider and wider. And then over the course of tens of seconds, it seemed really quick that the darkness just kind of came over, and then the sun blinked out, and there was the corona.

Another rare event occurred the week after the eclipse when a Japanese reporter arrived at MLO. International visitors were common at MLO, but Robinson reported that this one may have been the first to arrive at the observatory by taxi.

Ralph Keeling Begins to Measure Oxygen at MLO (1991–Present)

Dr. Ralph Keeling of the National Center for Atmospheric Research (NCAR) visited MLO January 15–20, 1991, to begin a new cooperative air sampling program. The son of Charles David Keeling, who began the historic carbon dioxide program at MLO in 1958, Ralph Keeling was interested in the other side of the CO_2 equation. The oxidation of carbon that produces CO_2 removes oxygen from the atmosphere, and Keeling wondered if it might be possible to measure and track over time the slight reduction in the atmosphere's concentration of O_2 expected to accompany the upward trend in CO_2. In the late 1980s, Keeling had perfected a method for measuring atmospheric oxygen levels utilizing interferometry in the laboratory, and in 1989 his group began measuring changes in atmospheric oxygen concentrations in air samples collected at stations around the world.

By June 1991, air samples for Keeling's new project were being collected every two weeks. Eventually arrangements were made to place automatic air sampling equipment for Keeling's project at Cape Kumukahi on the extreme eastern coast of the Big Island, and permission was obtained to install sampling lines for the project high atop the vandalism-plagued lighthouse tower. In December 1995, Keeling visited the cape to check the sampling system and procedures and to recommend some changes. More changes were needed, for the ongoing vandalism made the area unsafe to visit.

Meanwhile, Ralph Keeling joined the Scripps Institution of Oceanography, University of California, San Diego, in 1992, where his research focused on atmospheric chemistry, the carbon cycle, and climate change. Charles Keeling had long been the leading investigator of the global carbon dioxide cycle, and Ralph Keeling had become the leading investigator of the global oxygen cycle. (The Keeling legacy at MLO is discussed more fully in chapter 13.)

The University of Denver FTIR System Arrives (1991)

The various trace gases in the atmosphere absorb narrow bands of infrared radiation found in sunlight. In May 1991, the University of Denver began a program at MLO to monitor trace gases by means of a sophisticated sun-tracking system installed in a telescope dome across from the AEC Building. The system is known by its acronym, FTIR (Fourier Transform Infrared Spectroscopy), and it precisely measures chlorine compounds, ozone, and other trace gases. Over the years the system would be significantly upgraded, and the FTIR system would eventually become part of the Network for the Detection of Atmospheric Composition Change and become a key tool for monitoring ozone, chlorofluorocarbons (CFCs), greenhouse gases, and air pollution from China that drifted over Hawaiʻi. Bob Uchida and Alan Yoshinaga of the MLO staff were assigned to the FTIR project. One of them had to rise at 2:30 a.m. at least once a week to drive up to the station and activate and warm up the instrument so it would be ready to measure changes in sunlight caused by various gases in the atmosphere as the sun peeked over the horizon. The special trips ended in September 1995 when a new automated system replaced the original one (David and Murcray 1996–1997).

Staff Matters

Like Howard Ellis before him, Tom DeFoor was nearly electrocuted while working on the high-voltage section of the laser in the MLO lidar system. Ellis' case was related in chapter 11, and in an interview on August 13, 2008, DeFoor clearly remembered his: "It about knocked me across the room," he said. In explaining the accident, he noted that "You don't have 100 percent of your mental acuity working at 11,000 feet."

On a more pleasant note, DeFoor and Elmer Robinson wrote a paper entitled "Stratospheric Lidar Profiles from Mauna Loa Observatory: Winter 1985–1986." The paper was accepted for publication by *Geophysical Research Letters* (DeFoor and Robinson 1987), where it became the latest of the hundreds of scientific publications that had emerged from MLO.

MLO had long attracted the attention of the international scientific community, and scientists from various countries visited the observatory to get ideas for establishing air-quality observatories in their respective countries. MLO had also hosted a number of visiting scientists over the years and played a role in the establishment of observing programs elsewhere. Elmer Robinson describes one such example of cooperation in several of the Activities Reports:

> John Chin has received an invitation from the People's Republic of China, Academy of Meteorological Science, to visit China and assist them in establishing their CO_2 sampling program. The visit is scheduled to begin August 15. He will work with Mr. Ding Guo An who was a visiting foreign scientist at MLO in 1984. (Robinson 1988c)

Chin left for Beijing on August 9 and remained there until August 24. He assisted the Academy of Meteorological Science with its CO_2 program and also visited four other scientific institutions. Robinson devotes a page in the August 1988 Activities Report to Chin's description of his trip:

> All of these facilities are equipped with up-to-date electronic apparatus and with U.S., Japanese and Chinese computers. . . . For weather forecasting, they have a SA 3 weather satellite which provided satellite photo coverage twice daily from India to the west and Japan to the east.
>
> The PRC [People's Republic of China] CO_2 project is in its infant stage. Both the reference gas and air sample were passed through a moisture preparation apparatus to have them both at a dew point of +4°C before entering the analyzer. This system was changed at the suggestion of John Chin. A glass freeze-out trap according to GMCC design was made by a local glass blower. In the future, both air sample and reference gases will pass through this freeze-out trap with a dry ice and alcohol mixture as a coolant [to remove all water vapor that would otherwise cause serious measurement errors]. The Academy needs reference gases badly. Presently they have only two working reference gases, W1 = 330 ppm, W2 = 350 ppm. . . .
>
> Most of the PRC institutions were very excited about Chin's visit because they have known of Mauna

Loa Observatory as a very famous observatory and this was the first time they actually were able to meet a member of the observatory staff at their institutions. (Robinson 1988d)

Elmer Robinson Retires (1991)

By the end of 1991, MLO director Elmer Robinson had served longer than any previous director. Major construction projects had been completed by the time he arrived, and during his tenure MLO continued its tradition of generating many scientific papers. The November 1991 Activities Report carries Robinson's farewell note:

> After almost seven and a half years as MLO Director, the time has come to step down and turn this job over to a successor. On December 1st, Dr. Russ Schnell was appointed as the new MLO Director, and he will be in Hilo to take over this post on January 6, 1992. I will spend my last official day in the office on December 13. The time has gone fast, and I wish Russ all the luck in the world in this position. I also express my thanks to the MLO staff and all of my colleagues. (Robinson 1991e)

Robinson recently described the construction of the lava barrier as a highlight of his time at MLO. He also remembered the solar eclipse of July 11, 1991 (Robinson, telephone interview with author, April 9, 2008). John Chin briefly filled in again as acting director of MLO after Robinson's departure, which he and Judy Pereira describe in the next Activities Report:

> Among many other things, he was instrumental in seeing to the completion of a lava barrier which was constructed above the observatory to protect it from any new volcanic activity. Elmer came to us from the University of Washington at Pullman, and has now taken a position at the University of Hawaii at Hilo on a part time basis teaching beginning meteorology. All of us at Mauna Loa will miss him, and wish him the very best for a great retirement! (Chin 1992)

Judy Pereira (1996) remembers both Kinsell Coulson and Elmer Robinson as father figures during their long tenure at MLO:

> When Elmer decided to retire in 1991 after 7 1/2 years at MLO it was like losing a family member. We didn't think we'd know how to behave with another boss. It was a long and difficult hiring process to find a replacement, and we all sort of expected another post retiree to take over the reins. Boy, were we surprised! We received word that a Boulderite that we knew only slightly was coming to be Director: Dr. Russ Schnell.

MLO Director Dr. Russell Schnell (1992–1997)

Russell Schnell was born in the farming community of Castor in rural Alberta, Canada. He attended universities in Canada, Wales, and Hawai'i and earned degrees in biology from the University of Alberta and chemistry from Memorial University, Newfoundland.

He then attended the Department of Atmospheric Science at the University of Wyoming, where he earned his masters and doctorate. Schnell's doctorate was a cross-disciplinary study that helped prepare him for the wide variety of experiments he would eventually direct at MLO. His faculty advisor suggested that he explore the nuclei of hailstones to better understand their origin, and in less than three months he discovered that ice nuclei were being made by *Pseudomonas syringae* bacteria, a finding that led to a string of publications in leading scientific journals (Schnell 1976; Schnell and Vali 1976; Vali et al. 1976).

Schnell (video interview with author, Boulder, Colorado, April 27, 2007) explains how his rural upbringing merged with his scientific training and led to his discovery:

> To me, coming from a rural farming community and watching hailstorms every summer, it was very obvious . . . you could see when hailstorms form, how the air just rises underneath them, and obviously that is where they are getting their nuclei. I mean, it was a no-brainer.

> Why you would see that is because where I grew up there would be hawks . . . soaring, and then in the springtime a crow would come chase the hawk . . . away from its nest. And then behind the crow, you would have a blackbird, and behind the blackbird, you would have a kingbird, and behind the kingbird, you would have a sparrow, and behind the sparrow, you would have something else. You would have seven different species chasing each other, and they were spiraling under these clouds because they were all in the uplift. So, you would see this train going around and around and around and each bird was chasing other birds. It was fun to watch that, too, because obviously, the air was going up pretty fast.

After receiving his doctorate, Schnell worked for a year at the National Center for Atmospheric Research (NCAR) in Boulder. He then worked for the United Nations World Meteorological Organization (WMO) Global Atmospheric Watch (GAW) program to determine if Mount Kenya was a suitable site for an atmospheric baseline station to provide information about air quality and meteorology. Schnell and his wife Suan, who was from Singapore, arrived in Kenya a week before Suan gave birth to their daughter Alicia:

> She was born in an African public hospital, and my job during the birth was to kill the flies . . . because it is all open and there are no screens on the windows. (Schnell, video interview with author, Boulder, Colorado, April 27, 2007)

Some twenty years later, the Mount Kenya site became one of the six WMO stations and the only high-altitude (12,785 feet or 3,897 meters above sea level) baseline station on the equator (Rotich 2004).

Back in Boulder, Schnell began his own research program to study the phenomenon of thick haze over the arctic, an experience that provided firsthand experience in applying for research grants and organizing scientific expeditions (Schnell 1983). The experience would serve him well at the Mauna Loa Observatory, as he noted in the 2007 interview:

> I came back to Boulder and worked for the University of Colorado. I was a self-supported scientist, and most of my work eventually [evolved into] the Arctic Gas and Aerosol Sampling Program. I developed that program and funded it, because I did not really have a job. So I wrote proposals and developed the program and ran . . . an aircraft program studying pollution in the Arctic.
>
> So, every few years, we would take a NOAA P-3 aircraft, outfit it with equipment, and fly it across the whole Arctic Basin from Russia to the other side and map the pollution, and then come home and publish papers for a few years and do it again. We did that for 12 years.

When the directorship opening at MLO was announced, Schnell applied and was accepted. He and his family visited Hilo from September 16 to 30 for what Pereira describes as a traditional house-hunting visit. He was officially named director on December 1, 1991, and he was scheduled to arrive on the job in mid-January 1992.

Judy Pereira had long been the most senior member of the staff at MLO. Hired out of college in 1966, she had worked for directors Lothar Ruhnke, Howard Ellis, Rudi Pueschel, Ronald Fegley, John Miller, Kinsell Coulson, and Elmer Robinson. Now still another director was on the way:

> "What's he going to be like?" We all tried to reassure each other that he would work out fine, and the fact that he wasn't over 60 didn't mean a thing in terms of what he'd be like. For me it was just another chapter in my long career at MLO, and I figured I could work with anybody who wanted to be Director here. I was right. (Pereira 1996)

Schnell's Activity Reports were as lively as the man himself, and his first one begins with a bang. After explaining that he had been on the job for only two weeks and was still coming up to speed on the observatory's programs, he lists new progress at MLO. The tiny MLO bathroom was refurbished and painted, a new front door was installed on the main building, and a general cleanup of the MLO site was begun. Soon after Schnell's arrival, he and the MLO

FIGURE 12.4. The MLO staff shortly after the arrival in January 1992 of Director Russell Schnell included *(standing, from left to right),* Bob Uchida, Dr. Schnell, Judy Pereira, Melodie Keaunui, Steve Ryan, Alys Wall; and *(kneeling, from left to right),* Tom DeFoor, John Chin, Darryl Kuniyuki, and Alan Yoshinaga. *(MLO archive.)*

staff gathered for a group photograph at the Federal Building (Fig. 12.4.)

Schnell met shortfalls in the MLO operating budget head on by reporting that he was looking into charging cooperative and visiting research programs for the costs associated with their projects. This was a radical step, for cooperative programs had previously been hosted for little or nothing. In some cases, MLO staff donated their own time to maintain cooperative programs, the classic case being former MLO director Howard Ellis' work with the carbon dioxide program operated by the Scripps Institution of Oceanography. Schnell's years of experience running his own science programs provided some much-needed business sense that was about to be applied at MLO for the first time. All this was new to Judy Pereira (video interview with author, Boulder, Colorado, May 2, 2007):

> And, for me, it was difficult because he took being from a university situation a lot further than any of the others. He refused to accept the government paperwork and bureaucracy. It was a waste of time. He just didn't want anything to do with it. . . . But he was fun, because he traveled a lot, brought a lot of people to the observatory,

> made money for the observatory like nobody else. He would charge people for everything. Before you know it, all the cooperators were paying their fair share, which nobody else had ever asked for before that. He had people doing commercial videos up there. We had cars being run up and down the road [for TV commercials]. We had a [silver-painted] man who was running up the road for some commercial. . . . It was an interesting time, and it was a very lucrative time with Russ.

As for Schnell, he appreciated Judy Pereira's networking and her ability to organize and manage everything from paperwork to social functions. He remembers her as "an icon" and was impressed by the staff at MLO:

> The staff at Mauna Loa was unbelievable. If you just encourage them, they would work long hours and they would bring in their friends and everybody was related or friendly. And if they did not have a way of fixing it, if you just treat them nice and talk to them, they would get things done. We had unbelievable things done there. (Schnell, video interview with author, Boulder, Colorado, April 27, 2007)

Russ Schnell's business experience also made him a good communicator, a skill that proved handy when it came time for a major new addition to the observatory that would require more land than MLO's four acres:

> Glen Taguchi from the Dept. of Land and Natural Resources in Hilo has walked me through the procedure for acquiring the land on Mauna Loa. It is a potential six month process after the papers are accepted in Honolulu. (Schnell 1992a)

The Big Cleanup (1992–1993)

Like former MLO director John Miller years before (1976a, b), Schnell made cleaning up the observatory a major priority, and the cleanup and maintenance work on the mountain continued into February:

> The staff have been spending a few hours each mountain shift repairing, painting and reorganizing the observatory and its environs. Loose data and power lines have been covered with protective half-shell pipes, the gas inlet lines taken off the saw-horses, as the wind blew them over regularly, the outbuildings repaired, and some trim and door painting completed. (Schnell 1992b)

The observatory was rewired to bring it up to code and eliminate electrical hazards. The staff found a surprise while data cables were being rerouted in March 1993:

> The data cables which have been spread over the surface of the MLO site for decades are being systematically replaced with cables run through buried conduits. During this operation, two unused and long forgotten (but still hot!) power cables were discovered lying exposed on the site. (Schnell 1993b)

Considerable progress in the cleanup had occurred by April 1993:

> A 20-year accumulation of discarded shipping boxes, broken equipment, and related items deposited and forgotten at the observatory are slowly being removed at the rate of about one truck-load a week. This cleanup will probably last for another two months. (Schnell 1993c)

In the fall of 1993, some 160 feet (49 meters) of new wood walkways were built and installed to provide better and safer access to various instrument sites at the observatory.

From Hawai'i to Alaska and Back (1992)

Russell Schnell's extensive experience in arctic haze studies led to a special five-week project in Alaska only ten weeks after his arrival at MLO. NOAA assigned him to direct a project to explore a mysterious plume emerging from Bennett Island in the Arctic Ocean. He was scheduled to depart for Anchorage on March 28, the day after a five-day conference on the Mount Pinatubo eruption of 1991 at the University of Hawai'i at Hilo. Schnell and Tom DeFoor both gave papers at the meeting. Meanwhile, Schnell was finalizing plans for the research team he had organized and the rented Russian AN-26 aircraft that would make flights through and across the Arctic plume, while the scientists wore gas masks in case the plume was toxic. The plume turned out to be a harmless consequence of cold fronts passing by the island (Schnell, video interview with author, Boulder, Colorado, April 27, 2007).

Applying Business Principles at MLO (1992)

Water for hand washing and the single flush toilet at MLO came from the rain gutters installed along the eaves of the main building. The maximum storage capacity was around 1,000 gallons, and during dry months water had to be hauled from Hilo. Schnell's business experience influenced how he reported on the water problem:

> The observatory water tank was filled on 2/2 [1992] with 2 tank loads hauled up by Frank Grahek, NCAR. Each gallon of water, delivered to MLO, is worth $0.50, or, in more prosaic terms, each toilet flush sends $2.00 of water into the lava sump. (Schnell 1992b)

Schnell was serious about recovering expenses from cooperative observers:

> Financial: MLO is now in a position to recover vehicle, power, and telephone expenses from outside groups using MLO facilities. Donna Leise is establishing a MLO maintenance account such that non-government payments made out to NOAA, Department of Commerce, can be deposited in Boulder. (Schnell 1992b)

Most programs paid the fees they were assessed, but there was one major exception: the Scripps carbon dioxide project.

> That instrument was operated by us at no payment from Scripps. . . . We maintained it at our own expense. Scripps never paid for our time and material. (Schnell, video interview with author, Boulder, Colorado, April 27, 2007)

When the Coast Guard closed its station at 'Upolu Point, Schnell arranged for MLO to acquire much of the surplus property:

> Items included a Canon copier (for the mountain), 4 refrigerators to replace our decaying stock, a variety of hand and power tools, approximately 100 gallons of paint for the observatory, and a plethora of other items such as toilet paper, Xerox paper, cleanser, etc. The haul is valued at about $10K. (Schnell 1992c)

Somehow Schnell learned that the General Services Administration had been overcharging MLO for its chemistry lab in the basement of the Federal Building. The GSA agreed to a rebate of $70,000, which would amount to a savings of more than $80,000 over the next decade (Schnell 1993g).

Repaving the Observatory Road (1992)

The road to MLO had been a major issue since the summit observatory days. While some people found that working in a high-altitude environment was made easier by spending several days and nights at the observatory, overnight stays by the MLO staff generally ended after the power line was installed in 1966. But day trips do not allow time for the body to fully adjust to the elevation change. Moreover, having to negotiate a road sprinkled with giant potholes and sections where the pavement had disappeared slowed the trip to the observatory and tired the workers before they arrived. Russell Schnell wasn't shy about asking those who shared the MLO road to help pay for the repair of the many potholes.

The staff of MLO and the neighboring High Altitude Observatory generally drove only two vehicles a day to and from MLO. But construction projects and some major experiments caused a steep increase in traffic and accompanying dust and air pollution. They also took a toll on the road and what remained of its thin asphalt coating.

One of the most ambitious cooperative experiments at MLO was the Mauna Loa Observatory Photochemistry Experiment (MLOPEX). The project involved temporarily placing sophisticated instrumentation at MLO to measure ozone and other trace gases in the free atmosphere. The first MLOPEX expedition to MLO was during May–June 1988. There were then four one-month studies during each of the seasons of 1991–1992. Schnell wrote about the good and bad implications for MLO of the third MLOPEX visit in the April 1992 Activities Report:

> Activity at the observatory increased greatly in mid-April with the return of 30+ NCAR MLOPEX staff, associated vehicles, and tons of equipment. . . . We will learn a lot about the source of air at MLO from this data. The downside of this activity is that the MLO road has deteriorated rapidly in the past month from heavy and excessive traffic. Major repairs will be required soon. (Schnell 1992c)

The high traffic on the road continued into May and caused significant damage to the road's asphalt veneer.

> At the end of May, the bulk of the MLOPEX group (approximately 25 people) had departed leaving 3 scientists who will conduct measurements continuously until the return of the main group in mid-July. All the traffic associated with these and similar experiments has on occasions produced 30 vehicle trips per day on the road. The road has sprouted dozens of new potholes [over the] past two weeks. MLO has raised funds and donations of material and equipment from NCAR/MLOPEX, Hawaiian Telephone, the television station consortium, and we believe, the US Army, to repair the damage. (Schnell 1992d)

By June, Schnell had managed to secure enough funds to begin the repair work:

> The road to the observatory was repaved in June under the guidance of Alys Wall (MLO staff) and crew. In all, 40 tons of cold mix patch was added to the road "quilt" [which is what the road resembled]. Hawaiian Telephone paid $1800.00 cost of materials, the county hauled the mix, NCAR paid $2000.00 towards labor, and MLO paid the $800.00 balance. NOAA must consider a longer term solution to the road maintenance issue as the present road is past its original design life. With increasing traffic loads to be expected from future MLO and NDSC activities, the deterioration will be accelerating over the coming years. (Schnell 1992e)

Network for the Detection of Stratospheric Change (1992)

Russell Schnell's major MLO project was to work with David Hofmann to arrange for the construction of the largest building ever built at MLO. The building would include a large rooftop deck for the placement of MLO's many instruments that measure sunlight and atmospheric gases. The building would also house the NOAA and NASA lidars, and three separate rooms would be provided to support the rooftop ozone and ultraviolet instruments and NOAA's carbon cycle and halocarbon instrumentation. The latter two programs track the concentration of greenhouse gases and gases that damage the ozone layer. Also included would be a computer room, an electronics shop, two tiny bedrooms, a kitchen, and a bathroom.

The proposed new building would house many of MLO's projects planned to support the Network for the Detection of Stratospheric Change (NDSC), a set of international research stations around the world that observed and analyzed the chemistry, aerosols, and the transport of air within the middle atmosphere. A key goal of the NDSC was careful monitoring of the ozone layer and anything that affected it. Another key goal was monitoring the relationship of the ozone layer with climate, the chemistry of the atmosphere and solar ultraviolet at the ground. Finally, the NDSC provided an independent means to verify the calibration of satellite sensors of ozone and other atmospheric trace gases (Chipperfield and Newman 2007).

Most of these scientific observations and measurements were being performed at MLO when Russell Schnell arrived, and his assignment was to consolidate many of them within a single structure, the proposed new NDSC Building. But first he had to find a place to put the new structure. The four-acre site occupied by MLO lacked space for the proposed building, and soon after his arrival at MLO Schnell had begun the lengthy process of acquiring an additional four acres east of and adjacent to the original site. In an Executive Order signed by territorial governor Samuel Wilder King on January 28, 1956, the original site had been "set aside for the uses and purposes of the United States of America. FOR a United States Weather Bureau Station Site, to be under the control and management of the United States Department of Commerce." Schnell's goal was to obtain an Executive Order with similar language granting to the Department of Commerce the use of the requested four new acres. He reported on progress in the monthly Activities Reports:

> Progress appears to be occurring as the "metes and bounds" survey has been completed and 3 members of the State Land Board made an unannounced inspection of the MLO site and observations on the second week of May [1992]. They appeared friendly, asked only a few questions, looked around, saw where the NDSC building would go, and left after a short site visit. . . . [They] never signed in. We are working on that! (Schnell 1992d)

The "State Land Board" was Hawai'i's Department of Land and Natural Resources (DLNR), one of the most powerful regulatory bodies in Hawai'i. The jurisdiction of DLNR includes the protection of natural resources, preparation for natural emergencies and disasters, fire protection on state land, preservation of historic sites, management of state parks, and many related functions. The DLNR was heavily involved in the lengthy approval process for the lava barrier project begun when Kinsell Coulson was director of MLO, and key decisions about Schnell's request for the additional four acres for the NDSC building would have to pass muster with the agency. Meanwhile, Schnell met with the architectural and engineering contractors in Hilo in June 1992 to discuss plans for the building:

> The building footprint was moved up the slope a few 10's of meters in order to provide better sunrise views. (Schnell 1992e)

The improved sunrise view was important for instruments designed to begin tracking the sun immediately after sunrise, especially during winter when the sunrise was blocked by the slope of the mountain. Drs. Schnell and James Peterson from the Climate Monitoring and Diagnostics laboratory in Boulder attended a DLNR public hearing about the land transfer on July 9, 1992:

> Questions from the land board were informational in nature and generally neutral or friendly. The overall impression from Jim and Russ is that the land board is not adverse to the land transfer request. A decision by the land board will be forthcoming by the end of the calendar year or sooner. (Schnell 1992f)

Meanwhile, the MLO staff worked on a contour survey of the building site and Tom DeFoor planned the utilities connections. NASA's Jet Propulsion Laboratory (JPL) was planning to move two large trailers to MLO for a new ultraviolet lidar system for probing the ozone layer at night. Russ Schnell held discussions with JPL's Stuart McDermid about this major new experiment, which would eventually occupy the west end of the proposed new NDSC Building.

While approvals required for the NDSC Building languished in Hawai'i's bureaucracy, NOAA's bureaucracy quickly approved the concrete pads for JPL's two lidar trailers:

> We thank all the NOAA/CMDL and MASC staff who worked on getting through the procurement for the JPL lidar trailer parking pad in an incredibly short time. The request for the pad came to MLO on Thursday of one week and by the following Thursday the 1200 sq. ft. concrete pad, 9" deep, was drying in the sun. The JPL

> [lidar] trailers were installed the following Tuesday and powered up that week. The 40' trailers traveled the MLO road with relative ease. (Schnell 1993e)

Back at the Hilo office in the Federal Building, simpler changes were taking place. The thirty-year-old carpet was replaced and furniture for the workroom was ordered. Finally, good news about the NDSC project arrived on August 27, 1993:

> On 8/27 the State land board voted to "set aside" the 4 acre NDSC site and transfer control of the land to NOAA at no cost to the Federal government. . . . Final transfer of the land will occur after one more "procedural" public hearing to be held in Kona in late October or early November. (Schnell 1993f)

As in the past, the good news was tempered with reality, and final approval of the proposed NDSC Building site did not come until March 7, 1995, when Governor Ben Cayetano signed Executive Order 3646 formally setting aside 4.05 acres of land adjacent to and east of the original 4.05 acres:

> FOR MAUNA LOA CLIMATOLOGICAL AND ATMOSPHERIC RESEARCH OBSERVATORY SITE to be under the control and management of the United States Department of Commerce, National Oceanic and Atmospheric Administration. . . .

The transfer of land for the original four acres was unconditional, but the new Executive Order included provisions for cancellation of the land transfer should NOAA not use the site in compliance with its 1992 permit.

While the State of Hawai'i had finally approved the land required for the NDSC Building, the cost of the building's foundation had escalated significantly, and the federal government had yet to act on requests for funding. In the March 1995 Activities Report, Schnell wrote, "NDSC—The silence is deafening."

Because of the increased expense of building the originally planned NDSC Building, in November 1995 plans for the facility were downsized, and the proposed building site was moved from the new 4.05 acres of open lava to directly across from and south of the main building on the original site.

Lessons from a Major Atmospheric Monitoring Failure (1991)

While MLO's NDSC Building project was underway, the World Meteorological Organization was closing down a global network of sun photometer stations established around the world under WMO's Background Air Pollution Monitoring Network (BAPMoN) program. Brent Holben et al. (2001) describe BAPMoN and its unhappy fate:

> Historically the most ambitious attempt to monitor background aerosol optical depth levels was that organized under the auspices of the WMO BAPMoN program, which officially operated from 1972 to 1992. The network was composed of 95 sites operated by member countries that provided processed data to the National Climate Data Archive in Asheville, NC, USA. The diversity of instruments, expertise, analysis methods and quality control led a WMO evaluation committee to recommend abandonment of the network and declare the data archive in its present form as unsuitable " . . . for scientific analysis of either short or long term changes in global aerosol optical depth."

The failure of the BAPMoN sun photometer network was well known to at least some of the data users before the network was ended. For example, Robert Charlson of the University of Washington was quite familiar with the high quality of sun photometer data collected at MLO, for he had worked closely with Kinsell Coulson and members of the staff during a NOAA contract that resulted in a paper in *Science* describing the possibility of the widespread hemispheric distribution of microscopic soot from industrial sources (Clarke and Charlson 1985). Before the WMO closed BAPMoN, Charlson (1990) reported on his disappointing experience with BAPMoN data:

> It is necessary to be severely critical of the bulk of these data [for the United States and the world] because of serious instability and drift in the sun photometers. Further difficulties exist with calibrating the instruments, especially at low values of optical depth. As a result of these instrumental difficulties, much if not most turbidity data is of suspect quality, even some of the data that are officially archived by the National Climatic Data Center at Asheville, NC. Thus it is necessary to utilize these data with caution, particularly data from sites with low aerosol loadings.

Findings like this raised a red flag, and NOAA's Atmospheric Research Laboratory (ARL 1991) soon assisted the WMO in assessing the viability of BAPMoN sun photometer network:

> ARL led an expert review of the operations of the global sunphotometer network run by BAPMoN, as a result of which the sunphotometer network has been shut down. In brief, the instruments used employed filters that drifted with time, and the calibration requirements were therefore too much for the network to handle. It has been widely concluded that the instrumentation was deployed before it was ready for prime time. Moreover, the sunphotometer approach is more suited to research-grade observatory stations than to routine operations by technicians rather than experienced and expert scientists.

BAPMoN's verdict was officially sealed when the WMO (1991) announced its findings following a 1991 WMO meeting of experts in Boulder to assess the program:

> Aerosol optical depth is the principle variable describing aerosol effects in classical radiative transfer calculations. . . . The lack of success with many previous efforts to accomplish this task, routinely, must be acknowledged and deficiencies corrected in any future effort to help assure satisfactory results.

As noted in the WMO report, aerosol optical depth (AOD) is "the principle variable" in the radiative transfer calculations that estimate the temperature change that can be caused by aerosols in the atmosphere. But twenty years of data were lost, and Dr. Bruce Hicks (2003) of ARL best summed up the bad news and what was to become the WMO solution:

> Global environmental monitoring is in sad shape. In particular, we lack a coherent system by which the quality of even the most basic environmental measurements can be assured. A fresh start has been made, with a new World Meteorological Organization's Global Atmosphere Watch (WMO/GAW) forming the framework for a nested network of regional and global observing stations. This new network subsumes and replaces the BAPMoN (Background Air Pollution Monitoring Network) and GOOS (Global Ozone Observing System) activities of earlier decades.

The 1991 ARL review (above) suggested that sun photometers designed to measure optical depth should be operated by "expert scientists." However, technicians at MLO routinely made high-quality sun photometer observations for many years, and some of their data was published by Charlson (1990: Table 24-11). Moreover, subsequent experience with the GLOBE program's sun photometer project (Brooks and Mims 2001) demonstrated that even trained high school students and their teachers can perform high-quality sun photometer measurements using simple, stable instruments based on designs perfected during several years of periodic calibrations at MLO by this book's author:

> Can measurements by 15–18 year undergraduate students be used to validate aerosol measurements by satellite instruments? . . . Some of the students' measurements have been directly compared to collocated professional measurements of aerosol optical thickness (AOT). As a proof of concept, we first characterised the LED-based GLOBE Sun photometer's performance relative to a full-automatic, sun-tracking Sun photometer at KNMI. Results from this comparison indicate that collocated measurements by the GLOBE instrument are within

> 0.02 AOT of the values retrieved from the SPUV. Subsequently, AOT measurements performed by undergraduate students in The Hague have been compared to results from the AERONET CIMEL-instrument in The Hague. The encouraging results from this comparison show that undergraduate students are capable of measuring AOT up to a reasonable degree of precision (0.03 AOT) and with no significant bias. (de Vroom et al. 2005)

In short, the failure of BAPMoN was not due to the instrument operators. Instead, it was caused by the degradation of the optical filters and the calibration problems cited in the ARL report. Regular calibrations of the sun photometers would have detected changes caused by drift in the filter specifications. But no information can be found that suggests BAPMoN provided calibration services during its twenty-year sun photometer program. Individual sun photometer operators could have independently calibrated their instrument using the Langley method, but this procedure required a very clear and stable sky for consistent results. In the end, WMO's Global Atmospheric Watch program replaced the BAPMoN sun photometer network with a much smaller network of precision instruments placed at MLO and a dozen or so other permanent observatories.

The MLO Museum Finds a New Home (1992)

Kinsell Coulson's meteorological museum included some extraordinarily rare meteorological instruments, but it received little publicity and very few visitors. MLO director Schnell needed more office space, and in April 1992 he decided it was time to look for a new home for the museum:

> After discussion with, and encouragement from Dr. Kin Coulson and Elmer Robinson, the contents of the MLO museum have been loaned to the Lyman Museum in Hilo. In 1990 and 1991 a total of only 5 people visited the museum in room 304 [in the Federal Building]. Lyman plans on setting up an MLO exhibit in the main museum and a display at the Hilo Airport lobby. MLO sponsorship will be given billing on the displays. (Schnell 1992c)

The Lidar Explosion (1992)

The ruby laser in the MLO lidar system required three sources of high voltage. One was provided by a bank of large capacitors that was discharged through the laser's flash. A second circuit generated a very high voltage pulse that ionized the xenon gas in the flash lamp so that it became conductive and allowed the capacitor bank to discharge through the lamp. The third powered an optical device called a Q-switch that blocked the release of light from the pink ruby rod until the optical energy resonating through the ruby reached a peak level and emerged through the Q-switch in a powerful burst lasting only twenty nanoseconds.

All these sources of high voltage were dangerous, especially in the ultradry air high on a mountain, which allows high-voltage electrical charges to build up inside the laser. Moreover, the high-voltage electronic components occasionally broke down. There was always a possibility that capacitors and flash lamps might explode, and even a solid ruby rod would blow apart if pumped with excessive light from a xenon flash lamp. While the lidar usually worked fine, the high voltages sometimes took charge, as Howard Ellis and Tom DeFoor learned on separate occasions when they were both nearly electrocuted.

Then there were more routine problems, as when the high voltage inside the lidar power supply zapped some of the components:

> On November 4 [1992] after 35 shots, a high voltage breakdown occurred in the bank power supply of the laser system which literally disintegrated a number of components. The components were identified and ordered from the laser manufacturer. After the components were received and installed the laser failed

> during initial test. This time one of the 6KV [6,000-volt], 100 uF bank capacitors split open, spreading oil inside the chassis and over the floor. (Schnell 1992g)

The 1993 New Year's Storm

The generally mild weather on the mountain took a holiday in January 1993:

> The new year was ushered in with two weeks of high winds, ice, and snow at MLO. On 1/4/93 the front door of MLO was found open in the morning with inside temperatures of 6° C. Check your data for the period 12/31/92 to 1/4/93. In the following days, winds of up to 80 mph were recorded at MLO followed by an ice and snow storm on 1/6. Staff were unable to reach the observatory on 1/7 due to the snow and ice; the truck slipped off the road and had to be retrieved the following day with a rented 4-wheel drive. The high winds tore off one quarter of the roof [from] the generator building and damaged the flashing on the Solar Variability Building. (Schnell 1993a)

John Barnes Replaces Tom DeFoor at MLO (1993–Present)

On June 29, 1993, Tom DeFoor made his last visit to the lidar at MLO before being reassigned to Boulder. He left Hilo with good memories of his work at MLO and praise for all three of the directors under whom he worked: Kinsell Coulson, Elmer Robinson, and Russell Schnell. He also left with a list of major scientific findings on his résumé, for he was the first lidar specialist to detect the giant El Chichón volcano cloud in 1982 (chapter 11) and the Mount Pinatubo cloud in 1991.

> I would say as far as getting to see some interesting data and interesting phenomena, I was pretty lucky for my 13 years there, because there were three or four major eruptions. . . . I really enjoyed working there. I loved living in Hawaii. I loved working for Mauna Loa Observatory. I loved the technical challenges of the job. I loved the people I worked with. It was hard leaving there. (DeFoor, video interview with author, Boulder, Colorado, May 3, 2007)

DeFoor was replaced as lidar scientist by Dr. John Barnes, who made his first working trip to the mountain on July 15. Barnes first became interested in science in the sixth grade. He received his bachelor's degree from St. Mary's University in Minnesota and then moved to Colorado State University for a master's in physics. He earned a Ph.D. in physics from the University of Minnesota while studying the spectroscopic and molecular properties of ozone.

Barnes first learned about the Mauna Loa Observatory from Dr. David Hofmann, director of NOAA's Climate Monitoring and Diagnostics Laboratory in Boulder. Barnes was intrigued by Hofmann's studies of the upper atmosphere using high-altitude balloons, and he was especially interested in Hofmann's studies of the Antarctic ozone hole. Like Barnes, Hofmann had received his doctorate from the University of Minnesota, and Barnes asked him if there were any openings in Boulder. There were none at the time, but Tom DeFoor's departure left an opening at MLO:

> So Dave called me and asked if I wanted to move to the rainiest city in the United States. I told him I definitely was interested. (Barnes, video interview with author, Hilo, Hawai'i, December 2006)

For the first time since Barry Bodhaine had returned to Boulder, two members of the MLO staff held doctorates. Barnes' background wasn't merely academic, for he had hands-on experience building lidar systems. He also liked the challenge of designing and building a new system, getting it to work, and then collecting data with it. So he arrived at MLO with plans for a major upgrade from the old ruby lidar to a much more sophisticated system designed around a neodymium-doped YAG laser that would emit an invisible beam of near infrared superimposed over a brilliant green beam produced by a frequency doubling crystal. Barnes noted in the same interview that while building the new lidar, he continued to operate the old ruby system:

> [I] kept operating the ruby because it had this long history [and] had recorded two large volcanic eruptions, the biggest ones in more recent times—El Chichón in Mexico and Mt. Pinatubo in the Philippines. And meanwhile I started building the new one.

Tom DeFoor spent November 1993 in Hilo helping Barnes automate the old ruby system so that a personal computer could fire the laser and change the voltages applied to the photomultiplier tube that detected the light scattered back to the ground from aerosols high in the sky. The modified system also allowed the computer to automatically record the data (Barnes, personal communication with author, June 3, 2008).

Working on his own, Barnes assembled the new lidar, which was completed and in operation around nine months later. Dave Hofmann spent several days working on the lidar project with Barnes in February 1994.

Darryl Kuniyuki took over DeFoor's computer server and network responsibilities, an assignment that was to soon lead to major changes in the way the MLO data were acquired, processed, stored, and sent to scientists on the mainland and around the world.

The Personal Computer Revolution Expands at MLO (1993)

When he assumed the directorship of MLO, Russell Schnell's first priority was how best to modernize the data acquisition process, and in an interview in Boulder, Colorado, on April 27, 2007, he reflected back on major changes he initiated:

> I said, "Man, we got a lot of work to do here. The observatory is in a great place but we are still putting stuff on strip charts and people are making records and measurements on data sheets. Then they go down to Hilo and get copied, if you can find a copier, and then they get mailed. Man, we better do something about this, electronically and physically. It would be really ideal to send this data in real time." . . . There was no Internet at that time, but at least we started getting computers. . . . So we started buying computers, and Darryl Kuniyuki started buying computers and teaching us how to use them. . . . We started getting old computers at first to put on the data systems, and then we revolutionized how data was collected.

Judy Pereira played a major role in the computer acquisition process, Schnell notes:

> She was incredibly valuable and good in getting things done. I'm not a detail-person, I never had been and to my detriment, too many times. So I would just say, or Darryl would come in and say, "We need a new computer." I would say, "Okay, I will find some money." And Judy would figure out how to do it.

Schnell and Kuniyuki sometimes negotiated with one another about which computer to buy.

> [Darryl] is a very excellent engineer, and he is very innovative and enthusiastic. . . . His computers were always the best and highest priced one you could get. Mine were always the ones that would just marginally do the job. So we would settle in between.

New personal computers began to appear both in the Hilo office and the mountain station in 1993:

> New Wordperfect V6.0 was installed and DOS 6.0 upgraded on the 486 PC in the Hilo office. A new 486 and a color printer have been ordered. (Schnell 1993g)

During November 1993, e-mail arrived at MLO, and each staff member was assigned an individual address. A new PC with a host of new features also arrived:

> We can now read data on cd-rom media with the arrival of our new 486 pc with cd-rom drive. Also this month we received an HP Deskjet 550c color printer which is being enthusiastically used by MLO staff. (Schnell 1993h)

On the mountain, the old CAMS data logging system was replaced by more sophisticated systems. The NOAA and JPL lidar systems on the mountain were each connected to the Internet via optical fiber cables. Electrical engineer Darryl Kuniyuki was MLO's resident computer expert, and he played the major role in networking the new systems. In November 1993 he attended an Internet engineers' workshop in Honolulu.

Russell Schnell was pleased by the transformation that had taken place with the introduction of computers and the Internet, some of which was made possible by creative financing:

> It made it easier to have better measurements and more measurements . . . staff were not spending a lot of time writing stuff down and looking at strip charts. They were calibrating more, and we introduced a lot better and more rigorous calibration and maintenance of equipment. And if the staff wanted a new piece of equipment, if it was needed, we would find a way to get it. Maybe it was not through the normal budget way, but we would get an agency or somebody to do it.
>
> And we started to charge people to come to Mauna Loa. When I went there, Mauna Loa staff was doing everything free, and people would come and it was kind of an ad-hoc system. So we formalized it and made them pay for it, and now it is a business. We make [money], and then we take that money that we used for the user fees and buy better equipment and better computers. . . . A lot of the programs are run by scientists in different parts of the world. They get on the Internet, and they work on their equipment in real time right on the mountain. It does not have to go through us, so we just maintain it and keep it running. (Schnell, interview with author, Boulder, Colorado, April 27, 2007)

Since the earliest days at MLO, the staff had been intimately involved in breaking the data from chart recordings, but automated data acquisition sharply reduced the number of chart recorders after Schnell took over. Michael O'Neill (video interview with author, Boulder, Colorado, May 3, 2007) of the Global Monitoring Division reflects on the relative advantages and disadvantages of chart recorders and automatic data logging:

> The digital format is nice in that we have more data coming in faster. I have a UV spectral radiometer at Mauna Loa that I can set up automatically to record data at predefined times. And I can set that up for every minute, every eight minutes or ten minutes. I have a choice.
>
> The strip chart recorders are nice in that they are continuous; there are no breaks unless the chart recorder breaks down. The problem comes in, like the chart that we just looked at, if you start having too many traces over the same spot, they become obscured. If you start getting a blotchy pen on a strip chart recorder, your trace becomes very wide. Where do you mark the line? How do you mark the line? What are your zeroes? Your zeroes can drift. [A digital system is best], because we just take a number and assign that to memory, and that is our measurement. So there is good and bad [for] both. One is full time; one is selective. One has problems with the recording medium; the other one usually does not.

With the arrival of the Internet and modern data acquisition technology, the MLO staff suddenly found itself out of its own data loop:

> The familiar monthly summary of MLO measurements [begun in 1987] will not appear for another month or so. This is so because recent changes in the manner in which solar radiation and meteorology programs record their data no longer allows MLO direct access to the data. We are developing procedures for having the data sent to us from Boulder via Internet on a regular basis. Look for the graph to be back in March. (Schnell 1994a)

Second-Generation MLO Scientists

Over the years, several second-generation MLO–related scientists developed connections with the observatory.

Charlie Garcia supervised the High Altitude Observatory for many years, and his son Tom had worked at MLO for five years. Charles Keeling's son Ralph expanded on his father's measurements of carbon dioxide by measuring oxygen. Glenn Shaw's son Joseph expanded on his father's measurements of aerosols by studying the polarization of the sky over MLO. Then there was Kevin Ryan, who arrived at MLO on June 23, 1992, for the purpose of "helping dad Steve run the station." Kevin was fifteen months old at the time.

Visitors (1992–1993)

Environmentalist Steve Schneider of NCAR visited MLO on August 10, 1992. MLO veteran Bernard Mendonca, who was CMDL's chief of observatory operations, visited on October 19, 1992, and signed in "Great to be back!" He was welcomed again on May 20, 1993, with an honor so special that he neglected to sign in:

> Bernard Mendonca, NOAA/CMDL, visited and waxed nostalgic about the many years he worked at MLO. We rewarded him with the honor of allowing him to help load a truckload of junk. (Schnell 1993d)

Former MLO director Elmer Robinson, who still lived in Hilo, visited with twelve members of the Waikoloa Marine Life Fund on November 28, 1992.

Steve Ryan Monitors Sulfur Dioxide (1994)

Sulfur dioxide is among the principle gases emitted by volcanoes, and Hawai'i's Mauna Loa and Kīlauea are no exception. Bright yellow crystals of sulfur can be seen at Sulfur Banks near the Kīlauea Crater, and sulfur dioxide gas forms the volcanic haze known locally as vog.

Carbon dioxide is also released by volcanoes, and sulfur dioxide provided a convenient tracer for determining if excess CO_2 detected at MLO was leaking from Mauna Loa. Monitoring sulfur dioxide (SO_2) also provided a means for tracking Asian pollution events, for the combustion by-products from coal-burning power plants include the gas.

Steve Ryan, who was very much aware of the sulfur cycle, initiated a program to continuously monitor SO_2 on the mountain, and he describes the project in an addendum to the June 1994 Activities Report. Instead of using government funds, Ryan borrowed unused equipment from MLO and other labs to build his SO_2 monitoring system. Ryan continues to monitor SO_2 to 2011, and he describes his findings in a series of scientific publications and presentations (1995, 1997, and 2001).

Leslie Pajo (1994–2006)

Leslie Pajo joined MLO as a half-time data clerk in November 1994. Her choice fit well with Russ Schnell's approach to running the observatory, for Pajo had a degree in business administration with a major in accounting.

> Over the coming months many of you will learn to recognize her cheery voice on the telephone. She is a welcome addition to our staff. (Schnell 1994b)

Judy Pereira (video interview with author, Boulder, Colorado, May 2, 2007) relates how Leslie Pajo earned her position at MLO:

> I met Leslie before we even had a position open for her, or for anybody, really. She had come into the office with a résumé, and I talked to her a little bit, and I immediately said, this girl is special, she's definitely special. She's bright; she's willing to do just about anything. . . . I told her, if anything comes up that's permanent or in the direction of becoming permanent, I'll give you a call. And, sure enough, one position opened up, and I called her, and I told [Russell Schnell] please don't let this girl go; she is something special. She had a degree in accounting . . . and she was very good with numbers, as we all found out. . . . We hired her, [and] she learned everything that anybody wanted to teach her.

> She was like a sponge, she absorbed everything. We all loved her. She was special to everyone at the observatory, especially me, I think. When I left, she took over without skipping a beat. She just fit right in [and] did way more than I could do before, because she was computer literate, mostly because of [Darryl Kuniyuki]; he taught her a lot.

Darryl Kuniyuki (video interview with author, Boulder, Colorado, May 2, 2007) started from scratch with Pajo, who had accounting expertise but no formal science or computer training:

> When I first met her, she held up a floppy disk to me and said, "What is this?" She didn't know what that was. So, throughout the years since then, she learned . . . how to use computers and everything during her work time at Mauna Loa.

Leslie Pajo would go on to become a proficient computer user and play an important role in the day-to-day work at MLO. She even learned the mountain routine and occasionally worked at the observatory with the day crew, where she was assisted by *Leslie's Mountain Bible,* a thick notebook of operating instructions that she assembled for virtually every instrument at MLO.

Today new employees use an updated version of *Leslie's Mountain Bible* when they are working on the mountain.

More Rocks in the Mail (1994)

In December 1994, an anonymous visitor sent a box of lava rocks and a message to MLO, and her note was appended to the December Activities Report:

> My ex-husband took these rocks from your beautiful island many years ago. I do not want to be the victim of any more bad luck. I was very upset when I found that he had taken them.
>
> I hope that in returning them to you, my life will be filled with good things.
>
> I hope that he will experience some of the unhappiness that has come my way.

Road Hazards (1994)

On May 28, 1994, a motorcyclist participating in the international Mauna Kea Endura Race was killed when he crashed in the lava off the Mauna Loa road while racing downhill. A long, black skid mark is still visible on the road.

A new kind of road hazard occurred in March 1995 when Bob Uchida was driving to the observatory:

> In mid-month a group of armed pot growers were surprised and arrested by a 20+ unit of state police for growing the plant along the Mauna Loa Road. They had a system of cultivating the plants in pots set in depressions in the lava. They watered the plants with 5 gallon pails transported on a flatbed. Bob Uchida surprised them one morning but carried on as if he had not seen a thing. Now the hunters with automatic rifles and assault guns are our major worry on the road. (Schnell 1995a)

NASA's AERONET Comes to MLO (1994)

The Mauna Loa Observatory had long played a significant role in the calibration of instruments used elsewhere, especially sun photometers that measured various wavelengths of sunlight to determine the optical depth of the atmosphere caused by aerosols and instruments that inferred the total abundance of ozone in a column through the atmosphere. Because the sky conditions are often pristine, sunlight instruments calibrated at MLO became standards against which other instruments were compared.

In 1993 NASA's Goddard Space Flight Center (GSFC) began developing a global network to automatically monitor the optical depth of the sky and the total column water vapor using automated sun photometers. A key objective of the network was to provide ground truth validation of observations that would be made by a planned series of remote sensing Earth satellites. The project quickly blossomed into AERONET (AErosol RObotic NETwork). The network employed identical weather-proof robotic sun photometers that scanned the sun and sky at programmed intervals. Direct sunlight was measured at seven wavelengths ranging from 340 nanometers (nm) in the ultraviolet to 1,020 nm in the near infrared. The instruments were solar powered, and the data they collected were transmitted to GSFC via a satellite link (Holben et al. 2001).

Initial calibration and operational checks of CIMEL photometers began at MLO in 1993. In 1994, a permanent AERONET site was established at MLO on an old concrete pad southwest of the main building (now known as the Keeling Building), and the site regularly hosted two or three CIMEL radiometers. AERONET provides an important source of data for validating the optical thickness of the atmosphere measured by satellites, including NASA's Terra, Aqua, and AURA. AERONET also provides important data for routine atmospheric monitoring when not providing validation data during satellite overpasses.

AERONET (http://aeronet.gsfc.nasa.gov/) continues in operation today, and project head and principal investigator Brent Holben (personal communication, September 6, 2007) credits the Mauna Loa Observatory for the success of the global network that he manages:

> Measuring AOD [aerosol optical depth] by sun photometry has a checkered past at best, largely due to poor calibration. MLO offers the most stable, accessible and reliable platform on the planet for calibrating sun photometers using the Langley plot technique. This method is employed by AERONET reference instruments at MLO to calibrate the UV to SWIR [near infrared] bands and, thus, serves as the world reference for all 300 globally deployed AERONET sites. All AERONET data from 1993 to the present is traceable back to a MLO Langley calibration. Collaboration between NASA and NOAA has allowed a nearly continuous presence of AERONET on MLO since 1993. As the program has grown so has the collaboration and necessity of increasingly accurate measurements, testing and analysis as the number of instruments needed to operate the network increases. The reliable and extremely low AOD of 0.01 to 0.02 at 500 nm [green light near the peak response of the human eye] from MLO provides morning Langley plots year round and assessment of instrument performance that is routinely possible at no other place on Earth. Without the geographic location of MLO and naturally low aerosol loading combined [with] the infrastructure of MLO, there most certainly would be no AERONET program.

Climate Change (1995)

Considerable Asian dust drifted over Hawai'i in the spring of 1994. Hilo experienced forty-two days with rain that fall, while the Kona side of the Big Island remained relatively dry. Then there was an abrupt change in the weather pattern, and Russ Schnell comments on it in the March 1995 Activities Report:

> The dry weather pattern produced by El Niño continues on the east side of the island. Now the Kona Coast has the rains normally associated with Hilo. Three cheers for climate change! (Schnell 1995a)

In the December 1995 Activities Report, Schnell reported that the rainfall measured in Hilo during 1995 "was 8 feet below that recorded in 1994. That is correct, 8 feet!"

New Zealand Comes to MLO (1995)

Because of its location relative to Antarctica, portions of New Zealand are sometimes subjected to higher than normal levels of ultraviolet when sharply diminished amounts of ozone sweep across the country during the spring ozone hole season. For this and other reasons, Dr. Richard L. McKenzie and his colleagues at New Zealand's National Institute of Water and Atmospheric Research (NIWA) had long been interested in monitoring the ozone layer and solar ultraviolet.

In July 1995, McKenzie, Paul Johnston, and NOAA installed at MLO a research-grade ultraviolet spectroradiometer developed by NIWA. Barry Bodhaine was assigned as project leader until his retirement in 2001, when Mike O'Neill took over.

The instrument measures the full sky irradiance from 285 to 450 nm with a resolution of 0.8 nm by means of a double monochrometer that increases the precision of measurements by greatly reducing noise caused by stray light within the instrument. The high resolution of the spectroradiometer permits the irradiance across specific wavelength bands to be integrated into various action spectra (e.g., the erythemal action spectrum that causes sunburn, the DNA action spectrum that inactivates cells, the vitamin D action spectrum, and so forth). The narrow wavelength resolution also permits the instrument to detect ozone, sulfur dioxide, and other atmospheric trace gases that absorb UV.

The UV measurements made by the NIWA spectroradiometer have continued to the present and are viewed as among the best anywhere. UV measurements made by other instruments at the station were also considered high quality, and the scientists on the mainland responsible for them were very particular about how they were maintained by the MLO staff. When NOAA UV expert Dr. John De Luisi learned a UV specialist was planning a visit to MLO, he asked him to check the cleanliness of his instruments at the station. They were spotless.

Expansion of MLO Facilities (1995)

During Russell Schnell's tenure, a considerable amount of junk and trash was hauled off the mountain. Also during his time, a number of new structures were installed. The first were a pair of large trailers for the JPL lidar system. In April 1995, the prefabricated Global Oscillation Network Group (GONG) facility was installed on the east edge of the MLO site just south of the road to the main building. The instrument became operational in June. GONG was designed to study the internal structure and dynamics of the sun by means of interferometry, and the project was managed by the National Solar Observatory (NSO).

In July 1995, NASA's Microwave Ozone Building was hauled up the mountain and installed near the southeast corner of the original MLO land grant. The MLO Visiting Scientist Building was built in Kea'au and installed just north of the old AEC Building in August 1995.

Also in July, the MLO staff constructed the Tank Building, a room attached to the west end of the original main building. The Tank Building was used for a brief experiment by the US Air Force, after which it became a storage room for gas cylinders and shipping boxes for the flasks used to collect air samples.

While MLO did not hold title to the 1960s-era AEC Building, the staff had long been responsible for the building's maintenance and remodeling. Schnell reported on the building's status in the August 1995 Activities Report:

> On August 30, about a dozen DOE staff and contractors inspected the AEC and camera (radon/SO2) buildings which they still own and which are on long term loan to MLO. Everything appeared to be in order and they departed happy. No, they do not have any maintenance responsibilities for these buildings. And yes, we asked. (Schnell 1995b)

In only three years, Russell Schnell had transformed the appearance of MLO. Many truckloads of junk and rubbish had been removed, the dangerous underground electrical cables were repaired, and the arrival of an array of new structures had transformed MLO into a little science village along the lower side of the observatory's original four acres. All was well at MLO with but one exception: Plans for the long-awaited NDSC Building languished as the fourth decade at MLO came to an end.

References

ARL. 1991. Aerosol optical depth—atmospheric turbidity. Air Resources Laboratory. (Accessed at http://www.arl.noaa.gov/research/programs/aod.html.)

Bodhaine, B. A., J. M. Harris, J. A. Ogren, and D. J. Hofmann. 1992. Aerosol optical properties at Mauna Loa Observatory: Long-range transport from Kuwait? *Geophysical Research Letters* 19: 581–584.

Brooks, David R., and Forrest M. Mims III. 2001. Development of an inexpensive handheld LED–based sun photometer for the GLOBE program. *Journal of Geophysical Research* 106: 4733–4740.

Brooks, David R., Forrest M. Mims III, and Richard Roettger. 2007. Inexpensive near-IR sun photometer for measuring total column water vapor. *Journal of Atmospheric and Oceanic Technology* 24: 1268–1276. (Accessed at http://ams.allenpress.com/perlserv/?request=get-abstract&doi.)

Charlson, Robert. 1990. Existing conditions and historical trends for climatic effects of aerosols. In John C. Trijonis, William C. Malm, Marc Pitchford, and Warren H. White, *Visibility: Existing and Historical Conditions—Causes and Effects.* Report 24, National Acid Precipitation Assessment Program, 76–79.

Chin, John. 1992. December 1991 activities report, January 10. In *Geophysical Monitoring for Climatic Change,* Mauna Loa Observatory, Monthly Reports, Hilo, Hawai'i, 1992.

Chipperfield, Martin P., and Paul A. Newman. 2007. Public access to the network for the detection of stratospheric change database. World Climate Research Programme. (Accessed at http://www.atmosp.physics.utoronto.ca/SPARC/News16/16_Chipperfield.html.)

Clarke, Antony D., and Robert J. Charlson. 1985. Radiative properties of the background aerosol: Absorption component of extinction. *Science* 229 (July 19): 263–265.

David, Shelle J., and Frank J. Murcray. 1996–1997. Ultrahigh resolution infrared solar spectroscopy from Mauna Loa Observatory: Update, Climate Monitoring and Diagnostics Laboratory. Summary Report No. 24. (Accessed at http://www.al.noaa.gov/gmd/publications/annrpt24/david.htm.)

DeFoor, Tom, and Elmer Robinson. 1987. Stratospheric lidar profiles from Mauna Loa Observatory, Winter 1985–1986. *Geophysical Research Letters* 14: 618–621.

de Vroom, J. P., K. F. Boersma, and D. R. Brooks. 2005. Undergraduate student measurements to validate satellite-derived aerosol optical thickness. *Geophysical Research Abstracts* 7: 03497, SRef-ID: 1607-7962/gra/EGU05-A-03497. (Accessed at http://www.cosis.net/abstracts/EGU05/03497/EGU05-J-03497.pdf.)

Espenak, Fred. 1991. Path of the total solar eclipse of 1991 July 11. NASA Eclipse Home Page. (Accessed at http://sunearth.gsfc.nasa.gov/eclipse/SEpath/SEpath1951/SE1991Jul11Tpath.html.)

Fegley, Ronald W. 1972. September–October 1972 activity report, November 15. In *Geophysical Monitoring for Climatic Change,* Mauna Loa Observatory, Monthly Reports, Hilo, Hawai'i, 1971–1973.

Hamilton, Kevin, Steven C. Ryan, and Wataru Ohfuchi. 2008. Topographic effects on the solar semidiurnal surface tide simulated in a very fine resolution general circulation model. *Journal of Geophysical Research* 113, D17114, doi:10.1029/2008JD010115.

Hicks, B. B. 2003. Department of Energy support for operations of the WMO/GAW Quality Control/Science Activity Center for the Americas. Air Resources Laboratory, Reference DOE DE-A102-94ER61753 MACOOORRI B55WM. (Accessed at http://www.osti.gov/energycitations/servlets/purl/821794-o8ZD2S/native/821794.PDF.)

Hofmann, David. 1997. *Climate Monitoring & Diagnostics Laboratory: Science Review, Mauna Loa Observatory.* 40th Anniversary Commemoration and Dedication of the NDSC Building, November 20–21, Hilo, Hawai'i.

Holben, Brent. 2005. AERONET background. Goddard Space Flight Center. (Accessed at http://aeronet.gsfc.nasa.gov/F_Info/system_info_additional.html.)

Holben, B. N., D. Tanre, A. Smirnov, T. F. Eck, I. Slutsker, N. Abuhassan, W. W. Newcomb, J. Schafer, B. Chatenet, F. Lavenue, Y. J. Kaufman, J. Vande Castle, A.

Setzer, B. Markham, D. Clark, R. Frouin, R. Halthore, A. Karnieli, N. T .O'Neill, C. Pietras, R. T. Pinker, K. Voss, and G. Zibordi. 2001. An emerging ground-based aerosol climatology: Aerosol optical depth from AERONET. *Journal of Geophysical Research* 106: 12 067–12 097. (Accessed at http://aeronet.gsfc.nasa.gov/new_web/PDF/AERONET_climo.pdf.)

McPeters, Richard D., and Walter D. Komhyr. 1991. Long-term changes in the total ozone mapping spectrometer relative to world primary standard Dobson Spectrometer 83. *Journal of Geophysical Research* 96: D2, 2987–2993.

Mendonca, Bernard G. 2006. Former MLO staff information sheet. MLO archive.

Miller, John. 1976a. July 1976 activities report. In *Geophysical Monitoring for Climatic Change,* Mauna Loa Observatory, Monthly Reports, Hilo, Hawai'i, 1976.

———. 1976b. August 1976 activities report. In *Geophysical Monitoring for Climatic Change,* Mauna Loa Observatory, Monthly Reports, Hilo, Hawai'i, 1976.

Mims, F. M. III. 1992a. How to measure the ozone layer. *Science Probe* 2 (November): 45–51. (Accessed at http://www.sas.org/tcs/weeklyIssues_2007/2007-02-02/project1/index.html.)

———. 1992b. Sun photometer with light-emitting diodes as spectrally selective detectors. *Applied Optics* 31, 33: 6965–6967.

———. 1993. Satellite monitoring error. *Nature* 361: 505.

———. 1999. An international haze-monitoring network for students. *Bulletin of the American Meteorological Society* 80: 1421–1431 (cover article).

———. 2002. An inexpensive and stable LED sun photometer for measuring the water vapor column over south Texas from 1990 to 2001. *Geophysical Research Letters* 29: 20-1 to 20-4.

Mims, F. M. III, and John E. Frederick. 1994. Cumulus clouds and UV-B. *Nature* 371: 291.

Pereira, Judy. 1996. Mauna Loa Observatory: A 30-year personal overview. (Accessed at www.mlo.noaa.gov/pdf/memorabilia/JUDY%20BKLT.doc.)

Robinson, Elmer. 1986. January 1986 activities report, February 7. In *Geophysical Monitoring for Climatic Change,* Mauna Loa Observatory, Monthly Reports, Hilo, Hawai'i, 1986.

———. 1987a. December 1986 activities report, January 9. In *Geophysical Monitoring for Climatic Change,* Mauna Loa Observatory, Monthly Reports, Hilo, Hawai'i, 1987.

———. 1987b. April 1987 activities report, May 13. In *Geophysical Monitoring for Climatic Change,* Mauna Loa Observatory, Monthly Reports, Hilo, Hawai'i, 1987.

———. 1987c. July 1987 activities report, August 21. In *Geophysical Monitoring for Climatic Change,* Mauna Loa Observatory, Monthly Reports, Hilo, Hawai'i, 1987.

———. 1987d. August 1987 activities report, September 11. In *Geophysical Monitoring for Climatic Change,* Mauna Loa Observatory, Monthly Reports, Hilo, Hawai'i, 1987.

———. 1987e. September 1987 activities report, October 19. In *Geophysical Monitoring for Climatic Change,* Mauna Loa Observatory, Monthly Reports, Hilo, Hawai'i, 1987.

———. 1988a. February 1988 activities report, March 16. In *Geophysical Monitoring for Climatic Change,* Mauna Loa Observatory, Monthly Reports, Hilo, Hawai'i, 1988.

———. 1988b. March 1988 activities report, April 21. In *Geophysical Monitoring for Climatic Change,* Mauna Loa Observatory, Monthly Reports, Hilo, Hawai'i, 1988.

———. 1988c. June 1988 activities report, July 13. In *Geophysical Monitoring for Climatic Change,* Mauna Loa Observatory, Monthly Reports, Hilo, Hawai'i, 1988.

———. 1988d. August 1988 activities report, September 16. In *Geophysical Monitoring for Climatic Change,* Mauna Loa Observatory, Monthly Reports, Hilo, Hawai'i, 1988.

———. 1988e. October 1988 activities report, November 14. In *Geophysical Monitoring for Climatic Change,* Mauna Loa Observatory, Monthly Reports, Hilo, Hawai'i, 1988.

———. 1989a. December 1988 activities report, January 13. In *Geophysical Monitoring for Climatic Change,*

Mauna Loa Observatory, Monthly Reports, Hilo, Hawai'i, 1989.

———. 1989b. January 1989 activities report, February 10. In *Geophysical Monitoring for Climatic Change*, Mauna Loa Observatory, Monthly Reports, Hilo, Hawai'i, 1989.

———. 1989c. February 1989 activities report, March 14. In *Geophysical Monitoring for Climatic Change*, Mauna Loa Observatory, Monthly Reports, Hilo, Hawai'i, 1989.

———. 1989d. March 1989 activities report, April 14. In *Geophysical Monitoring for Climatic Change*, Mauna Loa Observatory, Monthly Reports, Hilo, Hawai'i, 1989.

———. 1989e. *April 1989* activities report, May 15. In *Geophysical Monitoring for Climatic Change*, Mauna Loa Observatory, Monthly Reports, Hilo, Hawai'i, 1989.

———. 1989f. October 1989 activities report, November 16. In *Geophysical Monitoring for Climatic Change*, Mauna Loa Observatory, Monthly Reports, Hilo, Hawai'i, 1989.

———. 1989g. November 1989 activities report, December 14. In *Geophysical Monitoring for Climatic Change*, Mauna Loa Observatory, Monthly Reports, Hilo, Hawai'i, 1989.

———. 1990a. January activities report, February 20. In *Geophysical Monitoring for Climatic Change*, Mauna Loa Observatory, Monthly Reports, Hilo, Hawai'i, 1990.

———. 1990b. March activities report, April 11. In *Geophysical Monitoring for Climatic Change*, Mauna Loa Observatory, Monthly Reports, Hilo, Hawai'i, 1990.

———. 1990c. November activities report, December 13. In *Geophysical Monitoring for Climatic Change*, Mauna Loa Observatory, Monthly Reports, Hilo, Hawai'i, 1990.

———. 1991a. March activities report, April 12. In *Geophysical Monitoring for Climatic Change*, Mauna Loa Observatory, Monthly Reports, Hilo, Hawai'i, 1991.

———. 1991b. April 1991 activities report, May 13. In *Geophysical Monitoring for Climatic Change*, Mauna Loa Observatory, Monthly Reports, Hilo, Hawai'i, 1991.

———. 1991c. May 1991 activities report, June 5. In *Geophysical Monitoring for Climatic Change*, Mauna Loa Observatory, Monthly Reports, Hilo, Hawai'i, 1991.

———. 1991d. July 1991 activities report, August 14. In *Geophysical Monitoring for Climatic Change*, Mauna Loa Observatory, Monthly Reports, Hilo, Hawai'i, 1991.

———. 1991e. November 1991 activities report, December 12. In *Geophysical Monitoring for Climatic Change*, Mauna Loa Observatory, Monthly Reports, Hilo, Hawai'i, 1991.

Rotich, J. K. 2004. Mt Kenya GAW station setup and initial results. Kenya Meteorological Department, Nairobi. (Accessed at http://www.esrl.noaa.gov/gmd/publications/annmeet2004/pdf_2004/rotich.pdf.)

Ryan, Steve. 1990. Diurnal CO_2 exchange and photosynthesis of the Samoa tropical forest. *Global Biogeochemistry Cycles* 4: 69–84.

———. 1991. July 1991 activities report, August 14. In *Geophysical Monitoring for Climatic Change*, Mauna Loa Observatory, Monthly Reports, Hilo, Hawai'i, 1991.

———. 1995. Quiescent outgassing of Mauna Loa volcano 1958–1994. In *Mauna Loa Revealed: Structure, Composition, History, and Hazards*. Geophysical Monograph Series vol. 92: 95–116. Washington, DC: American Geophysical Union.

———. 1997. The wind field around Mauna Loa derived from surface and balloon observations. *Journal of Geophysical Research* 102: 10,711–10,726.

———. 2001. Estimating volcanic CO_2 emission rates from atmospheric measurements on the slope of Mauna Loa. *Chemical Geology* 177: 201–211.

Ryan, Steve, E. J. Dlugokencky, P. P. Tans, and M. Trudeau. 2006. Mauna Loa volcano is not a methane source: Implications for Mars. *Geophysical Research Letters* 33: L12301, 10.1029/2006GL026223.

Schnell, R. C. 1976. Bacteria acting as natural ice nucleants at temperatures approaching 1°C. *Bulletin of the American Meteorological Society* 57: 1356–1357.

———. 1983. Arctic haze and the Arctic Gas and Aerosol Sampling Program (AGASP). Cooperative Institute for Research in Environmental Sciences, University of Colorado, Boulder. (Accessed at http://pdf.aiaa.org/preview/1983/PV1983_439.pdf.)

———. 1992a. January 1992 activities report, February 21. In *Geophysical Monitoring for Climatic Change,* Mauna Loa Observatory, Monthly Reports, Hilo, Hawai'i, 1992.

———. 1992b. February 1992 activities report, March 19. In *Geophysical Monitoring for Climatic Change,* Mauna Loa Observatory, Monthly Reports, Hilo, Hawai'i, 1992.

———. 1992c. April 1992 activities report, May 14. In *Geophysical Monitoring for Climatic Change,* Mauna Loa Observatory, Monthly Reports, Hilo, Hawai'i, 1992.

———. 1992d. May 1992 activities report, June 12. In *Geophysical Monitoring for Climatic Change,* Mauna Loa Observatory, Monthly Reports, Hilo, Hawai'i, 1992.

———. 1992e. June 1992 activities report, July 12. In *Geophysical Monitoring for Climatic Change,* Mauna Loa Observatory, Monthly Reports, Hilo, Hawai'i, 1992.

———. 1992f. July 1992 activities report, August 12. In *Geophysical Monitoring for Climatic Change,* Mauna Loa Observatory, Monthly Reports, Hilo, Hawai'i, 1992.

———. 1992g. November 1992 activities report, December 15. In *Geophysical Monitoring for Climatic Change,* Mauna Loa Observatory, Monthly Reports, Hilo, Hawai'i, 1992.

———. 1993a. January 1993 activities report, February 22. In *Geophysical Monitoring for Climatic Change,* Mauna Loa Observatory, Monthly Reports, Hilo, Hawai'i, 1993.

———. 1993b. March 1993 activities report, April 9. In *Geophysical Monitoring for Climatic Change,* Mauna Loa Observatory, Monthly Reports, Hilo, Hawai'i, 1993.

———. 1993c. April 1993 activities report, May 12. In *Geophysical Monitoring for Climatic Change,* Mauna Loa Observatory, Monthly Reports, Hilo, Hawai'i, 1993.

———. 1993d. May 1993 activities report, June 11. In *Geophysical Monitoring for Climatic Change,* Mauna Loa Observatory, Monthly Reports, Hilo, Hawai'i, 1993.

———. 1993e. June 1993 activities report, July 20. In *Geophysical Monitoring for Climatic Change,* Mauna Loa Observatory, Monthly Reports, Hilo, Hawai'i, 1993.

———. 1993f. August 1993 activities report, September 16. In *Geophysical Monitoring for Climatic Change,* Mauna Loa Observatory, Monthly Reports, Hilo, Hawai'i, 1993.

———. 1993g. October 1993 activities report, November 30. In *Geophysical Monitoring for Climatic Change,* Mauna Loa Observatory, Monthly Reports, Hilo, Hawai'i, 1993.

———. 1993h. November 1993 activities report, December 30. In *Geophysical Monitoring for Climatic Change,* Mauna Loa Observatory, Monthly Reports, Hilo, Hawai'i, 1993.

———. 1994a. December 1993 activities report, January 30. In *Geophysical Monitoring for Climatic Change,* Mauna Loa Observatory, Monthly Reports, Hilo, Hawai'i, 1993.

———. 1994b. November 1994 activities report, December 30. In *Geophysical Monitoring for Climatic Change,* Mauna Loa Observatory, Monthly Reports, Hilo, Hawai'i, 1994.

———. 1995a. March 1995 activities report, April 18. In *Geophysical Monitoring for Climatic Change,* Mauna Loa Observatory, Monthly Reports, Hilo, Hawai'i, 1995.

———. 1995b. August 1995 activities report, September 19. In *Geophysical Monitoring for Climatic Change,*

Mauna Loa Observatory, Monthly Reports, Hilo, Hawai'i, 1995.

Schnell, R. C., and G. Vali. 1976. Biogenic ice nuclei: Part I. Terrestrial and marine sources. *Journal of Atmospheric Science* 33: 1554–1564.

Vali, G., M. Christensen, R. W. Fresh, E. L. Galyan, L. R. Maki, and R. C. Schnell. 1976. Biogenic ice nuclei: Part II. Bacterial sources. *Journal of Atmospheric Science* 33: 1565–1570.

WMO. 1991. Report of the meeting of experts to assess available data and define the aerosol component for GAW-BAPMON stations. Boulder, CO, December 16–19. WMO Global Atmosphere Watch Report No. 79. Geneva: WMO Secretariat.

CHAPTER THIRTEEN

The Fifth Decade (1996–2005)

In 1963, MLO was nearly closed by major federal budget cuts, but in December 1995 there was not even a budget to cut. The fifth decade at the Mauna Loa Observatory began with no budget. A politically charged budget battle between the president and the congressional majority shut down the federal government in mid-December. The MLO staff was officially placed on furlough, along with most of the rest of the government. Fortunately for the sake of science, MLO director Russell Schnell (personal communication with author, September 11, 2007) managed to keep the observatory's key functions running during the furlough:

> During the furlough we operated with a skeleton staff with one person going to the mountain each day and one doing the balloons and the Kumukahi flask samples. We maintained our computer operations and core data gathering and archiving. Some of the non-core NOAA measurements may have been neglected a bit, but immediately after the furlough we put extra staff on the mountain to bring everything back to full operation.

The furlough lasted from December 16, 1995, to January 6, 1996. The two personnel assigned to keep the observatory functioning were Bob Uchida and Darryl Kuniyuki. Uchida was selected because of his knowledge of the electrical power distribution at MLO. Kuniyuki (personal communication with author, June 9, 2008) was chosen because of his role in setting up and maintaining the computer network on the mountain.

Schnell described a few casualties of the furlough in his December and January activities reports:

> The mid-month furlough reduced MLO activities in some areas, especially ozonesonde launches, Cape Kumukahi flask sampling, and CO_2 calibrations. During the furlough, shipments of flasks and related materials kept arriving and built up to a point that the Post Office threatened to return 20+ boxes which had accumulated in the receiving area. (Schnell 1996a)

> The January furlough of Federal employees reduced the normal care and maintenance on the venerable SIO [Scripps Institute of Oceanography] CO_2 equipment, which suffered a malfunction during the early days of the year. The system was revived by John Chin in mid-month and it has since operated well. (Schnell 1996b)

Back in Washington, DC, a compromise of sorts was reached on January 5, 1996, and most of the federal government returned to work. Ironically, Washington was promptly shut down again by a blizzard of historic proportions that became known as "The Great Furlough Storm." Meanwhile, work at MLO resumed a normal schedule.

On February 13, 1996, former MLO director Rudolf Pueschel of NASA visited MLO. Pueschel had taken over from Howard Ellis thirty years earlier, and he had been Schnell's postdoc mentor at NOAA in 1975 when Schnell first learned about MLO. Schnell reported in the February Activities Report that Pueschel's influence led to his goal of someday becoming director of the observatory. He then added, "Mission accomplished."

David Hofmann Becomes Director of CMDL (1996)

Dr. David Hofmann had served as chief scientist of the Climate Monitoring and Diagnostics Laboratory (CMDL) since his arrival from the University of Wyoming in April 1990. When Eldon Ferguson retired as director of CMDL in 1995, Hofmann became acting director. He became permanent director in April 1996, a position he was to hold until December 31, 2007.

The Aging Carbon Dioxide Analyzer (1996)

Visitors to the main building at MLO in the 1990s were watched closely if they wandered too close to Charles Keeling's venerable carbon dioxide analyzer. The instrument, which retained the traditional sensitivity to temperature and vibration it had exhibited since 1958, had been installed in a cabinet with a transparent acrylic door. Inside the cabinet was an incandescent lamp that was automatically switched on and off by a thermostat to regulate the temperature of the air inside the cabinet. Opening the cabinet's door would likely cause the recorder trace to quickly drift.

The staff had little good to say about what by then had become the world's most famous environmental monitoring instrument, and who could blame them? The data it produced almost perfectly matched that from NOAA's more modern analyzer, which was much easier to maintain and service and was not sensitive to temperature. Yet it was the close agreement between the old APC Model 70 and the newer instrument that justified keeping the Scripps carbon dioxide program alive at MLO. The close agreement of the measurements made by the two instruments since 1974 enhanced the value of the Model 70 data that went all the way back to 1958.

Because the old instrument produced high-quality data in spite of its many problems, it's of interest to review some of the difficulties experienced with it that reached a level worthy of recording in the monthly Activities Reports during a typical year.

> [April 1996] The thermal control unit of the Applied Physics CO_2 analyzer was unable to keep up with the room temperature changes on 3/22, causing drifting data traces on the chart recorder. Replacing tubes 2050 and 12AX7, the amplifier tubes, improved the situation. SIO was informed and asked to purchase more vacuum tubes. (Schnell 1996d)

> [June 1996] The thermal control unit in the analyzer was down from 6/6 through 6/13. The problem was corrected after replacing the amplifier tubes and the oscillator tubes. The main analyzer amplifier malfunctioned on 6/26, but after replacing five 6AK5 amplifier tubes and rephasing the amplifier unit, the system was back in daily operation. (Schnell 1996f)

> [August 1996] The thermal controller in the IR analyzer developed frequent problems during the month. The 2050 regulator tubes were replaced on 8/2, and the 12AX7 amplifier tubes needed to be replaced at the end of the month. The 6AK5 amplifier tubes for the analyzer were replaced on 8/26. (Schnell 1996g)

> [October 1996] The thermal controller in the Applied Physics CO_2 analyzer malfunctioned on 10/3, 10/15 and 10/21. After replacing the amplifier tubes and/or regulator tubes in the thermal controller in separate incidents,

the analyzer was back to normal operation. (Schnell 1996h)

[December 1996] The thermal controller in the Applied Physics IR analyzer malfunctioned on 12/2 and 12/30. After replacing the electronic tubes, the system was back to normal operation. The data acquisition unit was down on 12/9. The unit was back on line after reset and restart were performed. On 12/11 to 12/13 the downslope intake line, L3, was checked by disconnecting it from the base of the sampling tower to sample air at 2 ft level. The intake line seemed to be okay. (Schnell 1997a)

VOGNET Sniffs Volcanic Air Quality across Hawai'i (1996)

Since Kīlauea began erupting in 1983, visitors to the Big Island of Hawai'i are often surprised by the significant volcanic fog, or "vog," over parts of the island. The haze, which mainlanders often equate with smog, is especially bad on the *kona* (leeward) side of Hawai'i, where it tends to accumulate as a consequence of the trade winds blowing the stuff around the southern flank of Mauna Loa from its source on the opposite side of the island.

The vog originates from sulfur dioxide gas emitted from the lava emerging from the active Kīlauea vents on the southeastern side of the island. The gas quickly evolves into microscopic sulfate particles upon which water vapor condenses to form tiny droplets of acidic vog. People with asthma are especially sensitive to vog.

Because of the health and visibility problems associated with vog, there was a need to better understand its distribution around the Big Island. Physical scientist Steve Ryan had long been responsible for MLO's continuous measurements of sulfur dioxide (SO_2), a program designed to monitor volcanic activity and to also detect occasional bursts of SO_2 arriving from Asia. Ryan was well aware of the lack of knowledge about the distribution of vog, but the MLO staff was far too busy to take up the task. So Ryan developed a student-scientist partnership with a group of schools around the Big Island under the leadership of Deighton Emmons, a science instructor at the Hawai'i Preparatory Academy in Waimea. MLO director Russell Schnell introduced the program in the MLO September 1996 Activities Report:

> A new cooperative effort between five Big Island public and private high schools and MLO was begun by Steve Ryan this month. The program, called VOGNET, has students making regular observations of volcanic aerosols from a network of five locations evenly spaced around the island in order to study the distribution and relative levels of "vog" in Hawaii. The first organizational meeting was held on Saturday, 9/28. Science teachers from the five schools were given a seminar on how condensation nuclei and aerosol optical depth are measured, and each school was given a Gardner CN counter and a two-wavelength sunphotometer. These instruments had been in storage in Boulder, and were sent to Hilo to be cleaned up and refurbished prior to the meeting.
>
> Each school will take hourly observations and will periodically submit these to the lead school, Hawaii Preparatory Academy, which will maintain a database and distribute copies back to all the schools. Aerosol optical depth and CN measurements from MLO, along with radiosonde data from the Hilo weather service will be included in the database. Each school will be free to use the data in science classes and for research projects of their choosing. Every few months, a one-day conference will be held in which all the participants get together to discuss results and participate in an instrument intercalibration session. Students will present talks or posters on their research projects, and outside experts in aerosol chemistry, meteorology, and volcanology will be invited to give presentations.
>
> Besides being an educational tool, VOGNET will hopefully become a scientific resource. There are no existing monitoring networks in place, despite years of concern over the public health effects of vog. Over time, network measurements could answer questions about how vog is distributed as a function of the prevailing wind, how it evolves throughout the day, which sites have the greatest/least average and extreme levels, how

> ground concentrations are related to column amounts, etc. The students will be part of a real environmental monitoring project which could benefit the people of Hawaii. (Schnell 1996g)

VOGNET became a notable example of a successful student-scientist partnership, and environmental scientists and public health officials in Hawai'i learned far more about vog than previously known. Moreover, the hands-on aspect of the research taught the participating students far more about real environmental monitoring than could be taught in a classroom. Besides their regular measurement of vog, some of the students helped Emmons calibrate sun photometers at MLO. These instruments were then used to measure the optical thickness of the haze formed by vog. Some students pursued serious science fair projects based on their vog research. For example, in January 1999, MLO's Dr. John Barnes judged the Hawai'i District Science Fair, where he presented a senior research prize ($50) to Robin Ness for "How the wind direction influences vog" and a junior research prize ($50) to Steven Inouye for "Vog and you."

In December 1999, Steve Ryan traveled to Washington, DC, to receive a Department of Commerce bronze medal in recognition of his role in establishing VOGNET. While VOGNET is no longer in operation, the new information it provided greatly enhanced the scientific knowledge of the evolution and transport of vog around the Big Island (Ryan 2000).

VOGNET was only one of Steve Ryan's many projects, and former MLO director Russell Schnell (personal communication with author, September 11, 2007) describes Ryan as "an excellent scientist [with] a great mind":

> He would very easily be a PhD if he had felt the need or the interest, but he went a different route. . . . He does very innovative science, and he has recently had a paper published that suggests that methane on Mars is a non-issue from volcanoes because you cannot see any methane coming out of Mauna Loa. And then he took data [about] the CO_2 that is coming out of Mauna Loa volcano. . . . So, one man's garbage is another man's gold, and he has done a lot of good work by taking stuff that is peripheral to our mission and turning it around and making a very good science out of it.

Preparing for the Next Mauna Loa Eruption (1996)

Those who have spent nights at MLO sometimes discuss what to do should the volcano erupt during their stay. Because Mauna Loa signals its eruptions in advance with swarms of earthquakes and puffs of sulfur dioxide and carbon dioxide, it's unlikely that anyone would be trapped on the mountain. But it could happen.

A visitor who was calibrating instruments at MLO once asked MLO director Russell Schnell what he should do in the event of an eruption. Schnell replied that the visitor should retrieve the Dobson spectrophotometer, place the huge instrument under an arm and head for safety straight down the slope of the mountain! Schnell was jesting about how to save a valuable instrument that weighs about 150 pounds (68 kg) and is 5 feet (1.5 meters) long. But evacuation plans were definitely on his mind and remained so when he later became the overall director of all of NOAA's climate monitoring observatories. In 1995, Schnell devised a plan to retrieve all of MLO's key instruments should an eruption be imminent:

> Mauna Loa Mountain is still inflating and carbon dioxide (CO_2) gas releases from the summit caldera persist which means that an eruption may occur within a few years. Therefore, an escape plan and an equipment removal list have been drawn up. In essence, most equipment valued over $10,000 per item will be removed when it is predicted that lava will inundate MLO within 12 hours. (Schnell et al. 1995)

A helicopter landing area on the newly acquired four acres east of the original four acres provided for very limited emergency evacuation of personnel—but not necessarily MLO equipment. One morning a film crew and a geologist arrived to make preparations to film an unusual sulfur deposit discovered on the north slope of Mauna Loa.

Because of the thin air, the helicopter seemed to labor as it rose up toward the altitude of MLO and landed on the lava pad. After the helicopter took off, the pilot steered the craft sharply down the slope of the mountain toward denser air.

Russ Schnell experienced firsthand the difficulty of landing a helicopter at the observatory during a practice rescue attempt during September 1996:

> The County of Hawaii conducted a test of an emergency rescue of MLO staff from MLO due to a volcanic eruption. The helicopter was airborne in 9 minutes and on site after 18 minutes of flight time. An attempted landing at MLO was aborted due to overweight conditions in the helicopter as the MLO Director was onboard along with the pilot, a rescue crewman, and a full load of fuel. In future tests, and for the real thing, the fuel load will be lowered and two less bodies carried up the mountain. We did get some nice photos of the observatory. (Schnell 1996h)

Basic helicopter emergency training for the MLO staff was provided in December:

> Early in the month, four MLO staff were trained in the fine art of entering, being a good passenger, and exiting the County of Hawaii rescue helicopter in preparation for the next Mauna Loa eruption. (Schnell 1997a)

Later the remainder of the staff received helicopter training (Kuniyuki, personal communication with author, June 9, 2008). On January 16–17, 1997, Schnell reviewed evacuation plans and responsibilities with fire and civil defense officials:

> From this exercise it became clear that MLO has to be better prepared for attending to site and road accidents, attending to heart attack victims, handling fires, etc. MLO is now preparing plans for everything from hurricanes to bomb threats. In addition, all the mountain staff will have to take first aid courses and yearly refreshers. (Schnell 1997b)

Where is MLO? (1996)

In 1996 the approximate location of the Mauna Loa Observatory could be found on some maps, and the observatory could be seen in satellite images. Today MLO can be seen online in high-quality aerial photographs at Google Earth (www.googleearth.com). But what are the exact geographic coordinates of the site? It is well known that Mauna Loa inflates with lava prior to an eruption, but the dimensional changes of the mountain are minuscule (a few cm per year in 2007) when compared with GPS measurements made in 1996, which showed substantial differences from earlier surveys. The GPS measurements were a surprise, for a precise knowledge of the elevation and geographic coordinates of MLO is important to some of the observation programs at MLO—particularly those that require an exact knowledge of the position of the sun in the sky. Russell Schnell et al. (1997) explain the details:

> Recently a number of GPS measurements (two different systems, two different times) of elevation at MLO have produced data which is 70m (210 feet) off from the conventional and generally accepted rod and level survey height of MLO. To address this discrepancy, MLO arranged funding from NASA to obtain an elevation survey using a new benchmark established for the recent resurvey of the Saddle Road. This new survey determined that the Stair benchmark is about 12 feet too high, and that the GPS elevations are 200 feet out.
>
> With respect to the GPS measurements, apparently the normal mainland algorithms do not account for the geoid-ellipsoid differences introduced by the mass of Mauna Loa and Mauna Kea which distorts the globe in this part of the world. Because of this local distortion of GPS, the NOAA National Geodetic Survey will be coming to MLO this summer to obtain enough data to produce a Hawaii Island algorithm for GPS measurements. In so doing, they will produce MLO elevation information accurate to within a few centimeters. In the meantime, the elevations of various locations at MLO based upon the most recent rod and level survey are: Stair marker = 11,128 ft.; floor of the MLO main

building = 11,141 ft.; . . . These elevations are considered to be accurate within +/– 1 ft.

Google Earth gave an elevation of 11,129 feet for the Stair marker when checked in 2008, which is within the 1-foot range reported by Dr. Schnell.

Canada Comes to MLO (1997)

Environment Canada designed the Automated Brewer Ozone Spectrophotometer, commonly known as the Brewer, as a modern and fully automated replacement for the older Dobson spectrophotometer. Both the Dobson and the Brewer measure the total column abundance of ozone in the atmosphere. The Brewer also measures the sun's ultraviolet wavelengths with high accuracy. Thus, the Brewer can measure almost simultaneously both the ozone layer and changes in solar ultraviolet owing to changes in the column abundance of ozone, haze, and clouds. The Brewer's spectrometer system scans the wavelength range of 290 to 325 nm with a resolution of half a nanometer, which allows the instrument to distinguish the narrow absorption bands of sulfur dioxide and measure its abundance in a column through the atmosphere.

The Canadian Atmospheric Environment Service (AES) installed Brewer No. 009 at MLO on March 24, 1997. This instrument is equipped with an all-sky camera that records an image of the sky every ten minutes. Brewer No. 119, which is equipped with a double monochrometer that greatly reduces internal light scattering and improves accuracy, was added in November 1997. This instrument has an extended UV response of 286.5 to 363 nm, the lower wavelength being at the very edge of the UV that penetrates the atmosphere. Both Brewers were installed on the solar deck of the NDACC Building at MLO (described below) and coupled by cables to individual computers in a room inside the building from where the instruments can be controlled. Both Brewers can also be controlled over the Internet by AES scientists.

The two Brewers provide an important comparison of total ozone measured by one another, the station Dobson (No. 76), and other permanent and temporary instruments at MLO that also measure the ozone layer. By 1998, some 150 Brewers were in operation around the world, primarily in Northern Hemisphere locations. The two Mauna Loa Brewers remain in service today.

The Lightning Blowout of 1997

Lightning strikes are uncommon at MLO, but they cause considerable damage when they occur. Some of the power poles along the MLO road are surrounded by long splinters blown from them by lightning bolts. Several lightning rods protruded from the roof of the NDACC building, each connected to heavy cables designed to divert electrical charges to ground. But the electrical grounding properties of a mountain composed of lava are inadequate for the proper functioning of lightning rods, and the station remained vulnerable to lightning strikes.

MLO engineer Bob Uchida describes an electrical storm that occurred while he was at work in the Keeling Building. When the storm arrived at MLO, the interior of the building and its many metal features were illuminated with a bluish electrical discharge. This and similar anecdotes explain why the MLO staff sometimes fled from the station during a lightning storm.

On the morning of June 27, 2001, the author, who was staying at MLO to conduct sun photometer calibrations, ceased work when clouds from an approaching thunderstorm blocked the sun. Soon the station was shrouded in thick fog, and booms of thunder were growing louder. Suddenly there was a flash of lightning accompanied by a sharp, very loud clap of thunder. A pair of red taillights could be dimly seen through the fog as the staff fled downslope and away from the impending storm that brought a heavy downpour (Plate 49).

Four days later a more powerful storm arrived. MLO was pounded by rain and small hail, which roared louder than the accompanying thunder as it struck the 1956 building's metal roof and transformed the lava outside from black to white. Several lightning bolts struck at or very near the station. While storms such as this replenish

MLO's water supply, the vast increase in the number of computers and electronic data acquisition systems at MLO has made the station much more vulnerable to lightning strikes than in the past. Fortunately only minor damage occurred during this storm. Russell Schnell et al. (1997) described the consequences of a particularly devastating lightning strike several years before:

> The August 1997 lighting strike caused the worst damage and expense from any single natural event ever recorded at MLO, probably even more than the volcanic eruption of 1984. Thirteen of the MLO-supplied computers on the mountain were affected. Two network hubs and the router box were burnt out, as were the cordless telephone, fax machine, and two phone switch boxes. Most repairs were completed quickly, but some systems took months to repair or replace. The above losses do not include damage to equipment belonging to CMDL Boulder, NDSC, and MLO cooperative programs.

The NDSC Building Is Dedicated (1997)

CMDL director David Hofmann's work included strategic planning for the worldwide sampling and measurement of atmospheric trace gases having the potential to cause climate change and ozone depletion. Key to this planning was the construction of the Network for the Detection of Stratospheric Change (NDSC) building at MLO. (In 2005 the name was changed to the Network for the Detection of Atmospheric Composition Change, or NDACC.) Planning and construction of the building dominated the MLO agenda throughout 1996–1997, and Hofmann (personal communication with author, August 28, 2007) summarizes the project:

> Mauna Loa was named as a primary station of the NDSC (then Network for the Detection of Stratospheric Change) in the early 1990's. NDSC was an International program run with NASA leadership to monitor, from the earth's surface, ozone and anything that might affect the ozone layer, but also anything else important in the stratosphere. I lobbied to have the Mauna Loa lidar included as a primary instrument, which was accepted, as were the Dobson ozone instrument and ozonesondes. That's why we [became] heavily involved in the program.
>
> The need for a building to house many of the instruments (both NOAA and non-NOAA) was clear, and after about 3 years, $1.6 M was scraped together. That would have been enough to put up just a foundation in the lava rock just to the east of the present site. The dilemma was solved by deciding to go with a modular building on the present site.
>
> Eight 12' x 36' mobile units were constructed in Oregon and put together to form a 96' x 36' building. I went out to Oregon and inspected it when it was put together. All the plumbing, electrical, etc. was in. The individual units were then separated and trucked (they had regular mobile home dual-axle wheels) to the coast where they were shipped to Honolulu and barged to Hilo. They were then trucked up the mountain (the road had to be leveled in tight turns) and set into a pre-poured concrete frame by a large truck crane. Within a few weeks the building was put back together and operating. It came in at just $1.6 million.

Before the foundation for the NDSC Building could be installed, it was necessary for work crews to excavate trenches for the utility lines. MLO director Schnell describes the ability of the construction crew to perform their task at the high-altitude site:

> The ground preparation crew are excellent and careful workers who have not broken one underground line or surface cable. They put in long days, rain, snow, or shine. . . . The rock removed from the building foundation is being used to fill a depression north of the lidar trailers to extend the parking area. . . . The same crew will level and expand the helicopter landing area when they get a break from their present activities. (Schnell 1997c)

While the ground crew did its job well, they could not avoid creating clouds of engine exhaust and dust and vibration from jackhammering, which interfered with the adjustment of the lidar receiver mirror and both ozone and carbon dioxide measurements.

Construction of the NDSC Building transformed the appearance of MLO (Plate 40). The building was completed in November 1997, just in time for an international gathering of scientists in Hilo for a two-day Climate Monitoring and Diagnostic Laboratory conference. The meeting included the dedication of the NDSC Building and observance of the fortieth anniversary of MLO (Plates 41 and 42). A month-long exhibit on the history of MLO and its scientific instruments was featured at Hilo's Lyman Museum.

A highlight of the fortieth anniversary celebration was the naming of the original MLO building as the Keeling Building in honor of Charles David Keeling, whose original carbon dioxide analyzer had been installed in the building in March 1958. The aging vacuum-tube instrument (Plates 43–45) was still in operation during the ceremonies. Keeling, who first visited MLO in November 22, 1958, was all smiles as he posed for photos with John Chin on either side of a plaque affixed to the outside of the original MLO building that had just been named for him (Plate 46). The plaque features the famous Keeling curve depicting both the annual cycle and the overall increase in carbon dioxide since March 1958. The text reads: "Keeling Building.

Named in honor of Professor Charles David Keeling, Scripps Institution of Oceanography, who initiated continuous CO_2 measurements at this site in 1958. November 1997."

John Chin had operated Keeling's CO_2 equipment at MLO for more than thirty years, and as he and Keeling stood outside the building during the dedication, hard at work inside were noisy pumps that sucked air into the old vacuum tube gas analyzer while a pen traced a green chart of the CO_2 concentration onto a slowly moving roll of paper. The old battles between Keeling and NOAA were briefly forgotten, but one year after the dedication, Keeling (1998) published a memoir in which he related the old conflicts.

After the NDSC Building was completed, the MLO staff invested considerable time moving instruments into the building and onto the rooftop observation platform and connecting them to electrical power, the computer network, and air sampling lines. The NDSC Building then became the sensor platform and nerve center for most of the key climate change measurements made at MLO.

While the new NDSC Building greatly improved the facilities at MLO, nature was unimpressed. Occasional snowstorms occurred as before (Plate 47). And it wasn't long before the new building was pummeled by hail and shaken by thunder from nearby lightning bolts (Plates 48–50). The hailstorms often filled the new water tank behind the building with water collected from the roof (Plate 51).

Colorado State University Comes to MLO (1997)

As noted in chapter 12, the Mauna Loa Observatory proved essential as a calibration site for the worldwide AERONET network of robotic sun photometers. The US Department of Agriculture's UV-B Monitoring and Research Program (UVMRP) operates another network of sunlight monitoring instruments that also benefits from MLO's alpine location. The MLO instruments in this network were installed on the northeast side of the solar deck over the NDSC Building. They included a Yankee Ultraviolet Radiometer that measures erythemal flux, the band of UV wavelengths that causes sunburn. Also included were two Yankee Multifilter Rotating Shadowband Radiometers, both of which measure global (full-sky) solar irradiance and the sunlight arriving directly from the sun. One version measures solar UV and the other measures visible and near-infrared light. A diffuser disk on the instruments scatters sunlight to seven filtered light detectors inside. A rotating metal shadow band shades the diffuser every fifteen seconds, during which time the instrument receives only sunlight scattered from the sky and clouds. Subtracting the instrument's signal when the diffuser is shaded from the signal when it is unshaded provides the direct sun irradiance.

These instruments remain in operation at MLO today, and details about the UV-B Monitoring and Research

Program and data it has collected at MLO and from across the United States are available at http://uvb.nrel.colostate.edu. Dr. James Slusser, the program's former director, summarizes the network and MLO's role (Slusser et al. 2007):

> In 1990 the USDA had become concerned about the decreasing stratospheric ozone over North America and the potential for increasing levels of crop- and biosphere-damaging ultraviolet radiation. They initiated the UVMRP in 1992 through a grant to Colorado State University in Fort Collins, Colorado, to provide information on the geographical distribution and temporal trends of ultraviolet radiation in the United States, which is critical to the assessment of the potential impacts of increasing ultraviolet radiation levels on agricultural crops and forests (Bigelow et al. 1998). The program consists of both a research component and a climatologic network [with thirty-six sites across the United States]. Specifically the monitoring program:
>
> - Provides information to the agricultural community and others about the climatological and geographical distribution of UV-B irradiance;
> - Furnishes the basic information necessary to support evaluations of the potential damaging effects of UV-B to agricultural crops and forests;
> - Supplies ground truth for satellite measurements of UV-B, and basic information for radiation transfer model calculations;
> - Establishes long-term records of UV-B irradiance necessary to assess trends.
>
> The UVMRP instrument suite was installed at MLO on 11 November 1997. . . . The UVMRP uses the MLO facility for three tasks: (1) to provide data as part of the climatologic network, (2) for development of the Langley calibration technique, and (3) for intercomparison of the shadowband instruments with other UV and visible radiation instrumentation, such as Dobson, Brewer and NIWA spectroradiometers, CIMELs and other atmospheric monitoring instruments operated by the variety of agencies using MLO. It is essential to be able to distinguish between atmospheric effects and instrument effects, and the near-constant atmospheric conditions at MLO allow the UVMRP to validate in-situ (Langley) calibrations and detect systematic biases due to instrument aging.

MLO Fully Enters the Computer Era

New computers proliferated at MLO during the Russ Schnell years. No longer was most data inscribed on paper by old-fashioned chart recorders or by early computers that saved data on floppy disks. Principal investigators on the mainland could access their instruments and data files via the Internet, and visits to MLO's site on the World Wide Web skyrocketed. Dr. Schnell included an update of MLO's computer status in the 1996–1997 report about the science at the CMDL stations:

> A major shift in MLO operations and activities over the past 2 years has been brought about by the absorption of Internet technology and direct computer control and communications into all but a few relic programs. Only a few years ago, MLO data were mainly recorded on disks and paper tapes, with a few programs accessible to off-site principal investigators (PIs) by using a dial-up telephone modem. Today MLO has about 50 [57 at the end of 1997] computers with most data going by way of the Internet direct to PIs located around the world. . . . Two years ago MLO installed a camera overlooking the mountain site and made the images available on the Internet. In early 1996 only a few dozen people per week logged onto the site, whereas by late 1997 this grew to a peak of 25,000 per month. At present the primary efforts of two MLO staff are required to keep the computer/Internet systems operating and updated. (Schnell et al. 1997)

Each month, Judy Pereira typed the Activities Report into a desktop personal computer instead of a typewriter. The reports were then transmitted to Boulder over the Internet.

The hard copy reports that were once bound in handsome volumes and placed on a shelf in a book room at CMDL in Boulder became a thing of the past.

The Cape Kumukahi Lighthouse

A unique feature of the high-altitude Mauna Loa Observatory is its relatively short distance from the sea, where measurements can be taken at or near sea level for comparison. When a team from the US Exploring Expedition spent three weeks making measurements from the summit of Mauna Loa during 1840–1841, teams aboard their ship in Hilo Bay and at a temporary observatory near the water on Waiākea Point also made measurements. For half a century, the US Weather Service had made meteorological measurements near sea level in Hilo while similar measurements were made on the mountain at MLO. Similarly, for decades MLO had collected air samples for NOAA and the Scripps Institution of Oceanography at both MLO and the Cape Kumukahi Lighthouse near sea level at the easternmost tip of the Island of Hawai'i.

Over the years, the Cape Kumukahi Lighthouse had been challenged by volcanic activity and human vandalism. The site's significance in atmospheric monitoring makes these challenges a story worth telling.

In 1928 the US government paid $500 for fifty-eight acres on Cape Kumukahi. The land was acquired to build an open-frame steel tower to serve as a lighthouse. Because of earthquakes, the tower was installed atop a double foundation. The lowest layer of the foundation was a heavy concrete slab installed in the bottom of a pit excavated in the lava. A thick layer of sand was placed over the first foundation, and a second layer of concrete was then poured. The 125-foot (38-meter) tower was anchored to the upper concrete pad, and the layer of sand helped isolate the structure from vibrations caused by earthquakes (Dean 1986).

In January 1960, a massive lava flow from the Kapoho vent threatened the lighthouse. During Kīlauea's third-largest eruption of the twentieth century, lava quickly covered virtually all of Cape Kumukahi and pushed most of its shoreline as much as 2,500 feet (762 meters) into the Pacific. While the lava destroyed all the US Coast Guard living quarters and support buildings, the molten rock flowed around the lighthouse tower and left it unscathed (Kakugawa 1999).

During the 1970s, MLO placed an instrumented trailer at the site to collect various kinds of atmospheric data. The trailer was eventually vandalized so thoroughly that only its frame remained, a predictable fate for a site that was littered with trash and abandoned cars. After a flask sampling program was established at the cape, vandalism was a regular occurrence, and a typical incident is related by Russell Schnell (1996b):

> Cape Kumukahi flask sampling pump box system was vandalized again in mid-month with entry gained by use of a hefty bolt cutter. Bob Uchida and John Chin have welded a new lock system onto the box which will require a cutting torch to bypass. That should last about 2 months based on prior experience!

Locks and fences did not solve the problem, so Dr. Schnell tried a new approach several months later:

> In an attempt to get the Puna rowdies and hippies to stop vandalizing the Cape Kumukahi facilities, Russ Schnell will be giving a talk on our MLO and Kumukahi activities at a "Gathering of the Clans and Bikers" at Kahena Beach, Pahoa, on April 14. Russ has gained fairly high visibility and recognition in the area for having convinced the Coast Guard to block that portion of the lighthouse beam that formerly shone onto land and into the rooms of a few hundred area residents. Maybe that explains why the $10,000 beacon has not been shot out in the last six months! (Schnell 1996c)

Schnell (personal communication with author, September 11, 2007) made the presentation without incident, and he recalls that "vandalism disappeared for a decade after we did that community involvement."

In 2004, the primitive air sampling facilities at the lighthouse were greatly improved by installation of a sturdy new air-conditioned building that measured about 8 x 20 feet (2.5 x 6 meters). A large sign on the door of the facility explained its purpose:

> **Alliance for Long-term Observations of the Hawaiian Atmosphere "ALOHA"**
>
> A joint project between the University of Hawai'i, Scripps Institution of Oceanography, National Oceanic and Atmospheric Administration and the National Aeronautics and Space Administration to Monitor Local and Global Air Quality.
>
> This site at Cape Kumukahi has some of the cleanest air in the northern hemisphere under typical trade wind conditions. This air from over the Central Pacific contains trace amounts of gases and particles that originated from pollution on the continents. By monitoring the variability in these amounts, we can estimate the influence of human activity upon global pollution and the air quality of the Hawaiian Islands. Similar measurements in the upper atmosphere at the top of Mauna Loa provide one of the world's most important data sets on CO_2 and the link to human activity, fossil fuel combustion and global warming. This site at Kumukahi will provide similar data for the atmosphere near the ocean surface.

A chain-link fence topped with barbed wire protected the building, but vandals managed to get inside the fenced area and decorate the building with graffiti. Meanwhile, MLO staff continued to make weekly visits to the cape to collect flask samples at ground level and via sampling tubes located on the lighthouse tower.

Vandalism returned to the lighthouse station in April 2005 when an expensive Cimel robotic sun photometer placed atop the building by NASA's Goddard Space Flight Center as part of the AERONET global network was totally destroyed by vandals wielding concrete blocks. AERONET director Brent Holben (personal communication with author, September 6, 2007) describes what happened:

> Our vandalized site ran from March 23 to April 29 when it was pummeled with cinder blocks I suspect by the local pot growers who probably thought the Cimel was a surveillance camera. Vandalism occurred at 20:30 UTC April 29th when our transmitter stopped and data collection ceased.

The Flask Sampling Ritual

The earliest form of flask sampling, which was instituted by Charles Keeling, is still in use at MLO and Cape Kumukahi. The procedure involves an almost ritualistic protocol using an evacuated, five-liter glass sphere wrapped with protective tape and having an access tube fitted with a stopcock. The sample is collected by holding the flask overhead while slowly walking into the wind. The stopcock is opened with one hand while a small rod held in the other hand is moved back and forth inside the aperture of the stopcock, the idea being to stir away any residual air that might have accumulated in the opening. Should tourists unfamiliar with this protocol happen to see MLO's Steve Ryan collecting an air sample, they would have good reason to stare, smile, and wonder what is taking place (Plate 52). Is this an ancient Hawaiian ritual? What is the man holding over his head? Why is he walking toward the ocean? What is that loud hissing sound?

Cynthia McFee (video interview with author, Boulder, Colorado, April 27, 2007), who took plenty of flask samples during her time in Antarctica and Hawai'i, reflects on the protocol that she had learned from Dave Keeling at Scripps:

> I actually have a photo of myself . . . my arms spread out [holding a sample flask]. . . . Yes, it does look like a ritual to the gods, the CO_2 gods or something.

Today flask samples are also collected using modern automated systems. One such system is housed in a suitcase-sized enclosure that includes support electronics and an extendable sampling tower (Plate 53).

Air is also sampled high in the stratosphere to measure its ozone content. This task is performed by an instrument known as an ozonesonde that is strapped to a standard radiosonde and flown from a weather balloon (Plates 54–55).

MLO Visitors and Guests

Scientists and visitors from around the world signed the *Visitors* book during Russell Schnell's tenure as director of MLO, and he often led them on tours (Plate 56). Sometimes he didn't wait for them to arrive at the observatory, and, like MLO directors before and after, he gave public lectures and visited schools in Hilo and elsewhere (Plate 57).

The author of this book first visited MLO on three days in August 1992 to calibrate a variety of atmospheric instruments. The author and Dr. Derek Chignell of the University of the Nations (U of N) returned in June 1993 with more instruments and two van loads of international students who were taking a short course on atmospheric science.

> Forrest Mims, Seguin, Texas, recent winner of a Rolex Award for scientific innovation, was at MLO calibrating some of his small ozone meters. He brought along a group of students from the University of the Nations. These 26 students came from countries such as Germany, Switzerland, South Africa, Netherlands, Canada, Burkina-Faso, England, Australia, and Tonga to name a few. (Schnell 1993)

Schnell describes another visit from U of N in 1996:

> MLO hosted 35 visitors, noteworthy for the 11 countries they hailed from. On 5/14 Forrest Mims, Seguin, Texas escorted a group of University of the Nations graduate students to MLO for a lesson in environmental measurements. This group of 24 claimed citizenship from 8 countries as varied as South Africa and Indonesia to Switzerland. (Schnell 1996d)

Schnell was surprised by one of the U of N visitors, a tall gentleman who was on the Big Island to attend the author's course:

> Visitors to MLO in May numbered 52 with 30 students representing 12 countries from the University of the Nations comprising the bulk of the count. Included in this group, was James Watt, former Secretary of the Interior in the Reagan Administration. You just never know who may show up at the most unexpected time! Other visitors of note were Al Gostewski, NOAA/ETL on 5/7 and Forrest Mims, Seguin, TX on 5/22. (Schnell 1998)

The annual tours by students from the U of N, which have continued to the present, were popular with the students. They especially enjoyed meeting Russell Schnell, for he had previously visited many of their countries. Without speaking down to them, he made the measurements program understandable by using an inflatable globe to illustrate the thinness of the atmosphere and how pollution is transported around the world. Schnell (personal communication with author, September 11, 2007) explains why he went out of his way to host the tours:

> Oh, man, because that was fun . . . if you could get young people excited about science or excited about something, then they will never forget it. And I liked talking to the kids. Some of them were really sharp. . . . And it is just fun to get people excited.

As noted in previous chapters, some visitors experience altitude sickness at MLO, a problem that is exacerbated by being bounced about in the back of a car or van for an hour or so during the drive along the very rough road leading up the mountain from Saddle Road. The condition can also be brought about by diminished oxygen when a group of visitors is packed together in a building to observe an instrument or see charts displayed on the wall, and on one occasion two student visitors fainted and were caught by their fellow students. MLO has long kept an oxygen tank in the Keeling Building for visitors who experience altitude sickness, and a few minutes inhaling the gas generally helps victims survive their visit. But not everyone who needs oxygen is willing to take it, as Russell Schnell (1996e) relates:

> On May 8, NBC Channel 8 conducted a live report from MLO on ozone depletion and UV starring John Chin and Russ Schnell. This 5 minute segment of the 6:00 p.m. news was the first ever accomplished from a remote on MLO direct to the on air newscast which included 2-way dialogue. The anchorman forgot a few of his lines and stumbled over some words due to oxygen deprivation. The fact that he and the cameraman had only 3 hours of sleep the night before and were on MLO from 0900 until the 1810 broadcast probably did not help. Next time he will take the oxygen he was offered.

As always, visitors continued to arrive after Schnell's time as director, and several groups of them are shown in Plates 58 to 61. From the earliest days at MLO, visitors have included many scientists who conducted various projects and measurements at the observatory, several of whom are shown in Plates 62 to 66. One of the many media teams to visit the observatory is shown in Plate 67.

MLO guests might best be described as long-term or even permanent visitors. During the 1960s, many Atomic Energy Commission scientists and technicians were guests at MLO. The longest guest program began in 1965 with the establishment of the Mauna Loa High Altitude Solar Observatory, whose founding was described in chapter 9 (Plate 68). From shortly after sunrise to shortly before sunset 363 days of the year, the observatory's telescope stared at the sun under the watchful attention of permanent guests from the National Center for Atmospheric Research (Plate 69). The second major guest program, the lidar program of NASA's Jet Propulsion Laboratory, was begun in 1993 and is described in chapter 12 (Plate 70).

John Barnes Replaces Russell Schnell as MLO Director (1998 to Present)

Near the end of his six-year tenure as director of MLO, Dr. Russell Schnell gathered with the observatory's staff for a group photo at the Federal Building (Fig. 13.1). He returned to Boulder in August 1998, where he reported to CMDL director Dr. David Hofmann as physical science administrator with responsibility for the operation of MLO and the other Climate Monitoring and Diagnostics Laboratory's baseline observatories. Schnell also became editor of the biannual *CMDL Summary Report,* a valuable resource for those who track observations of the key atmospheric parameters that influence climate and the ozone layer. In one of his final Activities Reports, Schnell (1998) wrote, "Life on the mountain continues at a steady pace with few major disruptions, which is good for both science and the staff."

Russell Schnell had led MLO out of the old strip chart era into a fully computerized, Internet-connected observatory. He had transformed MLO into a science village. And he had found innovative ways to raise new money to support the observatory.

MLO lidar scientist Dr. John Barnes, who was introduced in chapter 12, was appointed to replace Schnell and was assigned the new job title of MLO station chief. MLO computer specialist Darryl Kuniyuki was assigned various administrative duties to assist Barnes in running the station. Barnes inherited a close-knit, experienced staff with whom he had worked both in Hilo and at the mountain station.

Major Progress with the Lidar Program

Probing the stratosphere with a ruby laser lidar system had begun at MLO after the arrival of Ronald Fegley in 1972. Lidar was then an emerging technology, and the original ruby laser and its sensitive light detection system were plagued with problems. Fegley and Howard Ellis invested considerable time on the mountain at night using the system to try to detect the presence of aerosols in the stratosphere. Later Tom DeFoor replaced Ellis, and he in turn was replaced by John Barnes in 1993.

While the unofficial lidar observations at MLO date back to Fegley's time, the official record dates back only to 1980. John Barnes (personal communications with author,

FIGURE 13.1. The MLO staff about 1997 included *(left to right)* John Barnes, Steve Ryan, Judy Pereira, Bob Uchida, MLO director Russell Schnell, Alan Yoshinaga, Darryl Kuniyuki, John Chin, and Leslie Pajo. *(MLO archive.)*

August 30, 2007, and June 4, 2008) explains why the early data are not part of the official record:

> I think I would say that the record starts in December 1974, and it does capture the Fuego eruption. But I believe the very low background values have too large an error to be useful. I know that the data took a lot of work to get. But the low [backscatter] values are just too low to be believable. They are below the 1980, 1990 and 1997 low points in the curve and also below the German and NASA lidars that started at the same time. The [volcanic] eruptions were probably measured pretty well. . . . You could say that there was a substantial learning curve. Lidar was very new. The A/D [analog-to-digital computer interface] electronics for the fast binning of the data was new, and a lot of home-built electronics was needed to work with it. The profile had to be acquired in altitude segments, because the A/D electronics didn't have enough significant bits. Each shot was recorded separately, so an observation generated a lot more data than any other project. They had to work with the guts of the ruby laser and 25,000 volt power supply, because a service call was very expensive.

Barnes used the ruby lidar regularly until the system was retired in 1998 and replaced by a more sophisticated lidar system he had previously developed that employed a powerful near-infrared YAG (yttrium-aluminum-garnet) laser with a frequency doubler. This enables the laser to produce a dual-wavelength beam in the green and near infrared (see chapter 12). Like all critical projects at MLO, the old ruby lidar wasn't immediately retired. First, backscatter measurements made with the ruby lidar were carefully compared with those from the new YAG system. When the ruby laser eventually broke down, the YAG system was prepared to take over:

> I had about 50 overlap nights to join the records and the YAG data was so much better that I didn't have a good reason to fix the ruby. I did figure out later that the detector was damaged. The YAG has both 532 and 1064 nm, and I can interpolate to 694 nm. When I did that with the 50 overlap nights I got within 2% difference. (Barnes, personal communications with author, August 30, 2007)

The high-voltage discharges of the lidar systems at MLO emitted considerable electromagnetic noise, so after the JPL ultraviolet lidar was moved from its trailers into the new NDSC (NDACC) Building in May 2001, John Barnes and the JPL team switched on their systems at the same time to check for problems:

> On 5/10 the lidar was run while the JPL ozone lidar was also operating. Tests were performed to look for possible noise or interference between the two systems. No problems were found. (Barnes 2001)

Aerosols propelled high into the stratosphere by major volcanic eruptions, like El Chichón in 1982 and Pinatubo in 1991, were easily detected by the MLO lidar and received the most notice. Yet lidar is quite capable of detecting evidence of much smaller eruptions. The original ruby lidar detected the Fuego eruption of October 1974. In November 2002, the lidar detected aerosols from a volcano that erupted in South America:

> Lidar—Observations were made on 12/5, 12/19, and 12/31. . . . A clear peak at 20 km was seen above the normal background layer and most likely can be attributed to the Reventador eruption in Ecuador. The layer has a clear top at 37 km which is the highest altitude ever observed. . . . The normal reference range of 35 to 40 km used in all previous analysis was changed to 38 to 45 km for these observations. (Barnes 2002)

During weekly lidar sessions at MLO, the brilliant lidar beam forms a spectacular green shaft in the sky that ends in the stratosphere over MLO (Plate 71). While this spectacle is playing overhead, John Barnes sits at his desk watching a computer monitor that displays the presence of any aerosols high in the sky (Plate 72). A chart of the lidar data since the earliest days is shown in Plate 73. This chart, which includes the pre-1980 data, begins with an upward spike caused by the eruption of Fuego (1974). The second two spikes were caused by aerosols from two major volcano eruptions, El Chichón (1982) and Pinatubo (1991).

MLO's Ozone Monitoring Program

JPL's powerful ultraviolet lidar, the Canadian Brewer spectrophotometers on the solar deck, and the station Dobson spectrophotometer in its protective dome (Plate 74) provided the means to monitor the ozone layer both day and night at MLO. Even before this expanded capability, MLO's ozone measurements fulfilled Robert Simpson's dream for the observatory he founded, for the MLO ozone record had earned a positive reputation in the scientific community, in large part because MLO was the world's highest ozone station. The simultaneous measurements made from MLO and from space by various satellites were of special interest.

Nimbus 7 was a NASA satellite loaded with sophisticated instrumentation for monitoring sunlight, the Earth, and the ozone layer. The latter measurement was made by an instrument known as TOMS (Total Ozone Mapping Spectrometer). Nimbus 7 was launched in late 1978, and during eight summers from 1979 to 1989, NASA's Goddard Space Flight Center contracted with the Climate Monitoring and Diagnostics Laboratory to ship the world-standard Dobson spectrophotometer, Instrument 83 (I83), to Hawai'i so that its measurements of the ozone layer could be compared with those made by TOMS as it flew over MLO each day near noon. By 1989, this research had found a drift of almost 7 percent in the ozone measured by the satellite (McPeters and Komhyr 1991), so the program was extended in order to provide a means for correcting the satellite ozone retrievals. Besides establishing MLO as an important site for the ground validation of a major satellite instrument, the program provided much needed funds for CMDL and an opportunity to perform annual calibrations of 183 from the best site in the world.

Each summer, the CMDL sent a trained Dobson operator to Hawai'i to conduct the 183–TOMS ozone comparisons. The work was rigorous, for the operator had to rise early in order to arrive at MLO shortly after sunrise. Because cirrus clouds sometimes interrupted the measurements, the operator had to take advantage of every clear day, which could mean driving up and down the mountain road every day for a week or more.

Kevin McKee did the Dobson calibrations during the summer of 1980, arriving on June 8. He signed out on August 7 and wrote in the *Visitors* book, "Worshipper of Pele. Spent the summer on this fascinating mountain. (I'd like to return.)" As noted previously, MLO staff conducted some of the other Dobson calibrations.

Grae Roth, a biology student at the University of Colorado, was hired by CMDL to perform the calibrations during the summer of 1991. She returned for the same purpose in the summer of 1992 (Plate 75), and in early August drove the author of this book to MLO for three of her calibration sessions. This permitted the calibration of the author's two homemade TOPS (total ozone portable spectrometer) instruments that measured the ozone layer (Mims 1992a) and a Russian M-124 ozone spectrometer. The 1992 TOPS ozone measurements at MLO affirmed a new drift in ozone retrievals from TOMS aboard NASA's aging Nimbus-7 satellite, a finding that became a short paper in the British science journal *Nature* (Mims 1993).

Roth's assistance had other spinoffs, for the trips up the mountain that she provided eventually led to the writing of this book. They also permitted the calibration of a new kind of sun photometer that measured the optical depth of the atmosphere and total column water vapor using light-emitting diodes operated as spectrally selective, filterless photodiodes (Mims 1992b). The latter instrument became the model for the GLOBE program sun photometer (Mims 1999; Brooks and Mims 2001) and infrared-hygrometer programs (Mims 2002; Brooks et al. 2007). The calibrations begun with Roth's help in 1992 led to annual calibrations at MLO each year to the present and a major study of solar ultraviolet in Hawai'i. The ozone layer plays a major role in regulating solar UV, and in the spring of 1994, sixteen miniature solar ultraviolet monitors were placed around the Big Island in a study of the UV environment of Hawai'i. While the purpose of the study was to investigate the difference in UV between sea level and MLO (around 20 percent), a surprising finding was that scattering of sunlight from cumulus clouds in the pristine sky over MLO caused increases of up to 15 percent in solar UV-B (Mims and Frederick 1994).

Grae Roth was followed by physicist Mark Clark, who performed the annual 183 ozone measurements each summer from 1993 to 2002 while listening to radio talk show programs (Fig. 13.2). During the summers of 1994–1996, Clark spent nights in the AEC Building, sometimes working twenty days in a row. Each afternoon he sped down the mountain for his daily swimming appointment with the Pacific at Honoli'i Beach near Hilo. Brooke Walsh conducted the 183 measurements from 2003 to 2006 and became a close friend of Leslie Pajo, who conducted ozone measurements with the station Dobson (Plate 76).

Visitors from Space (1998)

MLO had long provided grand views of meteor showers, comets, and a solar eclipse, and the Leonid meteor shower of November 1998 was an especially memorable visual treat:

> The Leonid meteor shower gives a high rate of meteors every 33 years, and this year was one of those exceptional years. On the night of November 16–17, John Barnes did a lidar measurement and checked the skies over MLO every few hours during the night for meteors. He saw many bright meteors that left glowing trails. That morning, Steve Ryan came up to the lower observatory road with family and friends, and watched the skies from 3:30

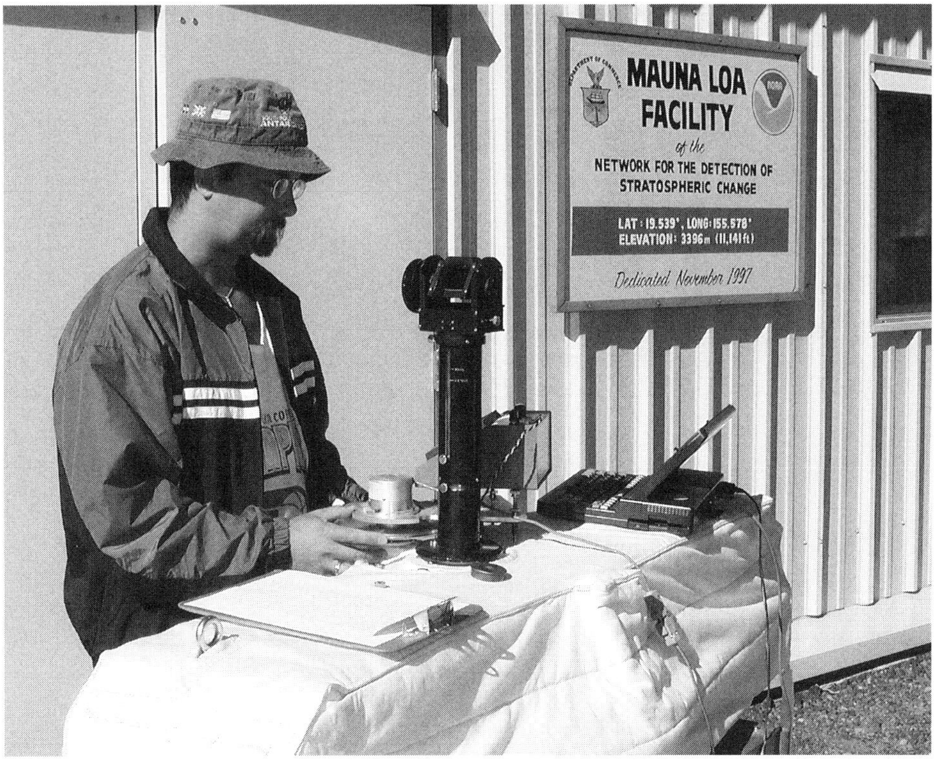

FIGURE 13.2. Mark Clark measures the ozone layer on July 9, 2002, using the world-standard Dobson spectrophotometer during its annual visit to MLO.

> a.m. to dawn. During the hour between 3:30 and 4:30, 174 meteors were seen, with about a third being bright and leaving glowing trails that persisted from less than a second to about 5 seconds. (Barnes 1998)

A second kind of outer space event during November was installation of an interplanetary dust collector. The principle investigator was Ken Farley from the Division of Geological and Planetary Sciences at Caltech. The apparatus consisted of a concave dish about 1.5 meters across. The dish was filled with water that was circulated by a pump, and antifreeze was added to prevent the water from freezing on cold nights. A magnet was placed at the bottom of the dish to attract any metallic dust that happened to fall into the water. The program was ended in October 2003, and Dr. Farley (personal communication with author, September 25, 2007) summarizes the findings as follows:

> We collected weekly samples from Mauna Loa for several years and analyzed the magnetic matter for extraterrestrial 3He. We detected the presence of cosmic dust in many of those samples and documented fluctuations in the cosmic dust flux over several annual cycles. I have correlated the MLO results with similar data from Pasadena [and] there are some interesting temporal relationships.

While the apparatus was intended to collect cosmic dust, it also collected insects that were blown up to MLO during upslope wind conditions, and it was quite interesting to visit the pool to see the daily catch.

In October 2003, Virginia Garrison of the US Geological Survey visited Hilo to discuss a possible sampling program for another kind of airborne visitor (Barnes 2003). Dr. Garrison had previously studied microorganisms that accompany Saharan dust that blows from North Africa to the Caribbean, where they were suspected of causing damage to certain kinds of coral, and she was interested in determining if Asian dust might also contain microorganisms. Garrison initially decided to do sampling at Cape Kumukahi. Later, John Barnes collected samples for her at MLO when he was there for his weekly lidar session.

Still Another Budget Crisis (2000)

Beginning with the 1963 budget crisis that nearly closed MLO, the observatory had periodically faced budget shortfalls. Occasional visits by officials charged with recommending or approving budget decisions had allowed the observatory to make its case, and one such visit occurred in August 1997:

> Congressman Hal Rogers, Chairman, of the Appropriations Committee for Commerce, Justice, State and Judiciary; along with Congressmen Tom Latham and James Moran, visited MLO on 8/11 along with two wives, a congressional aide and two Air Force escort officers. The group flew to the lower MLO helicopter pad on a National Guard helicopter where they were greeted by Dave Hofmann, Director, CMDL, who was on site to tell the group about CMDL science and to conduct a tour of the MLO facilities, including the new NDSC Building. . . . Overall, the congressional delegation visit (CODEL in bureaucratic language), went as smoothly as possible. They learned a lot about CMDL and NOAA science in pleasant surroundings; we had their attention (and control of their person) for about five hours. (Schnell 1997d)

Former MLO director Russell Schnell's application of basic business principles had raised some cash by charging fees to cooperative programs that maintained programs at the observatory. John Barnes continued the tradition and reported in the August 1999 Activities Report that a carbon monoxide study by the State University of New York and Caltech's cosmic dust project would be charged $5,000 per year if they continued their programs at MLO. The MLO road was sometimes used to film car commercials for television. In the August 1999 Activities Report, Barnes reports that MLO was charging $1,000 per day for use of the road for commercials and that $3,000 had been raised for road repairs.

The fees charged some of the cooperative programs provided only a minor boost to the budget needs of MLO. As the turn of the century approached, the funds to operate NOAA's baseline observatories were not keeping up with inflation, and there was a very real possibility that the Samoa station would have to be closed in order to keep the remaining observatories in operation. There was concern that monitoring operations at all the observatories might be seriously cut back. GMD director David Hofmann recalled the situation during an interview in Boulder, Colorado, on May 1, 2007:

> I think it was around the year 2000 when inflation and lack of adjustments to our base budget were really putting pressure on our operation and, in particular, one had to ponder whether we really needed to have all of the baseline observatories that we had. Could we afford to keep measuring at Samoa, for example? Only two people [were] there [and they had] a good record. But the observatory itself was in bad condition; it was going to require a lot of money to continue.
>
> I laid awake at night a lot thinking about what to do about the observatories. In those days, you approached problems like these through research initiatives. So I know of something we can do that is really fantastic; it is going to produce new information. But how do you do that for something that you have been doing for thirty years? How do you write an initiative that will pick up enough interest?
>
> We went ahead with the Baseline Observatory Initiative; that is what we called it. And we sold it on the fact that without these observatories we cannot have these long-term records. Do you want the carbon dioxide record at Mauna Loa to stop? Do you want the ozone records to stop? That is what will happen in the future if we do not figure out how to continue funding something

> that keeps costing more money every year just to do the same thing over and over.
>
> I spent a lot of time in Washington talking to people from, for instance, congressional staffs from various offices like Hawaii and Alaska. And most of them did not really know about these observatories; they did not know they existed. And once they did they thought, "My gosh! You mean we have that in our state? Something that important?" And so this initiative went through. It was not a lot; US $2 million. But in those days that was a lot of money, and it was enough to make sure that the observatories survived.

The $2 million that saved the observatories can be placed in perspective by a comparison with the cost of launching a scientific experiment into Earth orbit aboard NASA's Space Shuttle. At about $10,000 per pound (0.45 kg), the price for Hofmann's Baseline Observatory Initiative to keep the observatories open and their many dozens of crucial atmospheric measurement projects in full operation would pay for placing about 200 pounds (91 kg) into orbit.

Hofmann noted during the 2007 interview that the Baseline Observatory Initiative was "one of two key points in my tenure that really changed what we are doing now." The second point involved the role of North America as a carbon sink, and Hofmann explained how meetings with elected officials and their staffs at remote sites like GMD's South Pole station provided rare opportunities to have casual conversations that would never occur back in Washington:

> The South Pole is special in a certain way, because a lot of visitors come there. People like the notion of being able to go to the South Pole. So about '97 or '98, we had a lot of senators come down: Ted Stevens from Alaska, Pat Roberts from Kansas, and others. And you kind of have a captured audience at the South Pole. They [arrive] in the morning; they spend the day and fly out in the late afternoon. So once you showed them what you are doing . . . you are sitting around and you are able to talk to them like you never could in the US. You do not go to Washington and walk in a senator's office unless you are from the state and you make an arrangement to see him. But these people were all on the Commerce Appropriations Committee—high-level people.
>
> And in our little discussions, when I told them that, well, we are measuring carbon dioxide all over the world because it is important for climate, and we need to know more about the carbon cycle, because we do not really know what takes up carbon dioxide very well. We know the oceans take it up; we know the lands are important; we know trees and plants, but we do not know the details. And when I mentioned that, for example, soils are important in taking up carbon dioxide, crops are important, [Kansas Senator] Pat Roberts' eyes lit up, and he said, "You mean agriculture could be important in controlling carbon dioxide?" And I said, "Yes, it could." And when I got back from Antarctica, I got a call from his office; he wanted me to come up and talk to him. . . . I spent about an hour with him, and then he wanted me to testify before [the Senate Subcommittee on Production and Price Competitiveness of the Senate Agricultural Committee] about the importance [of sequestering carbon] in the soil and how crops are important . . . this was a key to getting funding later on for the North American Carbon Program to try to understand the carbon cycles.

On May 4, 2000, Hofmann appeared before the Senate Subcommittee on Production and Price Competitiveness to discuss how agricultural practices might enhance the sequestration of carbon in soil, thereby reducing the potential impact of carbon dioxide on climate. Sen. Roberts observed that "agriculture may have the ability to store 200 million tons of carbon annually or the equivalent of 307 million tons of coal" (GAP 2000). This cross-disciplinary dialogue on coal and agriculture that sprang from a meeting in Antarctica was reminiscent of many other such discussions that traced their origin to conversations at MLO.

Judy Pereira Retires (1999)

In the summer of 1966, newly arrived MLO director Lothar Ruhnke had just established an office for MLO at the Cloud Physics Building at the University of Hawai'i in Hilo. As related in chapter 10, Ruhnke was looking for secretarial help, and Howard Ellis recommended Judy Bright, who had just completed her sophomore year in speech therapy. Ruhnke quickly hired her for what became the first and only job of her working career. Judy Bright married Jerry Pereira three years later and had their three children while working at MLO.

During her thirty-three-year tenure, Pereira worked with every MLO director from Lothar Ruhnke to John Barnes. In the early days, Judy extracted data from chart recordings and typed MLO paperwork. Her duties expanded over the years so that she also functioned as the office administrator. She typed the drafts of all the scientific papers written by the MLO staff, helped launch ozonesondes, arranged tours, made hotel reservations for MLO visitors, hired summer workers, prepared purchase orders, helped in arranging repairs to the MLO road, worked closely with new staff and incoming directors, organized staff parties, and carried out a host of other duties. Persons planning to visit MLO were advised to contact Judy Pereira to work out the arrangements. As Tom DeFoor (video interview with author, Boulder, Colorado, May 3, 2007) recalls,

> She ran all the administrative stuff. If you had to figure out how to get something done, whether you were the low person on the totem pole or the director, you worked through Judy because she was amazing in her knowledge of how to get things done and her ability to work with people.

During the early years, Pereira would occasionally visit MLO, and she helped prepare the food during the twentieth anniversary ceremony in 1978. But thereafter altitude sickness kept her off the mountain. An account of Pereira's time at MLO is included in the *Climate Monitoring and Diagnostics Summary Report* (CMDL 2000–2001a):

> **Judy Pereira**
>
> In 1966, Judy, the reigning "Miss Aloha Hawaii" and a student at the University of Hawaii, Hilo, joined Mauna Loa as a part time worker. She soon became the permanent secretary and served as the heart and soul of Mauna Loa until her retirement in December 1999 after a 33-year association with the observatory. Over those years she worked for eight different directors, or rather eight directors worked with her, as she often knew much more about MLO history and daily operations than they did. She ran the observatory office in an efficient and professional manner with dashes of Hawaiian Island flavor and local culture mixed in. Because she was related to a large percentage of the people in the Hilo area, Judy could usually get repairs and services completed any time of the week, often free as a favor to a "family member." And, any member of the Mauna Loa staff was considered part of this extended family, with the requisite invitations to weddings, first-year birthday parties, Christmas celebrations, and family luaus.
>
> Judy's ready smile, enthusiasm, empathy, friendliness, and knowledge made her memorable to all she met. She continued her role as the official greeter for Mauna Loa and Hilo throughout all her years of service. Visitors and scientists who came to Mauna Loa from around the world remained her friends for life. Often these people would come back to the Big Island decades later just to renew her acquaintance and to introduce their spouses and children to her, a living embodiment of the "Aloha" spirit.

In 1996, Pereira wrote *Mauna Loa Observatory: A 30-Year Personal Overview,* a memoir about her time at MLO that provides important background for portions of this book. In January 2000 Pereira was given a retirement party at Hilo's Nihon Japanese restaurant. She and her husband Jerry then moved to Las Vegas to be near two of their children and grandchildren.

When Pereira (video interview with author, Boulder, Colorado, April 27, 2007) was asked if she would do it all again, she replied, "Absolutely, without any question." Asked to recall some of the names of those with whom she worked the longest at MLO, she mentioned several of the old-timers:

> Howard Ellis, he was the main one because he was the only one I knew, really. I also met Bernard Mendonca and

Barry Bodhaine, who was a grad student at the time, he was there. John Chin became my lifelong coworker. . . . Johnny and I kind of grew up together with Mauna Loa Observatory.

Reflecting on his thirty years working with Judy Pereira, John Chin (video interview with author, Hilo, Hawai'i, December 4, 2006) describes her as

> just like, I would say, a house mother for the observatory . . . everybody who had problems [went] to Judy, and she was the one that organized everything. . . . She took care of all the visitors. In fact, she was a very good thing for the directors.

Another transition occurred in 1999 when, on April 20, Mamoru "Al" Shibata of Hilo died at the age of seventy-two. Shibata had worked as an electronics technician at MLO from 1972 to 1990 and was very popular with his colleagues, including Alan Yoshinaga (video interview with author, Hilo, Hawai'i, December 4, 2006):

> Al Shibata was an ET [electronics technician] who came in from Las Vegas, I believe, at about the same time I did. And I enjoyed him, because he was a little older than me. We could talk stories about Las Vegas, and he would tell me all of his National Weather Service experiences in the South Pacific. And he was also a fantastic ET. If I needed help with any of my instruments, he was really willing to jump in there and help.

John Chin Retires (2000)

After thirty-seven years of service at MLO, John Chin retired in December 2000. An account of his service is included next to the one for Judy Pereira in the *Climate Monitoring and Diagnostics Summary Report* (CMDL 2000–2001b):

> **John Chin**
> John joined the Mauna Loa Observatory, U.S. Weather Bureau Research Station, in 1960 to work on the Scripps Institution of Oceanography (SIO) CO_2 program. He left in 1962 to broaden his skills at NASA and Lockheed before returning to Mauna Loa in 1964 to again conduct the SIO atmospheric CO_2 monitoring program along with other observatory duties. In 1966, John left for a brief time to work at Cloud Physics, University of Hawaii, studying warm rain projects. In 1968, he rejoined Mauna Loa and Federal service. In 1974 he helped in the expansion of the NOAA carbon cycle programs at MLO. As a physicist he continued to operate both SIO and NOAA CO_2 programs, in addition to many other observatory duties, through to his retirement in December 2000, completing 37 years of service at Mauna Loa.
>
> It has been calculated that during his tenure, John made at least 6000 round trips on the Mauna Loa Road, which is equivalent to driving around the earth 25 times at the equator.
>
> John is known for his reliability and predictability. He always arrived for work within 2 minutes of a set time, and his weekly lunch menu never varied. For 37 years, his lunch always consisted of a sandwich, tuna or chicken, a banana, carrot or celery sticks, and a small box of raisins that he ate exactly at 12 noon.
>
> When John first entered service at Mauna Loa Observatory in 1960, the annual mean atmospheric CO_2 concentration was 316 ppm. The annual mean was 370 ppm in the year he retired. The MLO CO_2 curve is recognized around the world as one of the longest and most important geophysical records in existence; John can be proud that he was instrumental in helping make it so.

The MLO staff celebrated Chin's retirement at a dinner party held at Hilo's Big Island Bistro on January 12, 2001. Chin was presented with a thick scrapbook filled with photographs and clippings that cover his entire tenure at MLO and provide several historical details useful to the organization of this book. The scrapbook also includes heretofore unpublished letters of tribute from several of Chin's former supervisors that provide colorful highlights about working at MLO. A letter from Charles Keeling (January 11, 2001) summarizes Chin's role in the carbon dioxide programs at MLO and Cape Kumukahi:

> Johnny has made it possible to keep the measurements of atmospheric carbon dioxide going without a significant break [since 1964]. He has patiently tended what is now probably the last instrument anywhere in the world of geophysical monitoring that runs on vacuum tubes. With Johnny's perseverance it has been possible to gather a record of breathing of the earth's carbon cycle with a single instrument based on a single principle.
>
> Johnny also pioneered the carbon dioxide sampling at Cape Kumukahi and has kept a watchful eye over the source of air at Mauna Loa as more and more programs came online, and more and more people visited and worked at the station.
>
> Someone at Mauna Loa has calculated the number of miles Johnny has traveled up and down the mountain top. Let me add that, according to my reckoning, Johnny has been responsible for about 3,892,104 measurements of carbon dioxide gas.
>
> With his cheerful dedication and loyalty to the Mauna Loa program, Johnny has been a wonderful resource and inspiration to all of us working on global environmental problems. We will all miss Johnny very much, but rejoice that he may now "rest from his labors," and yet his accomplishments will live after him.
>
> Charles D. Keeling
> Scripps Institution of Oceanography

Longtime Scripps engineer Dave Moss, who first signed in at MLO on July 23, 1972, wrote a letter (January 9, 2001) thanking Chin for introducing him both to Hawaiʻi and MLO. He also wrote,

> To survive traveling to and working on the mountain requires almost superhuman endurance and care. It seems that you possess both. If I am not mistaken, your career at Mauna Loa exceeds all others in duration. This is testimony to your strength of will. It is obvious that you have brought these strengths to bear on caring for the Scripps CO_2 program these many years. For this, you deserve undying gratitude.

Sam and Kay Oltmans lived at MLO for a month during the spring of 1972 while calibrating the world-standard Dobson spectrophotometer. Dr. Oltmans' letter (January 11, 2001) thanked Chin for a bag of fruit he gave him and his wife back then. He then wrote,

> Everyone including you were very worried about what we would do if the volcano erupted during the night since we had no car of our own at the observatory. There was an old army truck parked there that you told us we could release the parking brake and start rolling down the hill. I don't remember if the truck was supposed to start once we got rolling or [if] we were supposed to coast the whole way.

CMDL director David Hofmann also contributed the following letter, which begins with a relevant quotation:

> Webster's Dictionary: Monitoring—To check systematically or scrutinize for the purpose of collecting specified categories of data. To keep watch over.
>
> Since 1960 John Chin has been the monitor supreme. One can safely say that without his dedication for the past 40 years, the record of carbon dioxide measurements from Mauna Loa Observatory would not exist as we know it today. A record dating back over 40 years with *the same instrument* would not exist. Can you imagine watching television on a 1950's era TV? It would be a large wooden box with a very small oval screen, black and white of course! It would contain vacuum tubes, resistors and capacitors rather than transistors and integrated circuits. How many such TVs are in operation today? I would venture a guess that they could only be found in museums. Yet, that is exactly what John Chin has been working with these past 40 years—a 1950s vintage, vacuum tube operated infrared absorption carbon dioxide analyzer. And it still works!
>
> It was Charles "Dave" Keeling's foresight that continual measurements of carbon dioxide would be important because carbon dioxide, which is emitted whenever carbon-bearing fuel is burned, is a greenhouse gas that traps heat in the earth's atmosphere. And it was his

efforts that got the instrument running and in place on Mauna Loa volcano. But it was John Chin's daily care of the instrument until his retirement that made the Mauna Loa carbon dioxide record, often referred to as the most important environmental record in existence, possible.

Now John has retired. What will become of this instrument and this record? Measurements of carbon dioxide will continue at Mauna Loa Observatory with new state of the art instruments, as has been done since NOAA was created in 1970. But it is not likely that Dave Keeling's old instrument will continue to operate, at least not as well as when John was the caretaker. We will miss John's dedication to his work, but just as importantly, we will miss his kindness and friendship. Bon voyage John, and the best of luck in retirement. . . . You earned it!

Dave Hofmann, Director
Climate Monitoring and Diagnostics Laboratory
Boulder, Colorado
January 2001

After he retired, John Chin stopped by the MLO office every Wednesday to check his e-mail and have lunch with his former coworkers and the new staff. His lunch menu remained unchanged: a chicken salad sandwich, carrots, and a banana.

Replacing Judy Pereira and John Chin

Leslie Ann Pajo, who was introduced in chapter 12, took over from Judy Pereira as MLO's administrative secretary. The transition was smooth, for the two women had worked closely together since Leslie arrived in 1994. Longtime MLO chemist Alan Yoshinaga (video interview with author, Hilo, Hawai'i, December 4, 2006) reflects back on the contribution that both women made to the observatory:

> We have been really lucky with Judy and with Les. . . . They do their work, and yet they are friendly. That is an extra that you do not ask for in the [job] qualification: "Are you friendly?" They just had that personality where they just made people at ease. They took care of them. If people had extra problems, they went out of their way to help them. They took them out to dinner or took them out to lunch, introduced them to other people. They just went out of their way to help whoever came to Mauna Loa to do some work or whatever. . . . That is the nice thing about Mauna Loa—we had good people; we have good people.

The Mauna Loa Observatory Employee Association (2000)

In January 2000, Darryl Kuniyuki, John Chin, Leslie Pajo, and Alan Yoshinaga organized the Mauna Loa Observatory Employee Association (MLOEA). Kuniyuki was named president, and he and the others were named board members. All present and former MLO staff were named regular members. The MLOEA was organized as a nonprofit organization designed "to assist the Mauna Loa Observatory's employee's activities. The association also promotes outreach and provides the cash award for the local Science fair."

The MLOEA raises money for its outreach activities through the sales of soft drinks, snacks, shirts, hats, and other items. Since 2000, the MLOEA has provided a cash award to an outstanding junior and senior project at the Hawai'i County Science Fair related to environmental science, physics, chemistry, or computers.

Paul Fukumura-Sawada and David Nardini Arrive (2001)

After John Chin retired, MLO veteran Steve Ryan took over the carbon dioxide duties on the mountain and at Cape Kumukahi. In February 2001, electronics engineer Paul Fukumura-Sawada was hired to fill Chin's vacated slot and take over Ryan's projects. Fukumura-Sawada was a Honolulu native who did well in mathematics in high school and was president of his school's Math League. Besides taking over Ryan's responsibilities, his MLO duties included maintaining the solar radiation instruments and other tasks on the mountain three days per week. Fukumura-Sawada was accustomed to high altitude, for he had worked four-day shifts at NASA's infrared telescope on the summit of Mauna Kea for eighteen months prior to moving to MLO. The only altitude symptom he reported was the loss of memory so common at alpine elevations: "Maybe you go get a tool, and you forget which one you were looking for" (interview with author, Hilo, Hawai'i, December 5, 2006). Fukumura-Sawada enjoyed his new MLO assignment:

> Down here [in Hilo] it can be raining, and up there it's sunny. . . . Even when there's nothing to do, there's something to do. We have no maintenance person, so we basically take care of that ourselves.

Hilo native and electronics specialist David Nardini joined MLO as a research/operation associate in 2001. Nardini is a NOAA veteran with extensive experience working around the Pacific. When he was working for the National Weather Service (NWS) in Samoa in 1991, category 5 typhoon Val slowly passed directly over the island, causing extensive damage, dropping 17.5 inches (444.5 mm) of rain, and soaking some of the instruments he managed. The eye of the storm lingered for a day, and the storm battered the island for more than three days. In 1992, Nardini and his NWS colleagues received NOAA's Silver Medal for their service during and after the storm.

Nardini was originally hired under an MLO contract. Then, like the majority of today's MLO staff, in 2005 Nardini was hired as a contract employee with the Research Corporation, University of Hawai'i (RCUH). During a typical week he was assigned three trips to the mountain, where his responsibilities ranged from checking the rain buckets to measuring the ozone layer using the station Dobson spectrophotometer.

Nardini (video interview with author, Hilo, Hawai'i, December 5, 2006) reports that he enjoyed working in science and learning more about the impact on climate science of the measurements being made at MLO. The long drives to and from the station and working at high altitude were negatives, as was the lack of job security in the new era of MLO's increasing reliance on contract workers.

The EPA Mercury Program Arrives (2001)

While MLO had long monitored particulate matter and dozens of gases in the atmosphere, the observatory did not monitor mercury until Dr. Matthew S. Landis of the US Environmental Protection Agency (EPA) pioneered a new cooperative mercury monitoring program in 2001. Airborne mercury can be either naturally occurring or a by-product of human activity, and it can be found in the air in its gaseous form or included as compounds in particulate matter. Dust and the emissions from coal-burning power plants contain mercury and its compounds. Asia is a source of major dust storms, power plants in China are believed to be major sources of airborne mercury, and air from China often passes over Hawai'i as it blows toward North America, especially during spring.

The objective of the EPA program was to quantify the presence of mercury over the center of the Pacific Ocean and to determine its sources, and the best place to locate the EPA's sophisticated new instruments was atop the old wood tower at the east end of the Keeling Building. The solar instruments formerly located on the tower had long since been moved to the solar deck over the NDACC Building, and Bob Uchida was assigned the task of preparing the tower for the mercury project. Uchida was a jack-of-all-trades who could do everything from repairing a delicate instrument to designing and assembling a complete building. For this assignment, his initial tool was a sledge hammer, which he deftly used to demolish the upper portion of the old tower while balanced on what remained. He then constructed a new platform for the EPA instruments. As Uchida (video interview, Hilo, Hawai'i, November 30, 2006) later explained to the author, who had watched him at work, he had to work rapidly, for there had been a communications mix-up:

> When the principals for that project came over, they had expected to see the tower ready for them to put up their project. Well, I was not told that that project was coming in and when it was coming in and where it was going. So, when that guy got here, he was very, very unhappy that . . .

> it was not ready for him to set up. What he told me was that he had a lot of powerful friends in Washington and he was going to call Washington and tell them what was going on. . . . And so, I told him I'll do it for him. I'll get it done. And that is what you saw me doing.

The lab equipment for the mercury project was placed at the southeast end of the Keeling Building where MLO staff once slept when they spent nights at the observatory. Darryl Kuniyuki was placed in charge of the administrative side of the EPA project, and chemist Alan Yoshinaga was assigned as the project's scientist.

The EPA project introduced many new systems with sophisticated names, including two mercury analyzers, a reactive gaseous and particulate collector, an aerosol collector, an aethalometer, a sequential fine particle sampler to collect halides, sulfur dioxide and surface ozone analyzers, and a precipitation collector (Kuniyuki, personal communication with author, June 9, 2008). The time consumed in maintaining the project would eventually require the addition of two new positions on the MLO staff (2005).

AMiBA Comes to MLO (2003)

Thanks in part to Dr. Russell Schnell's entrepreneurial approach to help support MLO, in July 2003 the Academia Sinica (Latin for "Chinese Academy") of the Republic of China (Taiwan) signed a contract with NOAA for the use of a portion of the northwest section of the new 4.05-acre expansion of MLO for a highly specialized radio telescope. Regulations limit the construction of new telescopes on Mauna Kea, and Mauna Loa provides an alpine site above most of the water vapor that absorbs the gigahertz signals that such telescopes are designed to receive. In December 2004, the telescope's six-legged mount and receiver platform arrived at MLO. The system was formally dedicated on October 3, 2006, and AMiBA (Array for Microwave Background Anisotropy) went to work providing new kinds of imagery based on radio emissions from Jupiter and distant galaxies.

Outreach Activities in Hilo (2003 to Present)

The Mauna Loa Observatory staff have long helped judge Hilo Science Fairs and participated in mentoring students for fair projects and summer science programs. Steve Ryan's VOGNET provided an opportunity for students at several Big Island schools to conduct significant science.

A visit to the MLO office in Hilo by the Mountain View Elementary School in October 2003 led to MLO's program to develop interactive displays for both schools and the general public. In 2005 MLO participated in Hilo's Earth Day Fair and has participated each year since. In 2008 MLO began participation in Hilo's annual AstroDay.

The MLO outreach program has continued to expand due to the public's increasing interest in climate change. Station Chief Dr. John Barnes continues the longtime tradition of giving talks at Big Island schools and organizations.

Remembering Nobuichi Yamanouchi (1912–2004)

Beginning with Robert Simpson's founding of the Mauna Loa Summit Observatory in 1951, a cross section of Hilo businessmen, diesel mechanics, Lions Club members, prisoners, and others provided valuable assistance to the observatory and its staff. One of the most memorable of these MLO boosters was Nobuichi "Nobu" Yamanouchi, the owner of the Yamanouchi Garage on Kaumana Drive near where Saddle Road began. In the early days, MLO vehicles received a pounding every trip up and down the unpaved lava road, and breakdowns of all kinds were frequent. As related in chapter 8, Yamanouchi often worked late into the night repairing MLO's damaged vehicles.

Nobuichi Yamanouchi died on October 26, 2004, at the age of ninety-two (*Honolulu Star-Bulletin* 2004). While "Nobu" never made a scientific measurement on the mountain, for many years he enabled those who did by serving as the observatory's de facto mechanic. As Bernard Mendonca (personal communication with author, May 6, 2008) recalls,

> He was definitely one of the most unsung heroes of the early days of the observatory. I cannot imagine how many days the observatory would have been left unmanned and unattended if not for his dedication and service to our vehicles. I miss his friendly smile on our return from the mountain to this day. . . . May he rest in peace.

Remembering Charles David Keeling (1928–2005)

By far the longest running and most significant cooperative research program at MLO is the nearly continuous measurement of carbon dioxide begun by Dr. Charles David Keeling of the Scripps Institution of Oceanography in March 1958 (Plates 77–78). Early in his career, Keeling became the world's expert at measuring CO_2 in the atmosphere. He also became the first scientist to accurately measure the annual cycle of CO_2 and its relentless year-after-year increase. Most of the attention given to the famous Mauna Loa carbon dioxide curve—the chart of the CO_2 concentration measured at MLO since 1958—centers on the warming potential of the ever-increasing concentration of the gas in the atmosphere. But there's much more to this curve, for, as described best in Keeling's memoir (1998), it is a visualization of the earth's respiration:

> The maximum concentration at Mauna Loa occurred just before plants in temperate and boreal regions put on new leaves. At Mauna Loa the regular seasonal pattern almost exactly repeated itself during the second year of measurements. . . . We were witnessing for the first time nature's withdrawing CO_2 from the air for plant growth during the summer and returning it each succeeding winter.

Charles Keeling was a perfectionist who insisted that no changes be made to his CO_2 measurement program, especially its ancient Applied Physics Corporation infrared gas analyzer at MLO. He was highly demanding of those who maintained his instrument at MLO, especially Howard Ellis, who in the mid-1960s invested thousands of hours of his own time and spent many nights alone on the mountain to care for the analyzer and make sure the generators that powered the observatory were running.

In January 1964, shortly after the observatory was placed on a starvation budget and the staff reduced to only Ellis, Bernard Mendonca, and Malcolm Keyes, Keeling blamed the staff for shutting down the APC analyzer after it experienced a major breakdown. More than four months of data were lost, but the data gap provided Keeling with the opportunity to hire John Chin to care for the carbon dioxide program at MLO.

Keeling had little tolerance for those whom he believed did not fully understand the difficulty of maintaining an accurate, continuous, and permanent record of the atmosphere's most important anthropogenic greenhouse gas. He developed a reputation as a tenacious scientist who did not hesitate to confront government bureaucrats and scientists over funding issues and the significance of his measurement program. He did battle with the National Science Foundation when his research was classified as "routine" and no longer eligible for funding. He also clashed with NOAA over funding and NOAA's establishment of a parallel CO_2 monitoring program.

Yet in the end, Charles Keeling was recognized for what he had become: the world's leading authority on the measurement of carbon dioxide. When Ellis, Mendonca, and Keyes visited MLO in 2007, they all agreed that the success of the carbon dioxide record was due to Keeling's perseverance. The chart in Plate 79, which is often referred to as the Keeling curve, affirms their evaluation, for the Scripps measurements of CO_2 merged seamlessly with NOAA's measurements that began in 1978 and have remained in close synchronization ever since.

Keeling made comparatively few visits to MLO over the years, and his last was certainly the most memorable. After decades of wrangling with NOAA and its predecessor agencies and after requiring the MLO staff to maintain his ancient vacuum tube CO_2 analyzer that was so sensitive to temperature changes that it responded when a door was opened, in 1997 NOAA honored Keeling by naming the original MLO building after him.

While Keeling's sometimes cantankerous side was well known, there was another side to him that has received little publicity, and that was his willingness to share credit for his accomplishments. The key portions of the first published paper that announced the increase in carbon dioxide measured at MLO (Pales and Keeling 1965) were almost certainly written by Keeling. Yet he gave first authorship of the paper to MLO director Jack C. Pales, who wrote the opening portion of the paper and had personal responsibility for the operation of the instrument. Similarly, Keeling gave lead authorship of the first paper announcing the CO_2 results from the South Pole to the station director. When John Chin retired (Plate 80), Dr. Keeling wrote a letter to MLO (see above) praising his role in producing the historic carbon dioxide record. In an epilogue to his 1998 memoir, Keeling credits some of those who helped with his carbon dioxide program:

> I have tried to be as objective as I could be in this narrative, but I apologize if, in my recollections, I have unfairly recalled, or never properly understood, activities recounted here. I hope that my text makes clear that I had a great deal of help in trying to stay in carbon dioxide research. Much of this help came from personnel below the management level in the agencies opposing my research. I also received crucial assistance from interveners, not all of whom I have had space to mention. On a more personal note, I express my appreciation to John Harte, who after suggesting that I write this narrative, helped me in this understandably difficult task, given the necessity to criticize the actions of others, however well meaning their intentions may have been. I also appreciate reviews of my text by Robert Socolow, Devendra Lal, and Ralph Keeling. Contributing to my research at the Scripps Institution of Oceanography, besides those mentioned in the text, are David Moss, who as the project engineer for 32 years, assured that my program had the best possible instruments within budget, and Alane Bollenbacher and Peter Guenther, who carried out many of the measurements in the laboratory for 20 and 30 years, respectively. Also, John Chin of NOAA, recruited for work at Mauna Loa, who has taken care of our equipment at Mauna Loa Observatory ever since the budget crisis of 1963, and is probably more directly responsible than anyone else for creating the Mauna Loa CO_2 curve that has been credited to me. I would like to close by expressing my deep gratitude to Louise Keeling, my wife for almost my entire career, whose patience, sympathy, love, and understanding gave me much needed support in my pursuit of the carbon dioxide molecule in all of its ramifications.

When Charles Keeling died on March 20, 2005, even those in the scientific community with whom he had done battle acknowledged the man's genius as an experimental scientist. The Scripps Institution of Oceanography obituary notice announcing Keeling's death includes these reflections by NOAA administrator Conrad C. Lautenbacher:

> Charles David Keeling, the world's leading authority on atmospheric greenhouse gas accumulation and climate science pioneer at Scripps Institution of Oceanography, University of California, San Diego (UCSD), died Monday, June 20, 2005, while at his Montana home, of a heart attack. He was 77 years old. Keeling has been affiliated with Scripps since 1956.

"Dr. Keeling will be sorely missed by the NOAA family," said Retired Navy Vice Adm. Conrad C. Lautenbacher, Ph.D., under secretary of commerce for oceans and atmosphere and NOAA administrator. "He has left us with the eternal gift of his vast knowledge, but more importantly he left us and future generations with the gift of inspiration. His legacy will inspire future generations to follow in his footsteps in the quest for scientific discovery" (Scripps 2005).

Continuing the Keeling Legacy at MLO

Charles Keeling's carbon dioxide work was continued at MLO under the direction of his son, Dr. Ralph Keeling, who pioneered the long-term measurement of the concentration of oxygen in the atmosphere introduced in chapter 12. Keeling was inspired by his father, and some have written that he followed in his footsteps. But the son was determined to make his own mark, and he quickly did so by staking out his own scientific territory at MLO and elsewhere that extended well beyond his famous father's shadow. While it was the elder Keeling who discovered that the concentration of carbon dioxide in the atmosphere was increasing each year, it was the son who discovered that the concentration of oxygen was declining. In view of the half-century Keeling legacy at MLO, by far the longest of the observatory's cooperative programs, this father-son story is worth telling.

The second of five children, Keeling relates how his father would take him and his older brother Andrew on hikes from the time they were very young. The hikes inspired an interest in the natural world, and when he was in the fifth grade, Ralph and a friend, whose father was also at the University of California at San Diego, brainstormed about perpetual motion machines and related topics. The elder Keeling was an accomplished pianist, and Ralph recalls the family's musical pursuits (this and excerpts following are from Ralph Keeling's personal communication with author dated August 28, 2007):

> He had a program, in a sense, to turn his children into [classical] musicians and even to develop ensembles by the fact that he had enough children to fill different parts in groups. So I started on piano at . . . age five.

Keeling promoted Ralph to the violin at age eleven, and he eventually considered becoming a professional violinist. But his first interest was science, which he considered much more exciting and challenging:

> I was not interested in being in mediocre science. And until I went to college and really could test myself against the [required] standards . . . I was not sure that it was really something I was going to be able to [succeed] in. . . . I guess I had my father as a consultant through that process.

Keeling attended Yale, where he initially majored in chemistry but could not see the value of performing canned experiments in lab sessions:

> I grew weary of all the labs that I had to go to. I viewed them as not worth the time investment, and I would still defend that perception. I do not think I was just trying to . . . find the easy route. I really found the lab work to be somewhat like trying to bake cookies as fast as you could with a recipe and not really learn much from the process. I thought it was too structured and artificial of a learning process for me. I would have much rather gone in and repaired a helium leak detector. That would have taught me something.

Keeling learned that he could avoid the boring lab sessions by switching his major to physics. After receiving his bachelor's degree in physics, he began a graduate program in applied physics at Harvard in 1979 and earned his doctorate in 1988. Along the way he spent time with his family in Switzerland, where he was inspired to pursue a thesis project that was to significantly impact his career and establish his scientific reputation:

> I was inspired by my father's work probably earlier than this, but the first time I really remember was when I was in college looking at some of his data, and a friend of mine from college who was visiting said, "Wow. That is really cool stuff," and I was thinking the same thing. He said, "That looks like very clean data." That was his reaction because his father was also in an experimental [field, and he] recognized that data often looks a little bit scattered, and this record did not look scattered. It just looked like [it was] tracking something. My father was explaining his work to me at various times, so I was gathering an appreciation for his work. Also, because of my schooling, I was developing a better ability to understand it at some depth.
>
> So he was explaining the CO_2 record and I remember him explaining that there was a problem unsolved about where the carbon dioxide was going—that part of it that was not staying in the air, some fraction, something like half, was going into some of the reservoirs. He mentioned that it would be useful to measure oxygen changes in order to distinguish between, say, an ocean and a land sink for this extra carbon.

It became clear to me that—not then but over time—the idea of measuring oxygen was something that he had planned on pursuing himself very early on and never got around to. So the idea of measuring oxygen was old. I do not know who it really came from. It is never clear whether it was his idea or someone else's.

. . . [An oxygen] paper was published by Lester Machta [Machta and Hughes 1970]. That was the first one to propose it in kind of a gray literature, but it was clear [from] talking to my father that the idea did not originate at that point in time. So I do not know the genesis of the idea. I did not come up with it.

So I had this idea, but I was in graduate school, and my interests were continuing as much in physics as in earth science. . . . So I did not have a clear idea of what I wanted to do when I went to graduate school.

At some point, though, about the first year I was there, and I think this was before I went to Switzerland to join my father who was on sabbatical, I remember going on a walk thinking, "Wow, I really should work on earth science because there is something about studying the earth in this era that is more important than breakthroughs in fundamental physics." It just seemed like it mattered more that some new theorem in non-equilibrium thermodynamics could wait 20 years, but if we wait 20 years, the earth will be a different place. So it really mattered. We needed to get going, and I felt like I should join in that effort. I was clearly inspired by my father's work in that thought process. . . .

I went back to Harvard, in part still not quite sure what I wanted to do, but at that point I decided I did want to do experimental work. . . . [During a return visit to Switzerland] we were staying in . . . [an old restored farm that] had extremely low beams in the doorway. It is probably a very old building from when people were shorter, and I continually was bumping my head. . . . One time, I bumped it hard enough that I decided to lie down on the bed, but I was thinking all along this time about what I wanted to do with myself. I had not really got a project that I was ready to sink time into. I was exploring.

As I was lying on the bed, I do not think the head bump had anything to do with it, but it was a moment of contemplation, I somehow put together the idea of these precision measurements that I had learned about and this old idea that had been circulating in my head of measuring changes in oxygen, and I realized there had to be a way to apply this technology to this problem and get something that is precise enough to measure these changes.

After he returned to Cambridge, Keeling researched early measurements of oxygen at the Harvard and Massachusetts Institute of Technology libraries and became convinced he could devise a new optically based method to measure oxygen with high precision. The measurement would be much more challenging than measuring carbon dioxide, for it is much easier to measure large changes in a gas like CO_2 that represents a very small fraction of the atmosphere (about 0.0394 percent in 2011) than small changes in a gas like oxygen that constitutes more than a fifth of the atmosphere. But his thesis advisor, Dr. Jim Anderson, reacted positively:

> It was a project . . . with less potential for me to make my own mark. But Jim Anderson immediately realized the interest in this oxygen idea and was behind it. He did not have to reflect; his reaction was, "Oh gosh. We've got to try to do this." It became my thesis project.

During graduate school, Keeling built two versions of his apparatus to measure with higher precision than ever before the concentration of oxygen in air. With help from engineer Nate Hazen, he settled on a third design, and the resulting instrument remains in use at Scripps Institution of Oceanography, where he has been affiliated since 1992. Between graduate school and Scripps, Keeling was a visiting scientist at the National Center for Atmospheric Research (NCAR) in Boulder. His connection with Scripps developed during his NCAR time when he needed a source of compressed air for calibrating his oxygen monitoring apparatus:

> Going to Scripps really accelerated that, because . . . my father had an apparatus for pumping air so you could fill up compressed air [tanks]. . . . La Jolla, where Scripps is at, is on the ocean, and we get on occasion air [from a] very clean sector of the Pacific Ocean, you can get very clean air pumped into a compressed air tank. There are very few places in the world that can do that, because in most places, their compressors are not necessarily upwind and everything else is polluting. So I had access to very good air.

Dr. Ralph Keeling became a professor of geochemistry at Scripps, where, in addition to directing his father's carbon dioxide project, he became the leading investigator of measuring changes in the concentration of oxygen in the atmosphere (R. Keeling 1993, 1996). He also began a new program to measure the concentration of argon in the atmosphere (2004). Because the solubility of argon in water is governed by temperature, the atmosphere's concentration of argon measured at MLO and elsewhere could be a proxy for monitoring the temperature of the world's oceans.

NOAA's Global Monitoring Division (2005)

The Mauna Loa Observatory was begun by the US Weather Bureau and has always been under the Department of Commerce. Over the years there were changes in the agencies responsible for the station, and on October 1, 2005, the Climate Monitoring and Diagnostics Laboratory was merged into the Earth System Research Laboratory (ESRL) to become part of its Global Monitoring Division (GMD). Dr. David Hofmann, director of CMDL, remained as director of GMD until December 31, 2007.

> GMD's mission is to observe and understand, through accurate, long-term records of atmospheric gases, aerosol particles, and solar radiation, the earth's atmospheric system controlling climate forcing, ozone depletion and baseline air quality, in order to develop products that will advance global and regional environmental information and services. (NOAA/ESRL 2005)

The three major functions of the Global Monitoring Division are posted on the organization's Web site (http://www.esrl.noaa.gov/gmd/index.html):

> 1. Climate Forcing
> GMD conducts sustained observations and research related to global distributions, trends, sources and sinks of atmospheric constituents that are capable of forcing change in the climate of the earth. This research will advance climate projections and provide scientific policy-relevant, decision support information to enhance society's ability to plan and respond.
>
> 2. Ozone Depletion
> GMD conducts research on the depletion of the global stratospheric ozone layer and Antarctic ozone hole through global surface-based monitoring of total-column ozone, ultraviolet radiation, and ozone-depleting gases, including those regulated by the Montreal Protocol. Continued surveillance is necessary in order to verify the expected recovery of the ozone layer.
>
> 3. Baseline Air Quality
> GMD monitors levels of air quality elements such as tropospheric ozone, carbon monoxide and aerosol particles in non-source regions which may be affected by long range transport from distant sources of industrial pollution. This large-scale transport affects baseline air quality which must be monitored in order to determine the importance of regional sources that may impact the environment and public health.

Another name change occurred in the fall of 2005 when the Network for the Detection of Stratospheric Change (NDSC) was renamed the Network for the Detection of Atmospheric Composition Change (NDACC). The new moniker didn't change any of the science at MLO, but the NDSC Building became the NDACC Building.

Monitoring baseline air quality remained the function of MLO and GMD's other baseline observatories at Barrow, Alaska, the South Pole, Samoa, and Trinidad, California. All the baseline observatories were functioning well as the end of MLO's fifth decade approached. The various stations continued to churn out what was widely viewed as among the world's highest quality atmospheric data in spite of the usual number of equipment breakdowns, tropical storms, lightning strikes, blizzards, personnel transfers, retirements, and tight budgets.

New Staff at MLO (2005)

As new projects arrived on the mountain and old-timers retired or prepared to do so, responsibilities changed and a new generation of staff began arriving at MLO. Darryl Kuniyuki, who had extensive experience in maintaining most of the observatory's science programs over the years and who managed the computers and network for all the MLO sites, was assigned management responsibilities for the nonscientific functions of the observatory, including scheduling of workdays for the mountain teams, office space management, and acquiring new staff. His work with the EPA mercury project led to a new way of expanding the staff without the use of civil service employees. The mercury project required adding two positions to the MLO staff, but a federal hiring freeze blocked new civil service positions. So the new staff were hired through the Research Corporation of the University of Hawai'i's Joint Institute for Marine and Atmospheric Research (JIMAR). The new arrivals had skills that included chemistry, mathematics, computer programming, and Web services. Kuniyuki (video interview, Hilo, Hawai'i, December 5, 2006) explains that all the newcomers were expected to learn what was necessary to participate in the mountain observations:

> That is a requirement. It is a requirement that anyone at the Mauna Loa staff has to learn the obs [observations], because that is our lifeline. That is the most important thing of this whole observatory. So it does not matter what background you come from; whether you are a secretary, accountant, or engineer you have to learn all of the obs, because that is the most important thing. And we do not discriminate, and everyone seems to be really happy to do it. I mean, they seem excited that they get to participate in the observations.

Trevor Kaplan, who had a mathematics degree from Colgate, was hired through JIMAR on June 1, 2005, to work at MLO as a programmer and research assistant. Kaplan, who previously worked on a satellite project at NASA's Goddard Space Flight Institute, was well acquainted with MLO, for he had worked there part time since November 2002 developing software for John Barnes' lidar program. After he became a full-time JIMAR employee, Kaplan was made responsible for the MLO aerosols program and assisting with Web site development. He sometimes experienced unusual reactions to the altitude during his days on the mountain:

> My eyes; that is the biggest one. They have been getting really dry. . . . And then, just occasionally, you get spacey up there. You get back here, and you say, "Why did I do that?" when you forget to do something or you do it the wrong way. And then I have found that I get incredibly hungry on the mountain. I eat twice as much. So just eating and drinking a lot seems to help. (Kaplan, video interview with author, Hilo, Hawai'i, December 5, 2006)

Kaplan's altitude symptoms didn't keep him off the mountain:

> It is pretty; it is amazing. On a clear day, you will see the whole chain of islands. . . . It is just amazing. I was driving up there every week. I like doing it once a week a lot. It is like taking a little vacation to the moon.

Kaplan's talents attracted a new opportunity, and during his last trip to the mountain on July 29, 2008, he signed the *Visitors* book and added, "Signing out! It's been a pleasure to work for MLO since 2002. Good luck in the future and stay out of the lava zone!"

Aidan Colton, a Hilo native with an earth science degree from Boston College, joined MLO through JIMAR on December 1, 2004. He was hired to work on the EPA mercury project that he would eventually take over from

Alan Yoshinaga. One of his first assignments was to learn the operation of the EPA's ambient ion monitor instrument from Matt Landis of the EPA. He also assisted Landis in constructing a deck at the Hakalau Forest Reserve for an EPA–University of Michigan precipitation collector. Colton was assigned mountain duty twice a week, and he occasionally experienced the altitude sickness reported by most of those who have worked at MLO. Nevertheless, Colton (video interview with author, Hilo, Hawai'i, December 2006) reports that he liked the opportunity to work both in the office and on the mountain, "which makes the job very pleasant."

In August 2005, Dr. Nimmi Sharma of Central Connecticut State University began a one-year sabbatical at MLO working with Dr. Barnes on his new CLidar project.

Preston Sato, a Hilo native with a business degree from Pacific University, joined MLO as a JIMAR worker in May 2006. Sato specialized in multimedia studies in college and, in addition to mountain duty, he was placed in charge of digitizing some two thousand images in the MLO photography files and assisting Kaplan and Darryl Kuniyuki in redesigning the MLO Web site. Sato also worked on video conferencing and was the videographer for numerous interviews of past and present MLO and MLO–connected scientists and support staff conducted during the preparation of this book and as part of the observatory's fiftieth anniversary. He also used video to document the observatory and the drive along the MLO road. Sato became very familiar with the observatory's historical images, and a unique experience for him was visiting some of the sites around Hilo and Mauna Loa that played a role in the observatory's history. It was "like I was there back in 1960. . . . I could see everything where it was" (Sato, video interview with author, Hilo, Hawai'i, December 5, 2006).

The CLidar Project (2005)

The brilliant green beam of the MLO lidar is easily captured on photographic film or a solid-state image sensor (see Plate 71), and MLO station chief John Barnes developed the CLidar project to look for the presence of aerosols near the ground by photographing the lidar beam against the night sky (Barnes et al. 2007). Barnes (video interview with author, Hilo, Hawai'i, December 2006) explains the operation of the system:

> This is a new technique that I have developed in the last few years. It has been really a fun project. Lidars themselves have trouble measuring near the ground because you get too close to your own telescope and things do not focus properly. So that has always been a problem. . . . If you are interested in boundary-layer pollution, there is a lot of the material near the ground.
>
> So I had an idea way back when I was at the University of Michigan. I was up on the roof, and I just opened up the hatch and the laser beam was propagating up into the sky. And I remember watching it from about a hundred meters away, and you could see the little bits of sparkle when you would get aerosols going through the beam. And I thought, well, if I can see that with my eye I should certainly be able to set up a digital camera and record that.
>
> And it took many years for me to finally try. A few years ago I took that up, and it has been a very interesting project—take a wide-angle picture of the whole laser beam and then analyze that to get information about particulate [matter].

The CLidar project is still underway at MLO, and there's more to its story than simply taking snapshots of the laser beam. After sunset and shortly before the sky becomes totally dark, Barnes hikes upslope to a point a hundred meters or so away from the observatory to activate the digital camera and computer. He then returns to the NDACC Building and conducts the regular lidar program. An hour or so before midnight, he puts on a parka, finds a flashlight, and returns to the CLidar camera to shut down the system before retiring for the night in one of the two tiny bedrooms in the facility. Then there's the processing of the data acquired in the images, which is in itself time consuming. Barnes, a stickler for accuracy, also checks his CLidar findings against measurements made by particle samplers at MLO.

An Unexpected Inspection by Hawai'i's Land Board (2005)

Over the years, MLO survived visits by various fire, electrical, and safety inspectors. One such visit occurred on July 22, 2005, when a NOAA Environmental Compliance and Safety Assessment System (NECSES) team visited to inspect the facilities. A very different kind of inspection team arrived on September 21, 2004, when Hawai'i's Department of Land and Natural Resources (DLNR) made the first of two visits to MLO. As MLO's fifth decade drew to an end, its buildings, facilities, and instrumentation were all in good condition (Plates 81and 82). But the DLNR was drawing up a lengthy report that claimed some of the structures on the mountain lacked proper permits, even though in 1956 the Hawaiian territorial government had assigned the "control and management" of the site to the Department of Commerce. MLO faced a new crisis, and the outcome was anything but certain.

References

Barnes, John. 1998. November 1998 activities report. Mauna Loa Observatory, Monthly Reports (CD-ROM), Hilo, Hawai'i.

———. 2001. May 2001 activities report. Mauna Loa Observatory, Monthly Reports (CD-ROM), Hilo, Hawai'i.

———. 2002. December 2002 activities report. Mauna Loa Observatory, Monthly Reports (CD-ROM), Hilo, Hawai'i.

———. 2003. October 2003 activities report. Mauna Loa Observatory, Monthly Reports (CD-ROM), Hilo, Hawai'i.

Barnes, J. E., N. C. P. Sharma, and T. B. Kaplan. 2007. Atmospheric aerosol profiling with a bistatic imaging lidar system. *Applied Optics* 46: 2922–2929.

Bigelow, D. S., J. R. Slusser, A. F. Beaubien, and J. H. Gibson. 1998. The USDA Ultraviolet Radiation Monitoring Program. *Bulletin of the American Meteorological Society* 79: 601–615. (Accessed at http://ams.allenpress.com/archive/1520-0477/79/4/pdf/i1520-0477-79-4-601.pdf.)

Brooks, David R., and Forrest M. Mims III. 2001. Development of an inexpensive handheld LED–based sun photometer for the GLOBE program. *Journal of Geophysical Research* 106: 4733–4740.

Brooks, David R., Forrest M. Mims III, and Richard Roettger. 2007. Inexpensive near-IR sun photometer for measuring total column water vapor. *Journal of Atmospheric and Oceanic Technology* 24: 1268–1276. (Accessed at http://ams.allenpress.com/perlserv/?request=get-abstract&doi.)

CMDL. 2000–2001. Judy Pereira, Years of Service: 1966–1999. Climate Monitoring and Diagnostics Laboratory Annual Report 23.

———. 2000–2001. John Chin, Years of Service: 1960–1962; 1964–1966; 1968–2000. Climate Monitoring and Diagnostics Laboratory Annual Report 23.

Dean, Love. 1986. *The Lighthouses of Hawai'i.* Honolulu: University of Hawai'i Press. (Accessed at www.lighthousefriends.com/light.asp?ID=144.)

GAP. 2000. Monthly review: May 2000. Government Affairs Program (GAP), American Geological Institute. (Accessed at http://www.agiweb.org/gapac/review0500.html.)

Hamilton, Kevin, Steven C. Ryan, and Wataru Ohfuchi. 2008. Topographic effects on the solar semidiurnal surface tide simulated in a very fine resolution general circulation model. *Journal of Geophysical Research* 113: D17114, doi:10.1029/2008JD010115.

Hegger, Dawn T. 2006. Enforcement file no. ENF HA-06-40. State of Hawai'i Department of Land and Natural Resources, March 10.

Honolulu Star-Bulletin. 2004. Obituary: Nobuichi Yamanouchi. October 28. (Accessed at http://starbulletin.com/2004/10/28/news/obits.html.)

Kakugawa, Frances H. 1999. The 1960 Kapoho eruption of Kilauea Volcano, Hawaii. Hawaiian Volcano Observatory, USGS. (Accessed at http://wwwhvo.wr.usgs.gov/kilauea/history/1960Jan13/main.html.) See final map, Kapoho map after 1960 eruption. (Accessed at http://hvo.wr.usgs.gov/kilauea/history/1960Jan13/kapoho7.html.)

Keeling, Charles D. 1998. Rewards and penalties of monitoring the Earth. *Annual Review of Energy and the Environment* 23: 25–82. (Accessed at scrippsco2.ucsd .edu/publications/keeling_autobiography.pdf.)

Keeling, Ralph. 1993. What measurements of atmospheric oxygen can tell us about the global carbon cycle. *Global Biogeochemical Cycles* 7: 37–67.

———. 1996. Global and hemispheric CO_2 sinks deduced from changes in atmospheric O_2 concentration. *Nature* 381: 218–221.

———. 2004. Measurement of changes in atmospheric Ar/N2 ratio using a rapid-switching, single-capillary mass spectrometer system. *Tellus* 56B: 322–338.

Machta, L., and E. Hughes. 1970. Atmospheric oxygen in 1967 to 1970. *Science* 26 (168): 1582–1584.

McPeters, Richard D., and Walter D. Komhyr. 1991. Long-term changes in the total ozone mapping spectrometer relative to world primary standard Dobson Spectrometer 83. *Journal of Geophysical Research* 96: D2, 2987–2993.

Mims, F. M. III. 1992a. How to measure the ozone layer. *Science Probe* 2 (November): 45–51. (Accessed at http://www.sas.org/tcs/weeklyIssues_2007/2007-02-02/project1/index.html.)

———. 1992b. Sun photometer with light-emitting diodes as spectrally selective detectors. *Applied Optics* 31, 33: 6965–6967.

———. 1993. Satellite monitoring error. *Nature* 361: 505.

———. 1999. An international haze-monitoring network for students. *Bulletin of the American Meteorological Society* 80: 1421–1431 (cover article).

———. 2002. An inexpensive and stable LED sun photometer for measuring the water vapor column over south Texas from 1990 to 2001. *Geophysical Research Letters* 29: 20-1 to 20-4.

Mims, F. M. III, and John E. Frederick. 1994. Cumulus clouds and UV-B. *Nature* 371: 291.

NOAA/ESRL. 2005. NOAA/ESRL Global Monitoring Division. Earth System Research Laboratory. (Accessed at http://www.esrl.noaa.gov/gmd/index.html.)

Pales, Jack C., and Charles D. Keeling. 1965. The concentration of atmospheric carbon dioxide in Hawai'i. *Journal of Geophysical Research* 70: 6053–6076.

Pereira, Judy. 1996. Mauna Loa Observatory: A 30-year personal overview. (Accessed at www.mlo.noaa.gov/pdf/memorabilia/JUDY%20BKLT.doc.)

Ryan, S. 2000. The distribution of volcanic aerosols in Hawaii from VOGNET. Proceedings of the Geoscience and Remote Sensing Symposium (IGARSS 2000). *IEEE 2000 International* 4: 1652–1654.

Schnell, Russell C. 1993. June 1993 activities report, July 20. In *Geophysical Monitoring for Climatic Change*, Mauna Loa Observatory, Monthly Reports, Hilo, Hawai'i, 1993.

———. 1996a. December 1995 activities report. Mauna Loa Observatory, Monthly Reports (CD-ROM), Hilo, Hawai'i.

———. 1996b. January 1996 activities report. Mauna Loa Observatory, Monthly Reports (CD-ROM), Hilo, Hawai'i.

———. 1996c. March 1996 activities report. Mauna Loa Observatory, Monthly Reports (CD-ROM), Hilo, Hawai'i.

———. 1996d. April 1996 activities report. Mauna Loa Observatory, Monthly Reports (CD-ROM), Hilo, Hawai'i.

———. 1996e. May 1996 activities report. Mauna Loa Observatory, Monthly Reports (CD-ROM), Hilo, Hawai'i.

———. 1996f. June 1996 activities report. Mauna Loa Observatory, Monthly Reports (CD-ROM), Hilo, Hawai'i.

———. 1996g. August 1996 activities report. Mauna Loa Observatory, Monthly Reports (CD-ROM), Hilo, Hawai'i.

———. 1996h. November 1996 activities report. Mauna Loa Observatory, Monthly Reports (CD-ROM), Hilo, Hawai'i.

———. 1997a. December 1996 activities report. Mauna Loa Observatory, Monthly Reports (CD-ROM), Hilo, Hawai'i.

———. 1997b. January 1997 activities report. Mauna Loa Observatory, Monthly Reports (CD-ROM), Hilo, Hawai'i.

———. 1997c. March 1997 activities report. Mauna Loa Observatory, Monthly Reports (CD-ROM), Hilo, Hawai'i.

———. 1997d. August 1997 activities report. Mauna Loa Observatory, Monthly Reports (CD-ROM), Hilo, Hawai'i.

———. 1998. June 1998 activities report. Mauna Loa Observatory, Monthly Reports (CD-ROM), Hilo, Hawai'i.

Schnell, Russell C., and MLO Staff. 1995. Mauna Loa Observatory. In CMDL Annual Report 23, Boulder, Colorado. (Accessed at http://www.esrl.noaa.gov/gmd/publications/annrpt23/chapter1_1.htm.)

Schnell, R., J. Barnes, J. Chin, D. Kuniyuki, L. Pajo, J. Pereira, S. Ryan, B. Uchida, and A. Yoshinaga. 1997. Mauna Loa Observatory. In CMDL Annual Report 24, 1996–1997. NOAA Environmental Research Laboratories, National Technical Information Service, Springfield, VA.

Scripps. 2005. Obituary notice, climate science pioneer: Charles David Keeling, 1928–2005. Scripps Institution of Oceanography, June 22. (Accessed at http://scrippsnews.ucsd.edu/Releases/?releaseID=687.)

Slusser, James, John Davis, William Durham, George Janson, and Roger Tree. 2007. The UV-B Monitoring and Research Program. Unpublished manuscript submitted to author.

CHAPTER FOURTEEN

The Mauna Loa Observatory Today

When its fiftieth anniversary year arrived (Plate 83), the Mauna Loa Observatory had evolved from a small building and a dozen or so instruments into a little village of buildings, domes, and shelters that support a vast array of sophisticated atmospheric monitoring instruments. The contrast is clearly shown in Fig. 14.1, which compares a photograph of the original building taken by Ralph Stair in 1956 and the view of the building today. Today's MLO staff has a much easier schedule than that required of the old-timers, who coped with serious budget cuts, a sometimes impossible road, a pair of noisy diesel generators, and shifts of four or more days at the mountain station, which they manned every day of the year. And their experience was considerably less rigorous than that endured by the US Exploring Expedition during their three-week stay on the summit during the blustery winter of 1840–1841.

Sometimes the present MLO staff experiences a taste of what life on the mountain was like for their predecessors, as when a blizzard closes the road or lightning shuts down the power and destroys delicate electronic circuits. While the mountain crew may sometimes leave the station when a lightning storm approaches, the opposite applies when snow arrives. As Darryl Kuniyuki (video interview with author, Hilo, Hawai'i, December 5, 2006) explains, "Even if there are warnings, we go up to see if we can make it to the observatory." Kuniyuki speaks from personal experience:

> The worst snowstorm I ever saw at Mauna Loa was in 1990, July [see Plate 47]. That was when the huge snowstorm came in and the observatory was inaccessible for about a week. Steve [Ryan] and I were able to get about two miles below the observatory, and we hiked through the snow up to the observatory. If you stayed on the road, I think it was up to above your knees. But you could not tell where the road was . . . so we kind of tried to estimate where the middle of the road was and stayed on track. . . . I would lose my breath every once in a while and have to stop and rest. . . . It was the first time I ever hiked in snow that deep, so it was kind of exciting for me, actually, even if I was out of breath.

Station Chief John Barnes (video interview with author, Hilo, Hawai'i, December 2006) recalls an unexpected side of this policy that occurred during what began as a crystal-clear night at the observatory:

> The next morning I got up and there were five inches of snow on the ground. I got a call from the guys coming

FIGURE 14.1. The original building (now the Keeling Building) in 1956 *(top)* is now surrounded by a cluster of newer structures. *(Top: Stair family archive; used with permission.)*

> up, and they could not. They were about five miles down the road, and they could not get any farther. [The snow] filled in all the little ups and downs, and I was a little worried about driving down the road and not being able to tell exactly where the road was. Also, slipping off the road [was a possibility, and] it is not that hard to get stuck in lava. You do not get very good traction, but I had a little pickup, and so I piled snow in the back for weight. Because originally I'm from Minnesota, I have driven in snow a lot. In Hawaii, people drive up, fill up their vehicle with snow and bring it back down so the kids can play with it and build a snowman in your front yard. That is what I did with my kids . . . they had a great time, barefoot and shorts on and playing in the snow.

While Mauna Loa old-timers would not be surprised by the vagaries of the weather at the station where they once worked, those who visit MLO after an absence of a few decades or more are surprised by the growth of the observatory and the expansion of its programs. The one-building station has become a little science village perched high on the north slope of the mountain, and the original main building, now known as the Keeling Building, is part of a cluster of a dozen or so structures and domes dominated by the NDACC Building.

Still, Jack Pales and his original team would be very much at home with MLO's best-known monitoring projects, for half a century ago their colleague Dominic Conte of the Weather Bureau in Honolulu installed the first meteorology equipment when he was at MLO from December 4 to 8, 1956. During the summer and fall of 1957, Pales, Williams, and Kutaka added instruments to measure surface ozone, atmospheric transmission, and the ozone layer. In March 1958, Charles Keeling's infrared gas analyzer that continuously monitored carbon dioxide was added to MLO's list of major International Geophysical Year monitoring projects. The CO_2 project was the first in a long series of cooperative projects in which the MLO staff assisted in the operation and maintenance of monitoring instruments installed by scientists from academia, various US government agencies, and foreign organizations. For decades, the CO_2 instrument was among the oldest monitoring devices at MLO, and its days were numbered as the observatory's fiftieth anniversary approached. A major transition was also in store for MLO's home office in Hilo's Federal Building.

Farewell to the Federal Building (2006)

In July 1976, the Mauna Loa Observatory office in Hilo moved from the Cloud Physics Observatory at the University of Hawai'i to the second floor of the Federal Building. While the building was distinguished in appearance, the MLO offices were cramped and stuffed with file cabinets, desks, and tables. The electronics workshop in the basement was habitable, but the adjacent records storage area was humid and unpleasant. As reported in chapter 12, MLO's vehicles that were parked overnight at the Federal Building were subject to vandalism, and there were reports about drug trafficking in the area and even in the building itself. In the fall of 1995, Russell Schnell made plans to save vital records that were deteriorating from their twenty-year stay in the basement:

> Prior to . . . Christmas, MLO staff will be packing and shipping archives of old data (presently stored in the basement) to various CMDL groups and SIO [Scripps Institution of Oceanography]. These records are deteriorating from the high humidity, termites and ants that are a constant presence in Hilo. (Schnell 1995)

In March 2004, CMDL director David Hofmann inspected potential new office sites for the MLO offices. Finally, after thirty years in the Federal Building, the MLO office was moved to the newly constructed Kīlauea Financial Plaza at 1473 Kīlauea Avenue in January 2006. The new MLO office is either on or very near the route taken by the US Exploring Expedition in December 1840 on the way to its historic ascent of Mauna Loa and the establishment of Pendulum Peak, the first Mauna Loa Observatory. Charles Wilkes and a detachment from his expedition, accompanied by more than two hundred Hawaiians, marched from Observatory 1, their temporary meteorological station located near the mouth of the Wailoa River just west of Lihiwai Road, toward Kīlauea along the old volcano trail.

Darryl Kuniyuki planned the layout for the new office, lab, and shop facilities. During November 2005, the MLO staff installed more than 20,000 feet of network and telephone cable, made more than 3,700 wire connections, and installed more than 230 connection jacks (Kuniyuki, personal communication with author, June 9, 2008). The chemistry laboratory was moved to the new building in December. On January 10, 2006, regular work came to a halt when all the Hilo computers were taken off-line in preparation for the move. The office furniture, file cabinets, and some of the shop equipment were moved to the new building by a commercial mover on January 11–13. Because the elevator in the Federal Building was damaged by the movers, the remaining shop equipment was not moved until August 8, 2006.

MLO's new offices are a dramatic improvement over the Federal Building. There is a reception area with a couch and displays of many of the items from the former Mauna Loa Museum. Posters illustrate some of the measurements made at MLO. At one end of the reception area is the secretarial office space. At the other is a meeting room with a kitchen area where the staff has lunch each workday. The office of MLO station chief John Barnes is adjacent to the reception area. The MLO workshop and shipping and receiving area are located down the hall from the reception room. A large door facilitates the arrival of equipment. Additional offices and the chemistry lab are on the second floor. There is ample parking space for MLO vehicles and the personal vehicles of staff and visitors. The University of Hawai'i at Hilo is only blocks away. In short, the new office, lab, and shop facilities are the best the observatory has had since its founding.

Retiring the Original Scripps CO_2 Analyzer (1958–2006)

In 1958, the Applied Physics Corporation infrared gas analyzer that Charles Keeling selected to monitor carbon dioxide at MLO was a state-of-the-art instrument. But as Dr. David Hofmann asked in his letter to John Chin on the occasion of Chin's retirement in 2000,

> Can you imagine watching television on a 1950's era TV? . . . How many such TVs are in operation today? I would venture a guess that they could only be found in museums. Yet, that is exactly what John Chin has been working with these past 40 years—a 1950s vintage, vacuum tube operated infrared absorption carbon dioxide analyzer. And it still works!

And it worked so well that when the data from NOAA's series of CO_2 analyzers, the first of which was placed in service in 1974, were superimposed on a graph of the old Scripps analyzer's data and viewed on a computer monitor, there was no visually noticeable significant difference.

The graph of the fifty-year MLO carbon dioxide record has two names. The most widely used is the "Keeling curve," but NOAA prefers the "Mauna Loa CO_2 record." The former is a graph of only the Keeling data, while the latter depicts the Keeling data from 1958 to 1974 and the NOAA data thereafter. Ralph Keeling has suggested that Scripps and NOAA agree to a graph that depicts the data from both organizations on the same axis, an appropriate idea in view of the increasing attention the chart has received. (One version of a combined graph is shown in Plate 79.)

In view of the age of the Scripps analyzer and its sensitivity to temperature, the agreement of the Scripps and NOAA data is extraordinary, especially in an era of debate over how best to correct temperature records around the world for biases, errors, and instrument changes. The MLO carbon dioxide record is a testimony to Charles Keeling's stubborn insistence that his measurement methods be followed to the letter. As former MLO director Russell Schnell (video interview with author, Boulder, Colorado, April 27, 2007) observes,

> Keeling was not really a pain, but he had a very particular way of doing things, and his analyzer that he brought up to Hawaii in [1958] is the same analyzer that was there

> in 2006. . . . It was very clear when I came there that the system should have been replaced, because there were much better ways of doing it. But, to Dave's credit, he said, "I have made this measurement with this instrument since I started. I'm going to keep doing it, even though there are better ways."

While Keeling's old analyzer worked, it did so only with tender loving care. If a visiting scientist wanted to know how it was possible to keep the aging instrument working properly, the MLO staff on the mountain wasn't shy about discussing the latest breakdown or joking about the instrument's tendency to become a thermometer when its temperature changed.

For years the monthly Activities Report described in detail specific problems with the old analyzer. But beginning with his 2002 Activities Reports, Station Chief John Barnes generally limited his comments about the status of the instrument to "Normal operation during the month, which includes frequent changing of vacuum tubes." Some of the many problems with the analyzer have been described in previous chapters, and it is sufficient to observe here that the MLO staff was not fond of having to spend valuable time on the mountain maintaining an ancient, temperature-sensitive, vacuum-tube instrument in the integrated circuit era, particularly when the modern NOAA analyzer was much less subject to breakdowns and temperature drift.

Charles Keeling died on March 20, 2005, and as the fiftieth anniversary of the dedication of MLO neared, the retirement of his old instrument was rapidly approaching. Scripps installed a modern carbon dioxide analyzer in the Keeling Building during 2005, and the MLO staff knew that the cantankerous old instrument Jack Pales had hauled up the mountain in 1958 would be shut down after a period of comparison with its replacement. Finally, on Tuesday, February 7, 2006, Scripps notified MLO that, after forty-eight years of nearly continuous operation, it was time to shut down one of the world's most famous environmental monitoring instruments. Station Chief John Barnes, a man who chooses his words carefully when he devotes a sentence or two in the monthly Activities Report to the retirement of a staff member, devoted an entire paragraph in the February 2006 Activities Report to the retirement of the historic instrument:

> The old Applied Physics vacuum tube analyzer that had run at MLO since Spring of 1958 was turned off on 02/07 on orders from Dave Moss. Their Ultramat 6 analyzer, which has been running without problems for almost a year, will continue the Scripps record with . . . much less maintenance and much lower operating costs (reduced calibration gas consumption, reduced electrical consumption, no need for a temperature stabilized cabinet environment, fewer man hours). Dave Moss was at MLO the week of 02/21 to unhook the old analyzer and switch the system to exclusive use of the Ultramat 6. (Barnes 2006a)

Dave Moss was in Hilo and at MLO from February 20 to March 1 to disconnect the APC analyzer. Ralph Keeling was also at MLO when his father's famous instrument was being retired. He visited from February 21 to 23 to install a new air sampling line to the Scripps oxygen monitoring apparatus. Also at MLO during the transition to the new Scripps analyzer were Richard W. Spinrad, assistant administrator for NOAA's Office of Oceanic and Atmospheric Research (OAR), and David Hofmann, director of the Global Monitoring Division.

The March 2006 MLO Activities Report section on the Scripps CO_2 program reports nothing negative for the first time in years. The report simply reads, "Normal operation during the month." So long as Scripps continues to find funds to continue its work at MLO, the NOAA and Scripps CO_2 programs will continue their parallel measurements for the foreseeable future, for, as former MLO director Russell Schnell (personal communication of August 2007) has written,

> The NOAA and Scripps carbon dioxide analyzers operating together at Mauna Loa provide important redundancy of measurements, which is crucial in maintaining the integrity of the Mauna Loa carbon dioxide record.

The Department of Land and Natural Resources (2006)

On May 19, 2005, a team of six burly contract workers was installing a concrete slab for a new metal building just east of the old AEC Building. That morning a young woman arrived with the men and stood by watching as they methodically went about their work pouring concrete between the wood forms they had constructed. Periodically she looked at papers on a clipboard and took photographs of the crew at work. She walked slowly and deliberately and eventually found a place to wait while the work continued.

The next day the woman did not arrive with the work crew, and one of the workers explained why. She was their supervisor, and she was concerned about the fact that the men required more time to do their work on the mountain than down at sea level. As they were leaving for Hilo the previous day, she developed a bad case of altitude sickness, and they had to twice stop their vehicle when she became ill. This story had a happy ending, for the supervisor learned firsthand why the men were working more slowly at MLO than at sea level. She was so pleased that they were able to work on the mountain that she treated them to dinner.

Another story related to the inspection of the many construction projects on the mountain had a much more uncertain outcome. The government of Hawai'i has traditionally been very protective of its land and natural resources, and the US National Park service has followed suit. As discussed in chapters 4 and 5, MLO founder Robert Simpson met with strong opposition from the National Park Service in 1951 when he first attempted to gain permission to build the original summit observatory. When he was planning the present slope observatory, Ralph Stair bypassed the Park Service and went straight to Hawai'i's territorial governor, Samuel Wilder King, who in 1956 assigned four acres of land for the project.

In 1992, MLO director Russell Schnell and others worked closely with Hawai'i's Department of Land and Natural Resources (DLNR) to acquire an assignment of a second four-acre plot of land adjacent to and east of the original four acres. The new land was intended for the planned NDSC (now NDACC) Building, and during the second week of May 1992, three representatives from the DNLR made a surprise visit to MLO to inspect its facilities. There is no evidence that the DLNR was unfriendly or concerned by the status of the facilities and work at MLO during this time. Indeed, the DLNR agreed to MLO's land request, and the land was conditionally assigned to MLO by Governor Ben Cayetano on March 7, 1995. However, the situation changed dramatically between the 1992 DLNR visit and September 21, 2004, when another DLNR team made an unannounced inspection of MLO and surprised the day crew by announcing that the observatory lacked the required land use permits to enable it to carry out most of its mission. On seeing a team of contract workers installing a new gate across the road to the observatory near the public parking area, one of the inspectors shouted, "Cease and desist!" According to the MLO staff who were present, she threatened to fine the contractor $5,000 a day if the workers did not stop work. The warnings were unnecessary, for the gate installation had already received official approval.

The DLNR team photographed many of the MLO facilities during the 2004 visit. They returned on November 21, 2005, and took additional photographs. They also requested details about the nature and history of most of the facilities. During the visit, MLO's Bob Uchida informed the DLNR team about the Executive Order of 1956 that provided the original four-acre site to the Weather Bureau. Uchida also noted that the order was signed by territorial governor Samuel Wilder King before Hawai'i became a state and before the creation of the DLNR.

On March 10, 2006, the DLNR issued a report designated as an enforcement file and entitled "Regarding Alleged Unauthorized Land Uses Found at the Mauna Loa Observatory" by the US Department of Commerce through the Global Monitoring Division (GMD) of NOAA's Earth System Research Laboratory (Hegger 2006). The report begins with an introduction to the history and purpose of MLO and then describes the two visits by the DLNR site inspection team. While the report notes that "various facilities were under construction" during the initial visit in 2004, there is no mention that construction workers were ordered to cease work and that fines were threatened. The report continues with a detailed list provided by the observatory of the facilities at MLO. Various land use laws and rules are then cited and applied specifically to a list of "unauthorized land uses" that had occurred since 1964. The report concludes by recommending a fine of $26,000 for thirty-three alleged land use violations.

The DNLR considered twenty-two of the violations minor, including a wind sock, a National Weather Service thermometer base, two rain gauges, and a bare patch of lava devoid of human activity called the "All Sky Area." Eleven violations were deemed major, including various buildings and the forty-meter walk-up tower. The NDACC Building and its solar deck received separate fines. Even the original 1956 Keeling Building was fined. The report is accompanied by maps, photographs, and correspondence between Samuel J. Lemmo of the DLNR and Russell Schnell, director of NOAA Observatories and Global Network Operations and a former MLO director. The report, which appears to have required considerable preparation time by all concerned, has historical value, for it provides a history of the installation dates of many of today's structures at MLO.

The enforcement file against MLO was presented to a meeting of the Board of Land and Natural Resources on March 10, 2006, and a motion to withdraw the enforcement action was quickly made and unanimously approved (BLNR 2006). In response to questions about the matter, Samuel Lemmo explained in an e-mail that his understanding of the board's action was that NOAA is protected by sovereign immunity and not subject to the laws of Hawai'i. He also wrote that the traditional position of the DLNR is that permits are required for any use of state lands and that was the reason for the filing of the action.

Like the story about the contract supervisor who learned firsthand why her men worked somewhat more slowly on the mountain than at sea level, this story also had a happy ending, for in an e-mail dated May 23, 2007, Lemmo made it clear that he was not opposed to the scientific monitoring being conducted at MLO. When asked if a portion of his e-mail could be included in this book, Lemmo readily agreed:

> One of our greatest gifts is our ability to observe the universe because of our particular atmospheric structure and our location in the Galaxy. It would be tragic if this gift were ruined because of pollution.

Bob Uchida Retires (2006)

Electronics technician Bob Uchida (Plate 84) retired from MLO on January 3, 2006, after working thirty-two years for NOAA. Uchida first worked for the National Weather Service and had been with MLO since 1991. His MLO connections went back much farther, for in June 1965 he befriended Howard Ellis and other MLO old-timers when he worked on a cloud physics project on the mountain when the MLO office in Hilo was still a post office box. While Uchida was a skilled electronics technician, he also had considerable experience in construction and had even built the Cloud Physics Observatory building where MLO was headquartered from 1966 until the move to the Federal Building. Among Uchida's major assignments at MLO was assisting with the planning for the NDACC Building. When asked how he learned construction skills, Uchida (video interview with author, Hilo, Hawai'i, November 30, 2006) dryly replies,

> It does not take too much imagination to pound a nail or cut a piece of wood. You get a hammer and a nail, and you can hammer and nail. You got a saw, you can cut.

Like Howard Ellis, Steve Ryan, and other old-timers who climbed the towers and did physical labor on the mountain, Uchida was unfazed by the altitude. Several years ago when he was using a sledgehammer to dismantle the upper portion of the old wood tower adjacent to the Keeling Building in preparation for the EPA mercury monitoring project, he worked as if he were a young man down at sea level. When asked how he managed to do hard work on the mountain, Uchida shrugs and replies, "It does not bother me. Altitude does not bother me."

Former MLO director Russell Schnell (video interview with author, Boulder, Colorado, April 27, 2007) sums up Uchida's contribution to MLO:

> He could weld, do electricity. He could do science, and he had five or six brothers and sisters in Hilo, so if you needed a welder, and he could not do it . . . maybe on the weekend this would get done, and it got done.

Everyone who drives the Mauna Loa road at night, especially during foggy conditions, can thank Bob Uchida for helping guide the way with a squiggly line of white paint along the center of the road. The guideline sprang from his assignment to operate the University of Denver solar interferometer, a task that required arriving at MLO well before sunrise to switch on the equipment and allow it to warm.

> So for me to get everything set up, I had to leave town about 3:00 a.m. in the morning and drive up. There were so many days that the road would be all foggy where you had to stop . . . because there is no contrast between the road and the terrain. You had to stop and really get reoriented again, reorient yourself, and then continue on. And that happened to me so many times that I decided, well, I needed something to help me. So, I put that white line, that squiggly white line on the road. What I did was get a five-gallon bucket and then put little holes on that thing and poured the paint on and, with the hose, just drip the paint. . . . Five gallons only went for about a quarter-mile. (Uchida, video interview with author, Hilo, Hawai'i, November 30, 2006)

Although officially retired, Uchida's skills and experience are particularly important at MLO, where he has been hired to work at the mountain station two days a week. Why did he return to work? He answered in the same interview:

> You really want to know? Well, it gets a little boring sitting around doing nothing. And for me, I'm happier when I'm busy; I like doing things. You have to be constructive and be productive. Otherwise, you might as well be dead.

The Fiftieth Anniversary of the Mauna Loa Observatory (2006)

The Mauna Loa Observatory celebrated the fiftieth anniversary of the dedication of its original building, now the Keeling Building, on June 28, 2006:

> The morning included an informal video conference session between Boulder and the Hilo office. We then had a visit from a pre-school which sang songs for those in Boulder. We then had various activities for the children followed by a hotdog lunch. In the afternoon, we had an open house for everyone associated with MLO, and many previous MLO staffers came. The day ended with an informal barbeque dinner for the MLO staff and their families. (Barnes 2006b)

MLO physicist Steve Ryan added an unplanned and ironic twist to the history of the moment with the publication a week before the anniversary of a paper that used his measurements of methane at Mauna Loa to conclude that the volcanoes of Mars were not responsible for the methane in that planet's atmosphere (Ryan et al. 2006). The first scientific study conducted at MLO also involved Mars and was underway during the dedication of the original building in 1956. Carl and Hariett Kiess and Charles and Edith Corliss had set up housekeeping in the new building during their pioneering search for water vapor in the atmosphere of Mars a dozen years before the first temporary astronomical telescope was installed on the summit of Mauna Kea (Kiess et al. 1957). MLO director John Barnes led the June Activities Report with a note about Ryan's Mars paper:

> The paper used 13 years of continuous methane measurements to establish an upper limit of 9 metric tons per year of CH4 emissions from Mauna Loa volcano, and showed that volcanic outgassing on Mars is unlikely to be the source of the 10 ppb methane in that planet's atmosphere. (Barnes 2006b)

So it goes at MLO. Surely Ralph Stair and the Kiess-Corliss team would have felt completely at home if they had been present at the fiftieth anniversary activities, especially if they could have chatted with Steve Ryan about his Mars paper while visiting the Keeling Building, where they spent many nights while performing their pioneering research half a century ago. They might have also visited Professor Walter Steiger (Plate 85), who spoke at the dedication of MLO on June 28, 1956, and who held fond memories of his various trips to the remote mountain station. Steiger was professor emeritus of physics and astronomy at the University of Hawai'i at Mānoa, and though retired, he continued to work at the Caltech Submillimeter Observatory offices at the University of Hawai'i campus in Hilo.

Remembering Leslie Ann Pajo (1962–2006)

Judy Pereira had trained Leslie Pajo to take over her position as administrative secretary at MLO, and Pajo was fully prepared for the transition when Pereira retired in January 2000. John Chin, who had worked with Judy for thirty years, was pleased by her successor:

> Leslie is just like her, a very good organizer. So we hired Leslie, and she turned out to be, yes, she was just like another Judy. She even made all the contacts, and people who had problems just went to her. She is the one who set any appointments [for] visitors. . . . She took care of it. . . . We always thought that she was one of the sisters—our sisters. Actually, at Mauna Loa, we all operated just like a family. Yeah, everybody kind of knows each other well [and has] almost the same interests, and there is no in-house fight. (Chin, video interview with author, Hilo, Hawai'i, December 4, 2006)

Besides taking on administrative duties, Les Pajo was trained by Darryl Kuniyuki in various computer and networking responsibilities. In 2003 she and Kuniyuki received the NOAA Administrator's Award for the network, computer, and data development of the observatory (Kuniyuki, personal communication with author, June 9, 2008). She also compiled information and old newspaper clippings about the history of MLO that proved invaluable in the preparation of this book.

On October 5, 2004, a longtime visitor to MLO who was staying on the mountain during a ten-day instrument calibration exercise was surprised to see Leslie Pajo arrive with the daily work team for a day of duty on the mountain. She carried a thick three-ring binder with a white cover labeled *Leslie's Mountain Bible* that included details about the care and feeding of every instrument and project at MLO. After retrieving a clipboard from the Keeling Building, Pajo went about her duties, which included checking various solar instruments and measuring the ozone layer. This was nothing new, for Alan Yoshinaga (video interview with author, Hilo, Hawai'i, December 4, 2006) recalls how Pajo often worked at the station:

> She actually did observing. So she went up to the mountain numerous times. It is not as if she only did once or twice; it was numerous times. And then, also, when we had our cleanup days, she painted, she sanded, she scraped along with the rest of us, and then, on top of that, she made lunch for us.

On September 15, 2006, the MLO staff members who were not on the mountain were having lunch together in their new office in Hilo. Pajo, who was seated near one end of the table next to Yoshinaga, suddenly collapsed, and Yoshinaga caught her before she fell to the floor. Pajo was evacuated to the Straub Hospital on O'ahu, where she died three days later from a brain aneurysm. She was only forty-three. Leslie Ann Pajo, who was laid to rest wearing her Mauna Loa Observatory shirt, is survived by her husband, Patrick C. Pajo, and their daughters Jessica and Carissa. John Barnes wrote in the September 2006 Activities Report, "She was an extraordinary co-worker and a wonderful person, and will be greatly missed at MLO." A few months later, Barnes (video interview with author, Hilo, Hawai'i, December 2006) reflected back on Leslie Pajo's role at MLO:

> Leslie really made it fun to work at Mauna Loa. She was very energetic. Sometimes we do programs for visiting

> schoolchildren, and she was always into that. She had great ideas of how to create little demonstrations for the kids and what they would like to do. And she used to bake things, bring them in to eat. . . . A few times we went over to her house for karaoke singing and she liked that. She was just very personable and very, very capable. She kept really good records. She did many jobs, from personnel activities and purchasing things to arranging travel. And I think she had a degree in accounting, so she would make sure that the bills were being paid.

Several months after she died, Bob Uchida and Alan Yoshinaga reflected back on Leslie Pajo's time at MLO:

> Leslie was, I guess, one of a kind. She was very dedicated. I really have not seen too many people like her that would really put in extra time and be helpful all the time, never get upset. [Her death] was really a big shock to me. . . . She was special. (Uchida, video interview with author, Hilo, Hawai'i, November 30, 2006)

> She was something else. When she—I have to laugh—because when she first came to work . . . Darryl said, "Hey, Alan. This girl does not even know what a [computer] disk is!" . . . And first thing she did [was to buy] a computer, and she learned. And then back in those days, we did our own data reduction. We would pull the data from the computers and plot them. She came up to me after a while and said, "Hey, I can make this [chart] nicer, Alan." And doggone it, she did it better! And she did more than she [had to]. She was too good; she was too nice. After a while, she organized our little luncheons, our little gatherings, our cleanups at the mountain. If people did not . . . want to chip in and do some work, she was always here to egg us on and do it in such a way that you just felt that you were a part of the group. She was something else. She was—what can I say? I'll say she became the spirit of Mauna Loa. (Yoshinaga, video interview with author, Hilo, Hawai'i, December 4, 2006)

New employees at MLO use an updated version of *Leslie's Mountain Bible* when they are working on the mountain. Leslie Ann Pajo is also remembered by the many visiting scientists whom she assisted during her time at MLO. When this book was being finalized in spring 2011, they still heard her pleasant "Aloha!" greeting when they called the MLO switchboard.

Debra Hansen Fills in for Leslie Pajo (2006)

Leslie Pajo's role at MLO became more apparent than before after she was suddenly gone, and, back in Boulder, Debra Hansen quickly filled the vacuum. A US Navy veteran (1983–1991) who had worked in personnel assignments, Hansen began working for the Climate and Monitoring Diagnostic Laboratory in February 1993. In 2001, Hansen's duties were expanded to include administrative support for the Observatory Operations group of the Global Monitoring Division, which is when she began communicating by e-mail and telephone with Pajo on a weekly basis. She also trained Pajo in some new systems used for purchasing, timekeeping, and travel. After Pajo died, the MLO staff shipped the accumulated paperwork that she would have processed to Hansen in Boulder. This added to Hansen's workload, but she said she was glad to help. Hansen (video interview with author, Boulder, Colorado, May 2, 2007) remembers Leslie as "one of the most congenial people that I know, and just very easy to work with, and I just really miss working with her."

Fire, Smoke, and an Earthquake (2006)

Mauna Loa has not erupted since 1984, and the US Geological Survey maintains a close watch on the mountain and its slow, on-and-off inflation. Some of their instruments are installed at the new four-acre site at MLO adjacent to a flat area of lava designed for emergency use by helicopters. The question isn't if the great mountain will erupt again, for it is assumed that it will. The questions are when will it erupt and in which direction will the lava flow? While Mauna Loa remains quiet, the Kīlauea eruption has continued uninterrupted since 1983. Sulfur dioxide emitted from the Kīlauea lava occasionally reaches MLO, as it did in May 2009 during an annual calibration visit by the author.

Other natural events have also continued to influence MLO. On October 4, 2006, smoke from a wildfire on the Kona side of the island reached MLO. Besides influencing measurements made by the ultrasensitive instruments at MLO, the smoke could be seen and smelled (Barnes 2006c).

On October 15, 2006, a magnitude 6.7 earthquake occurred off the northwest coast of Hawai'i island and caused property damage in some areas. While the upper rail on the platform atop the old AEC Building was shifted about 6 inches north, the delicate GMD lidar system in the NDACC Building was undamaged and required no realignment of the laser and its receiving mirror. A power outage caused by the earthquake resulted in problems with some of the equipment.

These natural events had little impact on the daily operations at MLO. A new eruption of Mauna Loa, however, would be a very major development, particularly if lava cuts the power line, as it did in 1984, and especially if it blocks the road. Should lava flow toward the observatory itself, the lava diversion barrier will finally be tested. It is a test that everyone connected with MLO hopes will never occur.

Improving the Mauna Loa Road (2007)

The condition of the Mauna Loa road deteriorated significantly during the late 1990s. By 2006, portions of the first 4 miles (6.4 km) of the lower section of the road between Saddle Road and the Road to Nowhere were rapidly returning to lava clinkers as potholes expanded completely across the pavement. MLO station chief John Barnes, in an interview with the author in Hilo in December 2006, was asked about the poor condition of the road at that time:

> Right now the road is, in my time here, it is the worst that it has ever been. . . . We found out they were going to be repairing the Saddle Road and it was a large highway project. So at the groundbreaking, Russ and I got one of the highway people to come visit the observatory. So we bounced them up potholes in the road to make an impression. When they do these large highway projects, there is often a couple of million on the side that they can use for certain projects that involve federal sites like we have. So they committed to giving us some money to fix part of it.

In 2006, Paul Fukumura-Sawada and David Nardini of the MLO staff surveyed the MLO access road. They used a GPS receiver to accurately plot the location of every pothole and the various blind curves and hills on the road to assist in planning the first major resurfacing project in decades (Darryl Kuniyuki, personal communication with author, June 9, 2008). During 2007, the first 4 miles (6.4 km) mentioned above were given a thick new layer of asphalt that should last for a few decades. White stripes along the edges of the pavement made this section of the road much safer to drive at night and when the lower slope of the mountain is blanketed in thick fog. The downside to the newly paved section is the temptation it presents, for now drivers can travel the lower portion of the road at higher speeds than before. This can prove quite dangerous, for the road is still only a single lane wide, and several dangerous, blind curves remain.

The Mauna Loa Education Center (2007)

In 2007 the artifacts of the former Mauna Loa Meteorological Museum (MLMM) founded by former MLO director Kinsell Coulson were returned to MLO by the Lyman Museum in Hilo. The artifacts include rare meteorological and solar instruments, books, and charts dating back to the early twentieth century. The Mauna Loa Education Center (MLOEC) was organized by Darryl Kuniyuki to showcase the artifacts at the MLO office in Hilo and to inform the public about the observatory and its science. The MLOEC was also made responsible for organizing Earth Day and other public science exhibits and for providing cash awards for outstanding Science Fair projects using funds from the Mauna Loa Observatory Employees Association.

Alan Yoshinaga Retires (2007)

Still another major change at MLO occurred with the retirement of chemist Alan Yoshinaga (Plate 84) on January 3, 2007, after over thirty-five years of service. Like many other MLO staff, Yoshinaga is from Hilo. He was hired by Dr. Rudolf Pueschel and joined MLO on June 12, 1972. To date he holds the record as the longest continuously serving member of the MLO staff. Shortly before his retirement, Yoshinaga recalled the camaraderie of the staff at MLO that has long characterized the observatory (this and following excerpts are from a video interview with the author in Hilo, Hawai'i, on December 4, 2006):

> I remember coming in 1972 and working with what I would like to call the "old bunch," the "old boys," and that would include Howard Ellis, Bernard Mendonca, Barry Bodhaine, Johnny Chin, and Al Shibata, our late ET [Electronics Technician]. . . . Howard Ellis . . . started most of the solar radiation projects [and] the lidar projects. I really got along with him. He was a giant of a man, real mellow, real nice guy. He and Bernard Mendonca and Johnny Chin were my mentors. What I learned at the mountain, I learned from them. They were patient, persevering, and they also instilled in me . . . I guess, the inherent quality to do a good job and to do a good job not just for yourself, but for everyone at the observatory.
>
> Our projects were more or less cooperative among ourselves where if something went wrong we always tried to help each other out and get things going. It was not sort of a proprietary thing. So, that is what I really enjoyed about it at that time. If I needed help with any of my instruments, [Al Shibata] was really willing to jump in there and help. [They were] a good crew. I really enjoyed working with these guys, and I really appreciate them teaching me whatever I learned.

For a time Yoshinaga had to wake up at 2:30 a.m. each morning to drive up to MLO on his own or with Bob Uchida to set up the University of Denver's FTIR instrument so that it could begin tracking the sun at sunrise. He had no problem with the early hours:

> At that time we were still in that mode where you are doing an interesting job. You are doing something good for science. You are doing something good for the environment. It was still the adventure that the old-timers instilled in me.

As his retirement approached, Yoshinaga worked closely with Aidan Colton in the MLO chemistry laboratories in the Hilo office and on the mountain to prepare him for taking over his responsibilities. He remembers the way he was received when he arrived at MLO, and he passed on the tradition:

> I was taken in by Howard [Ellis], by Rudi [Pueschel], by Barry [Bodhaine], by Bernard [Mendonca]. . . . They just took me under their wing, because I was a young kid when I came in. And then as I became part of the group, then they treated me as an equal, which was really neat. They did not look down upon me. They did not give me the menial jobs. Then as time progressed [they would ask], "Kid, do you want to try this? Do you want to take this job?" It was just cooperation between our little family or *ohana*, just something I had never experienced before. . . .
>
> [Now] I have this young kid, Aidan [Colton], working with me . . . on the EPA's mercury project, and it is an intensive program. He is learning everything from scratch. And they had assigned him to me, and he was supposed to go to the mountain with me twice a week. But I made the point that I would let him stay in Hilo one day, and I said, "You stay in Hilo. I'll take your mountain day, and you work on the instruments." And then for about a month I was going to let him sink or swim, and I thought he would truly say, "Help me. What am I doing? How do I run the system?" He just sat there; he kept doing it, doing it. And every once in a while he would ask me a few questions. I just gave him some pointers, and then he just sucked it all in. And now, he is doing a bang-up job on his EPA mercury program. It is nice. I just saw what I could do to help him, sort of, do-it-himself and just grow and even [learn] troubleshooting. When Darryl and I interviewed him, I said, "Have you

> done troubleshooting?" He asked, "What is that?" Now, he knows what troubleshooting is. He has got over five different instruments; maybe four of them are our prototypes, nowhere else in the world. He has got them running. So he is great. He is good.

Colton (video interview with author, Hilo, Hawai'i, December 5, 2006), who became a federal employee in October 2007, appreciated the opportunity to work under a true MLO old-timer, and says, "I'm glad that I have somebody like him to kind of initiate me into this job." He also said that he learned more in two years working with Yoshinaga than in two years of college.

New Staff at MLO (2007–2008)

After Leslie Pajo died, Kalei Lau, a nursing student in Hilo, was hired as a temporary secretary to take care of basic office tasks and send the MLO bills to Debra Hansen in Boulder. At one point, former MLO director Russell Schnell asked Judy Pereira if she would consider returning to MLO. Pereira says she briefly considered the option, but she was too well established with her family and grandchildren in Las Vegas. From September to December 2007, Connie Craig was reassigned to Hilo from GMD in Boulder to become Leslie's temporary replacement. Having worked as a secretary for two of the GMD groups since 2001, Craig was very familiar with the paperwork and she knew the key people.

Meanwhile, Nash Kobayashi, a mechanical technician who formerly worked at the James Clerk Maxwell Telescope near the summit of Mauna Kea, joined the MLO staff in September 2007 as a JIMAR employee.

Lillian Kamigaki became MLO's new permanent secretary on May 12, 2008 and first signed in at the mountain station on October 15, 2008 ("First trip to MLO!"). Kamigaki had spent twenty-two years as an administrative assistant for the US Department of Agriculture. Her older sister Florence Miyamoto knew some of the MLO old-timers, for as a student she had worked for Charles Fullerton at the Cloud Physics Observatory when MLO had its offices there. Darryl Kuniyuki (personal communication with author, July 3, 2008) recalls that "Her younger brother Michael worked at MLO during a summer helping me fill potholes on the Mauna Loa road."

Thomas Davis, a chemist who taught a semester at the University of Hawai'i and who has his own consulting firm, joined the MLO staff as a JIMAR employee in May 2008. Davis worked with the EPA mercury program (Kuniyuki, personal communication with author, June 9, 2008).

An Old-Timer Visits MLO (2007)

On May 22, 2007, former MLO director Howard Ellis accompanied the author during Ellis' first visit to the observatory in more than twenty-six years. Ellis wore the same knit cap and clothing he had worn on his last workday on the mountain. He brought along snacks and fruit prepared by Lena Ellis, much as he did when he stayed on the mountain for days at a time. During the drive from Hilo along Saddle Road, Ellis commented on how its condition was so much better than he remembered. He was especially surprised by the thick layer of asphalt that had recently been laid over the shortcut section of the Mauna Loa road that he and Tom Vance had planned forty-four years earlier. As the car pulled up to the NDACC Building, Ellis looked around and exclaimed, "Wow! What a difference."

Ellis, who was completely unaffected by the altitude change, opened his door almost before the car came to a stop and headed straight for the Keeling Building. There he sat in a chair by the old Scripps carbon dioxide analyzer that had caused him so much grief when it failed in 1964 and created the famous four-month data gap (Plate 86). The Keeling Building became a time machine as he explained what had happened there forty-three years before. Back then he was worried about his future. But as he spoke, he displayed the same smile that appeared in many old pictures, including the time he was photographed with Ralph and Angie Stair while they were seated at the table having supper in what used to be the kitchen section of the building in 1966. When Ellis was handed the old CO_2 *LOG*, which was opened to the page of his last entry on October 23, 1963, he kept smiling but became silent as he stared at the timeworn page.

Later Ellis briskly walked up the boardwalk toward the solar wall. When asked where he used to place his sleeping bag when he was afraid to spend nights in the main building, he demonstrated by sitting down on the lava and lying flat on his back with his feet toward the Keeling Building. Ellis then headed for the Dobson dome, where Preston Sato was preparing to make an ozone measurement. Ellis, who had made countless Dobson measurements when he worked at MLO, instinctively helped set the periscope in place in preparation for the measurements (Plate 87). The deep blue sky was completely free of clouds and in perfect condition to reveal its ozone content. The Dobson dome became a virtual time machine as Ellis prepared the instrument.

During a visit to the NDACC Building, Ellis was excited to see posters on the wall about measurements made at MLO over the years. He was especially gratified when the sunlight atmospheric transmission curve, the longest of the International Geophysical Year experiments at MLO, was referred to as the Ellis curve, for he had devised the clever method for extracting the transmission fraction from the daily data. But his smile vanished as he looked at a graph of the lidar data that began in 1980. "What happened to the old data?" he asked. The lidar system he had worked so hard to operate during many chilly nights on the mountain had nearly electrocuted him and, later, Tom DeFoor. But the lidar signal in those days was not acquired with the sophisticated instrumentation in place today, and it lacked the high quality of the post-1980 data. After learning about Ellis' reaction, John Barnes carefully reassessed the old data and included it in a new version of the lidar chart (see Plate 73).

After looking around the observatory and making a second visit to the Keeling Building, Ellis signed the *Visitors* book, adding "What a difference." He then went outside for more photos, this time standing by the metal plaque that names the building after Charles David Keeling and features the famous carbon dioxide curve that Ellis had helped make possible. While Ellis had unpleasant memories about some of his interactions with Keeling, he smiled and commended Keeling for staying with the CO_2 program. During the trip down the mountain, Ellis shared the lunch Lena had packed for him and told stories about the early days at the observatory and what it was like to drive the road before it was paved. Along the way, he found remnants of some rusty water barrels that Tom Vance had placed along the road. He pointed out the location of the former workers' camp (Camp Jaggar) that burned when Vance was building the abandoned Kona-Hilo Road. And he talked about the dynamite cache Vance reportedly buried in the lava.

While returning to Hilo on Saddle Road, there was a car that had run off the road at high speed lying upside down on the lava. Its occupants were nursing their injuries while picking up their belongings and waiting for the police to arrive. Ellis wasn't surprised; he had seen many such accidents during his years driving Saddle Road, which is well known for its high accident rate.

There was one more stop to make back in Hilo: the old Air Cargo Building at the airport that served as the first MLO office for Jack Pales and the original Weather Bureau team. Back in 1964 when he became the physicist-in-charge, Ellis closed the Air Cargo office and trucked all the office and scientific equipment to the mountain and a storage site in Hilo. The building had changed very little since then, and Ellis beamed as he posed for photographs. It was a fitting way to conclude a day remembering the history of the observatory.

Celebrating the Fiftieth Anniversary of IGY Measurements at MLO (2007)

During November 2007, the observatory observed the fiftieth anniversary of the onset of the International Geophysical Year measurements that were begun by Jack Pales, Clifford Kutaka, and Robert Williams in November 1957 (MLO 2007). The anniversary celebration was dedicated to them and the other former staff members who kept MLO alive during financial hard times and when access to the mountain station was much more difficult than today.

The celebration was also dedicated to the late MLO secretary Leslie Pajo, who had done much to plan the event.

MLO station chief John Barnes and Darryl Kuniyuki did radio interviews about the occasion, and during most of the month the Wailoa Center in Hilo hosted a public exhibit and a series of presentations about MLO and its history. Special events of note included the launch of an ozonesonde by Steve Ryan, various school tours, and a presentation by Ryan about VOGNET, a network he pioneered that monitored sulfur dioxide emissions from the ongoing Kīlauea volcano eruption. November 27 was set aside for the MLO alumni gathering, a reunion of past and present MLO staff. Former staff who attended included Adeline Fujisawa, Malcolm "Mike" Keyes, Johnny Chin, Bill Cobb, Judy Pereira, Mark Goldman, Dr. Barry Bodhaine, Alan Yoshinaga, Lisa Wallace, Bob Uchida, Lynne Hanzawa, Dotty Korsey, and former MLO director Dr. John Miller. The day before the alumni gathering, MLO veteran John Chin returned from a week-long visit to China, where he gave an invited talk about the history of the observatory and its fifty years of carbon dioxide measurements.

Some alumni who couldn't make it to the reunion visited Hilo soon thereafter, including former MLO director Dr. Rudolf Pueschel (1970–1972) and his wife Brigitta and Bernard Mendonca and his wife Clara. Former MLO director Dr. Elmer Robinson (1984–1991) sent a letter that current MLO station chief Dr. John Barnes read at the reunion:

> November 9, 2007
> Dear Colleague:
>
> I wish to congratulate the present and past staff of MLO on the completion of 50 years of unmatched scientific service. In this period of intense interest of the public and the scientific community on the impact of the atmosphere on the earth's environment the MLO record plays a fundamental role. The MLO staff has played a major part in understanding this environmental crisis.
>
> While the almost countless trips by the staff up the Saddle Road from Hilo to the Observatory seemed to be routine at times, they all played an important role in our understanding of this atmosphere that is around us. I salute all of you who were part of the MLO program over this past 50 year period. Whenever you see a present day reference to "global warming" you can say to yourself "I have been part of trying to understand that problem."
>
> Aloha and thank you for your long period of service to MLO.
>
> Sincerely yours,
> Elmer [Robinson]

Dr. Barnes presented engraved clocks to MLO alumni at the reunion. Former MLO director Howard Ellis (1964–1966, 1968–1970, and 1978–1979) was unable to attend, and his clock was presented at his home in Hilo by Darryl Kuniyuki, Preston Sato, and Bernard Mendonca, Ellis' former MLO colleague.

Fiftieth Anniversary of the Global Carbon Dioxide Record Symposium (2007)

The month-long commemoration of the fiftieth anniversary of MLO's IGY measurements concluded on November 28–30 with the 50th Anniversary of the Global Carbon Dioxide Record Symposium and Celebration (CO_2 Conference 2007). The meeting was held 25 miles (40 km) due west of MLO at the Sheraton Keauhou Bay Resort and Spa, Kona, Hawai'i. The National Oceanic and Atmospheric Administration sponsored the meeting, which was cohosted by the Scripps Institution of Oceanography and Peabody Energy. Additional support for the meeting was provided by the World Meteorological Organization, the Science and Technology Corporation, Steelcoast, the Chicago Climate Exchange, and the National Marine Sanctuary Foundation.

Some forty scientists, officials, and corporate representatives addressed the conference, including Hawai'i's Senator Daniel Inouye (by video); Dr. Richard W. Spinrad, assistant administrator, NOAA; Dr. Ken Melville, deputy director for research, Scripps Institution of Oceanography; Timothy R. E. Keeney, deputy assistant secretary for oceans and atmosphere, NOAA, and Dr. Len Barrie, director, WMO Atmospheric Research and Environment Programme.

The welcoming remarks by Dr. Spinrad (2007) included a concise summary of the meeting's objectives:

> Climate research conducted by NOAA and other science organizations, many of whom are represented here at this conference, helps to frame the questions that need to be asked, helps to find answers, and helps policy makers make informed decisions that impact our world and our future.

Dr. Ralph J. Cicerone, president, National Academy of Sciences, delivered the keynote address: "Climate Change: It's Not Just for Scientists." Cicerone met Dave Keeling in 1978 when he joined the Scripps Institution of Oceanography as a research chemist, and he alluded to the careful work behind the Keeling carbon dioxide record:

> There is also the kind of story that some of you are just beginning to speak about on the panel, and that is the human story of the difficulty that was involved in achieving this kind of a record, the meticulous commitment to perfection, the mercury manometer on the wall that could only be viewed with a spyglass, the very careful taping of the glass flasks with adhesive tape in Dave's group. Everyone was sitting around on the floor putting the adhesive tape around the glass flasks, just like they were professional athletes' ankles. It was really fun to watch. Then, of course, NOAA's involvements. There are so many stories involved in this graph. (Cicerone 2007a)

Cicerone also mentioned the importance of tracking CFCs, another class of gases long measured with precision at MLO:

> In fact, there was a little noticed paper [Hansen et al. 1989] which showed something extremely stunning: if the growth in the worldwide chlorofluorocarbon industry had continued at the rate that it had actually achieved in the 1960s and '70s, by the year 1990, the radiative forcing due to the CFC's would have exceeded carbon dioxide. And the fact that the CFC growth did not continue at that rate is something that we can be happy about.

Dr. Ralph Keeling of Scripps described details about some of his father's historic carbon dioxide research at MLO, Antarctica, and elsewhere. Of special interest were slides of some of Dave Keeling's early notes. Keeling and conference organizer Dr. James Butler, who is now director of NOAA's Global Monitoring Division, presented John Chin with a plaque for the more than two decades he spent managing the APC gas analyzer that measured carbon dioxide for Scripps (Plate 88).

Dr. Susan Solomon of the NOAA Earth System Research Laboratory spoke on "A Climate Success Story: Reversing Ozone Depletion." Various exhibits provided information about climate change and related research conducted by various corporate, government, and academic laboratories and organizations. The Mauna Loa Observatory exhibit included the recently retired Applied Physics Corporation infrared gas analyzer that for some forty-eight years monitored carbon dioxide at MLO. The analyzer had apparently never been moved from MLO since its arrival in March 1958 and was returned to the observatory after the conference.

A dramatic moment during the conference occurred when Fredrick D. Palmer, senior vice president, Government Relations, Peabody Energy, showed a slide of an eight-story-high wall of coal looming high over an engineer and his vehicle:

> So I thought I would be provocative and give you a slide that shows an 80-foot wall of coal in Wyoming. Peabody owns more carbon than any company on Earth, more than ExxonMobil. I have personally been involved in the coal industry since 1980, for 27 years. I'm proud of what we do. I'm proud of the coal industry. I know the good that using coal does for the American people and for the world community. But we also recognize the concerns and the concerns that you all have, and that's why we're here. We start out with the very basic premise, and

> that is energy is key to all human activity. It is key to our being here today, key to our getting here. It is key to the way we live when we're at home. It's key to the growing world community and the economic phenomenon that is southeast Asia. It is a second industrial revolution. (Palmer 2007)

Palmer continued by describing the rapid industrialization and development occurring in China:

> China surpassed the United Sates this year [2007] as the largest emitter of greenhouse gases. And as sure as you are sitting in the room, they will continue to grow emissions. The utility executives, I met last week with a partner of ours, Huaneng Group, told me that China will be installing as much coal capacity in the next four years in the northern part of the country that exists in the United States today. Another billion tons of consumption per year in China will come on line by 2010 [or] 2011.

Palmer then discussed the goal of zero emission coal combustion:

> I know for a fact that we have to use more coal. And we can use it cleanly, and we can use it in a carbon-free way. If you look at our booth out there, you'll see the statement: "We are well on the path to zero emissions." And that's our goal. It's called carbon capture and sequestration. This is how we see a policy framework developing.

In a telephone interview after the meeting, Palmer (June 20, 2008) concisely summarized his energy vision:

> We have an aspirational goal to mine and sell every ton of coal we can, because we believe it's in the public interest.

Following Palmer's presentation, conference organizer Dr. James Butler, who is now director of NOAA's Global Monitoring Division, was visiting with his predecessor, the late Dr. Dave Hofmann, who asked why Palmer's talk was included in the meeting. Butler responded to the effect that scientists concerned about climate change must stay informed about what is taking place in the real world of the greenhouse gas industry outside the confines of their laboratories.

Professor Jay Socolow's presentation began with a passage from Dr. Charles David Keeling's autobiography in the *Annual Review of Energy and the Environment*, which Socolow had commissioned when he was the journal's editor. The Keeling quotation (1998) echoes sentiments similar to those that had been earlier expressed by Roger Revelle (1988), Keeling's mentor at Scripps:

> A safe approach is just to remain an interested observer of the unfolding scientific evidence of man-made global climate change and its possible significance to human welfare. Without risk one can comment dispassionately. . . . I believe, however, that a more prudent attitude would be to heed the rise in atmospheric CO_2 concentration as serious unless proven to be benign.

Socolow went on to challenge the conference attendees who had traveled to Hawai'i in aircraft. He asked them to reconsider their personal contributions to atmospheric CO_2 by adopting the prudent approach advocated by Keeling, a step followed by at least one NOAA person who privately questioned the propriety of holding the meeting in Hawai'i and declined to attend. (He later relented and made a separate trip to Hawai'i and MLO.) The room was quiet as Socolow (2007) stated,

> I very much want to emphasize that truly one of the things that we should be trying to do in carbon management is remove unwanted travel. Unwanted travel is when you really would rather stay home. And if you're not staying home, it's probably because of a failure in the communications industry, which can be removed with additional technology and investment and cleverness. And, of course, this can, also, deal with your airplane travel. We couldn't have had this meeting except face-to-face, I think, but an awful lot of conversations can happen because we met each other here with the help of information technology. . . .

> Flying a mile and driving a mile in a 30-mile-per-gallon car are about the same in their carbon consequences, with normal aviation, commercial aviation. So, in my case, and I'm afraid in many of yours, too, your carbon budget is completely dominated by your flying. Let's just admit it and then start thinking what we can do about it. . . .
>
> So the main message, obviously, is we are using a lot more than our share. The US average is five times the global average today, and I suspect in this room, if we worked it out, our shares would be at least another times two the average American.

The CO_2 meeting provided a unique opportunity for some of the world's leading atmospheric, oceanic, and earth scientists to exchange views with scholars, economists, policy planners, and representatives from the fossil fuel industry. The meeting also provided an opportunity for some MLO veterans to renew acquaintances (Plates 89–94). Space precludes a full review of the meeting, and the reader can find full details and transcripts of the presentations at the conference Web site (http://co2conference.org). An unexpected footnote to the meeting's Mauna Loa Observatory connection that was not recorded in the conference transcripts occurred when MLO veteran Malcolm Keyes met Louise Keeling, the widow of Dr. Charles David Keeling. Ms. Keeling listened politely as Keyes explained how he had attempted to repair the broken carbon dioxide analyzer at the mountain station during the 1964 data gap.

MLO Field Trip (2007)

The carbon dioxide conference was concluded on December 1, 2007, with a field trip to Mauna Loa Observatory. The evening before the field trip, the author, who spent the night at the observatory, watched a curious array of clouds over the Kona side of the Big Island, where the carbon dioxide conference was ending. To the north, several UFO–like lenticular clouds, tinted orange by the setting sun, hovered over the slopes of Mauna Kea. A mile (1.6 km) below the observatory a thick streamer of gray vog flowed through the Saddle between Mauna Loa and Mauna Kea. The eastern edge of the vog was met by puffy heads of cumulus that poked up from the orographic layer of clouds below. Wispy cirrus clouds sliced high across the eastern sky where the giant Mauna Loa cast its huge shadow. The apparition appeared exactly as Charles Wilkes first described it when he and his crew of tough explorers spent three weeks on the volcano's summit during the winter of 1840–1841:

> This evening [Jan. 1, 1841], at sunset, we had a beautiful appearance of the shadow of the mountain, dome-shaped, projected on the eastern sky: the colour, a light amethyst at the edges, increasing in intensity to a dark purple in the centre; it was as distinct as possible, and the vast dome seemed to rest on the distant horizon. (Wilkes 1845: 163; see chapter 3)

Sunrise on the morning of the tour revealed none of the dust and pollution bands from Asia that are often visible over the horizon during spring. A long cloud with straight sides and a flat bottom floated like a box top over the barren summit of the dormant Hualālai volcano, a prospect that was visible to early risers at the carbon dioxide conference far below the observatory. The temperature was 36 degrees F (2° C), and a cool breeze flowed down from the summit of the mountain and provided the pure downslope air that gives the best sampling conditions for the instruments that monitor carbon dioxide, methane, CFCs, and other trace gases. As the morning sunlight warmed the black *'a'ā* clinkers, half a dozen green aphids (Homoptera: Aphididae), an invasive species not native to Hawai'i, resumed sucking sap from the stems supporting the pretty yellow flowers of a hairy cat's ear *(Hypochoeris radicata),* another nonnative species that has somehow taken root far above its normal elevation (Starr 2008). Those golden medallions and the lime green aphids dining on their stems were as foreign to the barren black, brown, and orange slopes of Mauna Loa as the instruments scattered across the lava and inside the cluster of buildings in the silent scientific village perched high on the slope of the huge volcano.

By midmorning, conference attendees began arriving for their tour of the observatory. First-time visitor Emily Cloyd of the US Climate Change Science Program Office wrote in the *Visitors* book, "Perfect weather." The weather was indeed perfect, for there were no clouds in the deep blue expanse overhead. A robotic CIMEL sun photometer operated by NASA's Goddard Space Flight Center's AERONET program indicated that the atmosphere over MLO had an optical depth of only about 0.0074 at the green wavelengths of light to which the human eye is most sensitive. In plain English, 99.3 percent of the direct sun's green wavelengths arriving at the top of the atmosphere reached the observatory. The tour was occurring during one of the pristine mornings for which MLO is so desirable as a site for calibrating sun photometers.

By noon the temperature reached a pleasant 49 degrees F (9.4° C), and the barometric pressure hovered around 679 millibars (or hPa), which is two-thirds the pressure at sea level. The precipitable water over MLO measured by NOAA's GPS system rose to about 0.5 inch (13 mm), which is well above the 0.04 inch (1 mm) measured on the driest days. The high moisture level made the occasion more comfortable for the visitors and probably explains the prominent solar aureole around the sun. The bright sunshine piercing through the crystal-clear sky kept the visitors warmer than suggested by the temperature of the air. It also provided plenty of ultraviolet, which reached an index of about 10 at solar noon. While this is well below summer levels, it is enough to cause erythema (reddening) of unprotected, pale skin in about fifteen minutes (Vanicek et al. 1999).

During the tour, the pumps that sample air from the forty-meter walk-up tower continued humming away in the background. That month the fifty-year-old Scripps experiment measured an average concentration of 383.69 parts per million of carbon dioxide in the air at MLO, which was 21.5 percent higher than the 315.7 ppm measured when measurements were begun in March 1958.

Among the dignitaries and scientists who visited the observatory during the tour were MLO veterans Bernard Mendonca and Mike Keyes (Plate 95), both of whom had spent considerable time working on the mountain long before there was a paved road and an electrical power line. Mendonca's wife Clara was also present, and all three signed the *Visitors* book. Mendonca added, "Just like yesterday! Great!" He had often visited MLO after he was transferred to Boulder in June 1976 and was well aware of the many changes that had been made, but Keyes had not returned since the Weather Bureau sent him back to Honolulu in 1965, leaving only Howard Ellis and Mendonca to man the observatory during the budget crisis that had nearly closed the observatory. Both men were glad to be back at MLO, which was much smaller when Keyes was last there forty-two years before.

MLO veteran Dr. Barry Bodhaine (Plate 96) and his wife Helen Horton also visited. After his signature, Bodhaine added, "Déjà vu! It's great to be back." Dr. Michael Kurylo, accompanied by his wife Mary ("2nd visit—gorgeous day!"), was among the visitors (Plates 97 to 99), his fourth and final visit as a NASA representative.

A few days prior to the conference field trip, former MLO directors Dr. John Miller ("Changed a lot in 30 years!") and Dr. Rudolf Pueschel, accompanied by his wife Brigitta ("Congratulations MLO for good old times"), signed in at MLO. Former Global Monitoring Division director Dr. David Hofmann, who played a major role in planning and acquiring the NDACC Building and other MLO projects, also visited ("GMD Director for 34 more days!"). Ralph and Eric Keeling, sons of the late Charles David Keeling, also signed in.

Visiting conditions were perfect for the fiftieth anniversary tour. No one then knew that a winter storm much like the one experienced by the US Exploring Expedition in December 1840 was on the way, and that it would soon cover the mountain and the observatory with a thick blanket of snow and even break a power line.

Visiting with Howard Ellis (2007)

While Howard Ellis is viewed as a legendary figure by the current staff at MLO, none of them had met him prior to his May 22, 2007, visit to MLO, for he long avoided any contact with the observatory. During the preparation of this book, however, Ellis spent many hours cheerfully answering questions and providing details about the founding of MLO and his time there. The day after the MLO field trip, former MLO director Rudolf Pueschel and his wife Brigitta drove across the island specifically to see Howard and Lena Ellis at their home in Hilo (Plate 100). Prior to their reunion with the Ellises, Pueschel recalled the friendly manner in which Ellis had greeted him when he first arrived on the mountain in 1970. Pueschel, who coauthored the 1971 atmospheric transmission paper for *Science* with Ellis, knew of Ellis' history of being passed over as director of MLO after having served as interim director several times, and, when they met in Hilo, Pueschel (2007) expressed his appreciation to Ellis for the manner in which he transferred his authority thirty-seven years earlier. He then told him, "There's no one who did so much to preserve this station as you." The two couples then spent a pleasant afternoon sharing memories. The next day, Bernard Mendonca and MLO staffers Darryl Kuniyuki and Preston Sato visited Ellis to present him with an inscribed clock in recognition of his time at MLO. More than four decades had passed since Ellis and Mendonca kept MLO alive during the worst budget crisis in its history. Mendonca shook Ellis' hand and addressed him as "Boss" (Plate 101). Meanwhile, the regular work schedule continued as usual at the MLO office in Hilo (Plate 102) and at the mountain station (Plates 103 and 104) where they used to work.

Dr. James Butler Becomes Director of the Global Monitoring Division (2008)

The Program Committee for the 50th Anniversary of the Global Carbon Dioxide Record Symposium was chaired by Dr. James Butler, who was then deputy director of the Global Monitoring Division of NOAA's Earth System Research Laboratory (ESRL) and also acting deputy director of ESRL and its 650 scientists and support staff, a position that benefited from his cross-disciplinary background. Following the symposium, Butler visited MLO's new Hilo office and met with the staff "to discuss past events and the future of the observatory" (MLO 2007). Butler's visit was well timed, for he was soon to assume a new position at the GMD.

Dr. David Hofmann, who was assigned as chief scientist of the Climate Monitoring and Diagnostics Laboratory in Boulder in 1990, stepped down as director of the Global Monitoring Division on December 31, 2007. Hofmann played a major role in the establishment of the NDACC Building at MLO. He also devoted considerable time to securing funding to maintain the Global Monitoring Division and, especially, the atmospheric baseline observatories. He considered the latter to be one of the two most important contributions during his tenure, the second being his work in helping establish the importance of North America as a carbon sink. While he wrote many scientific papers during his tenure as GMD director, Hofmann's administrative duties limited the time he was able to devote to his passion of doing science, and he said that he looked forward to staying on at GMD as a research scientist. He subsequently visited MLO with his daughter Gretchen on October 21, 2008, where he signed the *Visitors* book and added, "ESRL/GMD Director Emeritus, Research Scientist . . . yay!"

Dave Hofmann was replaced as director of GMD by James Butler, and former MLO director Russell Schnell became deputy director. Butler's background illustrated the cross-disciplinary range of specialties at GMD, for his academic training included a bachelors in biology, a masters in fisheries with a radio chemistry minor, and a Ph.D. in chemical oceanography with a microbiology minor. His doctoral studies involved the study of trace gases in seawater, and he specialized in studying the exchange of gases between the atmosphere and the oceans. Butler (video interview with author, Boulder, Colorado, May 2, 2007), asked if he still studies the atmosphere-ocean connection, had the following response:

> I do some of that, because studying the atmosphere requires understanding the ocean. The ocean covers over 70 percent of the earth's surface. Gases are exchanged between the ocean surface and the atmosphere. Heat, energy momentum are all exchanged between the

> ocean-atmosphere. We like to think of it as an ocean-atmosphere system. . . . For many years I went out to sea about every year, year and one-half, to study the exchange of trace gases across the earth's sea interface.

While it's well known that the oceans absorb carbon dioxide, less well known is that they also absorb a number of other influential trace gases. Among them is methyl chloroform, which Butler points out "is both produced and consumed in the ocean." Methyl chloroform is important because it destroys ozone.

Jim Butler's expertise in the two-way transfer of gases that occurs at the ocean-atmosphere interface illustrates his versatile, cross-disciplinary approach to science at the GMD. In the same interview, he noted,

> A lot of my work has been to try to get the laboratory to integrate, work together, and get things done. But, I'm looking forward to focusing my efforts more on nurturing and assisting the Global Monitoring Division, because I think the long-term measurements that we make are critical to the decisions that society is going to have to make in the future.

Butler was asked if his administrative duties would slow his research:

> I feel really good, people ask me that a lot, and I can proudly say that I got a first authored paper out this year on very short-lived gases and how they influence the lower stratosphere and ozone depletion. I am involved in a considerable amount of work regarding these gases and others.

MLO and the News Media

The global warming controversy that had its genesis in the carbon dioxide findings at MLO has attracted far more media interest than the observatory itself. A few newspaper reporters were on hand during the dedications of the summit observatory in 1952 and the slope observatory in 1956. The MLO monthly Activities Reports describe various visits by both the print and television media over the years. Most were visits by individual reporters and writers, and some were by professional film crews making television documentaries broadcast in the United States, Japan, and elsewhere. Media interest in MLO has peaked during unusual events, such as the eruption of major volcanoes.

In spite of the occasional media attention, the MLO staff had long lived with the fact that most people in Hawai'i assumed they were somehow connected with the astronomical observatories atop Mauna Kea. The white telescope domes on Mauna Kea are visible from across the north side of the Big Island when clouds are not in the way, but it takes a trip along Saddle Road and a careful eye to make out MLO on the north slope of Mauna Loa. Images of distant galaxies and exploding stars attract much more attention than what some maps of Hawai'i label a "weather station" on Mauna Loa.

A dramatic photograph of the $183 million (NOAO 2008) Keck Telescopes atop Mauna Kea on the cover of the April 2008 *Smithsonian* magazine illustrates how MLO, arguably the world's most significant atmospheric observatory, is so often overlooked. The photograph, which shows a laser beam emerging from one of the Keck domes, illustrates an article about black holes (Irion 2008). But it also vividly depicts a far more important story that is not even mentioned by the magazine, for it shows very thick layers of haze hovering over the Big Island of Hawai'i. The dark layers were either sulfur dioxide vog from the Kīlauea volcano or dust and air pollution from China—or possibly both. The missed story, of course, is that a world-class atmospheric observatory on nearby Mauna Loa had been monitoring the quality of the atmosphere, greenhouse gases, and sunlight high over the Pacific for more than fifty years. The Mauna Loa Observatory also points powerful lasers high into the stratosphere, but their purpose is far more important to people since they measure the ozone layer, volcanic particles, and water vapor. While this research is not as glamorous as studies of black holes, it is far more relevant to understanding factors that influence earth's atmosphere and climate.

There is an ironic twist to MLO's identity problem, for the observatory was first proposed by Robert Simpson to Ralph Stair during a chance encounter at a solar observatory atop a mountain in New Mexico, a state that also has an identity problem. For many years, one of the most popular columns in *New Mexico Magazine* has been "One of Our 50 is Missing." Each installment relates anecdotes about clueless outsiders who assume that New Mexico is part of the country of Mexico, with which it shares a common border.

Overnighting in the NDACC Building

Projects by both US and international scientists have sometimes benefited by the presence of the two small bedrooms in the NDACC Building. When Station Chief John Barnes drives up to MLO to conduct his weekly lidar observations of the stratosphere between sunset and midnight, he often spends the night in one of the two tiny bedrooms in the NDACC Building. Occasionally visiting scientists spend nights there. Most scientists are involved with the various missions of the observatory, but a few stay for more specialized purposes.

Consider geologists Julie Herrick and Erika Ronchin, who spent a week collecting lava samples across the upper slopes of MLO in 2005. Ronchin, who was from the Universita degli Studi di Trieste, Italy, was completing her doctoral studies, and Herrick served as her guide. The two women left MLO early each morning in a surplus US Army vehicle, returned shortly before sunset, and spent nights in the NDACC Building (Mims 2005a,b).

Having company while overnighting is helpful, for, after the lights are out, inexperienced visitors are sometimes startled by the sounds produced by the shutters, fan motors, air sampling pumps, and ventilators that drone on continuously or switch on and off at intervals. As John Barnes has observed, "Yes, at night if you are up there alone, it gets a little spooky." First nights are sometimes difficult for visitors who have not fully acclimated to the high elevation, especially when the dry air makes breathing difficult and blocks sinus passages. Some relief can be provided by spraying saline solution into one's nostrils.

International Collaboration and Cooperation

Little known outside the atmospheric sciences community are the many international collaborations in which the Mauna Loa Observatory and other NOAA observatories have long been involved. MLO's operations have been studied by climate observatories in Korea, China, Taiwan, Japan, Indonesia, and various European countries (Darryl Kuniyuki, personal communication with author, June 9, 2008). Mauna Loa staff members have attended a variety of international atmospheric science conferences over the years, and the *Visitors* book is sprinkled with the signatures of scientists from around the world. There have also been scientific exchanges, as when Steve Ryan exchanged places with Laurie Porter of Australia's Cape Grim Baseline Station during the second half of 1994. In October 1995, then MLO director Russell Schnell visited China to give a talk on measurements of Chinese dust storms made at MLO (Schnell 1995).

An especially important cooperative project, which is described in chapter 13, began on March 24, 1997, when the Atmospheric Environment Service of Environment Canada installed a Brewer spectrophotometer for measuring the ozone layer and solar ultraviolet. The instrument was mounted on a concrete pad southeast of the main building, and its computer was housed in an adjacent dome. The Brewer was checked and its optical ports regularly cleaned by the MLO staff, and the computer that controlled the instrument was accessible from Canada via the Internet. After four and a half months, the Brewer system fell victim to one of Mauna Loa's powerful thunderstorms:

> The instrument operated nearly flawlessly until August 15, 1997, when the observatory took a direct hit from lightning. A surge from the lightning damaged the control computer, probably through the telephone port, but left the Brewer instrument undamaged. Observations resumed on November 18, 1997, when the computer was replaced and the instrument was relocated to the deck of the NDSC building. (Kerr 1996–1997)

A second major international collaboration is New Zealand's National Institute of Water and Atmospheric Research (NIWA) ultraviolet spectroradiometer, introduced in chapter 12. On a clear morning in May 1995, MLO director Russell Schnell was wandering around the observatory site with a transit. He periodically placed the tripod mount of the transit on the lava and peered upslope. When the author, who was at MLO to calibrate ozone instruments, asked the purpose of Schnell's observations, he explained that the NIWA team needed to know how much of the sky at MLO was blocked by the north slope of the volcano south of the observatory.

Some of the NOAA scientists who have been assigned to MLO have traveled extensively while doing science and giving talks, and, as noted earlier, several MLO scientists were assigned to the South Pole before moving to Hawai'i. Physicist Steve Ryan was assigned to the NOAA observatories in Antarctica, Samoa, Alaska, and MLO. Physicist Michael O'Neill, who began working for the Global Monitoring for Climatic Change laboratory (now GMD) in the late 1980s, has worked at MLO on various temporary assignments:

> At last count, I have been to 30 countries [and] all seven continents. I have spent over six years out of the past 20 out of the country on long-term projects. I have done everything from surface ozone to balloon ozone [and] total column ozone. . . . I was at the South Pole for a year as the technician. I have also been involved in some solar radiation, UV measurements, as well as automated MET systems; a lot of different programs. (O'Neill, video interview with author, Boulder, Colorado, May 3, 2007)

Former GMD director David Hofmann (video interview with author, Boulder, Colorado, May 1, 2007) cites the carbon dioxide monitoring network as a good example of international cooperation that sprang from the International Geophysical Year and MLO's legacy:

> There are well over 50 surface sites around the world where . . . local scientists [and] people who are not scientists help us in obtaining these samples, and they are all sent back here to Boulder where they are analyzed with the same calibration gases. And so without the global cooperation of many, many people we would not have the annual greenhouse gas index or a lot of other measurements that we make.
>
> And, also, most of our sites are part of the World Meteorological Organization Global Atmospheric Watch Program. So it is important because the world looks upon us not only to make good measurements but to provide calibrations for other countries. You well know about the ozone standards; we also have these carbon dioxide standard gases and some of the ozone depleting gases that are standards. So if someone in a country wants to start up a measurement program, we can provide them gases so that they will be on the same standard as the rest of the world. So this is very important.

The greenhouse gas monitoring programs described by Hofmann began in 1958 with Keeling's carbon dioxide project at MLO. The potential impact these gases have on Earth's temperature is determined by their *radiative forcing*, a key phrase in the lexicon of global warming.

Forcing the Issue

"Radiative forcing" is defined by the Intergovernmental Panel on Climate Change (IPCC 2007a) as follows:

> Radiative forcing is a measure of how the energy balance of the earth-atmosphere system is influenced when factors that affect climate are altered. The word radiative arises because these factors change the balance between incoming solar radiation and outgoing infrared radiation within the earth's atmosphere. This radiative balance controls the earth's surface temperature. The term forcing is used to indicate that Earth's radiative balance is being pushed away from its normal state. . . . When radiative forcing from a factor or group of factors is evaluated as positive, the energy of the earth-atmosphere system will ultimately increase, leading to a warming of the

system. In contrast, for a negative radiative forcing, the energy will ultimately decrease, leading to a cooling of the system.

The IPCC expresses forcing in watts per square meter (W/m^2) relative to the estimated forcing during the preindustrial year of 1750. According to the 2007 IPCC report, the primary greenhouse gases (leaving aside water vapor) are carbon dioxide, methane, nitrous oxide, the halocarbons, and tropospheric ozone. Together, according to the IPCC, these gases cause the greatest positive forcing, with smaller contributions from changes in solar irradiance and black carbon (soot) deposited on snow. The major negative forcing is from clouds, aerosols, and land use. The net forcing is positive and amounts to about 1.6 W/m^2 (IPCC 2007a). According to the IPCC, the single most important contributor (aside from water vapor) is carbon dioxide, which has been measured almost continuously at MLO since March 1958.

Dr. Pieter Tans, who spent a year working under Charles Keeling, directs the Carbon Cycle Group at NOAA's Global Monitoring Division in Boulder. He placed the 1.6 W/m^2 forcing in perspective during an interview in Boulder, Colorado, on May 1, 2007:

> You can say, "What is 1.6 watt per square meter?" I always compare it to the total amount of energy absorbed from the sun by the earth's surface, the atmosphere and the surface together. The earth absorbs, on the average, year round globally . . . 240 watts per square meter. . . . So it is a good fraction of 1 percent.
>
> So in a sense, for the climate system it is as if the sun has become brighter by 0.6 percent. I think that is quite significant. The scary part is that if we continue to burn carbon-based fuels and keep adding CO_2 [to] the atmosphere, the forcing due to CO_2 will continue to increase and might go as high as over 10 watts per square meter, several percent of what we absorb from the sun. And we have the capability of doing that. There is enough coal, mostly coal, but also oil and gas on earth to do this, and we have the technical capability of doing it.

Tans' concern about the potential impact of increasing carbon dioxide is reflected in the 2007 IPCC report.

> Most of the observed increase in global average temperatures since the mid-20th century is *very likely* due to the observed increase in anthropogenic greenhouse gas concentrations. This is an advance since the TAR's [IPCC *Third Assessment Report*] conclusion that "most of the observed warming over the last 50 years is *likely* to have been due to the increase in greenhouse gas concentrations." Discernible human influences now extend to other aspects of climate, including ocean warming, continental-average temperatures, temperature extremes and wind patterns. (IPCC 2007b)

Observations and Computer Models

Studies of climate change are driven by both observations and by models. Weather forecasters use computer programs known as models to make short-term predictions about the weather. Measurements of atmospheric pressure, temperature, humidity, and wind direction are entered into the model, which then generates weather maps depicting these parameters a day or more into the future plotted as lines and swirls across the country.

A climate model is a numerical simulation that attempts to represent the Earth's atmosphere over longer periods of time than weather models. Basic weather forecasting models have evolved into sophisticated analytical tools that permit the interaction of the various components of daily weather and long-term climate to be simulated. Models are used, with varying degrees of success, to make detailed weather forecasts, evaluate air pollution, predict the spread of pollen, smoke, and dust, and forecast everything from drought and floods to climate change.

While weather forecast models are designed to provide detailed predictions out to a week or so, global climate models attempt to represent the Earth's atmosphere over decades. The models divide the entire Earth into a grid of cells and its atmosphere into layers. They require inputs ranging from data about past climate to the latest measurements of surface and tropospheric temperature, wind, water vapor, aerosols, cloudiness, and other atmospheric parameters collected at MLO and from around the world. After these parameters are fed into the model, the effect on climate caused by changes in weather, air pollution, volcanic eruptions, greenhouse gases, and clouds can be simulated and depicted as color-coded images on computer monitors.

Climatologist Stephen H. Schneider has defined a model as "our best understanding of how the atmosphere works, how the oceans work, how ecosystems work, how chemical exchanges work." "Best understanding" is a caveat not always given the attention it deserves, and Schneider (2008) has argued that it is important to update models with new findings, for *climate models are only as good as the data fed into them:*

> I keep arguing, don't be too arrogant about the belief in your models; what you do is make projections, and then you crank a knob to try to avoid the more catastrophic outcomes or the outcomes that don't match your values, but we better be reflexive. That means we had better have what in the language of the systems guys is a "complex adaptive system." We need to always build in knobs so that as we get new information that changes our understanding of the structural bases of models, we can in turn crank up or down our degrees of policy control. But we rationalists, we systems analysts, think that's a great model of reality.

Global climate models were still a dream back in 1958 when MLO director Jack Pales unwound spools of chart recordings on the shiny red floor of the MLO building to examine the daily and weekly trends. He used a ruler and pencil to extract the data an inch at a time to create tables of results. Digital data acquisition systems and computers ended the need to spend hours tediously extracting data from chart recordings. Today, free computer spreadsheet programs permit students and scientists alike to quickly tabulate, analyze, and make charts of their data or data posted online. Climate models have also made important progress, but there was a time when computer modelers and their models were viewed with great skepticism by the experimentalists who collected and often processed their own data. The feeling was mutual, for the modelers sometimes mistrusted the accuracy of data they were provided. MLO veteran Bernard Mendonca (video interview, Boulder, Colorado, May 27, 2007) reflects on changes in this faceoff:

> In the beginning we didn't like modelers, because what did they know? . . . A data point is worth ten models any day. . . . In the beginning they didn't want to deal with us, and we with them. But now they are discovering that our information can make their models much better, and their models can make our interpretation much better.

While climate modeling has made significant advances, concerns remain about the inadequacy of the models to account for the temporal and spatial effects of clouds and aerosols, the primary negative forcing agents, with some asserting that models either overestimate or underestimate the impact of greenhouse gases. For example,

> The latest report from the Intergovernmental Panel on Climate Change assesses the skill of climate models by their ability to reproduce [past] warming over the twentieth century, but in doing so may give a false sense of their predictive capability. (Schwartz et al. 2007)

Another concern is the inability of models to accurately predict climate change at the regional level. For example,

> Global and regional climate models have not demonstrated skill at predicting regional and local climate change and variability on multi-decadal time scales. (Pielke 2006)

Pielke's concern and Schneider's caveat about the need to adjust climate models was demonstrated during the decade following the high temperatures measured during the 1998 El Niño. While a host of climate models predicted a continued warming trend, the global temperature flattened and even declined after about 2001.

Unless otherwise credited, all photographs taken by the author.

PLATE 83. MLO announces its fiftieth anniversary with a banner on its Hilo office.

PLATE 84. MLO's fiftieth anniversary was bracketed by the retirements of Alan Yoshinaga *(left)* and Bob Uchida.

PLATE 85. Prof. Walter Steiger of the University of Hawai'i spoke at the dedication of MLO on June 28, 1956.

PLATE 86. Former MLO director Howard Ellis visited MLO on May 22, 2007. Here he inspects the retired gas analyzer.

PLATE 87. During his visit to MLO, Howard Ellis helps prepare the Dobson spectrophotometer for an ozone measurement.

PLATE 88. At the 50th Anniversary Global Carbon Dioxide Record Symposium, James Butler *(left)* and Ralph Keeling *(right)* recognize John Chin for maintaining the Scripps CO_2 analyzer.

PLATE 89. From left, Bernard Mendonca, John Chin, and David Hofmann with the Scripps CO_2 analyzer that Chin maintained from 1964 to 1990.

PLATE 90. Ralph Keeling and his mother, Louise, at the 50th Anniversary Global Carbon Dioxide Record Symposium.

PLATE 91. Janet and John Chin and John Miller during the 50th Anniversary Global Carbon Dioxide Record Symposium.

PLATE 92. From left to right: Malcolm Keyes, Clara and Bernard Mendonca, former MLO director John Miller, and MLO station chief John Barnes during the 50th Anniversary Global Carbon Dioxide Record Symposium.

PLATE 93. MLO veterans who attended the 50th Anniversary Global Carbon Dioxide Record Symposium include *(left to right)* Malcolm Keyes, Rudolf Pueschel, Bernard Mendonca, Barry Bodhaine, and John Chin.

PLATE 94. Former MLO director John Miller *(left)* and John Chin with the Scripps CO_2 analyzer.

PLATE 95. On December 1, 2007, Malcolm Keyes *(left)* inspects the old Scripps CO_2 analyzer parts cabinet at MLO while Bernard Mendonca looks on.

PLATE 96. MLO veteran Barry Bodhaine points out the main air sampling port at MLO on December 1, 2007.

PLATE 97. From left to right: Bernard and Clara Mendonca, Helen Horton and Barry Bodhaine, and James and Sally Elkins in the NDSC Building at MLO during the 50th Anniversary Global CO_2 Record Symposium on December 1, 2007.

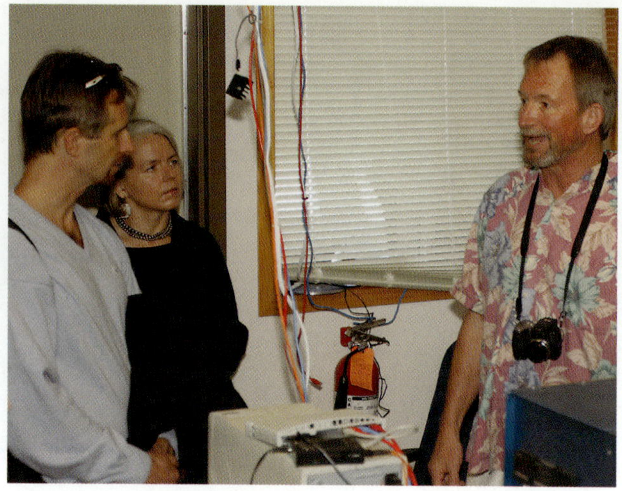

PLATE 98. James Butler *(right)* explains one of the air sampling programs at MLO during the 50th Anniversary Global CO_2 Record Symposium field trip.

PLATE 99. Group photograph of MLO field trip during the 50th Anniversary Global CO_2 Record Symposium. MLO station chief John Barnes is kneeling at left (yellow jacket).

PLATE 100. Brigitta and Rudolf Pueschel *(left)* and Lena and Howard Ellis at the Ellis home in Hilo on December 2, 2007.

PLATE 101. Howard Ellis *(left)* and Bernard Mendonca, who kept MLO alive during the budget crisis of 1964–1966, visited on December 3, 2007.

PLATE 102. John Barnes *(left)* and John Ogren prepare for a trip to MLO on December 3, 2007.

PLATE 103. Darryl Kuniyuki and the parts cabinet for the retired CO_2 analyzer at MLO.

PLATE 104. Paul Fukumura-Sawada cleaning the Eppley pyrheliometer at MLO.

PLATE 105. Dr. Robert Simpson receives NOAA's 2008 Environmental Hero Lifetime Achievement Award for founding the Mauna Loa Observatory and for his pioneering hurricane research. The award was presented on July 2, 2008, by Dr. Richard Spinrad, assistant administrator of NOAA's Office of Oceanic and Atmospheric Research. *(Daniel Parry, NOAA Office of Communications.)*

PLATE 106. The staff of MLO with the MLO centennial time capsule on March 10, 2009. Kneeling *(left to right):* Darryl Kuniyuki, Donald Mock, deputy director of ESRL, and Nash Kobayashi, who built the time capsule. Standing: Paul Fukumura-Sawada, Preston Sato, Tom Davis, Aidan Colton, Steve Ryan, Charles Mock, Lillian Kamigaki, Anita Colin, David Nardini, and Dr. John Barnes, MLO station chief. *(MLO archive.)*

PLATE 107. The MLO staff on March 10, 2009. From left to right: Preston Sato, Tom Davis, Darryl Kuniyuki, Steve Ryan, MLO station chief John Barnes, Aidan Colton, Nash Kobayashi, David Nardini, Paul Fukumura-Sawada, and Lillian Kamigaki. *(Donald Mock.)*

PLATE 108. Leslie Ann Pajo (1962–2006) on duty at MLO on October 5, 2004.

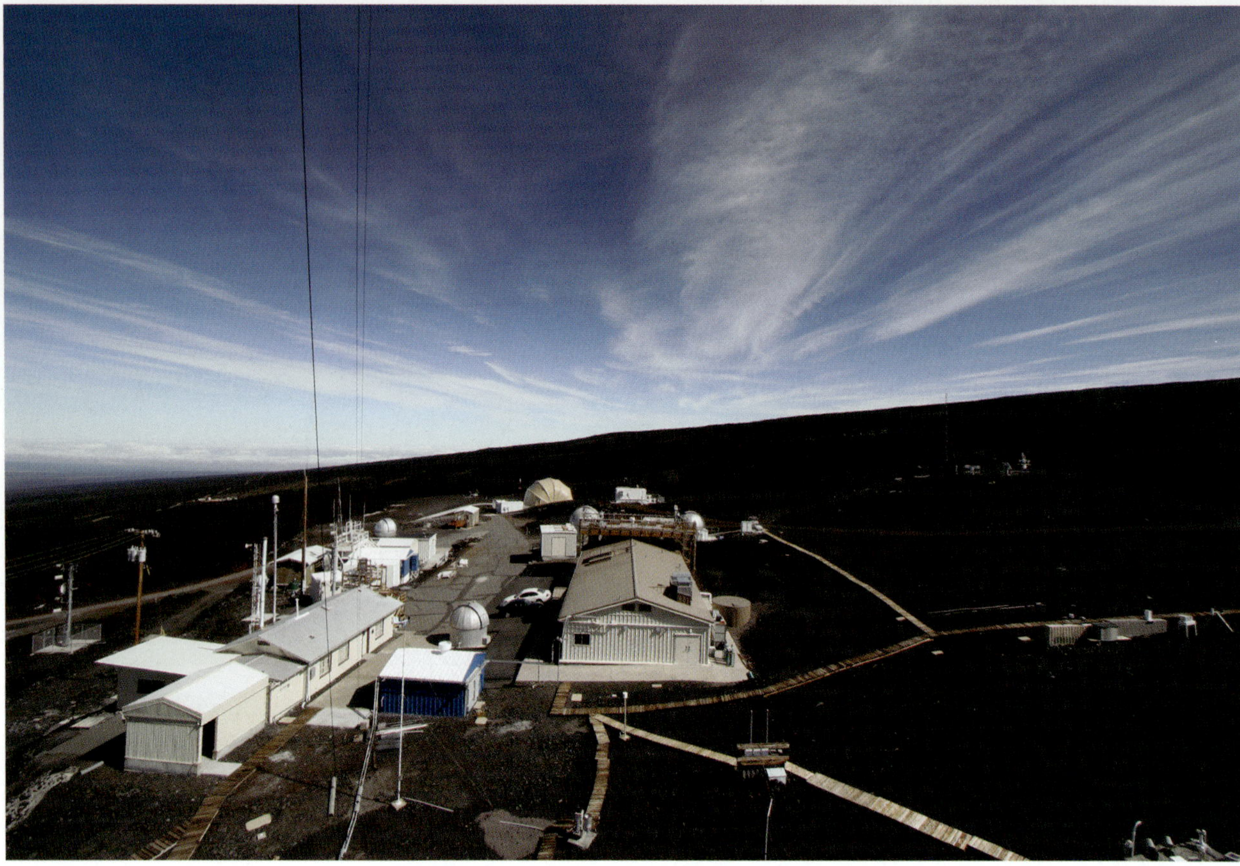

PLATE 109. The Mauna Loa Observatory as viewed from the walk-up tower on December 9, 2005. The original building is the tan structure left of the blue Butler Building. The large building *(center)* is the NDSC Building.

PLATE 110. The Keeling Building at sunset on May 23, 2007.

Meteorologists have noted that the winters of 2008–2009 and 2009–2010 resembled what occurred during the cooling of the 1970s.

The IPCC is well aware of both the capabilities and deficiencies of global climate models:

> There is considerable confidence that climate models provide credible quantitative estimates of future climate change, particularly at continental scales and above. This confidence comes from the foundation of the models in accepted physical principles and from their ability to reproduce observed features of current climate and past climate changes. Confidence in model estimates is higher for some climate variables (e.g., temperature) than for others (e.g., precipitation). Over several decades of development, models have consistently provided a robust and unambiguous picture of significant climate warming in response to increasing greenhouse gases. . . .
>
> Nevertheless, models still show significant errors. Although these are generally greater at smaller scales, important large-scale problems also remain. (IPCC 2007a)

Remote sensing satellites such as Terra, Aqua, Cloudsat, Aura, and a host of others have provided abundant new data about the role of clouds and aerosols that may improve future climate models. The interval between solar cycles 23 and 24, the longest in about a century, may provide new information about both direct and indirect roles of the sun in altering climate.

The Annual Greenhouse Gas Index

Former GMD director David Hofmann was long aware of the general public's inability to distinguish between the present state of model uncertainties and the high-quality data that for many decades had been collected at MLO and the other GMD observatories. In 2006, he announced the Annual Greenhouse Gas Index (AGGI), a new and easily understood method for bypassing the world of models by summarizing real, overall changes in the concentration of the major greenhouse gases in the atmosphere measured at MLO and elsewhere (Hofmann et al. 2006a,b). As explained by Hofmann (2007),

> Greenhouse gas changes since the industrial revolution are believed to be closely related to the change in climate that has been observed (IPCC 2007a). However, climate projections have model uncertainties which overwhelm the uncertainties in greenhouse gas measurements. In this work we sought an index that was directly proportional to the forcing of climate but with relatively small uncertainty.

During an interview on May 1, 2007, at GMD in Boulder, Hofmann elaborated on the AGGI and his reason for proposing it:

> I think when I first came here, and for the next ten years or so, our main scientific efforts were just to provide data for the IPCC reports and the ozone assessments. And when you talk to the man in the street or a policy maker, and you talked about why we have provided all this information for these assessments, I would not say their eyes glazed over, but they really did not understand what we were doing. And I think I felt it was important to come up with something that is a little easier to understand.
>
> So what we did was to use the concept of radiative forcing, [the] effect a molecule has on the radiative balance of the atmosphere, and calculate that for all the greenhouse gases, not just carbon dioxide; and then add up that radiative forcing. [The result] was in watts per square meter, which is still a term that the average person does not understand. I can tell them that a watt per square meter is approximately one of those little tiny Christmas tree bulbs [per] square meter; that is about what it is. But it does not really make it easy for them to keep track of what is going on.
>
> So what I decided to do is to normalize, that is, pin the total radiative forcing to the year 1990. . . . Susan Solomon, in our discussion with her, suggested, "Why not use 1990 to pin down the annual greenhouse gas

index, because that is the baseline year for the Kyoto protocol?" . . . So we pinned it there.

So effectively, each year we calculate the total radiative force of all the gases [and] then divide it by what it was in 1990, and that ratio is now up to 1.23 in 2006 [1.28 in 2009]. So the radiative forcing of all these gases has increased 23 percent since 1990, since the Kyoto protocol was designed. . . . And so to get the forcing back to those levels is going to be a major task, as you can see.

"The Best Measurements in the World"

Dave Hofmann closed his comments on the Annual Greenhouse Gas Index in the 2007 interview by distinguishing the work of the data collector from that of the modeler:

> We do not talk about what the effects of these changes are. You are getting into another area. I'm not a climate modeler; I'm not a climate scientist. I'm a person who makes atmospheric measurements, and the laboratory that I work for and the scientists in that laboratory pride themselves in making the best measurements in the world. So the numbers that we come up with are very significant in their own, and you can take those numbers and go where you want to with them as to what they mean for the future.

Hofmann's approach to environmental monitoring follows a long tradition in which experimental scientists strive to collect the best possible data while leaving the details of its implications to specialists, statisticians, and policy makers. As noted in chapters 2 and 3, this is very much in the tradition of early Mauna Loa scientists Archibald Menzies (1794), Charles Wilkes (1840–1841), and Henry Guppy (1897). NOAA's Dr. Susan Solomon, who was cochair of the UN Intergovernmental Panel on Climate Change (IPCC) Working Group I, echoed Hofmann's sentiments during a presentation at the 50th Anniversary of the Global CO_2 Record Celebration and Symposium in 2007. Solomon urged her audience to concentrate on their data and models while leaving public policy advocacy to environmental organizations. "Our role is not to be advocates," she said.

Solomon and other speakers at the CO_2 meeting also expressed concern about widely circulated statements in the media to the effect that global warming science is over and the debate about climate change has ended. She explained that *science is an ongoing process,* and that climate change research budgets might be negatively affected by such assertions.

Dr. Ralph Cicerone, president of the National Academy of Sciences, spoke at the same CO_2 meeting in Hawai'i that Susan Solomon addressed. In an interview with Bob Ryan, a former president of the American Meteorological Society and a television meteorologist (NBC Channel 4, Washington, DC), Cicerone, who first signed in at MLO on December 19, 1974, discussed the professional and public responsibilities of climate change scientists:

> What we need from scientists first and foremost is what we have always needed. They have to be absolutely open minded and tough minded in evaluating the evidence and conducting experiments and doing calculations. They have to let the chips fall where they may. . . . We need the rigorous, hard-nosed, extremely skeptical scientific attitude that is very independent minded. On the other hand, we're asking them to communicate across all professions and across all livelihoods at the same time. (Cicerone 2007b)

How the communication Cicerone spoke about can best take place among independent-minded scientists can lead to differences of opinion. For example, Dave Hofmann's passion for collecting world-class atmospheric data at MLO and other observatories, while leaving the implications to modelers and policy planners, once clashed with Pieter Tans' enthusiasm for spreading the word about the importance of slowing the increase in carbon dioxide. Hofmann was then director of NOAA's Global Monitoring Division (GMD), and Tans was leader of GMD's Carbon Cycle Greenhouse Gases Group. Both Hofmann and Tans have strong reputations in the atmospheric sciences community, and they didn't allow their differences to get in the way of their professional relationship. In December 2006, shortly after the controversy, which was publicized in the press, both men made presentations at the annual GMD meeting in Boulder, where they engaged in friendly conversation between sessions and sat together during some of the presentations.

This and similar controversies related to climate change research led to a study by the Government Accountability Office (GAO), which issued a report in May 2007 under a title that succinctly summarizes its principal findings: "Policies Guiding the Dissemination of Scientific Research from Selected Agencies Should Be Clarified and Better Communicated." The report (GAO 2007: 32–33) concludes that

> More and more of the major policy debates of the day hinge on the results of scientific research. Therefore, timely and thorough dissemination of research results within the research community and to the public at large is crucial. . . . Moreover, when disputes arise about an agency's decision regarding a dissemination request, few researchers are aware of the dispute resolution processes that exist within their agency.

The Department of Commerce revised its public communications guidelines in anticipation of the GAO study (US Dept. of Commerce 2007). The revised policy allows scientists to openly announce their findings, even controversial ones, so long as they do not claim to represent NOAA when making statements about policy matters. NOAA administrator Conrad C. Lautenbacher Jr. (2007) informed NOAA employees about the new policy, which is summarized on the agency's Web site (http://www.communications.noaa.gov/mediaguidance.htm) and encourages open communication:

> I am proud of the science that we pursue for the good of the American people, and I believe we should do whatever we can to ensure that the results of our research are widely shared. Let's continue telling our stories.

Remembering Dr. David Hofmann (1937–2009)

This book demonstrates that the former and current staff at MLO anticipated Conrad Lautenbacher's suggestion, "Let's continue telling our stories." Dr. David Hofmann, Dr. James Butler's predecessor as director of the Global Monitoring Division, played a major role in the development of this book, for he approved many of the extended visits to MLO by the author. While Hofmann did not work at MLO, he directed the observatory's expansion, guided it through the traditional budget gaps imposed from above, and wrote a series of papers about some of its key stratospheric ozone and aerosol findings. Hofmann died on August 11, 2009, as this book, which he had read, was being prepared for publication. In an announcement of the same date to the GMD staff, Butler wrote,

> You all know how valuable and important Dave was to this laboratory and as a friend to many of us. I cannot say too much about the quality of his research, which was always exemplary and meaningful. In many ways, Dave has represented what our work was about—long-term commitment, attention to detail, and focus on the important issues. He appreciated the value of science and was a wise and often fun person to spend time with.

Debating the Implications of the MLO CO_2 Record

Of all the measurements and studies conducted at MLO over the years, none have had the impact of the thin green line drawn on hundreds of charts by Dave Keeling's old infrared gas analyzer. For in spite of the observatory's relative obscurity, the graph depicting the carbon dioxide record collected at MLO has become world famous. It helped trigger a vast expansion of climate change research. And it became an icon in an unprecedented debate over climate change and its potential consequences that is very much a part of MLO's history.

As related in chapter 7, shortly before Dave Keeling's carbon dioxide measurements began at MLO, Roger

Revelle and Hans Suess (1957) of the Scripps Institute of Oceanography wrote about the potential impact of CO_2 on climate change:

> Human beings are now carrying out a large scale geophysical experiment of a kind that could not have happened in the past nor be reproduced in the future. . . . An opportunity exists during the International Geophysical Year to obtain much of the necessary information.

There was a pragmatic side to Revelle, which was displayed in a letter he wrote to Congressman Jim Bates on July 14, 1988, more than twenty-five years after Keeling had discovered the annual increase in CO_2. Revelle wrote that, while questions about the impact of climate change remained to be answered, there were practical benefits to be had by taking action to reduce greenhouse gases whether or not greenhouse warming materialized:

> The recent extreme temperatures and drought in the Middle West are not necessarily due to the greenhouse effect. Most scientists familiar with the subject are not yet willing to bet that the climate this year is the result of "green house warming." As you very well know, climate is highly variable from year to year, and the causes of these variations are not at all well understood. . . . My own personal belief is that we should wait for another ten or twenty years to really be convinced that the green house is going to be important for human beings, in both positive and negative ways. From this belief I conclude that we should take whatever actions would be desirable whether or not the green house effect materializes. A transition to nuclear power and development of publicly acceptable means for water and energy conservation are actions of this type.

The upper limit of "ten or twenty years" of waiting suggested by Revelle has passed, and the science of global warming has expanded far beyond the carbon dioxide studies begun by Keeling to include other factors having the potential to alter climate. They include the warming potential of water vapor, methane, and other greenhouse gases measured with precision at MLO. Also included are subtle changes in solar irradiance, the direct and indirect effects on the atmosphere of land use changes, regional and transoceanic air and dust pollution, deposition of black carbon and dust on snow and ice, aircraft contrails, changes in cloudiness, and urban and industrial heat islands, all of which can influence weather and climate to some degree, and all of which have been the subjects of many scientific studies.

The 2007 report of the Intergovernmental Panel on Climate Change (IPCC) addressed these broader issues related to global warming, some of which are the result of human activity and some of which are natural. The IPCC received worldwide publicity on October 12, 2007, when it was announced that the IPCC and former vice president Al Gore would share the 2007 Nobel Peace Prize:

> The Norwegian Nobel Committee has decided that the Nobel Peace Prize for 2007 is to be shared, in two equal parts, between the Intergovernmental Panel on Climate Change (IPCC) and Albert Arnold (Al) Gore Jr. for their efforts to build up and disseminate greater knowledge about man-made climate change, and to lay the foundations for the measures that are needed to counteract such change. (Nobel Committee 2007)

While the role of governments in the approval of IPCC scientific reports has been debated, questioned, and criticized, there is a growing consensus within the climate science community that global warming is exacerbated by human activity, and NOAA's Global Monitoring Division Web page recently summarized the potential consequences.

> What are the potential impacts of global warming and a changing climate? Our health, agriculture, water resources, forests, wildlife, and coastal areas are vulnerable to global warming and the climatic changes it will bring. The IPCC concluded that "climate change is likely to have wide-ranging and mostly adverse impacts on human health, with significant loss of life." A few degrees of warming increases the chances of more frequent and severe heat waves, which can cause more

heat-related death and illness. Greater heat can also mean worsened air pollution, as well as damaged crops and depleted water resources. Warming is likely to allow tropical diseases, such as malaria, to spread northward in some areas of the world. It will also intensify the earth's hydrological cycle. This means that both evaporation and precipitation will increase. Some areas will receive more rain, while other areas will be drier. At the same time, extreme events like floods and droughts are likely to become more frequent. Warming will cause glaciers to melt and oceans to expand. The IPCC projects that sea level will rise one half foot to three feet over the next century. This threatens low-lying coastal areas. Scientists are also concerned that warming could lead to more intense storms. (GMD 2008)

Dr. James Butler, the new director of the Global Monitoring Division, believes global warming is a significant threat and has shifted his responsibilities accordingly:

> My administrative duties are pushing more and more towards supporting the carbon dioxide efforts that we have to make. We've got a great global network today. . . . The kinds of decisions society is going to have to make over the next 30 to 50 years as they take CO_2 emissions, hopefully, down to zero by 50 years, from fossil fuel and from land-use change, if we could do that, then, it would be a great success for us. We're still going to have to live in a climatically changed world, but we will have stopped polluting ourselves continuously with this substance, which wasn't recognized so long ago as a pollutant. What we need now is a strong observational and analysis network that we can provide society's support for the decisions they make. (Butler, video interview with author, Boulder, Colorado, May 2, 2007)

Yet while many scientists are outspoken in their concerns about the dangers of unchecked global warming, there is a wide divergence of views about the details in the scientific community and within NOAA, which recognizes that "One of the most vigorously debated topics on Earth is the issue of climate change" (NCDC 2008). This became quite evident while researching this book and interviewing a wide cross section of government, academic, and retired atmospheric scientists with MLO connections. At one extreme, a respected MLO–connected climate scientist feels so strongly about the adverse consequences of global warming that he said Nuremberg-style trials should be held for political leaders who have failed to implement corrective action. Dr. Glenn Shaw, an equally respected climate scientist at the Geophysical Institute of the University of Alaska, who made important aerosol discoveries in the Arctic and during numerous visits to MLO, offered a very different perspective in a newspaper commentary:

> In the 1970s as a young scientist at the Geophysical Institute I wrote passionate letters complaining that for the first time in the geologic era man was changing the atmosphere of the planet. I argued that continued dumping of carbon dioxide into the atmosphere would be associated with a warming of the entire earth and pled for attention to this matter. The letters were ignored.
>
> They were ignored because in the 1970's *Newsweek*, the *Christian Science Monitor*, the *New York Times* and countless books and articles were warning of the dangers of global cooling.
>
> Things have changed. Global warming is now being noted, and I in the meantime have become a little skeptical about some of the claims being put forth. I'm skeptical despite the fact that "everybody knows that the science is in." The science isn't even close to being in.
>
> A significantly large fraction of the science being done on global climate change is perhaps not wrong, but not enough, a little naive, repetitive and incorporating only a fraction of the complexity required to base policy on. Though we scientists don't advertise it much, science is often muddled when it is working on difficult new problems and most especially when the problems start to become political. And the issue of global warming has become massively political. Special interests abound. Try getting funding while being a skeptic. (Shaw 2007a)

Shaw (personal communication, November 8, 2007) observes that, through his talks and papers about his discoveries at MLO and elsewhere, he had unwittingly contributed to what he describes as the "irrational proportions" of the public discourse over global warming"

> The hysteria over greenhouse [warming] is like nothing I ever saw before. Perhaps the overreacting during the cold war, what with the bomb shelters, the red baiters, etc., might qualify. Also, the tulip mania in Holland was [an] example of irrational, hysterical behavior.

The late Dr. Joanne Simpson (1923–2010), a famed meteorologist and hurricane scientist who was married to Robert Simpson, the founder of MLO, recently expressed her thoughts about the global warming debate in a privately circulated message that was posted online at Climate Science: Roger Pielke Sr. Research Group News, a portion of which follows:

> Since I am no longer affiliated with any organization nor receive any funding, I can speak quite frankly. For more than a decade now "global warming" and its impacts has become the primary interface between our science and society. A large group of earth scientists, voiced in an IPCC statement, have reached what they claim is a consensus of nearly all atmospheric scientists that man-released greenhouse gases are causing increasing harm to our planet. They predict that most icepacks including those in the Polar Regions, also sea ice, will continue melting with disastrous ecological consequences including coastal flooding. There is no doubt that atmospheric greenhouse gases are rising rapidly and little doubt that some warming and bad ecological events are occurring. However, the main basis of the claim that man's release of greenhouse gases is the cause of the warming is based almost entirely upon climate models. We all know the frailty of models concerning the air-surface system. We only need to watch the weather forecasts. However, a vocal minority of scientists so mistrusts the models and the complex fragmentary data, that some claim that global warming is a hoax. They have made public statements accusing other scientists of deliberate fraud in aid of their research funding. Both sides are now hurling personal epithets at each other, a very bad development in Earth sciences. . . .
>
> What should we as a nation do? Decisions have to be made on incomplete information. In this case, we must act on the recommendations of Gore and the IPCC because if we do not reduce emissions of greenhouse gases and the climate models are right, the planet as we know it will in this century become unsustainable. But as a scientist I remain skeptical. (Simpson 2007)

Then there's the retired climate scientist who once operated a sophisticated experiment at MLO and who strongly disagrees with his colleagues who appear before government committees to speak about global warming. He compared the process with Lysenkoism, an ideology from the former Soviet Union in which scientists were required to support a state-sanctioned view of agricultural science or be declared enemies of the state. "They get these guys to dress up in suits and talk about only one side of the issue; it's a huge war that our field is caught up in," he said when interviewed for this book. "The good side is that it really might produce some results I'm happy with." The good side, he explained, is less dependence on fossil fuels and increased reliance on solar and wind power.

The global warming debate has gone well beyond the scientific community, for politicians, celebrities, pundits, and a host of bloggers have entered the fray sparked by the carbon dioxide measurements begun at MLO more than half a century ago. In his book, *An Inconvenient Truth*, former vice president Al Gore (2006) describes climate change as a "moral and spiritual challenge":

> The voluminous evidence now strongly suggests that unless we act boldly and quickly to deal with the underlying causes of global warming, our world will undergo a string of terrible catastrophes, including more and stronger storms like Hurricane Katrina, in both the Atlantic and the Pacific.

New York City mayor Michael R. Bloomberg spoke with reporters about global warming after a speech he gave before the General Assembly of the United Nations:

> "Terrorists kill people. Weapons of mass destruction have the potential to kill an enormous amount of people," Mr. Bloomberg told reporters after addressing the U.N. General Assembly, but "global warming in the long term has the potential to kill everybody. . . . No scientist knows for sure what's going to happen, but you don't want to wait to find out." (Avni 2008)

Public Television talk show host Charlie Rose asked Ted Turner (2008), founder and chairman of the United Nations Foundation, what will happen if global warming is not addressed immediately:

> Not doing it will be catastrophic. We'll be eight degrees hotter in ten, not ten but 30 or 40 years and basically none of the crops will grow. Most of the people will have died and the rest of us will be cannibals. Civilization will have broken down. The few people left will be living in a failed state—like Somalia or Sudan—and living conditions will be intolerable. The droughts will be so bad there'll be no more corn grown. Not doing it is suicide.

A 175-page minority report from the US Senate Environment and Public Works Committee (EPW 2007) offers an opposing view:

> Over 400 prominent scientists from more than two dozen countries recently voiced significant objections to major aspects of the so-called "consensus" on man-made global warming. These scientists, many of whom are current and former participants in the UN IPCC (Intergovernmental Panel on Climate Change), criticized the climate claims made by the UN IPCC and former Vice President Al Gore.

The views of those skeptical of the magnitude of the human impact on global warming received an unexpected boost during November 2009 when the controversy burst onto the Internet with the unauthorized disclosure of several thousand documents and e-mails from the Hadley Climatic Research Unit (CRU) at England's East Anglia University. The CRU and its collaborators played a key role in drafting the IPCC reports, and the leaked documents reveal an unseemly side of some of the world's most influential climate scientists (none of whom were associated with the Mauna Loa Observatory). Some of the e-mails garnered worldwide interest, because they appear to discuss the manipulation of climate data, efforts to subvert the peer review process, the blacklisting of some climate scientists with differing views, and the apparent withholding and destruction of documents, some of which had been requested under the British Freedom of Information Act. Investigations were quickly announced.

Some scientists and environmentalists brushed off the affair, which quickly came to be known as Climategate, as routine e-mails between busy scientists, some of whom were frustrated by Freedom of Information Act requests from several Internet climate bloggers who had published flaws in their research. Editorials in leading science publications asserted that the scientific findings related to global warming were affirmed by a large body of research that extended well beyond the Hadley CRU. Working Group I of the Intergovernmental Panel on Climate Change (IPCC 2009) issued a statement saying it "firmly stands behind the conclusions of the IPCC Fourth Assessment Report, the community of researchers and its individuals providing the scientific basis, and the procedures of IPCC Assessments."

But others declared that climate data should be transparently open, and that some of the leaked documents had damaged the reputation and integrity of climate science. Environmental journalist George Monbiot (2009), a leading believer in human-induced climate change, called the e-mails a "major blow" that "could scarcely be more damaging." Global warming believer Dr. James Lovelock (2010) told the *Guardian* that the reported content of the e-mails had left him feeling "utterly disgusted." "Fudging the data in any way whatsoever is quite literally a sin against the holy ghost of science," he said. "I'm not religious, but I put it that way because I feel so strongly. It's the one thing you do not ever do. You've got to have standards." Dr. Neil Frank (2010), who followed MLO founder Robert Simpson as director of the National Hurricane Center (1974 to 1987), wrote,

> Without question, there should be an immediate and thorough investigation of the scientific debauchery revealed by "Climategate." . . . If the climate models cannot get it right for the past 10 years, why should we trust them for the next century? . . . Climategate reveals how predetermined political agendas shaped science rather than the other way around. It is high time to question the true agenda of the scientists now on the hot seat and to bring skeptics back into the public debate.

Multiple investigations were announced, and one of the first was aimed at determining who leaked the e-mails. This criminal inquiry was led by the Norfolk Constabulary with assistance from various United Kingdom agencies, including the National Domestic Extremism Unit. The author was surprised to learn that his name had been added to the list of those the police wished to interrogate. In a February 2010 e-mail that requested an interview, Norfolk police officer David Irwin explained that he wanted to discuss my request for information from the University of East Anglia regarding the improperly released e-mails. During a subsequent twenty-minute interview, Officer Irwin seemed satisfied by the explanation that the information, which the university never provided, would be helpful for various writings about the affair (including this book). The police had not announced the results of their investigation when the book was being finalized in spring 2011.

While the police were conducting their criminal inquiry during 2010, several academic and scientific investigations were announced, conducted, and concluded. These investigations generally exonerated the principal players while chiding them for a lack of openness and suggesting the need for more transparency in the scientific process. Several major scientific societies and some leading scientists condemned the unauthorized release of the e-mails and endorsed the results of the investigations. But some commentators and scientists continued to raise questions about the matter. They were joined by assorted writers and Internet bloggers who questioned the impartiality of the investigations. Partisans on both sides engaged in the name calling that had troubled Joanne Simpson. Some leading believers in anthropogenic climate change referred to the skeptics as "deniers"; some of the latter responded by calling the former "alarmists." Many articles in the press and online sprang from the controversy, and a search of the World Wide Web will yield dozens of writings on both sides of the issue. A particularly revealing profile of one of the more prominent e-mailers in the affair was published in *Nature* (Adam 2010). Several books critical of the scandal were also published (Montford 2010; Mosher and Fuller 2010; Pearce 2010; and Sussman 2010).

Clearly the global warming controversy is far from over, and prominent scientists have weighed in on both sides, including a number of solar physicists who have predicted that an unusually long period of diminished sunspot activity between solar cycles 23 and 24 suggests the possibility of global cooling. Among them is Professor Henrik Svensmark of the Center for Sun-Climate Research, Technical Institute of Denmark. Svensmark (2007) and others have criticized climate models and proposed that the increase in cosmic rays that enter earth's atmosphere when the sun is quiet stimulates an increase in cloud cover that can lead to global cooling. Svensmark's cosmic-ray theory triggered a series of scientific papers by both critics and supporters.

All sides of the climate change issue will be tested over time, and the results may show that a combination of factors is at work. Meanwhile, MLO founder Robert Simpson (telephone interview with author, March 18, 2007) is concerned both by the sometimes rancorous debate within the scientific community and how best to respond to the science:

> Any time things as weighty as the questions about global warming come up, you always have people who play the role of devil's advocate, and controversies and questions and, all too frequently, personal insults are exchanged. I think it's too bad that people can't look at something as important as that and say, well, we don't need to know what's causing it. We need to do something about it and try to find some way to prevent disasters from meeting us earlier than we might have thought that they would.

In 1955 Harry Wexler endorsed Simpson's goal of establishing a new observatory on Mauna Loa, for he believed it would provide an ideal site for monitoring carbon dioxide during the International Geophysical Year. Back then neither man could have foretold that the measurements begun during the IGY would continue for more than half a century. Nor could they have predicted the major discoveries and research programs that would spring from the observatory and, especially, from Charles David Keeling's vacuum tube gas analyzer, much less the global controversies and debates that would accompany them.

Harry Wexler understood better than any scientist of his time both the potential warming that might accompany rising carbon dioxide and the sun's historical role in both cooling and warming. One can only wonder how he would have responded to today's debates over the human role in climate change and how much later the related discoveries and controversies would have occurred had Robert Simpson not met Ralph Stair at Sunspot, New Mexico, on a dusty day in June of 1955.

References

Adam, David. 2010. Climate: The hottest year. *Nature* 468: 362–364. (Accessed at http://www.nature.com/news/2010/101115/full/468362a.html.)

Avni, Benny. 2008. Mayor compares threat of global warming to terrorism. *New York Sun*, February 12. (Accessed at http://www.nysun.com/article/71103.)

Barnes, John. 2006a. February 2006 activities report. Mauna Loa Observatory, Monthly Reports (CD-ROM), Hilo, Hawai'i.

———. 2006b. June 2006 activities report. Mauna Loa Observatory, Monthly Reports (CD-ROM), Hilo, Hawai'i.

———. 2006c. October 2006 activities report. Mauna Loa Observatory, Monthly Reports (CD-ROM), Hilo, Hawai'i.

BLNR. 2006. Minutes for the meeting of the Board of Land and Natural Resources, March 10. (Accessed at http://hawaii.gov/dlnr/chair/meetings/minutes/060310-minutes.pdf.)

Cicerone, Ralph J. 2007a. Climate change: It's not just for scientists. Keynote address at the 50th Anniversary of the Global Carbon Dioxide Record, Symposium and Celebration. Kona, Hawai'i, November 28–30. (Accessed at http://www.esrl.noaa.gov/gmd/co2conference/transcripts/day1_cicerone.pdf.)

———. 2007b. Interview on climate change science. (Accessed at http://cwce.ametsoc.org/post?id=1735425.)

CO_2 Conference. 2007. Symposium report. 50th Anniversary of the Global Carbon Dioxide Record, Symposium and Celebration, Kona, Hawai'i, November 28–30. (Accessed at http://co2conference.org.)

EPW. 2007. Over 400 prominent scientists disputed man-made global warming claims in 2007. US Senate Environment and Public Works Committee (Minority), December 20. (Accessed at http://epw.senate.gov/public/index.cfm?FuseAction=Minority.SenateReport)

Frank, Neil. 2010. Climategate: You should be steamed. *Houston Chronicle*, January 2. (Accessed at http://www.chron.com/disp/story.mpl/editorial/outlook/6795858.)

Kiess, C. C., C. H. Corliss, Harriet K. Kiess, and Edith L. R. Corliss. 1957. High dispersion spectra of Mars. *Astrophysical Journal* 126: 579–584.

GAO. 2007. Policies guiding the dissemination of scientific research from selected agencies should be clarified and better communicated. Report by the Government Accountability Office to the Chairman, Subcommittee on Commerce, Justice, Science, and Related Agencies, Committee on Appropriations, US Senate, May. (Accessed at http://www.gao.gov/new.items/d07653.pdf.)

GMD. 2008. Frequently asked questions. Global Monitoring Division, Earth System Research Laboratory, National Oceanic and Atmospheric Administration. (Accessed at http://www.esrl.noaa.gov/gmd/infodata/faq_cat-1.html#28.)

Gore, Al. 2006. *An Inconvenient Truth: The Planetary Emergency of Global Warming and What We Can Do About It*. New York: Rodale Books.

Hansen, James, Andrew Lacis, Andrew Prather, and Michael Prather. 1989. Greenhouse effect of chlorofluorocarbons and other trace gases. *Journal of Geophysical Research* 94: 16,417–16,421.

Hegger, Dawn T. 2006. Enforcement File No. ENF HA-06-40. State of Hawai'i, Department of Land and Natural Resources, March 10.

Hoerling, M., J. Eischeid, X. Quan, and T. Xu. 2007. Explaining the record US warmth of 2006. *Geophysical Research Letters* 34: L17704, doi:10.1029/2007GL030643.

Hofmann, David J. 2007. The NOAA Annual Greenhouse Gas Index (AGGI). NOAA Earth System Research Laboratory, Boulder, Colorado. (Accessed at http://www.esrl.noaa.gov/gmd/aggi/.)

Hofmann, D. J., J. H. Butler, T. J. Conway, E. J. Dlugokencky, J. W. Elkins, K. Masarie, S. A. Montzka, R. C. Schnell, and P. Tans. 2006a. Tracking climate forcing: The annual greenhouse gas index, EOS. *Transactions of the American Geophysical Union* 87 (November 16): 509–511.

Hofmann, D. J., J. H. Butler, E. J. Dlugokencky, J. W. Elkins, K. Masarie, S. A. Montzka, and P. Tans. 2006b. The role of carbon dioxide in climate forcing from 1979–2004: Introduction of the Annual Greenhouse Gas Index. *Tellus B* 58B: 614–619.

IPCC. 2007a. FAQ 2.1, Box 1: What is radiative forcing? In S. Solomon, D. Qin, M. Manning, Z. Chen, M. Marquis, K. B. Averyt, M. Tignor, and H. L. Miller, eds., *Climate Change 2007: The Physical Science Basis. Contribution of Working Group I to the Fourth Assessment Report of the Intergovernmental Panel on Climate Change*, 101. Cambridge, UK, and New York: Cambridge University Press.

———. 2007b. Summary for policymakers. In S. Solomon, D. Qin, M. Manning, Z. Chen, M. Marquis, K. B. Averyt, M. Tignor, and H. L. Miller, eds., *Climate Change 2007: The Physical Science Basis. Contribution of Working Group I to the Fourth Assessment Report of the Intergovernmental Panel on Climate Change*, 10. Cambridge, UK, and New York: Cambridge University Press.

———. 2009. Statement by Working Group I of the Intergovernmental Panel on Climate Change on stolen emails from the Climatic Research Unit at the University of East Anglia, United Kingdom. Bern, December 4. (Accessed at http://www.ipcc.ch/pdf/presentations/WGIstatement04122009.pdf.)

Irion, Robert. 2008. Homing in on black holes. *Smithsonian* 39 (April): cover and 44–53.

Keeling, Charles D. 1998. Rewards and penalties of monitoring the Earth. *Annual Review of Energy and the Environment* 23: 25–82.

Keeling, Ralph. 2008. Atmospheric science: Recording Earth's vital signs. *Science* 319 (March 28): 1771–1772.

Kerr, J. B. 1996–1997. Operation of Brewer instruments at Mauna Loa Observatory. In CMDL Summary Report 24.

Kiess, C. C., C. H. Corliss, Harriet K. Kiess, and Edith L. R. Corliss, 1957. High dispersion spectra of Mars. *Astrophysical Journal* 126: 579–584.

Lovelock, James. 2010. Quoted in Leo Hickman, Humans are too stupid to prevent climate change. *Guardian*, March 29. Copyright by Leo Hickman/Guardian News & Media Ltd. 2010. Used with permission. (Accessed at http://www.guardian.co.uk/science/2010/mar/29/james-lovelock-climate-change.)

Mims, Forrest M. III. 2005a. Geological science on Hawai'i's Mauna Loa. *The Citizen Scientist*, June 17. (Accessed at http://www.sas.org/tcs/weeklyIssues_2005/2005-06-17/news2/index.html.)

———. 2005b. Geology update from Hawai'i's Mauna Loa. *The Citizen Scientist*, December 23. (Accessed at http://www.sas.org/tcs/weeklyIssues_2005/2005-12-23/news3/index.html.)

MLO. 2007. *Mauna Loa Observatory's 50 Years of Science and Observation*. Mauna Loa Observatory, November.

Monbiot, George. 2009. The Knights Carbonic, November 23. (Accessed at http://www.monbiot.com.)

Montford, Andrew. 2010. *The Hockey Stick Illusion: Climategate and the Corruption of Science*. Independent Minds.

Mosher, Steven, and Thomas W. Fuller. 2010. *Climategate: The Crutape Letters*. CreateSpace.

NAS. 1995. On being a scientist: Responsible conduct in research. Committee on Science, Engineering, and Public Policy, National Academy of Sciences, National Academy of Engineering, Institute of Medicine. Washington, DC: National Academy Press. (Accessed at http://www.nap.edu/readingroom/books/obas/index.html.)

———. 2003. Policy on committee composition and balance and conflicts of interest for committees used in the development of reports. National Academy of Sciences, National Academy of Engineering, Institute of Medicine and National Research Council. May 12. (Accessed at http://www.nationalacademies.org/coi/bi-coi_form-0.pdf.)

NCDC. 2008. Global warming: Frequently asked questions. National Climatic Data Center, National Oceanic and Atmospheric Administration. (Accessed at http://lwf.ncdc.noaa.gov/oa/climate/globalwarming.html.)

NOAO. 2008. Computation of the cost of a night of Keck Telescope observing. National Optical Astronomy Observatory. (Accessed at http://www.noao.edu/system/tsip/keck_cost.html.)

Nobel Committee. 2007. The Nobel Peace Prize for 2007. The Norwegian Nobel Committee. (Accessed at http://nobelprize.org/nobel_prizes/peace/laureates/2007/press.html.)

Palmer, Fredrick D. 2007. Coal, energy security and carbon—the path forward. Paper presented at the 50th Anniversary of the Global Carbon Dioxide Record Symposium and Celebration, Kona, Hawai'i, November 28–30. (Accessed at http://www.esrl.noaa.gov/gmd/co2conference/transcripts/day1_palmer.pdf.) (Note: The 80-foot-high exposed coal deposit in a slide shown at the CO_2 symposium was later revealed to have been digitally altered to stretch the image and to replace a murky sky with cumulus clouds in a blue sky as first reported at http://the-black-butterfly-effect.blogspot.com/2008/06/another-coal-bluff.html.)

Pearce, Fred. 2010. *The Climate Files.* Guardian Books.

Pielke, Roger Sr. 2007. Summary conclusions of climate science. Climate Science: Roger Pielke Sr. Research Group Weblog, September 2. (Accessed at http://climatesci.colorado.edu.)

Revelle, Roger, and Hans E. Suess. 1957. Carbon dioxide exchange between atmosphere and ocean and the question of an increase of atmospheric CO_2 during the past decades. *Tellus* 9: 18–27.

Ryan, S., E. J. Dlugokencky, P. P. Tans, and M. E. Trudeau. 2006. Mauna Loa volcano is not a methane source: Implications for Mars. *Geophysical Research Letters* 33 (June 22): L12301, doi:10.1029/2006GL026223.

Schneider, Stephen. 2008. Modeling the future: A talk with Stephen Schneider. Edge Foundation, Inc. (Accesed at http://www.edge.org/3rd_culture/schneider08/schneider08_index.html.)

Schnell, Russell. 1995. October 1995 activities report, November 16. In *Geophysical Monitoring for Climatic Change,* Mauna Loa Observatory, Monthly Reports, Hilo, Hawai'i, 1995.

Schwartz, Stephen E., Robert J. Charlson, and Henning Rodhe. 2007. Quantifying climate change—too rosy a picture? *Nature Reports* 2 (July). (Accessed at http://www.nature.com/climate/2007/0707/pdf/climate.2007.22.pdf.)

Shaw, Glenn. 2007. Impact of global warming looms on the horizon. *Fairbanks Daily News-Miner,* April 22. (Accessed at http://newsminer.com/2007/04/22/6603/.)

Simpson, Joanne. 2007. TRMM (Tropical Rainfall Measuring Mission) data set potential in climate controversy. Climate Science: Roger Pielke Sr. Research Group News, February 27. (Accessed at http://climatesci.org/2008/02/27/trmm-tropical-rainfall-measuring-mission-data-set-potential-in-climate-controversy-by-joanne-simpson-private-citizen/.)

Socolow, Robert H. 2007. Climate change mitigation under strong carbon constraints. Paper presented at the 50th Anniversary of the Global Carbon Dioxide Record Symposium and Celebration, Kona, Hawai'i, November 28–30. (Accessed at http://www.esrl.noaa.gov/gmd/co2conference/transcripts/day1_socolow.pdf.)

Spinrad, Richard W. 2007. Welcoming remarks. 50th Anniversary of the Global Carbon Dioxide Record Symposium and Celebration, Kona, Hawai'i, November 28–30. (Accessed at http://www.esrl.noaa.gov/gmd/co2conference/transcripts/day1_spinrad.pdf.)

Starr, Forest, and Kim Starr. 2008. *Hypochoeris radicata:* Hairy cat's ear, gosmore (Asteraceae). In *Plants of*

Hawai'i. (Accessed at http://www.hear.org/starr/hiplants/images/thumbnails/html/hypochoeris_radicata.htm.)

Sussman, Brian. 2010. *Climategate: A Veteran Meteorologist Exposes the Global Warming Scam*. WND Books.

Svensmark, Henrik. 2007. Astronomy and geophysics cosmoclimatology: A new theory emerges. *Astronomy & Geophysics* 48: 1.18–1.24.

Turner, Ted. 2008. Television interview conducted by Charlie Rose for the Public Broadcasting System, April 1. (Acccessed at http://newsbusters.org/blogs/brent-baker/2008/04/02/turner-iraqi-insurgents-patriots-inaction-warming-cannibalism.)

US Dept. of Commerce. 2007. Public communications. Number: DAO 219-1, Effective Date: 2007-05-14. (Accessed at http://dms.osec.doc.gov/cgi-bin/doit.cgi?204:112:f6c4d0b1f837467844b97280a1244c8f4e63956ea9862b66c099a79e1e443b3b:267.)

Vanicek, Karel, Thomas Frei, Zenobia Litynska, and Alois Schmalwieser. 1999. UV-Index for the public: A guide for publication and interpretation of solar UV-Index forecasts for the public. Prepared by Working Group 4 of the COST-713 Action "UVB Forecasting," Brussels. (Accessed at http://www.uv-index.ch/images/Leitfaden_COST-713.pdf.)

Wilkes, Charles. 1845. United States Exploring Expedition during the years 1838, 1839, 1840, 1841, 1842. Vol. 4. Philadelphia: Lea and Blanchard, 163. (Accessed at http://www.sil.si.edu/digitalcollections/usexex/follow-02.htm.)

CHAPTER FIFTEEN

The Next Fifty Years

On October 4, 1957, Jack Pales, Robert Williams, and Clifford Kutaka were approaching three months of duty at MLO when the former Soviet Union initiated the space race by launching Sputnik I, the world's first artificial satellite. The building where the three men worked is still in use. And, according to Sergei Khrushchev, son of the late prime minister of the USSR, Soyuz rockets are still launched from the same pad at the Baikonur Cosmodrome where the Sputnik was launched in 1957 (Zak 2007).

Similarly, Charles Keeling's carbon dioxide analyzer operated nearly continuously for forty-eight years at MLO. Several Dobson spectrophotometers have been used at MLO, all of them dating back to the early days of MLO. Several Eppley radiometers have been used to measure atmospheric transmission at MLO since November 1957, and they are all of identical construction.

What will happen at MLO over the next half century as today's instruments grow old and become antiquated? Will they be replaced with new technology? Or will they serve as long or longer than Keeling's famous CO_2 analyzer? What new kinds of measurements will be made at MLO? These are among the many questions that come to mind as MLO enters its second half century.

Maintaining the MLO Data Record

The future of the Mauna Loa Observatory has never been certain and has always been subject to the vagaries of federal budgets and the possible eruption of Mauna Loa. No matter how well their instruments are calibrated and maintained, and irrespective of the perceived importance of the data, the scientists connected with MLO know that the future of their measurement programs is ultimately beyond their control. In the end, the future of MLO and the Global Monitoring Division's (GMD) other observatories is dependent on NOAA's budget and the priority assigned to their measurements. Having served as a director of MLO and then all of GMD's observatories, Dr. Russell Schnell knows this as well as anyone:

> I think the priorities are maintaining the facility as a viable institution, and you might laugh, but there are always problems with, for instance, maintaining the road. It costs $100,000 a year just to fix the potholes. Electricity costs go up. Our electricity bills are hundreds of thousand dollars a year there, maintaining all the instruments and the measurements. And politically, this past year, Hawaii put some heavy fines on Mauna Loa because we did not, supposedly did not, have the correct

> permits. But in the end, we prevailed because we showed them that, in fact, a lot of this was grandfathered in and we did have the right.
>
> So, there are all kinds of political issues that could shut us down or shut the observatories down, funding issues. You know, maybe in the future, it is going to be difficult to get staff that is willing to work up there. John Chin worked there for [thirty-seven] years driving up every couple of days. He drove, somebody calculated, around the world twelve times just going up to Mauna Loa. That is tremendous dedication, and we may not have that in the staff. But the priorities are to keep measurements going, even measurements that you do not know have any value, because a long record always has importance. We measured species of gases up there . . . before the ozone hole was discovered. And once the ozone hole was discovered, people realized that [CFCs] were going up very fast, and you can see the turnover as soon as the Montreal protocols went in. It is a tremendous story. How fast it was going up and how fast it came down; unbelievably good story.
>
> . . . Nobody ever thought that the amount of [solar] energy coming into the earth would change, but it does because of volcanoes [and] the dust and pollution coming out of China. You could see the annual pollution events coming out of China. So . . . the measurements are all valuable. They've got to be taken well, and they've got to be taken for a long time, because we really do not know ten, twenty, fifty years from now how valuable those measurements will be. (Schnell, video interview with author, Boulder, Colorado, April 27, 2007)

Many scientists interviewed for this book share Schnell's views about the importance of maintaining the continuity of the MLO data record, and Dr. Ellsworth Dutton's remarks (video interview with author, Boulder, Colorado, May 1, 2007) are particularly relevant:

> For understanding climate and for being able to verify the predictions of climate through observations, we feel that it's absolutely necessary to maintain a continuity of measurements that otherwise won't allow us to accomplish that goal. The breaks in the record can be very detrimental to missing phenomena, interfering with the ability to analyze the data, interrupting the ability to maintain the long-term calibration consistencies. Although, in theory, if we had absolute measurement referenced standards, we could withstand some gaps in the record better than we can now.
>
> But the reality is, we don't have that type of capability, certainly the satellites don't have it. To keep the kind of records going to make sure that we understand the complete evolution of what's happening, the continuous high-time resolution data we feel are just absolutely necessary to grab the snapshot, because once it's gone . . . we can't go back and replay it. If it gets away from us, [it] will never be recovered.
>
> There [are] sophisticated techniques for filling gaps [with] various levels of sophistication success, but the true observation, once lost, will never be captured, and we try to absolutely minimize that. It's . . . one of our higher callings as a monitoring organization.

Sam Oltmans (video interview with author, Boulder, Colorado, April 27, 2007) stresses the fact that long-term measurements have provided a baseline against which the human impact on the atmosphere can be gauged:

> I think it's one of the critical [aspects of] atmospheric science. I think of the measurements that have been made by a number of the groups in our division, the early measurements in solar radiation, in greenhouse gases, ozone . . . their value has just grown immensely with time, because we have . . . a baseline to measure against when human impacts change the atmosphere. We know that we can trace it to human activity because we have measurements when human activity was having less of an impact on the atmosphere.

Deciding what to measure is not easy, especially when there are conflicting opinions and expectations. Oltmans provides a classic example in the form of missed measurements prior to the discovery of the precipitous annual drop in Antarctic ozone popularly known as the ozone hole:

> Probably the most frustrating part is that you see something that you think is very interesting scientifically, and maybe even has important social implications, and it's very hard to convince people that you should be making the measurements that'll solve that problem. That's why there were no ozonesonde measurements from the early 1970s until the ozone hole. People didn't think it was worthwhile to make ozone profile measurements. "Don't you know everything you should know about ozone in Antarctica?" As it turns out, we didn't.

Before a baseline of measurements can be initiated or continued, the scientific community and the administrators who disburse the available funds must agree on its relevance. Deciding which measurements must be cut back or even stopped so new ones can be initiated is a process not without flaws, particularly when decision makers at the top may not fully agree or understand that measurements may have global relevance. Russell Schnell (video interview with author, Boulder, Colorado, April 27, 2007) sums up the points that matter most:

> Monitoring is not a sexy science. It is a mundane science, and it is costly. It is very costly to monitor accurately, because you have to put in so many standards, and you have to keep your standards going, and you have to be able to check your standards and make sure your data ten years ago was as good as it is today and a hundred years from now will be just as good. And sometimes, our funding goes down and sometimes we are scrambling to keep things going.

Schnell is a realist about the future of the Mauna Loa Observatory. He found creative ways to find supplemental support from other government agencies and from users of MLO's facilities when he was its director and continued to do so after he was placed in charge of all of the Global Monitoring Division's observatories. But supplemental funds alone are not enough to keep the observatories going, and Schnell philosophized about the alternative in the same interview:

> And you always think, well, I always need a little extra money to do a little extra more. But then on the other side, if you are philosophical, you can say, it is unbelievable that taxpayers would support measurements for fifty years at Mauna Loa and keep the staff going and it is a nice job, so you always feel that you should have more money, but, again, you do not have any right to it.
>
> And scientists like to sometimes think they have a right to the funding . . . but you are always competing with other people's feelings and other people's ideas, and it is frustrating in that respect. But [it's best] to step back and say, "It is really great that we have been doing it for so many years, and we just hope it will maintain itself." But again, we do not have a right to those funds, even though we think we should get them. You have to realize there are a lot of competing issues out there and other people's feelings and needs. So, be thankful we could do it for so many years.

Satellite Validation

As reported in this book, data acquired by various instruments at MLO has long been used to validate satellite observations of the ozone layer, aerosols, and water vapor. Asked what he would add to the science at the observatory if unlimited funding were available, MLO veteran Bernard Mendonca (video interview, Boulder, Colorado, April 27, 2007) replies that satellite validation is more important than adding new measurement programs:

> If money was no object, what I would start to try to do is to use the measurements that we're making there as ground truth for satellite missions. I've always proposed that the next generation of what we've done is to get really integrated measurements. The way you do that now, of course, is with satellites. So, if I had a whole bucket of money to spend, I'd try to tie these two technologies together . . . because the real hardship with satellite observations has been they drift over time. [They are] hard to calibrate. So if you can get a tie-down to, like, all these stations—you know, we started in Algeria

> and Tibet and Indonesia and Argentina and Mauna Loa and South Pole—if you tie all this up to satellites, I think you get a lot of bang for your money. That's the way I'd spend unlimited amounts of money.

Mendonca is concerned that the current mix of ground and satellite measurements is often uncoordinated, as he noted in the same interview:

> The projects that have been going up in space looking down with the satellites have been almost like independent adventures. There's no tie-in. That's just the way our system was organized, I guess. I don't know why it's not corrected. We've struggled with that for a long time with the ozone measurements. When the ozone hole was first discovered, the satellite guys . . . got in trouble because their instruments were drifting. And it took so long, I thought, before they finally tied [their data] into ours. And now that they have, everybody is happy, and you're getting very good information.

Polishing MLO–Collaborator Relationships

Since the earliest years at MLO, there have been occasional instances when collaborating scientists at universities, NOAA, and other government agencies did not fully appreciate the difficulties experienced by people working on the mountain. Bernard Mendonca (video interview, Boulder, Colorado, April 27, 2007) is well aware of this problem and believes "it's still going on." He describes a typical situation involving a hypothetical, newly arrived instrument:

> It worked in the laboratory with all the kludges and clips and stuff, and they tried to bring that out to the field. Well, you can imagine what's going to happen. [When problems developed] the good principal investigators realized that and responded to the guys in the field [and] said, "Oh, I never thought of that. Yes, why don't we try this?" The ones who had the harder time insisted that their instrument worked in the laboratory. "Why can't you get it to work in the field, you so-and-so's?"

Mendonca believes there are lessons here for collaborators who are flexible and understanding, as opposed to those who complain that "you guys can't make it work":

> And it's a lesson that I think still is around. Unless you're in the field, it's like the GI in the battlefield, you know. You train him and train him, you get him on the battlefield, and things happen not the way you were trained, not the way your equipment is supposed to work, so you've got to adapt.

Detailed instructions for collaborators posted on the MLO Web site include advice learned over the years about how best to work with the mountain crew that will help maintain and service equipment and instruments:

> During the length of the project it is essential to keep in regular communication with MLO. Keep in mind, your project is a cooperative project and is only one of many for which the observatory staff is responsible. If your project is to be successful, you must take the responsibility of keeping it updated and provide a constant flow of necessary supplies. If at all possible we will arrange for you to present a discussion to the MLO staff of your research and your specific support requirements. This informal group discussion will be especially important if we are maintaining instruments in your absence. The more familiar we are with your program the better we can support it on a day to day basis. We would also appreciate receiving copies of papers and reports that result from your work at MLO. (MLO 2008)

Perhaps collaborators might also be referred to the model of cooperation exhibited by Glenn Shaw of the University of Alaska during his many visits to MLO, details of which are related in chapter 11.

Funding MLO's Future

Beginning with the federal budget crisis that nearly closed MLO in 1964, budget challenges have often threatened the scope of measurements and even the future of the Mauna Loa Observatory and the Global Monitoring Division's other observatories at the South Pole, Samoa, Barrow, Alaska, and Trinidad Head, California.

The MLO monthly Activities Reports often include comments that can be traced directly or indirectly to budget shortfalls. It's the weekly ozonesonde launch day, but there are no ozonesondes to be launched—or maybe no balloons. Segments of the road have more holes than asphalt, but there's no money to patch them. The ambitious plans for the NDSC (now NDACC) Building must be scaled back for lack of funds.

Budget uncertainties have become just another part of the job at many government laboratories. During the ESRL Global Monitoring Conference in Boulder, Colorado, in May 2007, a scientist asked David Hofmann, then GMD director, a question that pertained to the latest congressional impasse over the federal budget. Hofmann, who had grown accustomed to budget uncertainties, didn't editorialize about the problem. He calmly responded that everything was in the air until the budget was approved and moved on to the next topic.

Salesmanship can help save budgets, and when MLO has an opportunity to show off the amazing variety of science it conducts, it gladly does so. Consider former MLO director Russell Schnell's report about a visit by a group of dignitaries in 1996:

> Operations were normal for the month with the exception of a visit from three Congressional political appointees, who sit on committees that set the Department of Commerce budget, and three accompanying NOAA Legislative Affairs staffers. Alexis Mathios, Senior Legislative Assistant, Congressman Forbes (R-NY); Chris Mann, Minority Legislative Staff, Committee on Resources; William Testerman, Legislative Assistant, Senator Jeffords (R-VT); Brian Wheeler, Director, Legislative Affairs, NOAA; Brooke Davis, Senior Staff, Legislative Affairs NOAA; and Jim McCallum, NMFS were given a full day tour of MLO including a view of Mauna Loa from the summit of Mauna Kea. The group, as a whole, was rather knowledgeable in science. The legislative assistants did share that at the budget level they work setting the Commerce budget, the MLO budget would be 2 decimal places too small to appear in the accounting sheets! (Schnell 1996)

Then there is this enlightening anecdote in one of Russell Schnell's 1998 Activities Reports:

> Visitors—March was a quiet month with only 5 visitors in addition to the scientists mentioned above. One of these visitors was Dr. Richard Greenfield, Division Director, Atmospheric Sciences, National Science Foundation. He was pleased (amazed) at the high scientific output of MLO per $ invested. As he was saying this, he was looking at the $ billion [telescope domes] on Mauna Kea. (Schnell 1998)

Perhaps MLO budget planners might get more attention if they presented their requests in Space Shuttle dollars. Launching a payload into orbit aboard the Shuttle costs about $10,000 per pound (0.45 kg). How many pounds can be launched with MLO's comparatively tiny budget? GMD director James Butler (video interview with author, Boulder, Colorado, May 2, 2007) responds in terms of AIM, a $100 million satellite that carries just three instruments designed to monitor noctilucent clouds, the highest in the atmosphere:

> That's a cheap satellite. Our entire budget for our entire [GMD] division that measures 40 gases, [solar] radiation . . . 120 people [is] under $20 million a year. . . . I don't think we should give up on satellites. I think they do have real value. They have real value in terms of identifying vegetation, water resources, and things like that, which do play into these gas cycles. We have to remember, the whole thing is an earth system, it's not just the atmosphere.

Like Dave Hofmann before him, Butler waxes philosophical about budget uncertainties over which he has no control (quoting from the same interview):

> Sir Isaac Newton once said that what goes up must come down, but I always say that what goes down does come back up again, too. So, yes, we're going to have budget cycles. We have them now. We'll have them in the future. I think money goes where the need is. I think that's what people like to do; that's what societies do. Maintaining a long-term network is difficult in this kind of a situation. . . . But, in the past, we have very, very few gaps in our data because of the perseverance and dedication of individual scientists. The scientists here are not here to work a job; they're here because they believe in it. I've known that, I've seen that for twenty years, and I feel really proud to work with a group like this.

Former GMD director Hofmann (video interview with author, Boulder, Colorado, May 1, 2007) was asked what his priorities might be if funding was not an issue. He replied that he would establish a global monitoring network that would provide the data to fully understand the carbon cycle, the natural and the manmade sources, and sinks of carbon dioxide and methane:

> And it is going to depend on the observing system. The models are fine; they are beautiful. They take all this information. But [the process requires] a data assimilation model, which means you have [an array] of measurements and you have a model. Then you compare the model to the measurements, and if they differ, you correct the model . . . the amount of that correction tells you whether there was a source or a sink. And so that way you get this information on all the sources of CO_2 in sinks and you can determine if they are natural or if they are due to man [e.g.,] fossil-fuel emissions.
>
> And if we had unlimited funds, we would expand that observing system to a point where the data assimilation system Carbon Tracker could do exactly that, could tell the difference between the carbon properties of San Francisco and Sacramento, for example. And we think it will come in the future that the need for this will be so great that we will be able to expand this system and then we will look also at the globe. And so I envision someday a global network of tall towers, perhaps, and possibly aircraft, to collect the actual air samples so that you can do isotopic work that will be adequate to do really high-resolution source-and-sink work for the globe to help people manage carbon in the future.

Upgrading the MLO Infrastructure

While MLO has needs, the observatory's physical infrastructure is in its best condition since the original building was dedicated in 1956. This is a result of various major improvements, most notably the complete remodeling of the original building, now the Keeling Building, under former MLO director John Miller in 1976. Also important have been periodic cleanup campaigns to remove years of accumulated shipping crates, antiquated equipment, dangerous wiring, and so forth.

The expense of infrastructure upgrading has long been minimized by retaining some of the old structures left over from the days when MLO was used by the Atomic Energy Commission to observe and study nuclear tests in the Pacific. Two recently installed structures were donated by the Keck Observatory when they were no longer needed at Mauna Kea.

The NDACC Building is the hallmark structure at MLO, even though it's a scaled down version of what Dave Hofmann and his GMD staff originally planned. Longtime MLO engineer Bob Uchida describes the planned structure as a "fabulous building" and a "first-class facility." Though the budget was inadequate to construct what was originally planned, the NDACC Building was a major step forward for MLO.

Uchida played a key role in much of the recent construction at MLO, particularly in the assembly of several new metal buildings that are firmly anchored to concrete slabs. Uchida's construction experience goes back to the Cloud Physics Observatory building that he built for the University of Hawai'i and where MLO's Hilo office was located from 1966 to 1976. When asked what his priorities would be if he were in charge of MLO and had an unlimited budget, Uchida (video interview with author, Hilo, Hawai'i, November 30, 2006) responds,

> Unlimited budget? Well, number one is improve on the facilities [and] pick up more staff. And with bigger staff and better facilities we could, I think, have a much better scientific environment out there.

While working on the mountain a few days later, Uchida wistfully reflected on how much better the assortment of buildings that forms the MLO science village would look had more funds been available to upgrade the facilities that were scrounged, borrowed, inherited, or built with very limited funds.

Staffing MLO

The size of MLO's staff and the number who are civil service employees are determined by the budget. Recently MLO old-timer Bob Uchida discussed what some of the MLO staff feel is a need for more personnel at the observatory. Currently there are enough personnel to conduct the weekday observing program on the mountain while maintaining the Hilo office. (The mountain is not manned during weekends and holidays unless there is an unusual event to be monitored.) But the staff can be shorthanded during vacation time and when staff members are ill or attending conferences and meetings. During emergencies and unusual atmospheric events, such as a major volcanic eruption from somewhere around the world, additional help can be a bonus, as MLO station chief and lidar expert Dr. John Barnes (personal communication with author, October 9, 2007) explains:

> A single lidar operator would be more efficient if he/she ran both my lidar and the JPL lidar. That was in the original plan. I don't see a strong reason for more data during background conditions, but it would be good to get data more than once a week for the next eruption.

Another and perhaps more complex issue concerns the replacement of retired and transferred staff members. As budgets decline, so do funds available to hire government employees and provide them with their benefits. The result for MLO has been lost positions. When former MLO director Russell Schnell returned to Boulder to direct all the Global Monitoring Division observatories, John Barnes took over his responsibilities, and Schnell's former position was cut. When Judy Pereira retired, Leslie Pajo took over her responsibilities and Pereira's position was cut. Similarly, when Bob Uchida retired, he was not replaced (but his skills were so needed that MLO hired him back on a part-time basis).

Like some other federal agencies, NOAA's Global Monitoring Division increasingly relies on a cadre of non-government contract workers to keep their science moving along without adding to their complement of civil service employees. The Cooperative Institute for Research in Environmental Sciences (CIRES) was established at the University of Colorado in 1967, and NOAA's Earth System Research Laboratory (ESRL) is the major contributor to the CIRES budget. ESRL and CIRES collaborate on many scientific studies and projects, some of which are under the Global Monitoring Division and help support work at MLO. The ESRL–CIRES collaboration has resulted in many scientific findings and papers.

While many government and CIRES employees have worked together on NOAA projects in Colorado and around the world, some for more than two decades, the CIRES employees earn significantly less than their government counterparts. The University of Colorado is aware of the discontent this can cause, and a university finance official pointed out that the CIRES workers are paid at about the same scale as regular university employees. Raising the wages of the CIRES workers to match their government counterparts, he pointed out, would not sit well with the regular university staff. CIRES staff are free to apply for government positions, but few such positions are available. CIRES charges NOAA a 20 percent overhead for each employee it provides, but even this would not necessarily reduce the salary gap if CIRES employees worked directly for the government.

In the end, the disparity in civil service and CIRES compensation will likely continue. And as federal budgets tighten, the trend of replacing retiring civil service scientists with contract or cooperative scientists will also continue. This has already occurred at MLO. Of the present staff of nine employees, five are civil service and four are Joint Institute for Marine and Atmospheric Research (JIMAR) employees. JIMAR resembles CIRES in that it is a cooperative program begun by the University of Hawai'i and NOAA in 1977. When longtime Mauna Loa chemist Alan Yoshinaga retired in 2007, he was replaced by JIMAR chemist Aidan Colton.

The MLO staff is characterized by longevity, especially on the part of employees originally from Hawai'i, and a careful review of the observatory's history will disclose instances in which staff members having a decade or more of experience made an important difference during situations in which less experienced staff would have been unprepared. Longtime staff members are often able to service equipment and perform tasks outside their usual area of responsibility. They are also better able to cope with unexpected equipment breakdowns and changes in weather. As MLO enters its second half century, Darryl Kuniyuki (video interview with author, Hilo, Hawai'i, December 7, 2006) discusses how staffing the observatory will become an issue as current senior staff members retire:

> You have to convince other people to see why you need these persons. . . . In a time of budget cuts and staff reduction, especially in our headquarters, it is really hard to justify that. Right now that seems to be the hardest thing in the whole time I have been here because other things you can eventually fix. This is something kind of not in your control, so it is really hard.

Wishing and Hoping for the Next Fifty Years

In view of the long-term difficulty of finding new funding for MLO, it may seem pointless to include in this final chapter a scientific wish list for the next fifty years—but maybe not. And perhaps there are lessons to be drawn from former MLO director Russell Schnell's innovative and entrepreneurial approaches to finding additional funds. In view of MLO's fame, perhaps a benefactor might be willing to assist in expanding the science at what has become the world's most famous meteorological observatory. So with an optimistic view of the future, this history of the observatory closes with a potpourri of ideas for the future.

A Mauna Loa Science Zone

When Dr. Rudolf Pueschel directed MLO from 1970 to 1972, he invested considerable time in attempting to preserve the clean air at the observatory. Those were the days when a propane-fueled car was used to make the daily trips to and from MLO to reduce pollution on the mountain. Pueschel even campaigned against a temporary diesel generator across the Saddle at Mauna Kea when the first astronomical observatories were being constructed.

The monitoring instruments at MLO today are far more varied and sensitive than those in use during Pueschel's time, and the MLO staff has long been concerned about pollution from visiting vehicles and, worse, the possibility of new dust or pollution sources being constructed along the Mauna Loa road. The latter possibility became a special concern when Native Hawaiians succeeded in their movement to prevent new observatories from being constructed on Mauna Kea and astronomers began looking at Mauna Loa. In 2005, Mauna Loa Observatory's status as the lone observatory site on the big mountain ended when the State of Hawai'i approved the All-Sky Survey High Resolution Air-shower detector (Ashra), a specialized observatory sponsored by the University of Tokyo in collaboration with the University of Hawai'i at Hilo and Mānoa, the National Taiwan University, and a consortium of additional Japanese universities (Sasaki 2007). Ashra consists of two telescope arrays, one having twelve small telescopes and the other having four. The arrays are designed to detect very high energy particles arriving from space. The Ashra installation, which began service in 2007, is only about 1 mile (1.6 km) east of MLO on the south side of the Mauna Loa road. Based on a diagram of the facility, the site measures 260 x 230 feet (79 x 70 m), or about three-fourths of an acre.

Beginning especially with Pueschel, there have been efforts to establish a science zone around MLO to protect the site from any possible pollution, and the Ashra facility demonstrates why a science zone is far more important than ever before. Indeed, the Ashra Web site illustrates one such reason in a photograph of a large bulldozer raising clouds of dust when the site was being leveled. Because of the blocking of new science facilities on Mauna Kea, it is entirely possible that still more facilities will be allowed along the Mauna Loa road.

Darryl Kuniyuki has worked at MLO since 1982 and is well aware of the potential contamination problem. In his view, a science zone having a radius of at least 10 miles (16 km) should be established around the observatory, and all new construction within the zone should be approved by MLO (Kuniyuki, personal communication with author, August 24, 2007). The wind roses for MLO support Kuniyuki's concern. At night and through midmorning, the prevailing wind is from the southwest, south, and especially the southeast. This is the famous downslope wind that occurs as clean, cool air from the free atmosphere over Hawai'i flows down the slopes of Mauna Loa. During the day, the wind can come from any direction, but the prevailing wind is from the northwest, north, and northeast. In short, during a typical twenty-four-hour day, the air—aerosols or gases included—arrives at MLO from around the compass.

MLO enjoys a fine reputation within the scientific community, and it is reasonable to assume that scientists would not want to jeopardize its generally pristine air quality for the sake of new projects. Thus, in lieu of a formal science zone recognized by Hawai'i's DLNR, one solution might be for the international scientific community to simply establish an informal science zone around MLO. Dr. John Barnes (personal communication with author, October 18, 2007), the station chief of MLO, was asked his opinion about such an arrangement:

> Yes, it would be good to have some agreement process. Since Ashra wanted to be on our site, I had them fill out our cooperative project form. So I already could clear them of several potential problems. But in the next case I may not have that "inside" information. . . . I haven't seen any impact on our measurements. They have been very cooperative when I have asked questions like "What fluid do your air conditioners use?"

Dr. Barnes has a friendly demeanor, and his positive experience with Ashra demonstrates that the scientific community can work together to address and minimize potential interferences, just as the many cooperative observers within the boundaries of MLO have done for fifty years. In effect, Barnes has begun to establish an informal science zone with Ashra. Clearly there is merit to the idea of a de facto Mauna Loa Science Zone, and what remains is developing a flexible protocol, perhaps initially with the assistance of staff at the University of Hawai'i. The protocol would address issues such as notification of dust events, painting days, and chemical releases. Participants would be expected to avoid any dust or gaseous emissions when the wind blows from their site toward another. Because scientists from around the world tend to collaborate widely and on friendly terms, as often demonstrated over the years at MLO, the expense involved in developing a de facto Mauna Loa Science Zone might amount to little more than coffee or tea and perhaps a lunch or two. One might even envision cooperation within the Science Zone for sharing the expenses of road maintenance and repairs and for providing emergency assistance during snow events and the inevitable next eruption of Mauna Loa. An appropriate slogan for the Mauna Loa Science Zone that comes to mind was created by MLO founder Dr. Robert Simpson in 1957 when he signed in at MLO and wrote, "We must preserve this progress."

Tracking the Ozone Layer

In recent years climate change science has eclipsed ozone layer studies. Dr. James Elkins, lead author of the team that first observed and explained the reduced growth of ozone depleting CFCs in the atmosphere (Elkins et al. 1993), believes it is essential that anthropogenic ozone depleting gases, which are also greenhouse gases, continue to be carefully tracked. Elkins, who is group chief of the GMD Halocarbons and other Atmosphere Trace Species Group (HATS), is also concerned about reductions in his budget for studying these gases. When asked about his vision for the next half century at MLO, Elkins (personal communication with author, May 4, 2007) wrote,

> Fifty years from now we should see a remarkable recovery of the ozone layer and no more Antarctica ozone holes if scientists and public policy makers do the right thing. My hope is that NOAA or its successor agency continues to monitor the ozone depleting gases (CFCs, halons, chlorinated and brominated solvents) until they disappear from the atmosphere. In my opinion, it is a sad commentary of the times that many funding managers think that the problem is solved today and that the funding of this important time series may be reduced. These time series will be a valuable demonstration of the transport and chemistry of our atmosphere for generations to come, and the time series may hide the secrets to solving the problem of global warming.

Returning to the Summit

MLO founder Dr. Robert Simpson retains his enthusiasm for the original summit observatory atop Mauna Loa. When asked if he would like to see the reestablishment of a summit observatory atop Mauna Loa, he ponders the question for a moment before responding:

> I'm not quite sure how to answer that. I would probably not get much support for reestablishing something at the summit, but that would be my first goal—if money was not a problem and the politics of the situation were solvable.
>
> I think that the observatory where it is now is serving a great purpose, a grand purpose, and I hope it stays open and continues to offer the opportunities which I think are beyond what they're doing right now, if I understand it. But I suspect, considering the opposition that would be encountered to begin with, there would have to be a building block stage if it was ever reestablished at the summit I think, mainly because of the Department of Interior.
>
> That's still, as I understand it, that's still a wilderness area. They don't want any more people intruding into the wilderness area up there than they have to put up with. So I suspect if it was ever reestablished at the summit, and I wish it were, it would be a tedious, long job requiring a lot of very careful negotiations. But I would like to see it done. (Simpson, telephone interview with author, March 17, 2007)

Simpson's idea is neither new nor radical, for he and Ralph Stair had always planned for a major science observatory to be built on the summit. Indeed, most of the text of the program for the June 28, 1956, dedication of the present slope unit lays out the details:

> Recent increased interest in such problems as atmospheric circulation, the ozone content of the middle atmosphere and solar radiation in relation to intense tropical storms has focused the attention of meteorologists and radiation physicists on observations from mountain locations in the Pacific. This interest has led to the recent completion of the second or "slope" unit of the Mauna Loa Observatory at a point 11,134 feet above sea level by the National Bureau of Standards and the United States Weather Bureau, both of the U.S. Department of Commerce.
>
> The dedication of this unit does not mark the end of efforts to establish an adequate facility near the summit of Mauna Loa. Rather, it is another step toward the goal of developing a complete functional unit of the Observatory near the highest point on this enormous volcanic mountain. The realization of this aim is dependent on many things, including approval for future use of the existing site near the summit, the extension of the improved Mauna Loa Highway to that point and the erection and equipping of an appropriate building there.
>
> When these are accomplished, it is envisioned that most of the research programs of the Observatory will be carried out at the "summit" unit. The building being ded-

> icated today will then serve primarily as a staging point and living quarters for the scientific personnel. (NBS and US Weather Bureau 1956)

This half-century-old goal was proposed by the author during a talk at the 2007 ESRL Global Monitoring Conference in Boulder, Colorado (Mims 2007). The audience was informed that the Hawaiian Volcano Observatory installed a panoramic digital camera on the west side of the Moku'āweoweo summit caldera that provides a near real time, online view of the entire crater. The success of this arrangement demonstrates the reliability of today's electronic instruments, which are far more compact, reliable, and versatile than the primitive meteorological instruments installed on the summit from 1953 to 1954. Possible sites for a compact, unattended array of instruments include the concrete slab of the original summit observatory and the Mauna Loa Cabin on the east side of the summit caldera.

Following this presentation, then GMD director David Hofmann endorsed the possibility of returning to the summit. However, when the idea was discussed a few weeks later with some MLO staff working on the mountain, they pointed out that Mauna Kea's summit would be much more accessible than the summit of Mauna Loa. This is correct, for Mauna Kea has an all-weather road to the summit, and the old jeep trail to the summit of Mauna Loa is in very poor condition and would require major work. One source has reported that the National Park Service might eventually remove the summit road. As for Mauna Kea, permission to place instruments at the summit would have to come from one of the astronomical observatories, and that might be difficult in view of the restrictions now in place. Moreover, nearby observatory domes would shield much of the sky from solar irradiance instruments. Then there is the problem of increased dust and emissions from the vehicles that transport workers and visitors to the summit. In view of impending development near MLO and the fifty-year time series of data from the current Slope Observatory, returning to the summit of Mauna Loa offers intriguing research opportunities, especially for the measurement of full sky UV-B without the present blockage of the sky caused by the north slope of the mountain that rises above the observatory. Already the UV-B that is continuously measured at MLO is the highest measured anywhere on earth. With its unobstructed horizon and higher elevation, the UV-B at the summit must be even higher.

Returning to Sea Level

Scientific experiments and observations at MLO are conducted there because of the generally pristine conditions on Mauna Loa. There is another aspect of the site that has yet to be fully exploited, and that is the proximity of many sea level and intermediate sites from which observations could be made for comparison with those at MLO. This idea is not new, for when in 1794 Archibald Menzies made the very first scientific measurements from atop Mauna Loa with his thermometer and portable barometer, he compared his results with measurements he and colleagues made at sea level and in between. When the US Exploring Expedition ascended Mauna Loa in 1840 and camped on the summit for three frigid weeks, their hourly measurements of temperature and pressure and their observations of the wind were compared with hourly measurements made from the USS *Vincennes* in Hilo Bay and the temporary Observatory 1 set up among coconut palms at the mouth of the Wailoa River near old Hilo Town.

Jack Pales continued the tradition of making simultaneous mountain and sea level measurements from the current slope unit on the mountain and the original MLO office at the Air Cargo Building at the Hilo airport. With eighteen months of funding from the US Army Electronics Proving Ground at Fort Huachuca, Arizona, Pales and his men installed a row of four vertical posts in a grassy field extending from the southwest end of the Air Cargo Building. On each post they mounted an instrument designed to measure direct or full sky sunlight. They also installed instruments to measure surface ozone, aerosols, and meteorology. An identical suite of instruments was installed at the observatory.

Today some observations are made from both sea level and MLO. Flask samples are collected from both the observatory and near sea level at Cape Kumukahi. Total column water vapor is measured by NOAA GPS units near Hilo and MLO. Meteorological parameters are measured by the National Weather Service in Hilo and by MLO on the mountain. Missing are simultaneous sea level and MLO observations of optical depth, surface ozone, aerosols, solar ultraviolet, direct and diffuse sunlight, and other atmospheric parameters of high interest. Some of these observations would provide significant information about Hilo's urban heat island effect, the clarity of Pacific air, the impact of the Kīlauea volcanic emissions known locally as vog, and the differences in surface ozone and solar UV at sea level and at MLO.

Measurement Transects along the MLO Road

There have been a number of projects at MLO in which data were acquired by the mountain crew as they drove to and from MLO. A classic example, described in chapter 10, was Bernard Mendonca's study of the trade wind inversion, during which trips up and down the mountain were made while a psychrometer and an air sampling tube were held out an open window (Mendonca and Iwaoka 1969). Another important transect project, described in chapter 12, was Steve Ryan's deployment of nine miniature atmospheric pressure recorders across the Big Island for a study of pressure variations caused by the daily temperature cycle (Hamilton et al. 2008.) The author of this book and others have mounted UV sensors on cars and made measurements of total column ozone and particulate matter during drives up and down the mountain.

What was once the roughest segment of the MLO road now has a thick coat of smooth, dust-free asphalt, and the prospect of resuming transect observations has become more practical than ever before. An automated instrument rack could be mounted atop an MLO vehicle to collect data all the way from the Hilo office to MLO and back every workday. Besides providing an almost daily profile of the inversion layer, a transect program would provide new information on the variation with altitude of solar irradiance, particulate matter, and UV. These measurements would be of special interest during spring when layers of Asian dust sometimes pass over and below the observatory.

Suggestions From a Young Old-Timer

In years of service, electronics engineer Darryl Kuniyuki became the senior staff member at MLO when Alan Yoshinaga retired in January 2007. Today Kuniyuki serves as station manager of MLO under Station Chief Dr. John Barnes. In addition to participating in the routine observing schedule on the mountain, Kuniyuki's responsibilities include managing the MLO servers, network, computers, security systems, and program development. He also provides technical support for MLO staff and assists the station chief in administrative and supervisory duties.

In view of his vast experience at MLO, Kuniyuki was asked what his goals would be for MLO, and he responded with a list of what he described as "Realistic goals" and "If we could get whatever we want goals." Kuniyuki's first realistic goal is for a Mauna Loa Science Zone, as detailed earlier in this chapter. He also proposes a number of additional realistic goals, which are amplified upon here:

- MLO should be powered by solar and wind. There is ample space for solar cell arrays on the lava south of the NDACC Building.
- Measuring background mercury in the atmosphere should be included in the MLO monitoring program. There is increasing concern about the atmospheric transport of mercury in desert dust and smoke from coal-burning power plants in China. This program would supplement the current EPA mercury program.
- Larger bandwidth for the data network, since almost all data now goes through the network.
- Electronic check sheets for the mountain observations so that paper records will not need to be stored. (The monthly Activities Reports that were typed onto paper for decades have been entered into

a computer since 1990.) This suggestion would not provide a paperless environment, but it would simplify the storage and acquisition of the daily check sheets, which could even be sent to cooperative observers by e-mail or posted online.

- The station should have more public relations activities, especially with schools. A full-time PR person should give talks, tours, and promote the science being done. This suggestion has good potential, for MLO provides a significant resource for local high school and college students. And there's another angle of relevance. While the astronomical observatories atop Mauna Kea have become world famous for unlocking secrets of the universe over the past several decades, the Mauna Loa Observatory (MLO) is little known outside the scientific community and has long been Hawai'i's best-kept scientific secret. When a scientist happens to say he or she works at the Mauna Loa Observatory, the listener inevitably hears "Loa" as "Kea" and responds with comments or questions about giant telescopes and galaxies. This reaction occurs both on the mainland and the Big Island, where the white and silver telescope domes atop Mauna Kea are clearly visible from various points on the island. The Mauna Loa Observatory is visible only from Saddle Road and then only to those who know where to look.

In addition to the realistic goals listed above, Darryl Kuniyuki listed six "If we could get whatever we want goals":

- Realign the access road so it does not follow Mauna Loa's northeast rift zone, because there is significant risk that lava from a future summit eruption will cut off the present road to the observatory. Due to the risk of head-on collisions, some of which have occurred in the past, the road should be widened to two lanes and blind turns and blind hills should be eliminated.
- Redesign the observatory so the air intake lines are at the southernmost point (highest point) and all or most buildings are at the lower elevations.
- All buildings should be made of metal to reduce the need for maintenance. Even better, replace the many small structures with a single large building.
- The solar deck over the NDACC Building should be made sturdier. The present deck is not sufficiently rigid.
- Add a guest facility for visiting scientists. Presently there are two small bedrooms in the NDACC Building.
- [Build a] Monorail with substation at Saddle Road and an MLO Visitor Center. This would greatly reduce automobile traffic to MLO, but the track would have to be designed in a manner to minimize its visibility from Saddle Road.

Sky Imagery

When Saul Price made his first visit to MLO after Jack Pales and his Weather Bureau team moved in during the summer of 1957, he brought along an automatic camera designed to photograph clouds over MLO at predetermined times. Sky imagery at MLO is important for a variety of reasons, and the MLO Webcams and the all-sky Webcam installed in one of the Canadian Brewer spectrophotometers illustrate why. Prospective visitors can easily see if clouds are present prior to a day trip to MLO. Atmospheric scientists can use the images to study the effect of clouds on measurements of optical depth, solar UV, column water vapor, and so forth. Of special interest is the enhancement of sunlight, including UV, when cumulus clouds are near the sun (Mims and Frederick 1994).

An all-sky camera having higher resolution would provide even better results. A pair of cameras pointed at the sunrise and sunset region of the horizon would provide important information about the presence of dust and haze layers in the Pacific sky, especially before sunrise and after sunset. These images would be of special interest to those studying the passage of dust and air pollution from Asia and would also provide clues about the arrival of aerosols from major volcanic eruptions.

Interdisciplinary Research Opportunities

So many studies, experiments, and projects are at MLO that many cooperative observers and even some of the MLO staff know little or nothing about projects in which they are not involved. This information gap provides many unexploited opportunities for interdisciplinary research, data mining, and archiving by students, visiting scientists, and organizations. Because considerable MLO data are available online, many research studies can be conducted from anywhere Internet service is available without ever having to visit the observatory. Dozens of research opportunities are available, including those in the following discussion.

Ozone and Ultraviolet

The simultaneous measurement of the ozone layer and solar ultraviolet by multiple instruments at MLO has provided data for a number of scientific papers by MLO scientists. This research can be updated and expanded to include additional atmospheric parameters.

Water Vapor

Another research opportunity is a comparison of the column water vapor retrievals from NOAA's GPS unit at MLO with those from the various automated sun photometers.

The MLO Heat Island

An unintended consequence of the addition of more buildings, pavement, and concrete to MLO over the years is that the center of the observatory has become warmer than the surrounding lava. Especially notable is the warmth caused by the NDACC Building, the roof of which can reach 55 degrees Celsius on a typical sunny day. The expanded pavement that serves the NDACC Building has also enhanced the heat island effect. The heat island impact can be easily observed from the meteorology tower by pointing an infrared thermometer around the site. The black lava is consistently cooler than the pavement and buildings, perhaps because its porosity provides for air circulation that asphalt and rooftops do not. Steve Ryan has suggested that the blocking of the local air circulation by the buildings may also play a role.

ASTER (Advanced Spaceborne Thermal Emission and Reflection Radiometer) is a thermal imaging instrument aboard NASA's Terra satellite. An ASTER (2000) image of Mauna Loa and its summit crater reveals a cluster of warm pixels at and around the observatory. A much better thermal image of MLO's heat island could be obtained by flying a thermal camera over MLO or by making thermal images from an infrared camera carried to the top of the forty-meter tower.

The MLO temperature record is significant, and the observatory's heat island effect could be carefully reviewed to look for any impact it might have contributed to the temperature time series that goes back to 1958. The temperature measured at the top of the forty-meter tower since 1987 may be less affected than the temperature measured since 1958 from a site or sites very near the Keeling and NDACC Buildings. Thus, a comparison of the two temperature records might reveal any heat island effects, particularly after 1996 when work on the NDACC Building was underway. Should the data include a heat island signature, one possible way to preserve the entire record back to 1958 is to plot only the data acquired in the early morning hours during downslope conditions.

How best to minimize the heat island effect at MLO could form the basis of a related study. Some of the heat island effect can be mitigated by coating all structures and roofs with white paint. But this might also have an unintended consequence, since it would make the observatory much more visually obvious than it already is when viewed from Saddle Road far down below. Perhaps some mitigation could be achieved without enhancing the visibility of the observatory from Saddle Road by painting only the south sides of rooftops and walls of structures.

Flora and Fauna at MLO

A possible by-product of heat island warming at MLO is an unlikely assortment of a few dozen small plants growing primarily around the center of MLO's main building complex. The plants range from a few blades of grass emerging from cracks in the pavement to several flowering plants a foot or more high sprouting from the lava. Though few in number, the plants provide green, yellow, and brown splashes of color that are in stark contrast with the black lava clinkers on which MLO is constructed.

From one to three specimens of a dozen or so species are represented, along with some of their insect colleagues that probably floated in with the inversion layer. During December 2006 and June 2007, six or more green aphids (Homoptera: Aphididae), an invasive species not native to Hawai'i, were photographed dining on one of the stems of a plant with attractive yellow flowers growing from the lava immediately west of the NDACC Building. The plant was a hairy cat's ear *(Hypochoeris radicata),* which is as invasive as the aphids (Starr 2008a). The aphid population was matched by half a dozen or more specimens of a small flying insect congregating around the bright yellow blooms. It's not unusual to find ladybugs and other larger insects at MLO. Perhaps it's these insects that provide meals for the native wolf spiders (*Lycosa* sp.), which can exceed the diameter of a US quarter. These large black spiders are unnoticeable on the lava, and they can startle first-time visitors as they prowl the floors of MLO buildings.

Another MLO plant with attractive yellow blooms that has become naturalized in Hawai'i is fireweed *(Senecio madagascariensis)* (Starr 2008b), and for a time a very nice specimen could be found adjacent to the hairy cat's ear described above. For a few years an attractive basketball-sized plant with less attractive flowers grew by the sidewalk adjacent to the NDACC Building. This plant was a common mullein *(Verbascum thapsus)* (Starr 2008c), which is also a nonnative species. While all three of the most obvious plants at MLO are as nonnative as the buildings and pavement, they certainly embellish the bleak landscape when they are present.

The observatory is at least a few thousand feet (about half a kilometer) higher than the maximum elevation where the MLO plants are normally found, which suggests that warming caused by asphalt and concrete might explain their survivability. MLO physicist Steve Ryan has suggested that construction sand might also be a factor, for it may provide a more hospitable environment for seed germination and plant growth than the raw *'a'ā* lava at MLO. At places along the MLO road below the Kūlani Mauka junction, various plants proliferate along the edge of the pavement, presumably because of the rain runoff from the road. Rain runoff from roofs and pavement may similarly enhance the growth of the handful of plants at MLO.

Whatever the explanation, the author has observed plants at MLO since 1992, and there has been an increase since then, especially after the arrival of the NDACC Building. The plants provide an opportunity for alpine ecologists and biologists to study the ability of plants and insects to survive at an elevation significantly higher than their normal climate zone. Of special interest is how the plants survive warm days and freezing nights and how they show no visually obvious damage from the intense level of solar ultraviolet at MLO. How the plants and their associated tribes of insects arrive is also worthy of study, as is the possibility that the roots of the plants might be associated with mychorrizal fungi whose spores drifted up with the inversion layer and sifted down into the lava. Lee S. Motteler (personal communication with author, March 13, 2010) has suggested that

> Human transport (on boots, tires, etc.) would seem another possibility.

It seems highly unlikely that the very small number of plants at MLO has any impact on atmospheric monitoring, particularly in comparison with the exhaust of the several vehicles that may arrive and depart on a given day. But atmospheric chemists are able to detect trace gases in the air down to parts per trillion. So it's time for the ecologists to stake their claim and begin their studies before the chemists arrive with garden shears. Perhaps the former can hold off the latter by finding a few rare and endangered species growing in the MLO garden.

Preserving the Entire MLO Measurement Record

Although meteorological measurements were made at the original Mauna Loa summit observatory and along the road from 1952 to 1954, these data were unable to be located while researching this book. Original chart recordings may still be in storage at one or more locations, and GMD's Michael O'Neill and Thomas Mefford are concerned about the preservation of old data from MLO. O'Neill (video interview with author, Boulder, Colorado, May 3, 2007) makes his point with the help of a prime example of early MLO history from a few months after Howard Ellis arrived at MLO:

> An example of that is a strip chart that I have from Hawaii . . . which shows a solar radiation record from a pyrheliometer at [MLO] . . . for July 1961. . . . We are looking at this as being forty-six years old, and the concern for a lot of people like myself is that, by law, we only have to keep this data until it is fifty years old. Then they can be thrown away, and that is only four years from now. And we are not sure at this point if this data has been archived, if it has been reduced, if it is in a record and placed in electronic form. And these paper charts are actually becoming brittle and starting to deteriorate because of where they have been stored and how they have been stored for the past fifty years.

Old data like this is often in the form of rolled-up chart recordings that provide a visual look, but the information is not in digital form. Consider the old solar irradiance chart described by O'Neill in the same interview:

> In this particular case, we are looking at one from Mauna Loa in July of 1961 where we have clouds in the early morning, clear day for most of the day, and then clouds again in the late afternoon, which is fairly typical climatology for Mauna Loa. And I could say, probably, some of the artifacts that are in the data could be from shadows, could be from a recurring feature that we would have to look at day-to-day and see if there is something that is recurring. But since nobody has looked at this data, probably in forty years, we do not know that information.

In short, locating, preserving, and digitizing the entire MLO data record would seem to be a high priority if funds can be found to accomplish the task. Thomas Mefford (video interview with author, Boulder, Colorado, April 27, 2007) has a wish list about how best to format the old data:

> I would like to see it assembled as hourly files so that when people are at the observatories . . . anybody can easily download it. . . . I would like something like simple text files that can be easily analyzed.

MLO Universal Database

Scientists and university students who visit MLO are particularly interested in the wall charts hanging on the walls of the NDACC Building. Most of these charts include graphs of various kinds of data collected over the years by instruments the visitors also see. By far the longest such time series displayed on these charts are those that depict atmospheric transmission (since November 1957) and carbon dioxide (since March 1958). It would be useful if these two time series, which are respectively known as the Ellis and Keeling curves, could be depicted on a common axis. This would provide a foundation for the addition of additional time series plots. These could include measurements of the ozone layer, surface ozone, solar UV, temperature, pressure, total water vapor, lidar backscatter, particulate matter, optical depth of the atmosphere, photosynthetic radiation, sulfur dioxide, the various carbon-cycle gases, and the halocarbons. The Mauna Loa Solar Observatory could add the daily sunspot numbers for each of the solar cycles since 1965.

Placing all these data on a single chart would be difficult and time consuming, especially since the data are acquired under very different time schedules. Optical depth, for example, is measured every three minutes by the USDA shadowband instruments and every fifteen minutes by NASA's robotic sun photometers. The ideal presentation would be user selectable and would depict a daily mean, peak, or minimum value for each of the key measurement programs. For example, the total ozone might be the mean for the day, while solar UV might be the daily maximum.

In the end, making the historic MLO data available online would permit students and scientists to create their own charts and extend the value and public knowledge of the vast store of data accumulated at MLO since 1957.

MLO Library

Closely related to an MLO universal database is the need to create a comprehensive online library of all scientific papers, reports, and presentations that are based on data acquired at the observatory. Much of this library already exists in the form of several drawers of research papers in a file cabinet in MLO's Hilo office. While this is an important start, a random search through these files will quickly disclose the absence of many important papers. Ideally, papers based on MLO data should be freely available to anyone with Internet access. Some scientific journals have become much more open to this prospect in recent years, particularly those published by the American Meteorological Society. Perhaps scientists who write papers about their research at MLO can be encouraged to publish in journals that will allow their work to be made freely available.

MLO Photo Album

Thanks to Darryl Kuniyuki and the MLO staff, the observatory has assembled a comprehensive library of photographs related to the history of MLO, and Preston Sato has scanned thousands of images for this library. Digital images take up considerable memory space, so it is not currently practical to make all the images available online. But in view of the historical significance of MLO, perhaps the entire album of photographs could someday be placed online, complete with detailed captions about each image and, when known, the name of the photographer.

Visitors and Tourism

Ecotourists and adventurers looking for exciting places to visit may be unimpressed by the usual description of MLO on maps and brochures as a mere "weather station." Of course the observatory does far more than measure wind, temperature, and humidity. It is important for scientists and nonscientists alike to be aware of the significant knowledge about the atmosphere that results from research at MLO, but communicating this knowledge presents a dilemma. For as more people learn about the role of MLO in revealing new knowledge about the greenhouse effect, air pollution streaming across the Pacific to North America, and volcanic effects on climate, MLO could someday be as well known as the giant telescopes atop Mauna Kea and become a popular ecotourist destination.

The recent repaving of the lower portion of the MLO road is sure to attract tourists and other visitors who would not have risked driving across the gaping holes and loose lava they would have encountered before. Back in the 1950s and 1960s when the MLO staff was manned twenty-four hours a day, the staff was glad to receive occasional visits by tourists. They even shared their food with some of them. That changed after the power line was installed in 1966 when overnight stays became unnecessary. Bob Uchida (video interview with author, Hilo, Hawai'i, November 30, 2006) has observed that the daily drive to and from the observatory left little time for the staff to give tours to visitors while also accomplishing their duties on the mountain:

> Normally, because of the time limit that we have up on the mountain, we do not go up there to eat lunch or to sit down and talk story. We go up there to do the work, get the mission done and come back down.

Those who have observed the staff at work understand what Uchida means, for the staff begins their daily work schedule immediately after they arrive. Normally the work proceeds at a reasonable pace, but activity speeds up when unexpected repairs need to be made. There may be back and forth calls to Hilo or to a scientist on the mainland. Someone may dash into the electronics shop to retrieve tools and spare parts. Bob Uchida explained in the same interview why equipment repairs receive a high priority:

> Well, everything that is running up there is important, because if it is not, then it would not be there. So, if it is down you have to get it going. And, as far as I'm concerned, every downtime is zero time. And if you can be down for a long time, you do not really need [the malfunctioning equipment]. So, you better get it going.

Then there are the occasional tourists and hikers who become ill, who encounter car trouble on the road, or who try to drive the virtually impassible summit road that leads past MLO and up to the crater. One day Bob Uchida noticed a couple walking down the road to the summit and hoped they would not arrive at the observatory with bad news:

> But eventually, the missus came up first, and she was calling to the husband. The husband was dragging behind, taking a long time getting up there. And so when he finally got up, the guy was sitting there . . . ranting and raving how tired he was and how his vehicle was stuck way up there off road. So I asked him if he needed a ride to get back down because we had a lot of construction workers up on the mountain that if they needed a ride I could have arranged a ride for them. But he could not—this guy could not decide. He was basically sitting there just moaning how tired he was and how weak he felt and how he had to sleep overnight up on the hill. And so, I asked him if I could call somebody to come up and rescue him and his vehicle; and he could not make a decision. He was sitting there for a long time. I had to sit there and try to get him to decide whether to go down to Hilo, or wherever; off the mountain, anyway. By that time his wife was beginning to get upset with him. She wanted to kill him, because she was not happy sleeping overnight on the mountain. Anyway, he could not decide. (Uchida, video interview with author, Hilo, Hawai'i, November 30, 2006)

Uchida eventually persuaded the stranded visitors to call their insurance company for help. The story is related here to illustrate the problems that can occur when visitors and tourists encounter trouble in the vicinity of the observatory and distract the staff from their tasks.

Besides tourists, the improved road might also attract a return of the bicycle outfitters who once offered rides down the long road. Because of Mauna Loa's increasingly publicized role in monitoring carbon dioxide and other greenhouse gases, companies that offer ecology tours around Hawai'i and trips to the summit of Mauna Kea might begin offering tours to MLO. MLO has prior experience with formal bicycle tours:

> Within the last month [August 1993], commercial bicycle tours have begun operations coasting from the observatory to the Saddle Road. Initially, the riders were dressed in black and rode black bikes. The company has now agreed to put colored flags on the bikes and orange vests on the riders. Still, the riders will be hard to see. Drivers on the MLO road must use their headlights at all times and be aware that there is potential of meeting a fast moving cyclist coming down the middle of the road. We have no authority to stop the tours. (Schnell 1993)

A second bike tour company began using the MLO road in December 1993, and MLO director Dr. Russell Schnell contacted the company about safety concerns. Bob Uchida (video interview with author, Hilo, Hawai'i, November 30, 2006) is generally more impressed by cyclists than visitors who arrive by car: "I think the bicycle riders are physically fit, so I think they are mentally more fit also."

The major downside of increased visitors and tourism is the possibility of accidents. There is little or no room for evasive action should a driver traveling slowly uphill on the one-lane road suddenly encounter a fast-moving bicycle or motorcycle at one of the blind curves. If there is not a collision, the car or the bicycle or both could end up in the lava, which wrecks both tires and undercarriages of cars and in 1996 killed a motorcyclist. Placing large convex mirrors at these blind curves could reduce the possibility of collisions.

Another downside to recreational use of the road by tourists and tour companies is the increased air pollution that will occur if a constant stream of visitors begins the trek to MLO and the distraction that tourist traffic will pose for those who work at the observatory. It is rather tiring to spend a day working at altitude, and the staff is glad to leave for Hilo at the end of their shift. But the trip down the mountain can be more stressful than usual when it's necessary to constantly be on the lookout for vehicles arriving from below. Also, many times over the years MLO employees have given up time from their official duties to assist accident victims, people with altitude sickness, lost hikers, and people whose cars have broken down. While the staff is certainly willing to assist during emergencies, that is not why they are on the mountain, and they may have to spend more time at altitude when visitors run into difficulty.

The benefit to MLO from recreational use of the MLO road and visits by tourists is the opportunity to inform the public about the monitoring science conducted at MLO and the possibility of raising supplemental funds for MLO educational projects by offering merchandise to visitors. Charging tour companies for their use of the road can raise money for road maintenance. Should MLO wish to consider these opportunities, the staff need look no farther than across the Saddle to Mauna Kea to learn how their astronomer colleagues have responded to the influx of visitors. At least two of the astronomical observatories atop Mauna Kea allow limited, guided tours during the day. The tours depart from the Onizuka Center for International Astronomy Visitor Information Station at the 9,300-foot (2,800-meter) point on Mauna Kea road. The station sells snacks and drinks, plus T-shirts, jackets, coffee mugs, postcards, books, and other items, all with a Mauna Kea theme.

Tour groups often stay on the summit of Mauna Kea to observe the sunset, but they must depart soon thereafter. Each evening at MLO, the headlights of their vehicles can be seen slowly meandering down the steep Mauna Kea road. Most of the tour groups participate in the free stargazing program conducted at the Visitor Information Station every evening from 6 to 10 p.m.

An MLO Visitor's Center

Consider a different kind of visitor's station at a future, fully remodeled MLO. This visitor's station would be the Keeling Building, the original building at MLO. Visitors would leave their vehicles at the parking space below the observatory and walk along a sidewalk equipped with a handrail. On arriving at the MLO entrance, they would be greeted by a welcome sign that would instruct them to follow a painted walkway that would guide them directly to the Keeling Building. Arriving visitors would sign a logbook, just as visitors have done since June 28, 1956.

Visitors would see examples of old monitoring equipment from MLO, including Charles Keeling's original carbon dioxide analyzer. A photo of Keeling and John Chin standing by the plaque on the building when it was renamed the Keeling Building would be placed nearby. An open parts cabinet protected by an acrylic shield would display spare parts for the analyzer, including the stash of spare vacuum tubes that still remains on the mountain. Large wall charts like those in the NDACC Building would show the data collected by the observatory going back to 1957. A video might present the history of the observatory and explain its key measurements and findings. Back in the 1960s, Howard Ellis hung photos on the walls showing various scenes around MLO, and some of these old photos could be rehung. The photo of a smiling Ellis seated at supper with Ralph and Angie Stair inside the building in 1967 could be placed near where the kitchen was once located. A table and chairs placed in the same location would allow visitors to rest and enjoy snacks and water purchased in the building.

While visitors to Mauna Kea are treated to a star party each evening, visitors to Mauna Loa would have the opportunity to participate in the observing schedule by conducting real science. A sun photometer would allow visitors to measure the optical depth of the sky. An infrared hygrometer would permit them to measure the total water vapor. A handheld ozonometer would allow them to measure the ozone layer. An ultraviolet radiometer would allow them to measure the solar UV-B. The data could be entered into a computer or written on a worksheet just like those used at MLO. Literature about these kinds of devices could be distributed freely or for a fee. These ideas have been successfully tested each spring since 1993 when students from around the world visit both MLO and Mauna

Kea during annual field trips sponsored by the University of the Nations.

So long as MLO retains its present arrangement of buildings, the Keeling Building will be needed by the staff for day-to-day operations. While the NDACC Building houses many of the main instruments at MLO, the Keeling Building still remains the center of operations for the daily mountain crew. The crew reports to the Keeling Building when they arrive each morning. And they complete their paperwork, eat their lunch, and depart from there. While the MLO staff are certainly glad to know that visitors are interested in the work of the observatory, large groups of visitors can sometimes get in the way.

Thus, an alternative option might be a visitors' center modeled after the original building constructed at or near the junction of Saddle Road and the Mauna Loa road or perhaps at the junction of the Mauna Loa road and the uncompleted Kona-Hilo "Road to Nowhere" built by Tom Vance. The visitor's center could be made to look much like the actual building and fitted with some of the retired instruments. A small telescope would allow visitors to see MLO.

The measuring instruments made available for visitors to MLO or a visitors' center at a lower elevation could form a valuable resource for students at Hawai'i secondary schools and colleges who are interested in making various kinds of measurements near sea level where most people in Hawai'i live and also from MLO. Perhaps students could be allowed to sign out instruments on loan, much like checking books out of a library. The difference is that those who borrow instruments must agree to conduct a serious science project that is entered in a high school science fair or fulfills a requirement of a university course.

An Ecotourist Destination?

Whether or not MLO plans for more visitors and tourism, others are already making plans for the area. Their motivation is fueled by better access along Saddle Road, which connects the eastern and western sides of the Big Island and is the access to the roads to both Mauna Loa and Mauna Kea and their science installations. While the lower half of the MLO road was being paved, a major new stretch of Saddle Road between the Mauna Kea road and Mauna Kea State Park was opened in 2007. As this is written, major improvements are being made to Saddle Road between the Mauna Kea road and Hilo. The ultimate goal is to upgrade the most dangerous section of the road between the Mauna Kea State Park and the Hawai'i Belt Road (Māmalahoa Highway) south of Waimea. When this project is completed in 2014, there will likely be a dramatic increase in travel along Saddle Road.

This prospect has attracted the attention of students at the University of Hawai'i who worked in collaboration with the Sustainable Tourism and Environment Program at the University of Northern Iowa to produce a study entitled *Ecotourism Opportunities on the Saddle Road, Big Island of Hawaii* (STEP 2007):

> Four students from the University of Hawaii at Manoa: two from the School of Travel Industry Management and two from the School of Architecture have proposed a project and have worked together to fulfill their objectives. The architecture students predict that the reconstruction of the Saddle Road will cause unplanned and random development to occur. The project would propose a tourist route to enhance visitors' perception, provide services for their necessities, and minimize harmful intrusions and excess pressure on the environment. The purpose of this project is to promote and sustain ecotourism in the project corridor as well as facilitate tourist activities along the proposed route through design by the architecture students.

Saddle Road provides the sole land access to a number of significant locations, most of which are listed on a Web site named Saddle Road on Hawai'i's Big Island (http://www.saddleroad.com), devoted to construction progress on the road. The list is worth a review, for there is a significant omission:

> The Saddle Road, along with a portion of Kaumana Road, provides the only road access (some 53 miles) to Pohakuloa Training Area, Mauna Kea Science Reserve

> International Astronomical Observatory Complex, residential areas of Waikii Ranch, Mauna Kea State Recreation Area, Parker Ranch, Kilohana Girl Scout Camp, Kaumana City and Kaumana Caves County Park. Access to public lands and forest areas for hunting, gathering, ranching and recreation are also provided by Saddle Road. (Saddle Road 2005)

While the Mauna Loa Observatory is missing from the list, chances are that the highly improved road will eventually become well known and that tourists with even a passing interest in the environment will be much more likely to attempt the trip now that the lower portion of the road almost qualifies as a single-lane version of Saddle Road itself.

The Best Measurements in the World

While the greater scientific community refines its studies and models that address the causes and consequences of the gases and aerosols being pumped into the atmosphere by power plants, volcanic eruptions, vehicle engines, dust storms, biomass fires, and assorted other sources, the staff at the Mauna Loa Observatory will continue doing what they do best by measuring every aspect of the air that flows across Hawai'i. They will accomplish their mission using facilities and instruments that cost a small fraction of the giant telescopes atop neighboring Mauna Kea and the fleet of Earth resource satellites passing overhead. Their goal is to continue to produce what former GMD director David Hofmann (video interview with author, Boulder, Colorado, May 1, 2007) describes as "the best measurements in the world."

References

ASTER. 2000. ASTER instrument team. ml_tir.jpeg, image date: April 18. (Accessed at http://visibleearth.nasa.gov/view_rec.php?id=92.)

Elkins, J. W., T. M. Thompson, T. H. Swanson, J. H. Butler, B. D. Hall, S. O. Cummings, D. A. Fisher, and A. G. Raffo. 1993. Decrease in the growth rates of atmospheric chlorofluorocarbons 11 and 12. *Nature* 364: 780–783.

Hamilton, Kevin, Steven C. Ryan, and Wataru Ohfuchi. 2008. Topographic effects on the solar semidiurnal surface tide simulated in a very fine resolution general circulation model. *Journal of Geophysical Research* 113, D17114, doi:10.1029/2008JD010115.

IPCC. 2007. Summary for policymakers. In S. Solomon, D. Qin, M. Manning, Z. Chen, M. Marquis, K. B. Averyt, M. Tignor, and H. L. Miller, eds., *Climate Change 2007: The Physical Science Basis*. Contribution of Working Group I to the Fourth Assessment Report of the Intergovernmental Panel on Climate Change. Cambridge, UK, and New York: Cambridge University Press.

Mendonca, Bernard G., and Wayne T. Iwaoka. 1969. The trade wind inversion at the slopes of Mauna Loa, Hawaii. *Journal of Applied Meteorology* 8: 213–219.

Mims, Forrest M., III. 2007. Hawaii's Mauna Loa Observatory: The first fifty years. Paper presented at ESRL Global Monitoring Annual Conference, Boulder, Colorado, May 2–3.

Mims, Forrest M., III, and John E. Frederick. 1994. Cumulus clouds and UV-B. *Nature* 371: 291.

MLO. 2008. Information for visiting cooperative investigators and ESRL staff, Mauna Loa Observatory. (Accessed at http://www.mlo.noaa.gov/pdf/Information%20for%20Visiting%20Cooperative%20Investigators%20and%20CMDL%20Staff.pdf.)

NBS and US Weather Bureau. 1956. Dedication of the "slope" building, Mauna Loa Observatory, Thursday, June 28, 1956. Jointly sponsored by the National Bureau of Standards and the US Weather Bureau.

Saddle Road. 2005. Saddle Road on Hawaii's Big Island. Executive summary. (Accessed at http://www.saddleroad.com/05execsum.html.)

Sasaki, Makoto. 2007. All-sky survey high resolution air-shower detector. ICRR, University of Tokyo. (Accessed at http://ebhu.astron.s.u-tokyo.ac.jp/pdf/sasaki.pdf.)

Schnell, Russell. 1993. August 1993 activities report, September 16. In *Mauna Loa Observatory, Monthly Reports*, Hilo, Hawai'i, 1993.

———. 1996. August 1996 activities report. Mauna Loa Observatory, Monthly Reports (CD-ROM), Hilo, Hawai'i.

———. 1998. March 1998 activities report. Mauna Loa Observatory, Monthly Reports (CD-ROM), Hilo, Hawai'i.

Starr, Forest. 2008a. Personal communication, June 19. See also Forest and Kim Starr, *Plants of Hawaii, Hypochoeris radicata*, Hairy cat's ear, gosmore (Asteraceae), May 19. (Accessed at http://www.hear.org/starr/hiplants/images/thumbnails/html/hypochoeris_radicata.htm.)

———. 2008b. Personal communication, June 19. See also Forest and Kim Starr, *Plants of Hawaii, Senecio madagascariensis*, Fireweed (Asteraceae), May 19. (Accessed at http://www.hear.org/starr/hiplants/images/thumbnails/html/senecio_madagascariensis.htm.)

———. 2008c. Personal communication, June 19. See also Forest and Kim Starr, *Plants of Hawaii, Verbascum Thapsus*, common mullein (Scrophulariaceae), May 19. (Accessed at http://www.hear.org/starr/hiplants/images/thumbnails/html/verbascum_thapsus.htm.)

STEP. 2007. Ecotourism opportunities on the Saddle Road, Big Island of Hawaii. Saddle Road Project, STEP-UP, Sustainable Tourism and Environment Program, University of Northern Iowa. (Accessed at http://www.uni.edu/step/reports/saddle_road.pdf.)

Zak, Anatoly. 2007. Baikonur: Soyuz launch facilities. (Accessed at http://www.russianspaceweb.com/baikonur_r7_1.html.)

EPILOGUE

"We Must Preserve This Progress"

On November 10, 1957, Dr. Robert Simpson visited for the first time the new concrete block building high on the north slope of Mauna Loa that sprang from his meeting with Ralph Stair at Sunspot, New Mexico, in June 1955. Some of the International Geophysical Year instruments were in operation, including the Regener surface ozone project inside the building and the sun tracking Eppley pyrheliometer mounted on a post in the lava.

Simpson was primarily interested in the newly arrived Dobson spectrophotometer, for measuring the ozone layer was his main reason for helping Ralph Stair restart an observatory on Mauna Loa after the summit station was closed. The Dobson had arrived and was awaiting a power supply so ozone layer measurements could be initiated by Jack Pales and his team. Dave Keeling's carbon dioxide system that motivated Harry Wexler to approve the observatory project would be installed the following March.

After seeing that his dream for Mauna Loa was about to be fulfilled, Simpson signed the *Visitors* book and wrote, "A long step ahead. We must preserve this progress." The progress he saw was only the beginning of much more to come. And all the research projects that were underway—or soon would be—when Simpson visited the little observatory on Mauna Loa more than half a century ago remain in operation today, a fact that greatly pleases him.

After he left Hawai'i to manage the fledgling National Hurricane Research Project that provided the funds to build the slope building, now the Keeling Building, Simpson visited MLO again on April 18, 1963, and November 10, 1979. But his formal collaboration had ended long before, and his role as the observatory's founder was nearly forgotten.

That changed on July 2, 2008, when Simpson received NOAA's 2008 Environmental Hero Lifetime Achievement Award for founding the Mauna Loa Observatory and his pioneering hurricane research (Plate 105). The award was presented by Dr. Richard Spinrad, assistant administrator of NOAA's Office of Oceanic and Atmospheric Research.

"Where It All Began"

On a metal shelf in the Keeling Building at the Mauna Loa Observatory is a stack of self-adhesive address stickers designed by Russell Schnell. The stickers depict the famous carbon dioxide curve, and "Where It All Began" is printed in blue letters over the graph. The scientists and staff at MLO are well aware of the observatory's role in triggering the climate change research taking place around the world and from space. They are also aware of the implications of their work, yet they rarely discuss them. Instead, they continue to do what they and their predecessors have done best for the past fifty years by routinely and precisely measuring the key parameters of the atmosphere, including the concentrations of greenhouse gases, the clarity of the sky, the total ozone and water vapor over the observatory, and a host of other observations.

When the crew drives to and from the mountain, the conversation is likely to be about the weather and the condition of the road. The busy schedule on the mountain doesn't leave much time to ponder the implications of the data being meticulously collected there day after day. This suggests another meaning for those "Where It All Began" stickers, for many scientists have built their careers on data they personally acquired or that was acquired on their behalf at the Mauna Loa Observatory. Perhaps Bernard Mendonca (video interview with author, Boulder, Colorado, April 27, 2007) says it best when he reflects back on his years working on the lonely mountain in Hawai'i:

> To me the highlight [of my career] was I got to be a scientist at Mauna Loa. I learned and matured, I did my first publications there. It was a growth period for me. I never forgot that. It was where I realized I could do [science]. . . . And the other thing was knowing that we were breaking ground. I knew that. I think most of the people who were working up there really knew this hadn't been done before, and [knew] that we really were breaking ground. We were the ones providing the first information. At times we weren't sure where it was going, especially in the very early years, how this would end up. But, as the years passed, this confidence built up that we, hey, we're the guys now. They've got to come and ask us. . . . I think that, to me, was the thing that motivated me, and every time I went back it came back again, you know. This place, this is where we started, this is where I started. I get emotional here, but I don't regret it at all. It's a good career. I'd do it again in a heartbeat.

Howard Ellis, Bernard Mendonca, and Malcolm Keyes kept MLO alive during the budget crisis of 1964–1966. Their work was often boring and required long stays on the mountain, with Ellis staying as long as fifteen days. Had they failed, today there would not be a half century of atmospheric transmission measurements and the carbon dioxide record that jump-started the global warming research era. Ralph Keeling (2008) summed up the significance of long-term measurement series best when in an essay in *Science* he reflected on the fiftieth anniversary of the CO_2 measurements begun by his father in March 1958:

> The Mauna Loa story provides a valuable lesson on the importance of continuous earth observations in a time of accelerating global change. . . . The only way to figure out what is happening to our planet is to measure it, and this means tracking changes decade after decade and poring over the records. A point of diminishing scientific returns has never been realized in what is now known as the "Keeling Curve," the Mauna Loa CO_2 record.

A Recent Stay at Mauna Loa Observatory (2008)

The author continued annual visits to MLO begun in 1992 to calibrate a variety of atmospheric instruments. In 1999 the visits were extended to ten-day stays on the mountain to guarantee a sufficient number of clear days. These stays provided many opportunities to observe and better understand the world class atmospheric science conducted at MLO and to meet many of the scientists themselves.

The MLO calibration visits from 2006 to 2010 occurred during preparation of the book. Trips down to Hilo every few days to shower at a public beach or swimming pool and buy groceries provided many opportunities for interviews with MLO staff and veterans and reviews of the records and photographs in the MLO archive. One trip to MLO was devoted mainly to work on the book.

The 2008 calibrations at MLO took place during a particularly busy time for both events at the observatory and work on the book. Notes were kept to document the trip to Hawai'i, life on the big mountain, and visits with scientists. As during every previous trip to Hawai'i, the plane departed from San Antonio, the Texas city from where MLO founder Robert Simpson and his two daughters left for New Mexico in June 1955. The flight from Dallas–Fort Worth to Los Angeles passed over southern New Mexico, and the 275 square miles of gypsum dunes known as White Sands could be clearly seen from the right side of the aircraft. That snow-like patch contrasted sharply with the adjacent Sacramento Mountains, where Ralph Stair's measurements of sunlight were interrupted by a dust storm the day that Simpson drove up in his 1950 Buick, introduced himself to Stair and suggested that a mountain in Hawaii would make a much better observing site.

The purpose of the stay at MLO was to calibrate a variety of instruments, including two GLOBE sun photometers (Mims 1999), a Microtops ozonometer, a Microtops with light-emitting diodes (LED) as detectors (Mims 2009; see also Plate 64), and the first LED sun photometer, which was built in 1990 and first calibrated at MLO in 1992 (Mims 1992). During most weekends, only the Mauna Loa Solar Observatory operator reports for mountain duty. But early on Saturday, May 17, Phillip B. Russell and Yohei Sinizuka (see Plate 62) from NASA's Ames Research Laboratory arrived at MLO. Russell is among the world's leading specialists in measuring the optical depth of the atmosphere from aircraft, and the purpose of his visit was to complete a week-long calibration of a precision airborne sun photometer. They completed their work on Sunday, May 18, packed their heavy instrument in its custom packing crate, and signed out.

Monday, May 19, was a busy time for the MLO staff. The previous Thursday they had shut down many of the experiments because of a five-hour power outage planned for Friday, and after they arrived on Monday they quickly began powering up the instruments, including the Scripps and GMD carbon dioxide experiments and the Brewer spectrophotometers.

May 20 was another busy day. The morning began with a series of sun photometer calibration measurements. Then came a call from Dave Moss at the Scripps Institution of Oceanography, who reported that the reference gas for the Scripps carbon dioxide monitor had not yet been switched back on after the power outage. Moss talked the author through the procedures to open the valve on the gas cylinder and then switch on the gas line. A remarkable demonstration of the instantaneous communications between Scripps and the remote mountain observatory occurred as Moss moved the cursor on his computer screen to point out the impact of reactivating the gas. The movements of his cursor in California were instantly visible on the Scripps computer installed next to their CO_2 analyzer at MLO.

Later that morning, a group of communications specialists representing the State of Hawai'i and the US Army visited MLO to make plans for a future microwave installation. Three of the group had connections to the very earliest days at MLO.

The group included Robert J. Hlivak of Hawai'i's Telecommunications Services Branch, who said that Claude Kutaka reported to him when they both worked at the University of Hawai'i. Kutaka's father was the late Clifford Kutaka, one of the first three Weather Bureau staff to man the observatory twenty-four hours a day. Hlivak said Kutaka told him amazing stories about the experiences and hardships his father had endured when he worked at MLO back in the 1950s.

Bruce Judson and Pete Feenan of the US Army Information System Engineering Command, Fort Huachuca, Arizona, were also present. When MLO director Jack Pales was looking for new projects for the observatory nearly fifty years before, it was Fort Huachuca that responded with a project that used various instruments to measure sunlight from four posts installed in a grassy field adjacent to the Hilo office at the old Air Cargo Building and from MLO. The project also included the first measurements of ground-level ozone at MLO and Hilo.

The group also included Denny Greenwood and Bill Selvey, who represented the US Army's 311th Signal Command. Greenwood pointed out various US Army communications systems installed at MLO for coordinating military training activities in the Saddle between Mauna Kea and Mauna Loa. Some of the training activities visible or audible from MLO over the years have included aircraft bombing runs and the firing of heavy guns.

Later on May 20, Prof. Joseph Shaw (see Plate 63) of Montana State University and his wife Margaret arrived at MLO to begin preparations for a two-week experiment to be conducted by postdoc Nathan Pust and graduate student Andrew Dahlberg. The experiment was designed to test a sophisticated camera that provided a polarized image of the entire sky. Their equipment was loaded in several large wooden crates that MLO station chief John Barnes drove to the station several days earlier during the annual visit by students from the University of the Nations. Some of the students helped Dr. Barnes unload the heavy crates and slide them into the NDACC Building.

While Shaw began unpacking the crates, Barnes arrived to begin his weekly lidar session. That night Shaw met Pust and Dahlberg on Saddle Road and drove them up to the observatory, where the jet-lagged pair began assembling their apparatus. Shaw, who is an accomplished photographer, also made some spectacular long-exposure night images of the MLO lidar beam.

The next day, Pust and Dahlberg began making all-sky polarized images. To check the validity of their data, they produced a red-covered edition of former MLO director Kinsell Coulson's classic text, *Polarization and Intensity of Light in the Atmosphere* (1988) and found that in 1977 Coulson had measured polarization values at MLO remarkably close to what they were measuring. The difference was that Coulson had used a complex instrument that painstakingly scanned a portion of the sky a step at a time, while Pust's all-sky imager produced an image of the polarized sky in one step.

Like Coulson, Joseph Shaw's father Glenn made numerous sunlight observations at MLO during the 1970s and 1980s to calibrate his sun photometers and to measure the clarity of the atmosphere, including the measurements that led to his discovery of Asian dust in 1979 (see chapter 11). When Dr. Shaw learned that Joseph would be at MLO, he shipped two of his 1982 sun photometers for him to calibrate, and Joseph brought along one of his own. On two mornings, Shaw and the author calibrated the highly stable instruments, and the results were typical for the pristine atmosphere of MLO:

> The data are of excellent quality, with the rms deviation for the Langley plots being around the 0.998 level. Moreover, the two days agree with one another very closely, to about the 0.2% level. The grand average optical depth from aerosols was 0.013, 0.007 and somewhat less than 0.008 . . . for the 425, 500 and 790 nm bands. . . . These are incredibly clear conditions, probably the lowest . . . you can get on planet earth. (Shaw, personal communication with author, June 21, 2008)

On May 24, Prof. Bo Reipurth of the Institute for Astronomy at the University of Hawai'i arrived with a team of enthusiastic students to install a small dome for a new telescope designed to look for sunlike stars. They quickly assembled the frame of the assembly and returned late on May 25 to install the dome. Meanwhile, Joseph and Margaret Shaw departed for Montana, and Nathan Pust and Andrew Dahlberg stayed behind for a few weeks taking turns on the mountain to complete the sky imaging project.

During each of the ten days, one of the rotating staff of observers at the Mauna Loa Solar Observatory arrived shortly after sunrise to point their telescopes at the sun and begin imaging its disk. Although solar cycle 24 officially arrived in January 2008, solar activity was minimal. The sun ignored various forecasts made by experts that it would soon reawaken and slumbered on until 2010. The sun's nap between solar cycles 23 and 24 was the longest in nearly a century.

Other observatories and the SOHO satellite also image the sun. But MLO is uniquely equipped to monitor any effects of a quiet sun on the atmosphere and climate, for greenhouse gases warm the atmosphere, and history has shown that subdued or extended solar cycles are associated with cooling. MLO, of course, measures both greenhouse gases and solar activity. The extraordinarily long delay between solar cycles 23 and 24 and the fewer than expected number of sunspots during the first few years of solar cycle 24 led to speculation in the media about a possible period of global cooling and rebuttals by some climate scientists (Beal 2011). The instruments and solar telescopes at MLO don't engage in debates or speculation. They will simply measure with high accuracy whatever occurs.

In addition to new interviews and photographs for the book, the 2008 MLO visit related here provided 2.4 megabytes of data files, 889 photographs, and 72 video clips. The photos included 118 calibrated images of the glow around the sun known as the solar aureole that is a sensitive indicator of atmospheric dust. Similar amounts of data and photographs are acquired each year and new experiments are often added. For example, during the 2010 MLO visit, the author compared two US Department of Agriculture shadowband solar radiometers and his personal shadowband radiometer (with LEDs as photodiodes) with MLO and other instruments. New measurements of the tiny amount of water vapor over MLO were made with several infrared thermometers, and these data were included in a new scientific paper (Mims et al. 2011). These and other calibration data are crucial for the success of the author's long-term atmospheric measurements in Texas begun in 1990. But their significance is minuscule when compared with the overall volume of data that flows from MLO to scientists across the United States and various cooperating countries around the world.

Measuring the Future

The Mauna Loa Observatory has persevered through a nearly fatal budget cut and many smaller slashes. It has survived a major volcanic eruption, earthquakes, lightning bolts, hurricane-force winds, hailstorms, blizzards, power outages, and government red tape. While insiders ponder the future of the Mauna Loa Observatory and the outside world debates how best to respond to its discoveries about our changing atmosphere, the scientists and staff (Plates 106–108) at the lonely observatory on the world's largest mountain in the middle of the Pacific Ocean (Plates 109–110) will continue doing what former GMD director David Hofmann (video interview with author, Boulder, Colorado, May 1, 2007) said best:

> The laboratory that I work for and the scientists in that laboratory pride themselves in making the best measurements in the world.

The Mauna Loa Observatory will continue making "the best measurements in the world," for they are vital to understanding the thin blanket of air that provides our sustenance and regulates the climate of our planet. The words that MLO founder Robert Simpson wrote in the *Visitors* book in November 1957 are as relevant today as when he penned them:

> A long step ahead. We must preserve this progress.

References

Beal, Tom. 2011. Scientists ponder sun's next move from quiet phase. *Arizona Daily Star* (April 4). (Accessed at http://azstarnet.com/news/local/article_82ad0b62-c7e7-5211-89e7-135b9b92479a.html.)

Coulson, Kinsell L. 1988. *Polarization and Intensity of Light in the Atmosphere*. Hampton, VA: A. Deepak Publishing.

Keeling, Ralph. 2008. Recording Earth's vital signs. *Science* 319: 1771–1772.

Mims, Forrest M. III. 1992. Sun photometer with light-emitting diodes as spectrally selective detectors. *Applied Optics* 31: 6965–6967.

———. 1999. An international haze-monitoring network for students. *Bulletin of the American Meteorological Society* 80: 1421–1431.

———. 2009. LED sun photometry. *OPN: Optics & Photonics News* 20: 32–38.

Mims, Forrest M., III, Lin H. Chambers, and David R. Brooks. 2011. Measuring Total Column Water Vapor by Pointing an Infrared Thermometer at the Sky. *Bulletin of the American Meteorological Society* (in press).

APPENDIX

The Current Science Programs

The original International Geophysical Year measurements at the Mauna Loa Observatory were supplemented over the years by more than a hundred research projects. While some projects were temporary or have achieved their purpose, many others remain active and have lifetimes measured in decades. This appendix examines some of the current science programs at MLO conducted and facilitated by the various research groups of the Global Monitoring Division (GMD) of NOAA's Earth System Research Laboratory (ESRL). The scientists who lead these groups are also introduced.

The mission statement of the Global Monitoring Division summarizes the broad suite of measurements made at MLO and GMD's observatories at Samoa, South Pole, Barrow, and Trinidad Head:

> To acquire, evaluate, and make available accurate, long-term records of atmospheric gases, aerosol particles, and solar radiation in a manner that allows the causes of change to be understood. (GMD 2010)

MLO station chief John Barnes (video interview with author, Hilo, Hawai'i, December 2006) discusses the importance of making multiple kinds of measurements that can be compared with one another and gives some specific examples:

> It is easy to measure CO_2 globally because it . . . spreads all over and it is quite uniform. But aerosols are given off mainly from industrial areas [and deserts], so [they are] more localized. They have a very short lifetime of days or weeks before they get rained out. So you need a combination of surface sampling and some upper air sampling and then pulling it all together with satellite measurements and then a theory to really tie it all together. So you can take the theory and then put it into climate models and see what effect they are going to have.

MLO also services many cooperative programs, some of which are closely tied to GMD projects and activities. The various research programs are divided into six major groups below, some of which can be traced to the very beginning of full-time science at MLO in 1957–1958 and will be familiar to the reader who has read the previous chapters about the history of the observatory. Much more information about the various programs can be found by visiting the Web sites of the GMD (www.esrl.noaa.gov/gmd/index.html) and MLO (www.mlo.noaa.gov/home.html).

1. Carbon Cycle and Greenhouse Gases (CCGG)

The CCGG Group is headed by Dr. Pieter Tans, who studied physics at the University of Groningen in the Netherlands. His studies were influenced by the MLO carbon dioxide record, which stimulated his interest in measuring isotopes of carbon in tree rings to learn about changes in atmospheric carbon dioxide:

> At that point I was pretty heavily involved in theoretical solid state physics, and it was a little bit too abstract for my taste. I wanted to do something that was a little bit more related to real life, to the earth and policies and future fate of the earth, because I was somewhat worried already about sustainability of our civilization. So I ran into a little book in a bookstore in 1972, and the title was *Inadvertent Climate Modification* [MIT 1971]. . . .

> A lot of the Mauna Loa record were in the book. There were lots of stuff about aerosols and other possible influences of human action on climate. I read this, because I was intrigued, and I was convinced right then that it was going to be an important problem. (Tans, video interview with author, Boulder, Colorado, May 1, 2007)

After receiving his doctorate, Tans moved to Scripps Institution of Oceanography to spend a year working with Charles David Keeling. He enjoyed the speakers and visitors at Scripps, but he was disappointed by Keeling's inflexible style of doing science, as he noted in the same interview:

> He knew extremely well what he wanted to do and how he wanted to do it. I tell you, when I set up that extraction line he did not give me any leeway. It had to be done exactly the way he wanted it. I found it a little bit oppressive, actually. It contributed to my not wanting to stay too long.

While Tans was impressed by Keeling's long-term measurements of CO_2, he didn't understand why Keeling was unwilling to modernize his apparatus. Shortly after moving to NOAA in 1985, Tans attended a meeting of the world's carbon dioxide experts at the World Meteorological Organization. In the 2007 interview, he noted that

> Dave Keeling had actually just changed to digital recording of the data, and he was singing the praises of that and selling it to us during the presentation he gave, and we were just looking around and [thinking] "Oh, my God, we have been doing this already for a decade." So he had just caught on to that. It was funny.

What Tans and his colleagues at GMD share with Keeling's legacy is high-quality measurements. For Tans, this involves precise measurements of the principle carbon cycle greenhouse gases that absorb some of the infrared radiation that is naturally emitted by substances that have been warmed by sunlight, including vegetation, soil, rocks, water, roads, and buildings. All these things are heated during the day by the visible wavelengths of sunlight, most of which are freely transmitted by greenhouse gases. It is the infrared radiation that these warmed objects emit that is absorbed by greenhouse gases. Leaving aside water vapor, the most significant natural greenhouse gas, all the specific carbon cycle greenhouse gases of most interest have both natural and human sources. They include carbon dioxide (CO_2), methane (CH_4), and carbon monoxide (CO). Together with nitrous oxide (N_2O) and two of the chlorofluorocarbons (CFC-11 and CFC-12, which are discussed below), these gases account for "about 97 percent of the direct radiative forcing by long-lived greenhouse gases since 1750" (Hofmann 2007). The purpose of monitoring these gases is to better understand the seasonal distribution and the trends in the concentration of greenhouse gases.

Clean, dry air consists of nitrogen (78 percent), oxygen (21 percent), argon (0.9 percent), and various trace gases. Dry air is ordinarily specified, because ambient water vapor is highly variable in time and space and can be substantial in volume (from near 0 to as much as 4 percent). The residence time of a water molecule in the atmosphere is much briefer than that of carbon dioxide, methane, and other greenhouse gases. A water molecule may remain in the atmosphere from minutes to weeks, while other greenhouse molecules may remain from days to centuries.

The carbon cycle greenhouse gases constitute a tiny fraction of the atmosphere by volume. It is estimated that carbon dioxide, the major carbon cycle trace gas, comprised about 0.028 percent of the atmosphere at the beginning of the industrial age (Biutner et al. 1986). Today's concentration is about 0.039 percent. If we assume that the increase in CO_2 since the beginning of the industrial age is due entirely to people, then about a fourth of the CO_2 in the atmosphere today is a by-product of human activity, mainly from the combustion of petroleum, natural gas, and coal. The remaining three-fourths is primarily from vegetation, biomass fires, animals, and volcanoes. There is much less methane in the atmosphere, but methane is a much more potent greenhouse gas than CO_2. Its sources include rice fields, rotting vegetation, biomass fires, termites, livestock, and leaks from natural gas plants and transmission lines.

MLO provides key data for the Carbon Cycle Greenhouse Gases Group supervised by Tans. The CCGG Group also collects air samples from ships, aircraft, and various ground sites. The ground sites include tall towers that collect air from as high as 1,500 feet (457 meters) above the ground. As of this writing, five towers are in service and seven more are planned.

The best-known and still the most significant investigation at MLO is the long-term, continuous monitoring of the concentration of carbon dioxide in the atmosphere. This program began in March 1958 with funding from the Weather Bureau and support from the Scripps Institution of Oceanography under the direct supervision of Dr. Charles David Keeling, who supervised the program until his death in 2005. The Scripps program is now directed by Dr. Ralph Keeling, Charles Keeling's son. Keeling also directs the Scripps program that measures the oxygen content of the atmosphere at MLO and elsewhere. As noted earlier, a unique feature of MLO's CO_2 monitoring is that the program of continuous measurements begun in 1958 has since 1974 been duplicated by NOAA. The measurements made by the two independent monitoring systems since that time have been in very close agreement.

Pieter Tans is passionate about his science, and he is persuaded that rising greenhouse gases will lead to a powerful and serious impact on earth's climate:

> The climate forcing by CO_2 really took off starting in the seventies and eighties, [and] the rate of increase of CO_2 now is higher that it has ever been. Here is something to keep in mind, that half of the CO_2 emissions from fossil fuel burning, cumulative from all fossil fuels burned since the last century, half of that has taken place after 1981. Keep that in mind. (Tans, video interview with author, Boulder, Colorado, May 1, 2007)

While carbon dioxide has continued its inexorable increase year after year, in recent years the concentration of methane and the CFCs has stopped rising, and GMD director David Hofmann (2007) notes how this has made a significant impact on the Greenhouse Gas Index:

> While the radiative forcing of the long-lived, well-mixed greenhouse gases increased about 22% from 1990 to 2006 (~0.50 watts m^{-2}), CO_2 has accounted for about 80% of this increase (~0.40 watts m^{-2}). Had the ozone-depleting gases not been regulated by the Montreal Protocol and its amendments, it is estimated that climate forcing would have been as much as 0.2 watt m^{-2} higher [Velders et al. 2007], or about one-half of the increase in radiative forcing due to CO_2 alone since 1990.

While carbon monoxide (CO) is a minor greenhouse gas, it is measured at MLO and other GMD observatories for a variety of reasons. Of special interest is that biomass burning produces large quantities of CO; thus, CO is a tracer that provides an indication of fire and smoke events.

Flask Sampling of Ambient Air

Collecting and analyzing samples of air is key to the greenhouse gas program. The flask sampling program at MLO, which began with Keeling, had expanded to global coverage long before the observatory's fiftieth anniversary. In addition to continuous monitoring of CO_2, methane, and other gases, MLO collects air in specially designed flasks that are sent for analysis to GMD and Scripps. Flask samples are also collected just above sea level at the Cape Kumukahi Lighthouse.

As described in chapter 13, the oldest form of flask sampling, which is still in use, involves an almost ritualistic protocol using an evacuated, five-liter glass sphere wrapped with protective tape and having an access tube fitted with a stopcock. The five-liter samples are collected once a week at MLO and at Cape Kumukahi and shipped to the Scripps Institution of Oceanography. After the sampled air is removed and analyzed, the flasks are evacuated and returned to MLO.

NOAA's Global Monitoring Division in Boulder uses a modernized form of flask sampling in which smaller flasks are sampled automatically by means of portable briefcase-size systems. GMD maintains an inventory of some six thousand such flasks. The sampling case is placed flat on the lava, opened, and a telescoping sampling tube is pulled up above the head of the operator. The system is then switched on, whereupon it begins an automated sampling cycle in which two sample flasks are filled with ambient air.

The flask samples are stored in their shipping boxes until a set number have accumulated. They are then shipped to research chemist Thomas Conway's flask sampling unit at the Global Monitoring Division, where they are delivered to Molly Heller's lab. The flasks arrive in stacks of shipping boxes that are rolled into Heller's lab, where she unpacks the flasks and logs them into the system. The flasks are then sent to physical scientist Patricia Lang's lab, where they are mounted in analysis carousels and a portion of their air content is analyzed. After analysis, the flasks are returned to Heller, who sends them to the University of Colorado for further analysis of the remaining air. The flasks are then returned to Molly Heller, and what happens next is best explained by the flasks themselves:

> Back we go to Molly, who cleans us up to go out again. That's why I am waiting for evacuation now. That is, we need to have the previous sample that's been analyzed totally evacuated or cleaned out of us, and then we get tested for leaks or damage.
>
> When I get back from my next trip, and my air sample is analyzed, Molly will examine and evacuate me, put me back in a box and ship me out again, with my mate, #3332. (We really do travel in pairs!) I won't know where #3332 and I are going until the person at our destination opens the box. It might be one of the new sites we're working on, like along the Gulf Coast of Mexico or on Taiwan. Anywhere is fine with me; I'm just happy to be part of NOAA's careful, consistent, and continual worldwide air sampling network that shows scientists the composition of the atmosphere. Tracking trace gases in the atmosphere can help to explain warming of the climate, the movement of the atmosphere, and the source of greenhouse gases. (Knight 2007)

2. Halocarbons and Other Atmospheric Trace Species (HATS)

As noted above, the carbon cycle greenhouses gases constitute such a small fraction of the atmosphere that they are collectively known as trace gases. There are many other trace gases, including hydrogen, neon, helium, krypton, ozone, sulfur dioxide, nitrous oxide, and more than a dozen synthetic gases, some of which are both greenhouse gases and damaging to the ozone layer.

Dr. James W. Elkins of the Global Monitoring Division has long supervised NOAA's program to monitor halocarbons and certain other trace gases. Elkins' career was greatly influenced by an amateur scientist who two hundred years ago made daily measurements of the minimum and maximum temperature from his farm or where he happened to live or travel, including Philadelphia on July 4, 1776, and later his temporary residence in Washington, DC:

> Thomas Jefferson had written on his tombstone that the three greatest accomplishments of his career were author of the Declaration of Independence, author of Statute of Virginia for Religious Freedom, and the father of the University of Virginia. It is important to note that he did not mention on the obelisk marker at his beloved Monticello that he was the third President of the United States. Thomas Jefferson also founded one of the current NOAA organizations, the Survey of the Coast, in 1807 while he was President. He is credited today as a founding father of NOAA. Professor Robert H. Simpson, who founded our Mauna Loa observatory in the 1950s, began his first professorship at the University of Virginia in 1974, the year I graduated with a B.A. in physics. It is amazing to me how many individuals, ideas, and organizations were influenced by the contributions of one of our greatest Presidents, Thomas Jefferson. He kept weather observations wherever he traveled and had academic correspondence with some of the greatest scientists of his time. He was the first environmental President. (Elkins, personal communications with author, May 4, 2007)

Jefferson is Elkins' inspiration, and measuring trace gases in the atmosphere is his passion. He is so persuaded that greenhouse gases produced by human activity are causing significant warming of the planet that he has invested in a rooftop photovoltaic system that will power much or all of his home with sunlight. The gases that Elkins studies are measured at MLO by means of electron capture detection-gas chromatography in a program known as CATS (Chromatograph for Atmospheric Trace Species). Most of the gases are measured in concentrations of parts per trillion, and they include the following:

Chlorofluorocarbons:
- Trichloro(mono)fluoromethane CFC-11 (CCL_3F)
- Dichlorodifluoromethane CFC-12 (CCl_2F_2)
- Trichlorotrifluoroethane CFC-113 (CCl_2F-$CClF_2$)

Hydrochlorofluorocarbons:
- Chlorodifluoromethane HCFC-22 ($CHClF_2$)

Methyl Halides:
- Chloroform ($CHCl_3$)
- Methyl Bromide (CH_3Br)

Chlorinated Solvents:
- Carbon Tetrachloride (CCl_4)
- Halon-1211 ($CBrClF_2$)
- Methyl Chloride (CH_3Cl)

Other Measurements:
- Nitrous Oxide (N_2O)
- Methyl Chloroform (CH_3CCl_3)
- Carbonyl Sulfide (COS)

The CATS instrument is so sensitive that it can detect whiffs of gases in the list above that are artificially introduced into the air at MLO. This was driven home for a visiting scientist who was about to remove dust from the optics of an ozone instrument by means of a can of compressed gas. MLO station chief John Barnes happened to see the can and asked to inspect it before it was placed in use. He carefully studied the ingredients to make sure they did not include any of the CATS gases before passing it back to the visitor (the author of this book).

Measurements made at MLO and elsewhere reveal that certain of the CATS gases are declining while others are increasing. Those in decline include the CFCs, CH_3CCl_3, CH_3Br, and CCl_4. Their decrease is attributed to sharply reduced production following the adoption of the Montreal Protocol to stem the production of gases that harm the ozone layer (Elkins et al. 1993). N_2O, HCFCs, and HFCs are not controlled and are increasing.

The decline in some key CATS gases has lessened their priority in the face of the relentless rise of CO_2. This has resulted in funding shifts that have Elkins worried, for he believes that it is essential to continue monitoring all the CATS trace gases to properly monitor their trends over time.

3. Global Radiation (G-Rad)

The Global Radiation Group is headed by Dr. Ellsworth Dutton, who was fascinated by astronomy, storms, clouds, and weather while growing up in rural Iowa. Dutton's interest in science stayed with him as he grew up and led to, as he explains, a "scientific curiosity that went beyond the earliest years when every kid is a scientist." Dutton majored in aerospace engineering in college, but after graduating in 1971 he found that employment opportunities had been reduced by cutbacks in the space program. He returned to college to study meteorology and in 1978 was hired by NOAA's former Geophysical Monitoring for Climatic Change (GMCC) Division in Boulder. NOAA later financed his doctoral studies at Colorado State, and in 1995 Dutton received his Ph.D. His thesis was devoted to the climatic effects of the major volcanic eruptions of El Chichón and Mount Pinatubo, and he explains how this fits well with the objectives of the Global Radiation Group:

> We are interested in the radiation quantities as they relate to climate and climate change, particularly radiation at the surface, both solar and infrared. We are interested in fitting in with the rest of the work

> of the traditional GMCC, then CMDL, now GMD laboratory of changes in atmospheric composition and their impact on the radiation budget at the surface. We are interested in explaining the variability that we see in the radiation records as to whatever is causing it, changes in land surface, changes in atmospheric composition, aerosols, water vapor and clouds.
>
> I've concentrated largely on, actually, measurement techniques and methodologies, improving the measurements that take the science past where it was in previous decades where the interest was primarily in shorter-term applications, weather applications, not really at the level of the accuracy necessary for . . . climate change detection. (Dutton, video interview with author, Boulder, Colorado, May 1, 2007)

Dutton's latter objective includes a continuation of the Ellis curve of atmospheric transmission measurements begun at MLO in November 1958 and extracting improved column water vapor measurements from old data.

GMD maintains a full suite of sunlight measuring instruments at the Mauna Loa Observatory and the other GMD observatories, and several cooperative programs also have various sunlight monitoring instruments at MLO. The key instruments include radiometers that measure solar ultraviolet and robotic sun photometers that measure changes in the optical depth of the atmosphere caused by air pollution and volcano aerosols. Also included are multifilter rotating shadowband radiometers that measure both the total sunlight from the sky and the light directly from the sun at wavelengths from the blue to the near-infrared.

The oldest and arguably the most important solar radiation program at MLO is a pair of Eppley pyrheliometers that track the sun across the sky and measure its direct sunlight. This program was begun in November 1957 using a single pyrheliometer and is the earliest of the International Geophysical Year measurements begun at MLO. During the 1960s, Howard Ellis devised a novel, self-calibrating method for extracting the percentage of the sunlight transmitted through the atmosphere from ratios of direct solar irradiance measured by the pyrheliometer at four different sun angles. The resulting Ellis curve, which is discussed in chapter 10 and shown in Plate 21, clearly illustrates the effect on atmospheric transmission of sulfate injected into the stratosphere by several major volcanic eruptions. The dips in transmission caused by these volcanoes nicely match peaks in backscatter measured by the MLO lidar. The lidar provides the altitude of the responsible aerosol layers, and its two wavelengths (green and near-infrared) provide a means to estimate the size of the particles. A string of papers reporting on the half century of data has appeared since the first major paper by Ellis and Pueschel (1970). The latest such paper is an especially in-depth report on the first forty-two years of this unique record of atmospheric transmission (Dutton and Bodhaine 2001).

Since 1978, the diffuse sunlight scattered from the sky has been measured with two Eppley pyrgeometers equipped with a shading disk that blocks direct sunlight. Summing the direct sunlight from the pyrheliometers described above and the diffuse sky irradiance provides the full sky irradiance, a figure that is given in watts per square meter. G-Rad also makes continuous measurements of total sky thermal IR using a pyrgeometer:

> This is standard measurement of G-Rad at all our observatories and measures the combined downward natural and anthropogenic greenhouse radiation at the surface, which ironically is not expected to be increasing as rapidly at MLO because of its isolated near mountain-top location. (Dutton, personal communication with author, May 20, 2008)

In view of the possibility of future changes in the atmosphere over MLO, Dutton expresses concerns in the same communication about the fifty-year Ellis curve of atmospheric transmission, which, as he observes, is more correctly identified as "mean apparent transmission":

> Concerns that are now arising are related to minor flaws in the absolute stability of the technique. Not only could water vapor trends start to become significant at the level of long-term changes we are starting to see, but the transmission ratios are sensitive (and always have

been) to any nonlinearity that exists in the pyrheliometers. (The simple ratio method only removes the instrument calibration if the instrument response is perfectly linear. While the thermocouples in the instrument are assumed to be highly linear, they are not perfect.) Both of these matters were never an issue with the magnitude and length of trends in question in the early 1970s. The reality of now attempting attribution to something like a 0.005 change in fifty years is that both issues arise. But this is not MLO history; it is its future.

SURFRAD (Surface Radiation) Network

GMD has deployed a network of seven radiation monitoring sites across the United States to form a network known as SURFRAD. While there is no SURFRAD site at MLO, similar measurements are made at MLO by GMD or cooperative instruments.

Solar Radiation Calibration Facility

The Global Radiation Group operates a world-class laboratory for periodically calibrating GMD solar radiation and infrared instruments to assure that they are traceable to the World Radiometric Reference maintained in Davos and the World Radiation Center.

4. Ozone and Water Vapor Group (OZWV)

The Ozone and Water Vapor Group is responsible for NOAA's network of Dobson spectrometers that measure ozone across the United States and at MLO and the other GMD observatories. The group is also responsible for measuring surface ozone at MLO and thirteen other sites. The group also measures water vapor in the stratosphere by means of specialized instruments flown from high-altitude balloons.

Sam Oltmans heads GMD's Ozone and Water Vapor Group. Oltmans grew up in Minneapolis, Minnesota, and earned a degree in mathematics in Illinois. After he completed Air Force Officer Training School at Lackland Air Force Base, Texas, in 1964, the Air Force sent him to Texas A&M University for two years of meteorology studies. Oltmans was then assigned to the Strategic Air Command (SAC) as a weather forecaster and briefer for the SAC leadership. The SAC assignment went well beyond traditional meteorology and weather forecasting:

> It was significant because . . . [SAC aircraft were] flying in the lower stratosphere, not in the troposphere, and we were flying above things like cirrus clouds, and that sort of thing. So, that was part of the motivation for the interest I had in upper-atmosphere phenomena. (Oltmans, video interview with author, Boulder, Colorado, April 27, 2007)

As related in chapter 10, Oltmans gained hands-on experience learning about the ozone layer by calibrating Dobson Instrument 83 for Walter Komhyr at the Mauna Loa Observatory during the summer of 1972. Komhyr had recently come to NOAA from Canada, where he had established a calibration program for that country's Dobson network. Following its calibration at MLO, I83 became—and still remains—the world-standard Dobson instrument.

The simplicity of the Dobson, which was invented in 1924 by Gordon Dobson, helps explain its longevity. The instrument uses an arrangement of prisms to divide a beam of sunlight into a spectrum. Ultraviolet, with a wavelength less than about 320 nm, is absorbed by ozone, and the absorption increases as the wavelength is reduced. The Dobson instrument compares ratios of the intensity of sunlight in very thin slices of the ultraviolet portion of the spectrum to permit the amount of ozone between the instrument and the sun to be determined.

Measuring the ozone layer at MLO was one of the principle goals of Robert Simpson and Ralph Stair, the observatory's founders. Measurements of surface ozone at the observatory site and the entire ozone column were among the first scientific observations made at MLO after the Weather Bureau team began IGY observations in the summer of 1957. Ozone was of high interest for very different reasons, for persistently excessive ozone in ambient air can pose respiratory difficulties and damage vegetation, while reduced ozone in the stratosphere can allow potentially dangerous levels of solar ultraviolet to reach the surface.

Surface Ozone

Measuring surface ozone was one of International Geophysical Year (IGY) experiments begun by Jack Pales at MLO in August 1957, but, as discussed in chapter 7, the instruments then available were chemically based and unreliable. Many pages of patiently collected and transcribed data, some of which remain in the old MLO files, were marked as being unusable. Yet even with the limited data obtained back then, a strong diurnal pattern was observed in which the concentration of ozone in the morning, downslope air was consistently higher than that of air rising from below later in the day. By the end of the first year, an obvious seasonal pattern emerged in the data, with the concentration of surface ozone being highest during spring and lowest during fall. Because of the many instrument problems, the surface ozone program at MLO was eventually abandoned.

New instruments that rely on the absorption of ultraviolet light to detect ozone in ambient air were available by the 1970s, and Walter Komhyr decided it was time to end the fifteen-year gap in the surface ozone record at MLO. Komhyr was pleased with Oltmans' calibrations of I83 at MLO in 1972, and, after he earned his masters degree from the University of Colorado in 1973, Oltmans (video interview with author, Boulder, Colorado, April 27, 2007) began working for NOAA's Global Monitoring for Climate Change Group in Boulder:

> I was hired to set up a surface ozone measurement program. Nobody had really been making measurements of surface ozone in clean air. It was considered to be a pollutant that you measured in urban areas, and the idea that there might be an active chemistry in the clean atmosphere really hadn't been established at that time, and people didn't really have a good idea what surface ozone was like in clean locations. There had been a few measurements at Mauna Loa, there had been a few measurements made at some European sites. And, so, I think this was, again, the foresight of Walt Komhyr, [who] thought we should learn what tropospheric ozone is doing. . . . And, so, we set up surface ozone measurements at our observatories.

Today the record of surface ozone measurements begun in 1974 by Sam Oltmans at MLO and Barrow, Alaska, is among the longest in the world. The data provide evidence of occasional ozone intrusions from the stratosphere. Also in the record are high ozone events associated with polluted air arriving at MLO from China. An unusual surface ozone event occurred at MLO on July 6, 2003, when Asian smoke passed over the observatory while the author was calibrating various instruments. The smoke formed distinctly visible layers in the sky, and microscopic particles of black carbon (soot) were collected on a microscope slide. Ozone can be produced when gases entrained in biomass smoke are exposed to sunlight, and the surface ozone during the 2003 smoke event at MLO reached 70 parts per billion, which is considerably higher than the 40 ppb mean. The ozone concentration was highly correlated with the number of smoke particles in the air, with a correlation coefficient of 0.98 (Mims and Mims 2004).

The Ozone Layer

The Ozone and Water Vapor Group is responsible for daily measurements of the ozone layer at its observatories and across the United States. The ozone layer measurement series at MLO began on November 27, 1957, and would be one of the longest of the International Geophysical Year legacy experiments were it not for an accident that occurred during a morning calibration session on January 1, 1960. A wheel on the cart used to hold the heavy instrument came loose, spilling the 150-pound (68-kg) instrument onto the concrete pad below. The damaged Dobson was returned to Boulder for repair, and total ozone measurements at MLO did not resume until December 1963.

Total ozone measurements at MLO are also made as part of a cooperative arrangement with the Canadian Ozone and Ultraviolet Measurement Program. This program, which began in 1997 and is described in more detail below, uses two fully automated Brewer spectrophotometers to track the sun and measure both the sun's UV and the total ozone column. Both instruments are installed on the solar deck above the NDACC Building. Each instrument is connected to its own computer and Internet connection in a room below the deck. Besides providing an MLO calibration connection to Canada's ozone and UV network, the two Brewers provide an ongoing, long-term comparison of total ozone retrievals by one another, the MLO Dobson and occasional visiting ozone instruments.

Ozonesondes

An ozonesonde is a small electronics package flown from a weather balloon. It transmits to the ground a vertical profile of the ozone concentration from the surface to a typical altitude of about 115,000 feet (35,000 meters), which covers most of the ozone layer. Because of MLO's fortuitous location so near the ocean, ozonesonde flights that are launched from near sea level at Hilo provide important opportunities to compare the meteorological parameters on the mountain with those at the same altitude over the adjacent ocean.

The ozone sensor in an ozonesonde is simple in principle, but its design is the result of years of experimental trials and improvements. The sensor is an electrochemical cell divided into two compartments, each of which contains a platinum electrode and a solution of potassium iodide. The two compartments are bridged by a small opening that can pass an electrical current through the two solutions without changing the concentration in the two chambers. A small air pump draws in outside air and bubbles it through one of the chambers. Ozone in the sampled air immediately reacts with the potassium iodide to release iodine. When the two electrodes are connected together through an external circuit, a small electrical current flows that is proportional to the ozone concentration. Each ozone molecule causes a current of two electrons, and this current is transformed into a voltage that is transmitted to the ground (Science Pump Corporation 1999).

A weather balloon carries the ozonesonde into the stratosphere. It travels with a standard meteorological radiosonde, which transmits to a receiver on the ground the ozone concentration, pressure, temperature, and relative humidity during the flight of the balloon. The result is a profile of the ozone concentration from the surface to the altitude where the balloon bursts high in the stratosphere.

For decades, the MLO staff launched ozonesondes from the Hilo Weather Bureau balloon facility across from the Air Cargo Terminal at General Lyman Field (now Hilo International Airport) where the first MLO office was located. Today ozonesondes are launched from the new National Weather Service balloon facility across the runway from and east of the former location. The MLO equipment used to prepare the ozonesondes for flight was moved to the new facility during October 2006, and the first flight was made on October 26.

Once a week Steve Ryan drives to the Weather Service office to prepare and launch an ozonesonde. After the instrument is prepared, tested, and mated with a radiosonde, he carries it to the balloon building. After the balloon is inflated with hydrogen, a parachute is attached to a line tied to the balloon and the ozonesonde is connected to the parachute. After calling the airport control tower to request permission for the balloon launch, Ryan walks to the adjacent field while holding the balloon by its nozzle in one hand and the ozonesonde in the other. He then releases the balloon and, in turn, the ozonesonde. As the balloon rises, a mechanical reel called an unwinder attached between the ozonesonde and the parachute pays out string until the ozonesonde is around 130 feet (40 meters) below the parachute. This separation assures that the ozonesonde is exposed to free air and not air trapped in the wake of the balloon.

Frost-Point Hygrometer

The water vapor in the stratosphere constitutes only a very small fraction of the total column water measured from the surface. Water vapor in the stratosphere over Boulder, Colorado, has long been studied by Samuel Oltmans and David Hofmann, who use delicate, balloon-borne instruments called frost-point hygrometers to measure the water vapor in the stratosphere. From 1981 to 1994, they found a steady increase in stratospheric water vapor (Oltmans and Hofmann 1995). The increase continued thereafter, but it has lessened in recent years, a decline that may be related to a similar flattening in the abundance of methane in the stratosphere. Methane is emitted by termites, rice cultivation, biomass burning, the natural decomposition of biomass, and leaks from natural gas lines. The gas eventually drifts into the stratosphere, where water vapor is a by-product of its oxidation. It is thought that changes in natural gas production in the former Soviet Union may have contributed to the decline in methane. While frost-point hygrometry is mainly conducted from Boulder, during special projects the devices have been launched from MLO.

GMD Lidar

A powerful laser beam periodically probes the stratosphere over MLO to detect aerosols in the upper troposphere and stratosphere over the mountain. (The same lidar detects water vapor and temperature, as described below.) Aerosols in these levels of the atmosphere arise from volcanic eruptions or natural sulfate aerosols known as the Junge layer. The original lidar used a ruby laser that emitted high-power pulses of red light having a wavelength of 694 nm. The ruby lidar was installed in 1974 and used until 1998. A more advanced lidar system that uses a neodymium-doped YAG laser was installed by Dr. John Barnes in April 1994 and has since been in regular use. The YAG laser emits an invisible, near-infrared beam having a wavelength of 1064 nm. The beam is passed through a frequency-doubling crystal that halves the wavelength of a portion of the near-IR beam to 532 nm in the green. Thus, the lidar system probes the stratosphere with superimposed, highly collimated beams of green and near-IR light.

Aerosols in the path of the lidar beam scatter a portion of the laser light back toward the ground, where a telescope inside the NDACC Building intercepts a portion of the scattered light and reflects it to an ultrasensitive photomultiplier tube (PMT). The signal generated by the PMT is sent to a computer monitored by the lidar operator during the course of a late-night session at the observatory. The computer incorporates a program written by Barnes and Trevor Kaplan that clearly depicts the presence of any aerosol layers in the stratosphere over the mountain. Layers below the tropopause—the boundary between the moist troposphere and the ultradry stratosphere—might be cirrus clouds or aerosols. Layers in the stratosphere are most likely aerosols.

The GMD lidar program at MLO requires considerably more personal attention than the many automated data-collecting systems on the mountain. Lidar observations usually take place once a week, or more often when major volcanic eruptions occur or satellite comparisons are being conducted. John Barnes arrives before sunset to prepare the equipment for the session. The preparation involves opening doors on the roof through which the lidar beam emerges and which expose the telescope to the night sky. Occasionally Barnes climbs down to the roof of the NDACC Building from the solar deck to lubricate the moving parts of the door assemblies.

When the lidar system is switched on, a brilliant, pencil-thin beam of green projects from the roof of the NDACC Building. Barnes then hikes to a point south of and above the NDACC Building to activate a laptop computer that controls a special camera designed to photograph the green lidar beam during the session. This arrangement is known as Clidar (CCD Camera Lidar), a new technology developed by Barnes. It's a simple, economical method for supplementing conventional lidar studies by permitting the detection and analysis of particles much closer to the surface than the lidar system can detect (Barnes et al. 2007). Barnes operates the lidar system at two power levels over a period of several hours and then closes down the system. By then it is often close to midnight, so he usually spends the night in one of the building's tiny bedrooms. He analyzes the data from the session in his Hilo office, which is more conducive to such analytical work than the rarified atmosphere on the mountain.

5. Aerosols Group (AERO)

Aerosols are microscopic solid or liquid particles suspended in the atmosphere. Solid aerosols include mineral dust, sea salt, ice crystals, carbon (soot), pollen, fungal spores, bacteria, spider silk, and microscopic fragments of plant and animal matter. The most common liquid aerosol is microscopic droplets of water. Some aerosols consist of a solid core, such as a sulfate particle, coated with liquid water. At MLO aerosols are measured in the ambient air using various kinds of particle counters and in the sky high above by means of lidar (see the Ozone and Water Vapor Group above). Former MLO scientist Barry Bodhaine (personal communication with author, May 12, 2008), who was introduced in chapter 10, recalls the history of the Aerosols Group:

> The first routine aerosol measurements at MLO were established in 1967 by MLO veteran Bernard Mendonca using a Gardner Condensation Nucleus Counter. However, this instrument was not sensitive enough to detect aerosol concentrations at MLO under the cleanest background downslope wind conditions. Austin Hogan of the State University of New York at Albany installed a Pollak Condensation Nucleus Counter in 1974 which was sensitive enough and which operated routinely for more than 30 years. Also in 1974 I became the project leader for the GMCC aerosol program and installed a General Electric automatic condensation nucleus counter and a four-wavelength nephelometer at MLO. The General Electric counter was eventually replaced by an alcohol-based condensation nucleus counter, and the four-wavelength nephelometer was eventually replaced by a more sensitive three-wavelength nephelometer. I continued as aerosol project leader until 1995.

Bodhaine was replaced by physicist John A. Ogren. Originally from Troy, New York, Ogren received his bachelors and masters degrees in engineering from Harvey Mudd College of Claremont, California. His master's thesis was on the contribution of stratospheric ozone to the ambient air in the Los Angeles basin. In 1983 Ogren earned a doctorate in civil engineering at the University of Washington, where his dissertation was on elemental carbon in the atmosphere. From 1983 to 1991, Ogren served first as a research associate and then an associate professor in the Department of Meteorology at the University of Stockholm, Sweden. In 1991 he moved to Boulder, where he has since worked as a physical scientist, first at NOAA's former Climate Monitoring and Diagnostics Laboratory and then the Earth System Research Laboratory. Like most of the scientists and technicians interviewed for this book, Ogren (personal communication with author, February 21, 2008) is motivated by the challenge of his work:

> The challenge and societal relevance of the research that we are doing provide a large part of my motivation. I've always enjoyed building things, and I've progressed from building small things (individual instruments), to collections of things (an integrated set of instruments, like what we have at MLO), to a collection of those collections (a global network of long-term stations).

The Aerosols Group uses a wide array of instruments to detect or collect aerosols at MLO, the other GMD observatories, and various other sites, some of which are described below. At MLO the most interesting aerosols arrive from Asia, especially during spring. Aerosols may also arise from volcanic activity (vog) and occasional fires on the Big Island.

Particle Soot Absorption Photometer (PSAP)

This instrument continuously measures the amount of green light transmitted through a quartz filter while outside air is flowing through the filter. As soot is deposited on the filter, the transmission of light through the filter is reduced. The rate of change of the light transmission divided by the rate of flow of the air being sampled provides information about the concentration of the particles.

Integrating Nephelometer

This instrument measures the scattering of light by aerosols in a stream of sampled air. In operation, a beam of light is projected through the air stream. A very sensitive photomultiplier tube detects the light scattered toward it by particles in the sampled air.

Condensation Nucleus Counter (CNC)

Water vapor in the atmosphere may condense as a liquid film on microscopic particles known as condensation nuclei. These tiny particles are detected at MLO by passing sampled air into a chamber highly saturated with butanol vapor. When the butanol condenses on any particles that are present, they grow in size to a point that they can be detected and counted.

The Aerosol Group employs an array of additional instruments, and more about them and the aerosol measurements conducted at MLO can be found at the observatory's Web site (www.mlo.noaa.gov/programs/esrl/aero/aero.html). There is overlap in some measurements, for the Aerosols and Global Radiation Groups are both interested in the impact of aerosols on sunlight. The difference is that the Aerosols Group stresses the collection of particles in the atmosphere by means of instruments connected to the outside air with sampling tubing, while the Global Radiation Group detects particles indirectly in the open air by measuring the changes they cause in the intensity of direct and diffuse sunlight. Dr. Ogren (personal communication with author, February 21, 2008) has provided an example of how these two methods are merged together in a single project:

> The integration of the remote sensing (radiation) and in-situ (aerosol) measurements is best exemplified by a recent analysis that Bob Stone did of smoke aerosols over Alaska. He used the in-situ data to derive the aerosol optical properties needed by a radiative transfer model, and then used the model to explain surface and satellite measurements of radiative fluxes and aerosol optical depth.

6. Other NOAA and Cooperative Research Projects

The MLO and GMD Web sites provide a full listing of more than a hundred research projects. Among the key projects not covered in the discussion of the five research groups above are those that follow.

Water Vapor Programs

Water vapor drives weather and is the most important greenhouse gas. A variety of projects at MLO are designed to measure either the entire water vapor column or water vapor in the stratosphere. While carbon dioxide receives more attention due to its relentless increase, water vapor is the most important and least understood of all the greenhouse gases. This was driven home by the American Geophysical Union (AGU 1995) in a special report that has become a standard reference on the role of water vapor in the atmosphere:

> There are many atmospheric greenhouse gases, some naturally occurring and some resulting from industrial activities, but probably the most important greenhouse gas is water vapor. Water vapor is involved in an important climate feedback loop. As the temperature of the earth's surface and atmosphere increases, the atmosphere is able to hold more water vapor. The additional water vapor, acting as a greenhouse gas, absorbs energy that would otherwise escape to space and so causes further warming. This basic picture is complicated by important interactions between water vapor, clouds, atmospheric motion, and radiation from both the Sun and the Earth. There are some aspects of the role of water vapor as a greenhouse gas that are not well understood, again mainly because we lack the necessary observations to test theoretical models.

Human activity has the potential of altering the concentration and the distribution of water vapor in the atmosphere. For example, jet aircraft inject water vapor directly into the upper troposphere and lower stratosphere. Irrigation projects increase the concentration of water vapor over some regions. The IPCC (2007: 135) has concisely summarized the impact of human activity on water vapor:

> Water vapour is the most abundant and important greenhouse gas in the atmosphere. However, human activities have only a small direct influence on the amount of atmospheric water vapour. Indirectly, humans have the potential to affect water vapour substantially by changing climate. For example, a warmer atmosphere contains more water vapour. Human activities also influence water vapour through CH_4 [methane] emissions, because CH_4 undergoes chemical destruction in the stratosphere, producing a small amount of water vapour.

The Global Radiation Group and various cooperative instruments that observe direct sunlight during the day measure the total water vapor in a column through the atmosphere over MLO. This is known as *total column water vapor,* or precipitable water (PW), and it is expressed as the depth of the liquid water that would result from condensing the water vapor in a vertical column through the atmosphere. (The water vapor in the stratosphere and mesosphere over MLO is measured by other groups and programs discussed below.)

The air at the Mauna Loa Observatory is generally quite dry, especially during nights and mornings when downslope conditions prevail. During these times, the total water vapor column can on some days be less than a millimeter. The dryness can cause chapped lips, a stuffy nose, and potent electrical shocks when unsuspecting visitors slide out of a car upon their arrival. The column water vapor can reach several centimeters in the afternoon if the inversion layer arrives at the observatory.

Measurements of the total column water vapor over MLO were first conducted in the summer of 1956, when Carl C. Kiess and Charles H. Corliss and their wives Harriet and Edith used a spectrometer in an experiment to detect water vapor in the atmosphere of Mars (Kiess et al. 1957). They needed to know the water vapor over MLO so it could be subtracted from their observations of Mars. Harold Yates (1957) measured the water vapor between MLO and Mauna Kea in July 1956 and again in 1957. During April and May of 1957, MLO cofounder Ralph Stair and Russell Johnston measured the column water vapor over MLO with much better accuracy than the Kiess and Corliss team (Stair and Johnston 1958). These pioneering measurements of water vapor were followed by the installation of a Foskett infrared hygrometer at MLO in October 1960 (Price and Pales 1963).

For more than forty years, the Foskett hygrometer measured the water vapor in a column through the atmosphere over MLO. The instrument consisted of two parallel metal collimator tubes mounted on a motor-driven device that tracked the sun. One collimator contained an optical filter that transmitted a narrow band of sunlight centered at 935 or 940 nm in the near-infrared where water vapor is strongly absorbing (the filter wavelengths changed over the years). The second collimator tube contained a filter that transmitted a narrow band of near-IR centered at 860 or 880 nm, a wavelength region that is not absorbed by water vapor. The ratio of the signal from silicon photodiodes behind each of the two filters was related to the water vapor column between the instrument and the sun. An analysis of the water vapor over MLO measured by this instrument from 1978 to 1983 showed no trend (Dutton et al. 1985).

An unfunded study of some thirty years of the water vapor record from MLO is being conducted by Ellsworth Dutton. NOAA scientists Rebecca J. Ross and William P. Elliott studied carefully selected radiosonde humidity data from 1958 to 1995 to look for trends, and evidence of increasing water vapor was found across the United States. The data also show reduced water vapor during the cooling of the 1960s and 1970s and increased water vapor thereafter. However, the trends for Canada and Asia are much flatter and show little or no trends, and the data for Europe show a slight negative trend (Ross and Elliott 2001). Measurements from the Southern Hemisphere were too sparse to be included in the study. Column water vapor measured in Texas since 1990 shows a slight downward trend (Mims 2002). A team of scientists at the National Center for Atmospheric Research looked at satellite water vapor measurements and found considerable variability around the Earth (Trenberth et al. 2005). According to the IPCC (2007), satellite observations from 1988 to 2004 showed an overall global increase in atmospheric water vapor. These and other findings lend importance to the

MLO water vapor record, particularly because the data have been collected nearly as long as the fifty-year carbon dioxide record.

GPS-MET Total Water Vapor

The newest water vapor instrument at MLO was installed in November 2004 by the Ground-Based GPS Meteorology (GPS-MET) project of NOAA's Earth System Research Laboratory. Similar instruments were installed at the summit of neighboring Mauna Kea and near sea level at the Hilo International Airport. These instruments provide an unprecedented opportunity to compare the water vapor columns above MLO, Mauna Kea, and Hilo at any time of the day or night. This new system uses Global Positioning System (GPS) technology to infer the total water vapor in a column through the atmosphere. The method is based on the fact that microwave signals from GPS satellites are slightly delayed by water vapor as they travel from the satellites to the Earth. By measuring the signal delay from a cluster of typically six GPS satellites and mathematically adjusting each satellite so that it appears to be straight overhead at the zenith, the column amount of water vapor can be computed with high accuracy.

The GPS water vapor project at MLO is part of a network of more than three hundred such sites around the United States. The network is administered by Dr. Seth Gutman, who is contagiously enthusiastic about the combined simplicity, accuracy, and low cost of GPS water vapor monitoring. Especially intriguing about the network is that each site is merely an antenna and a GPS receiver. The MLO system resembles an upside-down, white dinner plate mounted atop an aluminum mast a meter or so high. The system is installed in the lava along a boardwalk south of the solar deck of the NDACC Building and is rarely seen by visitors. There is no requirement to track the sun and there are no optics to keep clean and eventually replace. Best of all, the GPS system functions twenty-four hours a day through clear skies, clouds, and rain, and the network's data are posted on an easily accessible Web site.

CIMEL Robotic Sun Photometer Total Water Vapor

The Goddard Space Flight Center (GSFC) AERONET program maintains a suite of at least two CIMEL robotic sun photometers at MLO. The CIMELs measure column water vapor using the same principle as the Foskett near-IR hygrometer. The data for MLO and numerous other sites are available through the GSFC AERONET Web site. As described below, the CIMELs also measure the optical depth of the atmosphere over MLO at five wavelengths of light.

GMD Lidar Stratospheric Water Vapor

MLO's powerful neodymium–YAG lidar system was described above in the section about the Ozone and Water Vapor Group. While aerosol detection was the original role for the lidar, the 532-nm beam from the laser has the ability to also detect water vapor. This is made possible by the fact that a small fraction of the green laser beam (532 nm) projected through water vapor is shifted to a red wavelength (660 nm). Some of this light is scattered back to the ground, where a narrow bandpass optical filter placed over a photomultiplier tube in the lidar receiving telescope detects the red photons and enables MLO station chief John Barnes to determine the amount of water vapor up to an altitude of 46,000 feet (14 km). Temperature is measured using the 532 nm wavelength in the upper stratosphere and mesosphere, 115,000 feet (35 km) to 230,000 feet (70 km) (Barnes and Voemel 2003). The technique uses the lidar's atmospheric density measurement, the ideal gas law, and assumes stable conditions in the atmosphere. Dr. Barnes' lidar research has been used to validate water vapor measurements made by NASA's Aqua and AURA satellites.

Water Vapor Millimeter Wave Spectrometer

High-altitude measurements of water vapor by frost-point hygrometers are limited by the maximum altitude of balloons, which is typically 115,000 feet (35 km). Lidar measurements of water vapor do not extend above about 46,000 feet (14 km). While there is very little water vapor above these altitudes, its presence plays an important role in the production of the hydroxyl radical that destroys ozone.

Since 1996, the US Naval Observatory and NASA have operated an instrument at MLO that measures water vapor from the upper stratosphere at 40 km to the mesosphere at 80 km. The instrument is part of a small network of receivers that detect the 22 GHz millimeter wave emissions of water vapor. The other sites in the network are Lauder, New Zealand (1992 to present) and Table Mountain, California (1992–1997, 2003 to present).

During the early 1990s, the water vapor spectrometers at Lauder and Table Mountain detected an increase in the water vapor in the middle atmosphere of about 2 percent a year. Since 1996, there has been no significant trend (Nedoluha et al. 2003). It is believed that half the increase in the early 1990s was caused by increasing methane. The other half of the increase remains to be explained. The MLO Water Vapor Millimeter Wave Spectrometer is at the highest location of the three-site network, which reduces interference caused by water vapor at the lower two sites. The tropical location of MLO reduces the seasonal impact on water vapor. Taken together, these advantages make MLO a prime site for observing any trends in middle atmospheric water vapor.

Cooperative Solar Radiation Programs

Several cooperative programs measure various wavelengths of sunlight at MLO, in most cases to detect and measure the presence of intervening aerosols or gases. The Brewer spectrophotometers measure both the total column abundance of ozone and full-sky ultraviolet.

NIWA Ultraviolet Spectroradiometer

Chapter 12 introduced MLO's precision ultraviolet spectroradiometer project. In July 1995, a research-grade ultraviolet spectroradiometer began making measurements of the full-sky UV irradiance from 285 to 450 nm. The instrument, which was developed by New Zealand's National Institute of Water and Atmospheric Research (NIWA), employs a double monochrometer that increases the precision of measurements by greatly reducing noise caused by stray light within the instrument. The instrument measures UV in discrete steps of less than a nanometer. This permits the irradiance at various wavelengths or clusters of wavelengths to be integrated into various action spectra that damage DNA, burn human skin, and synthesize vitamin D in vertebrates. The narrow wavelength resolution also permits the instrument to detect atmospheric trace gases that absorb UV, including ozone and sulfur dioxide.

Because of its location in the tropics and its high elevation, MLO receives some of the highest UV measured on Earth, and the NIWA instrument has occasionally measured a UV Index (UVI) exceeding 20. To place this in perspective, a UVI of 10 is considered extreme. The UV at the beach that lines scenic Hapuna Bay, 35 airline miles (57 km) from MLO, can reach 12 on a clear summer day. The difference between the UVI at sea level and at MLO is due to the intervening atmosphere (McKenzie et al. 2001).

While there is a significant seasonal variability in the UV at noon at MLO, and while in some years there is more or less UV than in other years, thus far the NIWA instrument has not measured any significant overall trends in the UV. A major volcanic eruption would be expected to cause reduced UV for several years, and any ozone decline would be expected to cause increased UV. How these two conflicting effects interact will be of great interest should there be another major volcanic eruption that injects a blanket of sulfur dioxide into the stratosphere.

Shadowband Radiometers

The UV-B Monitoring and Research Program managed for the US Department of Agriculture by Colorado State University employs a pair of shadowband radiometers that measure the total sunlight from the sky and the sunlight scattered by the sky alone. The MLO instruments serve as a reference for the UV-B network, which has thirty-one stations scattered across the United States. One of the radiometers measures seven wavelengths of ultraviolet (300, 305.5, 311.4, 317.6, 325.4, 332.4, and 368 nm). The wavelengths are carefully selected to permit the accurate measurement of the total column abundance of ozone. The UV instrument also provides important data about the magnitude of the UV-B wavelengths (the first four above) that can damage plants and cause skin cancer and eye damage in people. Also provided is data about the beneficial action spectrum near 300 nm that enables the synthesis of pre–vitamin D in human skin.

The second shadowband radiometer samples four visible wavelengths (415, 500, 610, and 665 nm) and two near-infrared wavelengths (862 and 940 nm). These measurements enable the measurement of the optical thickness of the atmosphere and, potentially, the total column abundance of water vapor. They also measure the blue (415 nm) and red (665 nm) wavelengths responsible for photosynthesis and the blue light responsible for the phototropic response of plants.

The USDA instrument suite also includes two broadband radiometers designed to measure the irradiance from the sun and the sky. One measures the UV-B action spectrum for erythema (sunburn) and the other measures the full photosynthetic action spectrum (PAR).

Cooperative Ozone Layer Programs

The long-term measurements of the ozone layer made by the Dobson spectrophotometer at MLO are complemented by observations from cooperative programs and by occasional calibrations and comparisons with a variety of visiting instruments.

Brewer Spectrophotometers

The Canadian Ozone and Ultraviolet Measurement Program has operated two Brewer spectrophotometers at MLO since 1997 (Kerr 1998), when they were installed under the direction of the Canadian Atmospheric Environment Service (see chapter 13). Every two years, the Meteorological Service of Canada (MSC) visits MLO to perform repairs and upgrades and to conduct calibrations of the Brewers. A unique feature of Brewer 009 is that it is equipped with an all-sky camera that captures an image of the sky once every ten minutes. The current sky image can be seen at the MLO Web site.

The two Brewers are installed on the solar deck over the NDACC Building, and they are controlled by individual computers in a room below accessible to MLO staff and, via the Internet, by the MSC staff in Toronto. When an occasional problem occurs, the MSC staff can quickly communicate with MLO by e-mail or telephone and provide instructions for any necessary maintenance. For example, in May 2007 the author noted that Brewer 009 was oscillating back and forth while emitting loud sounds when the instrument was attempting to rotate on its axis to track the afternoon sun at solar noon. E-mails were quickly exchanged, and the MSC staff suggested that rain might have entered the enclosure that houses the Brewer's tracking motor and gears. This was indeed the problem. The bolts on an access door were tightened, the problem disappeared, and the Brewer continued its automatic ozone and UV monitoring.

JPL Lidar

NASA's Jet Propulsion Laboratory (JPL) has operated a high-power ultraviolet lidar at MLO since 1991. The lidar system was originally housed in a pair of large trailers east of the Keeling Building. The system was moved to the west end of the NDACC Building after it was completed in 1997.

The JPL lidar program, which is directed by Dr. I. Stuart McDermid, measures stratospheric ozone from 15 to 55 km. It also measures the temperature from 15 to 80 km. The lidar employs high-power lasers that emit at 308 nm, which is absorbed by ozone, and 353 nm, which is unaffected by ozone. The ratio of the two wavelengths of UV scattered back to the lidar receiver is related to the concentration of ozone.

Cooperative Aerosol Programs

Several cooperative programs at MLO complement the measurements of aerosols conducted by GMD instruments. These include GSFC's robotic sun photometers and the

US Department of Agriculture's shadowband radiometers, the latter described above under Cooperative Solar Radiation Programs.

CIMEL Robotic Sun Photometers

As noted above, two or three robotic sun photometers in NASA's AERONET network are installed on a concrete pad southwest of the NDACC Building. They wake up every fifteen minutes, stare at the sun, and measure the direct sunlight at five wavelengths of light (340, 440, 675, 870, and 1020 nm). The photometers then scan the sky to measure the sunlight scattered from molecules of air and any aerosols that might be present. The direct sun measurements enable the instrument to determine the optical thickness of the atmosphere and the total column abundance of water vapor (see above). The data are transmitted by satellite to NASA's Goddard Space Flight Center. AERONET director Brent Holben attributes the high calibration accuracy of the entire AERONET network to the opportunity to maintain an AERONET site in the clean air of MLO.

7. Summary of Other NOAA and Cooperative Programs

Cooperative programs at MLO are those that generally require regular or occasional support by MLO personnel, and approximately fifty research programs at MLO are defined as cooperative. Some of the cooperative programs are closely related to the primary observations long conducted by NOAA at MLO, and some of these are described above. Others are specialized programs with MLO support ranging from simply providing a site to place an instrument (e.g., the GPS water vapor project) to regular instrument checks and basic maintenance (e.g., the two Brewer spectrophotometers). Many cooperative projects require Internet access (including the two just cited). A few require occasional overnight accommodations at MLO (e.g., annual GLOBE program sun photometer calibrations). Details about the various cooperative programs can be found on the MLO Web site. A list of the current cooperative programs follows:

Clidar (CCD Camera Lidar) (NOAA ESRL GMD / CCSU)
Radon-222 Monitoring (ANSTO)
Atmospheric Research—Greenhouse Gases (CSIRO)
US Climate Reference Network (CRN)
Global Atmosphere Passive Sampling (GAPS) Network (EC)
Aerosol Light Absorption Aethelometer (EPA)
Elemental, Particulate, and Reactive Gaseous Mercury Monitoring (EPA)
Mercury Precipitation Collection (EPA) (upcoming)
Ambient Ion Monitor (EPA)
Dichotomous Partisol Measurements (EPA)
Sequential Fine Particle Sampler (EPA)
Sulfur Dioxide Monitoring (EPA)
Surface Ozone Measurements (EPA)
National Satellite Test Bed at Mauna Loa (FAA)
Global Positioning System (GPS) Meteorology (FSL)
Groundwinds Lidar System (UNH)
Earthquake Seismometer (USGS / HVO)
Tilt and Strain Measurements (USGS / HVO)
IMPROVE (Interagency Monitoring of Protected Visual Environments) (UC–Davis)
AERONET (Aerosol Robotic Network) (NASA)
CERES (Clouds' and Earth's Radiant Energy System) (NASA)
Pohakuloa Training Area Range Surveillance System (PMRF)
Relationships between Mineral Dust and Radionuclides from Historical Nuclear Weapons Tests (NMSU)
Measurements of ^{14}CO in the NH Tropics (NSF)
GLOBE Sun Photometers (NSF / Drexel University)
Maximum / Minimum Temperature & Precipitation (NWS)
ROSA Maximum / Minimum Temperature & Precipitation (NWS)
Carbon Dioxide (CO_2) Measurements (Scripps)
Oxygen (O_2) Concentration Measurements (Scripps)
Chemical and Isotopic Studies of Atmospheric Aerosol Sources, Transport, and Evolution (UC–San Diego) (upcoming)

Measurements of Free Tropospheric Nitric Acid and Aerosols (UH–Mānoa)
High-Altitude Environment Corrosion Project (UH–Mānoa)
Comparison of Aerosol Organic Carbon Measurements and Models in the Free Troposphere (UH–Mānoa)
VYSOS (Variable Young Star Optical Survey) (UH–Mānoa)
Ultraviolet-B (UV-B) Monitoring & Research Program (USDA / CSU)
CRONUS—Earth (Cosmic Ray Produced Nuclide Systematics on Earth) (University of Arizona) (upcoming)
Long-Term Brewer Calibrations (NDACC / AES)
FTIR (Fourier Transform Infrared Spectroscopy) Monitoring (NDACC / U–Denver)
Lidar: Stratospheric Ozone, Temperature, and Aerosol Profiles (NDACC / NASA JPL)
Microwave Measurements of Ozone (NDACC / U–Mass / NIWA)
Bromine Oxide Monitoring (NDACC / NOAA ESRL GMD / NIWA)
Nitrous Oxide Monitoring (NDACC / NOAA ESRL GMD / NIWA)
Ultraviolet Spectroradiometer (NDACC / NOAA ESRL GMD / NIWA)
Water-Vapor Mm-Wave Spectrometer (NDACC / NRL / NASA)
Multi-Filter Rotating Shadowband Radiometer (MFRSR) Aerosol Monitoring Network (UH–Mānoa / Albireo Endeavors / PNNL)
Aeolian Dust Experiment on Climate Impact (MRI)
US Climate Reference Network (CRN)

8. Guest Programs

Guest programs at MLO are those that occupy space on the lava but are otherwise independent and require no scheduled maintenance by MLO staff. Guest programs are made possible by the infrastructure at MLO in the form of the access road, power, telephone, and Internet. The major guest programs are described next in the order in which they first began collecting data. The MLO Web site links to all these guest programs.

Mauna Loa Solar Observatory (1965 to Present)

The Mauna Loa Solar Observatory (MLSO), which is described in chapter 9 and shown in Plate 68, is the earliest and best known of the guest programs at MLO. The MLSO was established by the National Center for Atmospheric Research (NCAR) in 1965 and has been in operation ever since. It has completed observations through four complete eleven-year solar cycles. The MLSO is operated by NCAR's High Altitude Observatory (HAO).

A major purpose of the MLSO is to provide imagery of the solar corona, the ionized streams of gas ejected into space by the sun. The MLSO also observes sunspots and the solar disk at various wavelengths. On most days, an MLSO operator arrives an hour or so after sunrise to begin the daily observing program. Observations continue at three-minute intervals as long as the sun is not obscured by clouds. The high frequency of observations permits solar flares and other brief changes to be observed.

A unique aspect of the MLSO is its long-term working relationship with MLO. On a typical workday, the MLSO observer and the MLO day crew may never see one another due to their respective work responsibilities. But over the years, MLSO has worked closely with MLO to share some of the road repair responsibilities.

The Mauna Loa Solar Observatory is among the most obvious structures at MLO. The facility and its two telescope domes are located due south of the Keeling Building near the south edge of the original four-acre site. The MLSO was expanded in 1997 and may yet be expanded again. Still and moving images of the sun and its corona made through the MLSO telescope can be seen online at http://mlso.hao.ucar.edu/cgi-bin/mlso_homepage.cgi. When sunspots are present, they are clearly visible in these images.

JPL Lidar (1993 to Present)

As noted above under the Cooperative Ozone Layer Programs, NASA's Jet Propulsion Laboratory has operated a high-power ultraviolet lidar at MLO since 1993. The lidar system occupies the entire west end of the NDSC Building. When sky conditions permit, a JPL scientist arrives at MLO each weekday evening to prepare for several hours of lidar observations. The long-term nature of the JPL program and the almost daily visits by a JPL scientist make this among the most significant guest programs at MLO.

GONG (1995 to Present)

GONG (Global Oscillation Network Group) has maintained a solar observing instrument at MLO since 1995. GONG is an international program in which six identical instruments around the world stare directly at the sun in order to observe and record the sun's regular pulsations. The pulsations, which occur at about five-minute intervals, permit the internal dynamics of the sun to be studied by means of helioseismology:

> Helioseismology utilizes waves that propagate throughout the Sun to measure, for the first time, the invisible internal structure and dynamics of a star. There are millions of distinct, resonating sound waves, seen by the Doppler shifting of light emitted at the Sun's surface. The periods of these waves depend on their propagation speeds and the depths of their resonant cavities, and the large number of resonant modes, with different cavities, allows us to construct extremely narrow probes of the temperature, chemical composition, and motions from just below the surface down to the very core of the Sun. (Armet 2004)

All six GONG instruments are installed in identical white shipping containers. Sunlight enters the south end of the MLO GONG container via an external optical port that tracks the sun. Precision optical instrumentation mounted on an optical table inside the GONG container detects and measures the Doppler shift of sunlight and stores the data.

As one enters MLO along the access road, the white GONG facility is the first structure past AMiBA on the south (left) side of the road just inside the eastern boundary of the original four-acre MLO site. GONG, which can operate unattended for about a week, is managed by the National Solar Observatory, which is operated by AURA, Inc., under a cooperative agreement with the National Science Foundation.

GroundWinds (2002 to Present)

GroundWinds is a unique lidar project in which a UV (355 nm) laser beam probes the sky up to 20 km in the lower stratosphere to detect wind speed and direction. An interferometer compares the outgoing laser beam with the scattered return, and any Doppler shift between the two permits the wind speed to be determined as a function of altitude.

It is hoped that GroundWinds technology will eventually be launched into space aboard a satellite that can measure wind from orbit. GroundWinds is managed for NOAA by Dr. Berrien Moore III of the Institute for the Study of Earth, Oceans and Space at the University of New Hampshire. Project partners include NOAA, the University of Hawai'i, the Mount Washington Observatory, and the Michigan Aerospace Corporation. The GroundWinds facility is housed in a small white building. It is the first structure on the right (north) when entering MLO.

AMiBA (2006 to Present)

The most visually striking new facility at MLO is AMiBA (Array for Microwave Background Anisotropy). AMiBA is a project of the Academia Sinica (or "Chinese Academy" in Latin) of the Republic of China (Taiwan). It is among a new class of radio telescopes designed to experimentally explore topics in cosmology by means of a closely spaced array of gigahertz receivers that employ interferometry to observe objects ranging from the planet Jupiter to distant galaxies.

AMiBA was formally dedicated at MLO on October 3, 2006. Its objectives are much more akin to those of the telescopes atop Mauna Kea than to the traditional meteorology and atmospheric chemistry studies conducted at MLO, as can be seen in this description of the project:

> The AMiBA targets specifically the distribution of high red-shift clusters of galaxies via the Sunyaev-Zel'dovich Effect (SZE), as a means to probe the primordial and early structure of the universe. AMiBA will also measure the polarization properties of the Cosmic Microwave Background (CMB), which is sensitive to the ionization history of the universe and can be a potential probe for gravity waves. (AMiBA 2007)

AMiBA is built on the new four-acre expansion of MLO and is the first facility a visitor passes when entering the MLO site. During the day, the antenna array is protected by a large brown dome of flexible material. When AMiBA is in operation, the protective dome is folded down to reveal a large hexagonal structure supported by an array of six adjustable legs that resembles a creature from a science fiction movie, especially when illuminated. Currently AMiBA is equipped with seven receiving antennas. It is anticipated that discoveries made by AMiBA will rise as the number of receivers is increased over the next several years.

References

AGU. 1995. Water vapor in the climate system. Special Report, American Geophysical Union, December. (Accessed at http://www.agu.org/sci_soc/mockler.html.)

AMiBA. 2007. AMiBA: The Array for Microwave Background Anisotropy. Academia Sinica. (Accessed at http://amiba.asiaa.sinica.edu.tw/.)

Armet, Dave. 2004. Helioseismology. GONG, National Solar Observatory. (Accessed at http://gong.nso.edu/info/helioseismology.html.)

Barnes, John E., Nimmi Sharma, and Trevor Kaplan. 2007. GMD camera lidar (clidar), Mauna Loa Observatory. (Accessed at http://www.mlo.noaa.gov/programs/gmdlidar/clidar/gmdlidar_clidar.html.)

Barnes, John E., and Holger Voemel. 2003. Water vapor measurements by Raman lidar and balloon sonde at Mauna Loa Observatory. In Upendra N. Singh, ed., *Lidar Remote Sensing for Environmental Monitoring IV.* International Society for Optical Engineering.

Biutner, E. K., O. K. Zakharova, and A. G. Lapenis. 1986. On the estimation of the preindustrial carbon dioxide concentration in the earth's atmosphere. *Akademiia Nauk SSSR, Izvestiia, Fizika Atmosfery i Okeana* 22: 796–803. (Abstract accessed at http://adsabs.harvard.edu/abs/1986FizAO. .22. .796B.)

Dutton, Ellsworth G., and Barry A. Bodhaine. 2001. Solar irradiance anomalies caused by clear-sky transmission variations above Mauna Loa: 1958–1999. *Journal of Climate* 14: 3255–3262.

Dutton, E. G., J. J. DeLuisi, and A. P. Austring. 1985. Interpretation of Mauna Loa atmospheric transmission relative to aerosols, using photometric precipitable water amounts. *Journal of Atmospheric Chemistry* 3: 53–68.

Elkins, J. W., T. M. Thompson, T. H. Swanson, J. H. Butler, B. D. Hall, S. O. Cummings, D. A. Fisher, and A. G. Raffo. 1993. Decrease in the growth rates of atmospheric chlorofluorocarbons 11 and 12. *Nature* 364: 780–783.

Ellis, Howard T., and Rudolf F. Pueschel. 1971. Solar radiation: Absence of air pollution trends at Mauna Loa. *Science* 172: 845–846.

GMD. 2010. Global Monitoring Division: Mission. (Accessed at http://www.esrl.noaa.gov/gmd/about/aboutgmd.html.)

Hofmann, David. 2007. The NOAA Annual Greenhouse Gas Index (AGGI). NOAA Earth System Research Laboratory, Boulder, Colorado. (Accessed at http://www.esrl.noaa.gov/gmd/aggi/.)

IPCC. 2007. Summary for policymakers. In S. Solomon, D. Qin, M. Manning, Z. Chen, M. Marquis, K. B. Averyt, M.Tignor, and H. L. Miller, eds., *Climate Change 2007:*

The Physical Science Basis. Contribution of Working Group I to the Fourth Assessment Report of the Intergovernmental Panel on Climate Change. Cambridge, UK, and New York: Cambridge University Press.

Kerr, J. B. 1998. Operation of Brewer instruments at Mauna Loa Observatory. In *Summary Report No. 24 1996–1997,* Climate Monitoring and Diagnostics Laboratory, NOAA, December. (Accessed at http://www.esrl.noaa.gov/gmd/publications/annrpt24/kerr.htm.)

Khrushchev, Sergei. 2007. We shocked the world. *Air & Space* 22 (August): 32–34.

Kiess, C. C., C. H. Corliss, Harriet K. Kiess, and Edith L. R. Corliss. 1957. High dispersion spectra of Mars. *Astrophysical Journal* 126: 579–584.

Knight, Carol. 2007. Life of an air flask: NOAA celebrates 200 years of science, service and stewardship. (Accessed at http://celebrating200years.noaa.gov/magazine/life_air_flask/welcome.html#lab.)

McKenzie, R. L., P. V. Johnston, D. Smale, B. Bodhaine, and S. Madronich. 2001. Altitude effects on UV spectral irradiance deduced from measurements at Lauder, New Zealand and at Mauna Loa Observatory, Hawaii. *Journal of Geophysical Research* 106: 22,845–22,860.

Mims, F. M. III. 2002. An inexpensive and stable LED sun photometer for measuring the water vapor column over south Texas from 1990 to 2001. *Geophysical Research Letters* 29: 20-1 to 20-4. (Updated data are also available at http://www.forrestmims.org and http://sunandsky.org.)

Mims, Sarah A., and Forrest M. Mims III. 2004. Fungal spores are transported long distances in smoke from biomass fires. *Atmospheric Environment* 38: 651–655.

MIT. 1971. *Inadvertent Climate Modification: Report.* Workshop sponsored by the Massachusetts Institute of Technology (MIT) and hosted by the Royal Swedish Academy of Sciences and the Royal Swedish Academy of Engineering Sciences. Cambridge, MA: MIT Press.

Nedoluha, Gerald E., Richard M. Bevilacqua, R. Michael Gomez, Brian C. Hicks, James M. Russell, and Brian J. Connor. 2003. An evaluation of trends in middle atmospheric water vapor as measured by HALOE, WVMS, and POAM. *Journal of Geophysical Research* 108 (D13): 4391.

Oltmans, S. J., and D. J. Hofmann. 1994. Increase in lower-stratospheric water vapour at a mid-latitude Northern Hemisphere site from 1981 to 1994. *Nature* 374: 146–149.

Price, Saul, and Jack C. Pales. 1963. Mauna Loa Observatory: The first five years. *Monthly Weather Review* 91 (October): 665–680.

Ross, R. J., and W. P. Elliott. 2001. Radiosonde-based Northern Hemisphere tropospheric water vapor trends. *Journal of Climate* 14: 1602–1612.

Science Pump Corporation. 1999. *Operator's Manual: Model 6A ECC Ozonesonde.* Camden, NJ: Science Pump Corporation.

Stair, Ralph, and Russell G. Johnston. 1958. Some studies of atmospheric transmittance on Mauna Loa. *Journal of Research of the National Bureau of Standards* 61: 419–425.

Trenberth, Kevin E., John Fasullo, and Lesley Smith. 2005. Trends and variability in column-integrated atmospheric water vapor. *Climate Dynamics* 24: 741–758.

Velders, G. J. M., S. O. Andersen, J. S. Daniel, D. W. Fahey, and M. McFarland. 2007. The importance of the Montreal Protocol in protecting climate. *Proceedings of the National Academy of Sciences* 104: 4814–4819.

Yates, Harold W. 1957. Recent atmospheric transmission studies. Paper presented at 42nd Annual Meeting of Optical Society of America, Cleveland, Ohio, October 17–19. (Abstract in *Journal of the Optical Society of America* 47: 1054.)

Subject Index

Page numbers in **boldface** *refer to photographs.*
Some See also *references are to entries in the Name Index.*

Name Index

Page numbers in **boldface** *refer to photographs.*
Some See also *references are to entries in the Subject Index.*

About the Author

Forrest M. Mims III is the most widely read electronics author in the world. His sixty-five books have sold more than 7.5 million copies and have twice been honored for excellence by the Computer Press Association. His many hundreds of articles and photographs have appeared in some seventy magazines and science journals, including *Nature, Scientific American, Science, Popular Photography, Geophysical Research Letters, New Scientist, Applied Optics, Sky & Telescope, Popular Mechanics, Physics Today, Electronics, PC Magazine,* and *IEEE Spectrum.* Mims cofounded the company that developed the Altair 8800, the computer that began the personal computer era, in Albuquerque, New Mexico. While he was writing the Altair's operating manual in 1975, two young programmers named Bill Gates and Paul Allen were creating a version of the BASIC computer language for the Altair. They soon moved to Albuquerque, where they named their partnership Microsoft. In 1993, Mims received a Rolex Award for a global ozone measuring network that used instruments based on his design. He has been calibrating those instruments and many others at the Mauna Loa Observatory at least once each year since 1992 and has published scientific papers about his findings there. Thanks in part to his research at Mauna Loa, in 2008 Mims was named by *Discover* magazine as one of the "50 Best Brains in Science" and one of the "50 Most Important, Influential, and Promising People in Science."

Production Notes for Mims / HAWAIʻI'S MAUNA LOA OBSERVATORY

Interior design by Mardee Melton, in 9.5-point Goudy Hawaiian, with display type in Stempel Garamond.

Composition by Mardee Melton

Printing and binding by Sheridan Books, Inc.

Printed on 60# House White, 444ppi